Organic Chemistry

*the text of this book is printed
on 100% recycled paper*

ASSISTANT EDITORS

David E. Adelson
Shell Development Company

Paul Allen, Jr.
Stevens Inst. of Tech.

J. Allen Baker
Simpson College

Virginia Bartow
University of Illinois

Robert A. Berntsen
Augustana College

W. W. Binkley
Ohio State University

Luther B. Bolstad
Floodwood, Minnesota

Carl Bordenca
Southern Research Inst.

Sam Brelant
Transparent Package Corp.

E. M. Burdick
Texsun Citrus Exchange

C. A. Burkhard
General Electric Company

George B. Ceresia
General Aniline & Film

John Chittum
College of Wooster

W. A. Cook
University of Akron

R. R. Coons
American University

E. J. Crane
Ohio State University

H. W. Davis
Univ. of South Carolina

Brian L. Dehoff
Westvaco Chlorine Prod. Co.

John DeVries
Calvin College

A. A. Dietz
Toledo Hosp. Inst. Med. Res.

Ralph E. Dunbar
North Dakota Agr. College

Raymond F. Dunbrook
Firestone Tire & Rubber Co.

Gregg M. Evans
Yankton College

Fred Feasley
Socony Vacuum Oil Co.

M. H. Filson
Central State Teach. Coll.

Worth A. Fletcher
University of Wichita

John A. Froemke
Augustana College

Albert W. Fuhrman
Naugatuck Chemical Co.

Harry W. Galbraith
Galbraith Microanalytical Lab.

Harry D. Glenn
Naugatuck Chemical Co.

R. H. Goshorn
Sharples Solvents Corp.

G. C. Gross
E. I. du Pont de Nemours

H. J. Gryting
SWPPNOTS

Bernard Gwynn
Gulf Res. & Dev. Lab.

William S. Haldeman
Monmouth College

W. E. Herrell
Mayo Clinic

Carl W. Holl
Manchester College

Lloyd B. Howell
Wabash College

Maurice Huggins
Eastman Kodak Co.

Kenneth Johnson
Commercial Solvents Corp.

Irwin A. Koten
North Central College

John C. Krantz, Jr.
University of Maryland

H. P. Lankelma
Western Reserve Univ.

A. C. Magill
Southeast Mo. S. T. C.

Jerome Martin
Commercial Solvents Corp.

R. F. McCleary
Texas Company

J. W. Mench
Eastman Kodak Co., E.W.O.

Jack P. Montgomery
University of Alabama

C. A. Morey
Findlay College

V. N. Morris
Industrial Tape Corp.

Robert J. Myers
Resinous Prod. & Chem. Co.

Joe O'Brien
Purdue University

Durward G. O'Dell
E. I. du Pont de Nemours

A. R. Padgett
Humble Oil & Refining Co.

John N. Pattison
Batelle Memorial Institute

Harold A. Price
Michigan State College

David Rankin
Novadel-Agene Corp.

Francis E. Ray
University of Cincinnati

R. Chester Roberts
Colgate University

Benjamin E. Sanders
Henry Ford Hospital Lab.

D. C. Sayles
Lowe Brothers Company

John W. Schick
Socony Vacuum Oil Co.

Raymond B. Seymour
University of Chattanooga

Ralph E. Silker
Chadron State Teach. Coll.

James M. Snyder
E. I. du Pont de Nemours

John A. Spence
California Research Corp.

C. A. Sprang
Emery Industries, Inc.

Thomas H. Stoudt
Merck and Company

Elise Tobin
Brooklyn College

L. G. R. Tompkins
Stetson University

Joe Tyree
Elizabeth, New Jersey

Walter O. Walker
The Ansul Chemical Co.

V. H. Wallingford
Mallinckrodt Chem. Works

Keith W. Wheeler
Wm. S. Merrell Co.

Richard Wistar
Mills College

Y. Yamada
Chicago, Illinois

Organic Chemistry

An Outline of the Beginning Course
Including Material for Advanced Study

ED. F. DEGERING, Ph. D.

*Quartermaster Research and
Development Command
Formerly Professor of Chemistry
at Purdue University*

and

Seventy-Six Assistant Editors

(See opposite page)

SIXTH EDITION

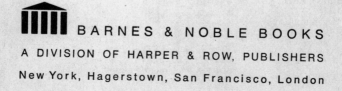

BARNES & NOBLE BOOKS

A DIVISION OF HARPER & ROW, PUBLISHERS

New York, Hagerstown, San Francisco, London

ABOUT THE AUTHOR

The late Dr. Ed. F. Degering received his Ph.D. degree from the University of Nebraska. For many years he was associated with Purdue University, where he taught organic chemistry, directed research, and wrote many scientific and educational articles and books. Subsequently he engaged in research for various industrial concerns, and held the position of chief of the federal agency studying chemical warfare.

Preface

The earlier editions of this Outline were so carefully checked and proofread that few changes were necessary in this sixth edition. A chapter on Organic Catalysis has been added, however.

As indicated by the subtitle, this book is an *outline* of Organic Chemistry. It is not intended as a textbook but as a supplement to any good text. It is the hope of the author that instructors will encourage the use of this Outline in lieu of requiring their students to take prolific lecture notes.

With something over one quarter of a million organic compounds already known, and new ones being described daily, it becomes apparent that the beginning student of Organic Chemistry is apt to become lost in a maze of factual knowledge if he is not provided with adequate guidance. The instructor endeavors to provide such guidance, but the task of taking notes often defeats the very objective of the lecture. It was with the hope that the practice of taking lecture notes could be largely abandoned that the preparation of this outline was undertaken.

The relative importance of material has been indicated by the use of different sizes of type. In beginning courses the material in eight point and six point type should doubtless be omitted from consideration. Some instructors may not care to give their students all of the material that the outline presents in the ten point type.

The authors plan to use the outline as a substitute for lecture notes rather than as a substitute for a textbook. Such a plan is recommended for consideration.

The outline represents a development of something over fifteen years. It appeared initially in a mimeographed form which was distributed to the assistant editors and used by them for a period of a year. Their corrections, criticisms, and suggestions were carefully edited and utilized in producing the first printed edition of the outline.

Among other things, the outline endeavors to emphasize organic terminology and to present it systematically. The nomenclature has been edited carefully by Professor Austin M. Patterson whose services are very gratefully acknowledged.

The late Professor Nelson and Professor Harrod rendered especially valuable service in the formulation of the general plan of the outline, and in editing and proofreading. The author wishes to express his sincere appreciation of the services rendered by the assistant editors, for without their cooperation this outline, doubtless lacking in some respects, would not have been as complete, accurate, and up-to-date as it is.

The authors wish to acknowledge services rendered by T. Boyd, E. Glusenkamp, L. F. Hatch, G. E. Hinds, H. C. Huffman, and F. H. Snyder, as well as by Professors D. Cottle, W. T. Hall, C. R. Kinney, H. P. Lankelma, W. T. Read, and John White. The authors are indebted to The Carter Laboratories, to E. I. duPont de Nemours and Co., and to The Hercules Powder Co. for certain items that were not available elsewhere.

It is hoped that the assistance rendered to students of Organic Chemistry will be in proportion to the time and effort expended by those who have so generously contributed.

ED. F. DEGERING

Table of Contents

TABLE OF CONTENTS

Organic Chemistry

CHAPTER I

INTRODUCTION

Definition:

Organic chemistry is the study of the compounds containing carbon.

Composition:

Many different elements may occur in organic compounds but the ones most commonly found, in the order of their relative occurrence, are: carbon, hydrogen, oxygen, nitrogen, phosphorus, sulfur, chlorine, bromine, and iodine.

Original Concept of Organic Chemistry:

It was formerly believed that organic compounds were exclusively of plant or animal origin and that vital processes were necessary for their formation. In 1824, however, Wöhler reported the preparation of the ammonium salt of **oxalic acid** by passing cyanogen into ammonia water, and in 1828 he reported the preparation of the organic compound urea by heating the inorganic compound ammonium cyanate: $NH_4OCN + heat \rightarrow H_2N \cdot CO \cdot NH_2$. This experimental work led to the development of the concept that animate processes are not essential for the formation of organic compounds.

Importance of Organic Chemistry:

Organic chemistry is involved in practically every activity of modern life. Anesthetics, anti-freezes, clothing, disinfectants, drugs, dyes, explosives, fertilizers, films, flavors, food, fuel, paints, perfumes, plastics, resins, rubber, and numerous other essential commodities are studied, analyzed, and synthesized with the aid of the basic principles of organic chemistry.

Number of Compounds:

More than a quarter of a million organic compounds are known. They out-number all other chemical compounds in the ratio of about ten to one.

Phenomenal Capacity of Carbon:

The existence of this large number of organic compounds is due to the unique capacity of carbon atoms to unite to form chains or rings of more or less complexity. A simple illustration is afforded by the following series of compounds which are known as the saturated hydrocarbons:

1

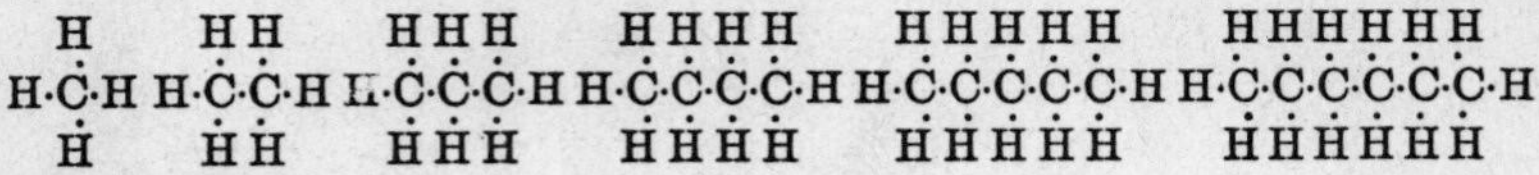

```
   H        H H       H H H      H H H H     H H H H H    H H H H H H
 H·C·H    H·C·C·H   H·C·C·C·H   H·C·C·C·C·H H·C·C·C·C·C·H H·C·C·C·C·C·C·H
   H        H H       H H H      H H H H     H H H H H    H H H H H H
methane   ethane    propane      butane       pentane        hexane
```

This continuous chain series includes known compounds containing sixty or more carbon atoms. In the following pages specific examples will indicate how any or all of the hydrogen atoms in a compound of a given series may be replaced, at least theoretically, by carbon groupings, by other elements, or by groups of elements. By the substitution of *hydroxyl groups* (OH) for hydrogen atoms in the members of the saturated hydrocarbon series the corresponding *alcohols* may be obtained:

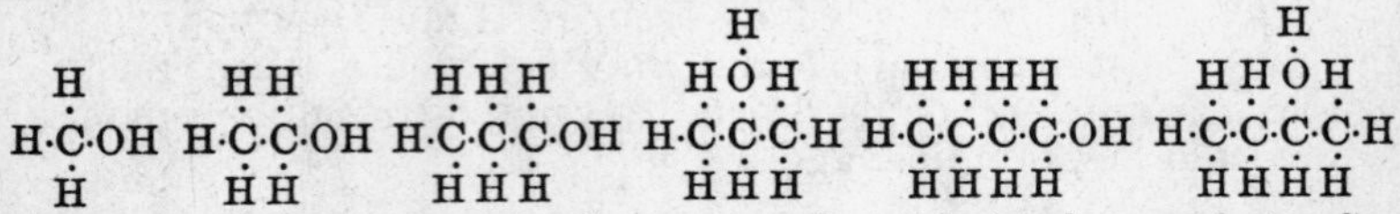

```
                             H                         H
   H        H H      H H H   H O H     H H H H      H H O H
 H·C·OH  H·C·C·OH H·C·C·C·OH H·C·C·C·H H·C·C·C·C·OH H·C·C·C·C·H
   H        H H      H H H   H H H      H H H H      H H H H
methanol,  ethanol,  1-propanol,  2-propanol,   1-butanol,   2-butanol,
methyl     ethyl     propyl       isopropyl     butyl        sec-butyl
alcohol    alcohol   alcohol      alcohol       alcohol      alcohol
```

Nomenclature:

As indicated by the names of the preceding alcohols, different systems of nomenclature are in use. The names shown in ordinary type conform to the recommendations of the *International Union of Chemistry* or to undetermined cases, and the names given in **bold type** represent common usage that does not conform to the I.U.C. recommendations. Throughout this outline, names recommended by the I.U.C. will be given precedence and shown in ordinary type; and common names will be given second place and shown, where the distinction is not obvious, in **bold type.** Where still another system of nomenclature has a limited use, it will be assigned to third place. In cases where no specific ruling has been formulated by the *International Union of Chemistry*, the names will be represented by ordinary type.

The trend in organic nomenclature is toward the adoption and use of the system recommended by the I.U.C., but names that have been established by common usage can not be ignored. *Those electing chemistry as a profession should master, therefore, both the system recommended by the I.U.C. and the common names.*

Terms and symbols used in the text without explanation will be found in the glossary.

Homologous Series:

In the first group of formulas given above, the hydrocarbon series, any given formula differs from the preceding and succeeding formula by a constant, "CH_2". Compounds that are structurally identical with the preceding and succeeding members of the group, except for the "CH_2" constant, constitute an *homologous series*.

Knowledge of the properties and reactions of one or two members of a given series will enable one to predict, qualitatively at least, the properties and reactions of the other members of the group. Homologous compounds react alike because they have the same *reactors* (characteristic groups). *The student should not attempt to memorize each equation but should concentrate on type reactions that are of general application to each homologous series.* This outline attempts to give these type reactions and a sufficient number of specific examples. The student should apply the ideas presented in the type equations to individual members of the group until he familiarizes himself with the characteristic properties of each series.

Where specific equations are not preceded in the same subdivision by a type equation, the equation should be interpreted as representing a specific reaction.

For tables of the homologous series, the student is referred to *Fundamental Organic Chemistry* and to the *Quadri-Service Manual, of Organic Chemistry*, both by Ed. F. Degering.

Formulas:

In the study of organic chemistry, a knowledge of the exact structure of a compound is of the greatest importance. 2-Hydroxypropanoic (**lactic**) acid may be represented by the following formulas:

$$
CH_2O, \quad C_3H_6O_3, \quad CH_3 \cdot CHOH \cdot CO \cdot OH, \quad \text{and} \quad H \cdot \overset{\displaystyle H}{\underset{\displaystyle H}{C}} \cdot \overset{\displaystyle O}{\underset{\displaystyle H}{C}} \cdot \overset{\displaystyle O}{C} \cdot OH \quad \text{or} \quad HO \cdot \overset{\displaystyle H}{C} \cdot \overset{\displaystyle O}{\underset{\displaystyle H}{C}} \cdot \overset{\displaystyle O}{\underset{\displaystyle H}{C}} \cdot H
$$

The first, which represents the simplest atomic ratio, is the *empirical* formula. The second, which is a simple multiple of the empirical formula and indicates the actual number of each constituent atom in the molecule, is the *molecular* formula. The third, which indicates the particular groups linked to each carbon atom, is a *constitutional* or *structural* formula. The last two, which indicate the spatial distribution of the atomic groups, are *graphic* formulas for the *d*- and *l*- forms of *lactic acid*. For most purposes, either the ordinary structural or constitutional formula is satisfactory for describing the structure of a compound, but the graphic formula aids the eye in the visualization of slight differences which distinguish isomeric forms, or one series of compounds from another series.

Sources:

The principal natural sources of organic compounds are: (1) plants, (2) animals, (3) coal, (4) petroleum, (5) natural gas, and (6) oil shale distillates, but laboratory syntheses afford an almost unlimited source for new compounds.

Distinguishing Characteristics:

The unique capacity of carbon to unite with itself gives rise to complex and isomeric structures not commonly encountered with the other elements. Organic compounds are, moreover, dissimilar to inorganic compounds with respect to the following:

1. They burn readily, some spontaneously, to give carbon dioxide, water, and, in some cases, other products.
2. They are usually less stable toward heat.
3. They are less soluble in water, but more soluble in organic solvents.
4. They are less ionic in their reactions.
5. They are, in general, less reactive and form compounds of greater complexity.

Identification:

1. *When a compound is known and described in the literature:*

The first step consists of purification. The purification of a solid is usually effected by one of the following methods: recrystallization from a suitable solvent, extraction, distillation, or sublimation. In the case of a liquid, purification is usually accomplished by rectification or fractional distillation. A sharp, constant melting point serves to indicate that a given solid substance is pure, and a sharp boiling point signifies a pure liquid unless, as sometimes occurs, a constant-boiling mixture is formed, or the constituents have nearly identical boiling points.

The identification of a compound involves one or more of the following: qualitative analysis, quantitative analysis, determination of the physical constants, and the formation of derivatives. The data obtained from these procedures, by comparison with the literature, serve to identify the compound. If the unknown is a solid, a mixed melting point is often determined for the purpose of confirmation. This is done by finding the melting point of a mixture of the unknown and the pure compound with which it has been identified. If the melting point of the mixture checks with that of the pure compound, the two substances are assumed to be identical.

2. *When the compound is not described in the literature:*

The establishment of the structure of a compound of unknown composition consists of (1) purification, (2) qualitative analysis, (3) quantitative analysis, (4) determination of the molecular weight, and (5) establishment of the atomic linkages in the molecule by synthesis, degradation, or the formation of known derivatives.

Nature of Chemical Bonds:

The attraction which binds the atoms in organic compounds is doubtless electrical. The basic concepts of the electron theory of valence are considered in the following chapter. The graphic representation of the electronic structures of methane, methanol (**methyl alcohol**), methanal (**formaldehyde**), and methanoic (**formic**) acid, is shown below:

$$
\begin{array}{cccc}
\text{H} & \text{H} & {:}\overset{..}{\text{O}}{:} & {:}\overset{..}{\text{O}}{:} \\
\text{H:C:H} & \text{H:C:O:H} & \text{H:C:H} & \text{H:C:O:H} \\
\text{H} & \text{H} & &
\end{array}
$$

From these examples it is observed that a pair of electrons functions in the capacity of each valence bond, and that the non-valence electrons are symmetrically distributed to complete the oxygen octets.

Classification:

The field of organic chemistry may be conveniently subdivided, according to the structure of the compounds, as follows:

1. Open-Chain, Aliphatic or Acyclic Compounds:

Homologous series of compounds with acyclic structures are illustrated by the following specific examples: $CH_3 \cdot CH_3$, ethane; $CH_3 \cdot CH_2 \cdot OH$, ethanol (**ethyl alcohol**); $CH_3 \cdot CHO$, ethanal (**acetaldehyde**); and $CH_3 \cdot CO \cdot OH$, ethanoic (**acetic**) acid.

2. Closed-Chain or Ring Compounds:

a. Alicyclic:

Examples are $H_2C{-}CH_2$ (cyclopropane) and $H_2C{-}CH_2$ (cyclobutane).

b. Aromatic or Carbocyclic:

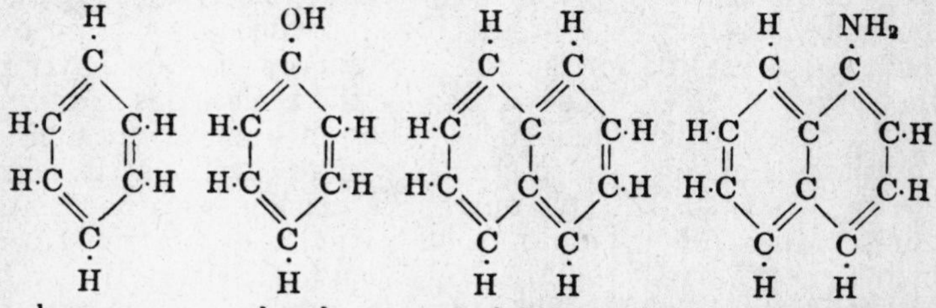

benzene,   phenol,   naphthalene,   and α-naphthylamine

c. Heterocyclic:

As the name indicates, these compounds contain constituents other than carbon in the ring. Examples are afforded by:

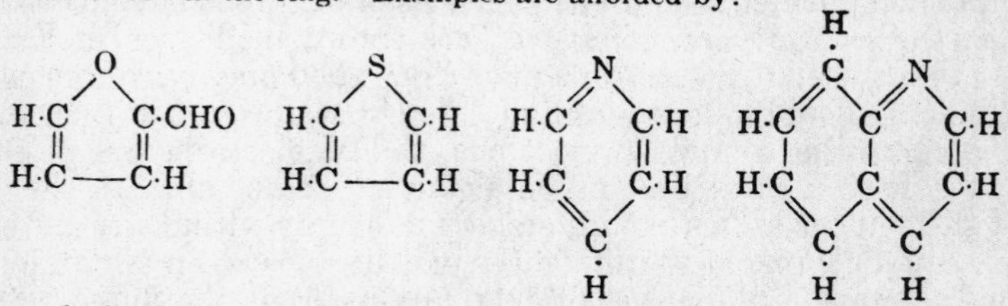

furfural   thiophene   pyridine   quinoline

CHAPTER II

THE ELECTRONIC BASIS OF VALENCE

An atom is a positively charged nucleus surrounded by negative electrons, the number of electrons being just sufficient to balance the positive charge on the nucleus. The positive charge of the nucleus of the uranium atom, the heaviest of atoms, is 92 times that of the hydrogen atom, the lightest of atoms. The nuclear charges of the other elements are multiples of the hydrogen nuclear charge; thus helium is 2, lithium 3, beryllium 4, and so on, up to uranium, 92. Elements are sometimes designated by means of the nuclear charge, which is, for this purpose, called the *atomic number*. Thus, the atomic number of hydrogen is 1, of helium 2, etc. The number of electrons outside the nucleus in neutral atoms likewise varies from 1, in hydrogen, to 92 in uranium.

Atomic Structure:

Atomic Structure deals with the arrangement of the electrons in atoms in relation to the nuclei. Since electrons are in rapid motion, representations of their positions by definite points in space are approximations only and are to be justified on the basis of convenience in describing their average positions. The electrons in atoms tend to group themselves into certain energy levels, or shells, around the nucleus. Those which are closest to the nucleus possess the lowest potential energy and are never more than two in number. They are called the K electrons and the shell they occupy is called the K shell. Above helium, atomic number 2, additional electrons, which are never more than eight in number, occupy the L shell. This number is reached with neon, atomic number 10. In heavier atoms, additional electrons must occupy higher energy levels, the M, N, O, P, and Q shells. Although the M shell is filled when eight more electrons are added (at argon, atomic number 18), it has the ability, after two have entered the N shell (following calcium, atomic number 20), to take on a total of 18 electrons. Beyond this point the electron arrangements are quite complicated. The energy levels of electrons in some atoms, i.e., the electron configurations, are shown in Table I. Except for the K shell, the electrons in any one shell may have somewhat different energies. For example, the L electrons are of two types, known as the 2s and 2p electrons, the M electrons are of three types, known as 3s, 3p, and 3d electrons, and so on. In the case of the smaller atoms the differences in energy within any one shell are small compared to differences in energy between two different shells. In the outer shells of the larger atoms the differences in energy between the different shells, and also between levels in any

TABLE I. ELECTRON CONFIGURATIONS OF SOME ATOMS

Element		K	L		M			N	
Symbol	At. No	1s	2s	2p	3s	3p	3d	4s	4p
H	1	1							
He	2	2							
Li	3	2	1						
Be	4	2	2						
B	5	2	2	1					
C	6	2	2	2					
N	7	2	2	3					
O	8	2	2	4					
F	9	2	2	5					
Ne	10	2	2	6					
Na	11				1				
Mg	12				2				
Al	13				2	1			
Si	14		Neon		2	2			
P	15		Configuration		2	3			
S	16				2	4			
Cl	17				2	5			
A	18				2	6			
K	19							1	
Ca	20							2	
Sc	21						1	2	
Ti	22						2	2	
V	23		Argon				3	2	
Cr	24		Configuration				5	1	
Mn	25						5	2	
Fe	26						6	2	
Co	27						7	2	
Ni	28						8	2	
Cu	29						10	1	
Zn	30						10	2	
Ga	31						10	2	1
Ge	32		Argon				10	2	2
As	33		Configuration				10	2	3
Se	34						10	2	4
Br	35						10	2	5
Kr	36						10	2	6

one shell, are less than for small atoms. Moreover, there may be overlapping between different shells.

There is a periodicity to atomic structures which is brought out in Table I. If only the outermost shells are considered, it is evident that with elements of similar chemical properties these shells have the same number of electrons. This number is one for the alkali metals, viz., lithium (3), sodium (11), potassium (19), rubidium (37), and caesium (55); two for the alkaline earth metals; six for the members of the oxygen family, viz., oxygen (8), sulfur (16), selenium (34), and tellurium (52); seven for the halogens; and eight for the rare gases (except helium), viz., neon (10), argon (18), krypton (36), xenon (54), and radon (86). The similarity in the chemical properties of the members of any one family is due to the fact that they have the same number of electrons in the outermost shell.   On this account it is called the *valence shell* and the electrons in it are often called the *valence electrons*, since they are largely responsible for the behavior of the element in chemical reactions. The part of the atom inside of the valence shell, that is, the nucleus and the other electrons, is often called the *kernel*.   The kernel carries a positive charge equal to the number of electrons in the valence shell.   For most of the common elements, the kernel charge is likewise the same as the column number or group in the periodic table.   Formulas are often written so as to show the valence shell of the atom.   In such a case the symbol of the element represents the kernel, for example,

$$\text{Li·, Na·, K·, Mg:, ·N:, ·P:, ·Cl:, ·Br:, :Ne:}$$

Stable electron groups are the *pair* and the *octet*.   The stability of the pair is indicated by (a) the presence of two electrons in the K shell of all atoms, except hydrogen; (b) the rarity of molecules having an odd number of electrons.   The stability of the octet is indicated by the nonreactivity of the noble gases, neon, argon, krypton, xenon, and radon, all of which have an outermost shell of eight electrons.

The tendency of many atoms to react may be ascribed to one or more of the following: (a) electrons tend to pair; (b) atoms of metals (alkali and alkaline earth metals, especially) tend to give up one or more electrons so as to form positive ions which have the octet structure of the next lower noble gas; (c) atoms of nonmetals (especially the members of the oxygen and halogen families) tend to acquire one or more electrons so as to form negative ions which have the octet structure of the next higher noble gas; (d) reactions tend toward a decrease in kernel repulsions (see kernel repulsions).

Types of Chemical Bonds:

There are two extreme types of chemical bonds, the *ionic* and the *covalent*.   The former results when there is a transfer, from the valence shell of one atom to the valence shell of another atom, of one or more electrons (proton transfer is discussed under kernel repulsions), the latter when electrons are shared between nuclei. Two typical examples of electron transfer are the reactions of sodium with chlorine and of barium with oxygen:

$$\text{:Na:· + ·Cl: → :Na:+ + :Cl:}^-$$

$$\text{:Ba:: + O: → :Ba:+ + + :O:}^-$$

The sodium easily loses one electron to chlorine to form positive sodium ion.   The chlorine, in taking up the electron, is converted into a negative chloride ion.   Sodium ion has the structure of the next lower noble gas, neon; chloride ion that of the next higher, argon.   A barium atom loses two electrons to the oxygen atom producing the doubly positively charged barium ion of the xenon structure and the doubly negatively charged oxygen ion of the neon structure. Compounds having the ionic type of bonding are salts and possess typical saltlike properties, such as high melting point, high boiling point, and ready solubility in water. They conduct an electric current both in the fused state and in solution, i.e., they are electrolytes. In solution they react rapidly with solutions of other salts. They are often called polar compounds, since they are built up of positively and negatively charged particles of small mass, the ions. These ions may be thought of as minute positive or negative electric poles.

The extreme covalent type of bond, produced by a sharing of electrons between atoms, is found in molecules where like atoms are joined together, as in H_2, Cl_2, S_8, diamond, etc. The extreme covalent type is discussed later in connection with atomic radii. The shared-electron bond, sometimes called the Lewis bond from G. N. Lewis, who first proposed it, is very common. It is formed whenever the sum of the valence electrons of two atoms in chemical combination is insufficient to allow each of the two atoms to have a separate octet of its own. The atoms may be alike or dissimilar, as in H_2, Cl_2, H_2O, PCl_3, SCl_2, CO_2, H_2SO_4: H:H, :C̈l:C̈l:,

$$H:\overset{..}{\underset{..}{O}}:H, \quad :\overset{..}{\underset{..}{Cl}}:\overset{}{P}:\overset{..}{\underset{..}{Cl}}:, \quad :\overset{..}{\underset{..}{Cl}}:\overset{..}{\underset{..}{S}}:\overset{..}{\underset{..}{Cl}}:, \quad :\overset{..}{\underset{..}{O}}::C::\overset{..}{\underset{..}{O}}:, \quad H:\overset{..}{\underset{..}{O}}:\overset{\overset{\textstyle :O:}{\displaystyle S}}{\underset{\textstyle :O:}{}}:\overset{..}{\underset{..}{O}}:H$$

The shared-electron bond is found also in complex ions, for example, in NO_3^-, CO_3^-, SO_4^-; i.e.,

$$:\overset{..}{\underset{..}{O}}::\overset{-}{N}:\overset{..}{\underset{..}{O}}:$$
$$:\overset{..}{\underset{..}{O}}:$$

Number of Shared Electrons:

For the usual shared-electron type of bond, the number of electrons shared between atomic kernels (not including the hydrogen nuclei) can be calculated, on the assumption that each kernel has a complete octet of electrons, as follows:

$$n = X - \Sigma$$

where n is the number of shared electrons, X is the number of electrons required if each kernel has a complete octet without any sharing at all, and Σ is the sum of the valence electrons of the atoms which make up the molecule, corrected, in the case of a complex ion, for the charge carried by the ion. The number of electrons shared between atomic kernels (excluding hydrogen atoms) in the above formulas is calculated as follows:

Cl_2, $2 \times 8 - (2 \times 7) = 2$ $\qquad CO_2$, $3 \times 8 - (4 + 2 \times 6) = 8$

PCl_3, $4 \times 8 - (5 + 3 \times 7) = 6$ $\qquad H_2SO_4$, $5 \times 8 - (2 + 5 \times 6) = 8$

SCl_2, $3 \times 8 - (6 + 2 \times 7) = 4$ $\qquad NO_3^-$, $4 \times 8 - (1 + 5 + 3 \times 6) = 8$

In shared-electron bonds, the number of shared electrons may be one, two, three, four, or six. The one-electron bond is rare; the best known example is found in boron hydride, B_2H_6, in which each boron is attached to one hydrogen by a single electron. The two-electron bond, the most common shared-electron bond, is the ordinary single bond of the organic chemist; the three-electron bond is rare; it is present in the oxygen molecule, in which the oxygen atoms are joined by a two-electron bond and two three-electron bonds; the four-electron and six-electron bonds are very common and are, respectively, the double bond and triple bond of the organic chemist. The four-electron bond is very common with first row elements, less common with second row, and rare with higher elements. The six-electron bond is rare except for the first row elements.

Atomic Radii:

Because of the repulsion which positively charged kernels exert upon each other, and also because of the mutual repulsions of electron shells, there is a limit to the closeness with which atomic nuclei can approach each other. The distance between any two nuclei in a compound, which is known as the internuclear (or interatomic) distance, can be determined in the case of crystalline solids by means of X-ray analysis, and in the case of gaseous molecules by the electron diffraction method or by an examination of the infra-red absorption spectrum or the Raman spectrum. As a result of measurements on numerous compounds, it has been possible to assign to many atoms definite values for their radii, so that, when two radii are added, the sum is the observed interatomic distance. But the situation is complicated because the radius of an atom in an ionic compound, i.e., the *ionic radius*, is not the same as the radius of the same atom when attached to another atom by a shared electron bond. For example, the ionic radius of sulfur in CaS is 1.69 Ångstrom units $(1 \text{ Å} = 10^{-8} \text{ cm})$, while the *covalent radius* of sulfur in S_8 is 1.04 Å. Whereas covalent radii are useful in organic chemistry, ionic radii are not important. On this account the latter are not discussed further. Table II lists the covalent radii of the atoms which are often present in organic compounds.

TABLE II. SINGLE NORMAL COVALENT RADII
Ångstrom Units $(1 \text{ Å} = 10^{-8} \text{ cm.})$

H	Be	B	C	N	O	F
0.29	1.07	0.89	0.77	0.70	0.66	0.64
			Si	P	S	Cl
			1.17	1.10	1.04	0.99
			Ge	As	Se	Br
			1.22	1.21	1.17	1.14
			Sn	Sb	Te	I
			1.40	1.41	1.37	1.33
			Pb	Bi		
			1.46	1.51		

The interatomic distance between two atoms joined by a double bond (two electron pairs) is less than when they are joined by a single bond (one electron pair), and still less when joined by a triple bond (three electron pairs). Since the attraction of the positive kernels for the shared electrons of the bond overcomes the repulsion between the two kernels (the kernel repulsion), it is evident that the greater the attraction the closer they are brought together. Four electrons exert more attraction than two, but the distance is not halved because the kernel repulsions increase rapidly with the closeness of approach. In general, the double bond distance is about 87%, and the triple bond distance is 78% of the single bond distance.

Resonance:

Whenever it is possible to write two or more equivalent electronic structures for a molecule, or for an ion, the particular atomic arrangement is more stable than would be expected. It is a consequence of the wave mechanical treatment of an electronic system that it is the more stable the greater its complexity, that is, the greater the number of possible arrangements which may be written for it. This condition is known as resonance and the substance is said to resonate among the different electronic forms. Thus, for carbonate ion, three equivalent electronic structures, I, II, and III, may be written:

$$\overset{\cdot\cdot}{:\!O\!:} \qquad\qquad \overset{\cdot\cdot}{:\!O\!:} \qquad\qquad \overset{\cdot\cdot}{:\!O\!:}$$
$$\overset{\cdot\cdot}{:\!O}\!::\!\overset{\cdot\cdot}{C}\!:\!\overset{\cdot\cdot}{O}\!: \qquad\qquad :\!\overset{\cdot\cdot}{O}\!:\!\overset{\cdot\cdot}{C}\!:\!\overset{\cdot\cdot}{O}\!: \qquad\qquad :\!\overset{\cdot\cdot}{O}\!:\!\overset{\cdot\cdot}{C}\!::\!\overset{\cdot\cdot}{O}\!:$$
$$\text{I} \qquad\qquad\qquad \text{II} \qquad\qquad\qquad \text{III}$$

Actually, the system has only one structure, which is sort of an average of all three, a hybridlike affair which cannot be represented by conventional electronic structures, but is reasonably well represented by the three structures given above; and, moreover, as a direct result of this, the system is more stable than it would be if it had the structure I (or II or III) above. The properties of the system are not exactly those corresponding to those for the various resonating structures; instead there is a shift in the direction of greater stability. Thus the carbon-oxygen distance is closer to the double-bond value (which corresponds to a stronger bond) than to the single-bond value. Whenever, in a resonating system, a given bond can resonate between a single and a double bond, or between a double and a triple bond, the observed distance is closer to the smaller value than would be expected from the relative contribution of each form. Moreover, the observed equality of all of the C-O bond distances (from X-ray analysis) and the three-fold symmetry of carbonate ion, i.e., the symmetrical arrangement of the three oxygen atoms about the central carbon atom, facts for which chemists have been unable to offer a satisfactory explanation on any other basis, are clearly understandable in terms of the resonating system. Two effects, that is, increased stability and a shortening of internuclear distances, are always observed whenever there is the possibility of resonance (see Table III).

TABLE III. RESONATING SYSTEMS

| Name | Formula (a) (b) | Internuclear distance ||||
| | | Bond a || Bond b ||
		obs.	sum of radii	obs.	sum of radii
carbonate ion	:O::C:O: :O:	1.31	1.24	1.31	1.43
nitrate ion	:O::N:O: :O:	1.22	1.18	1.22	1.36
vinyl chloride	H₂C::C:Cl: H	1.38	1.34	1.69	1.76
guanidinium ion	H₂N::C:NH₂ NH₂	1.36	1.28	1.36	1.47
acetate ion	O::C:O: CH₃	1.29	1.24	1.29	1.43
nitro group	O::N:O: CH₃	1.21	1.18	1.21	1.36

It is evident from Table III that the bond distance of (a), written as a double bond, is, in all cases, close to the calculated double bond distance, and that the bond distance of (b), written as a single bond, is, in all cases but one, also close to the double bond distance. In vinyl chloride, bond (b) has only a small amount of double bond character. That is, the contribution to the average structure of the electronic configuration

$$H:\ddot{C}:\underset{..}{C}::\ddot{C}l:$$
$$\ddot{H}\ H$$

is not very large.

Kernel Repulsions:

The positive kernels of atoms repel each other, but this repulsion is overcome, in molecules, by the attraction of the kernels for the shared electron pairs. There are thus two opposing tendencies, one, the kernel repulsions, tending to disrupt the union, and the other, the coulomb attraction, which creates the union. An approximation of the kernel repulsion between two atoms can

be obtained by the use of the expression $U = \dfrac{Z_1 Z_2}{r_o} \times 330$, where

U is the repulsion energy in kilogram calories per mole, Z_1 and Z_2 are the kernel charges, and r_o is the distance between atomic centers in Ångstrom units. The values are only approximate because other factors, such as resonance and the distortion of electron shells, modify them. However, for similar types of union, for example a single bond, and for atoms of similar size, as the first row elements, the calculated relationships are qualitatively correct. For rough comparison under these conditions, one may even use the product of the kernel charges. In comparing the relative stabilities of the three possible forms of nitrosyl chloride, NOCl, the sum of the product of the kernel charges in

$$:\overset{..}{N}::\overset{..}{Cl}:\overset{..}{O}: \text{ is 77, in } :\overset{..}{N}::\overset{..}{O}:\overset{..}{Cl}: \text{ is 72, and in } :\overset{..}{O}::\overset{..}{N}:\overset{..}{Cl}: \text{ is 65.}$$

It is evident, therefore, that the last form, in which the atom of smallest kernel charge is in the center, is the most stable. This is the correct formula of nitrosyl chloride. Although kernel repulsions are approximations and lack the accuracy of bond energies, they are useful in enabling one to select the most stable of two or more possible arrangements of a given number of atoms in a molecule.

In addition to the four tendencies which lead to chemical reaction, mentioned on page 8, should be added a fifth, viz.: (e) atoms joined by shared-electron bonds tend to react so as to form molecules in which they are joined to atoms of smaller kernel charge, or in which the kernel repulsions are smaller than in the reacting molecules. In some cases, especially in organic chemistry, only one kind of molecule is concerned in the chemical change (an intramolecular rearrangement), but in most cases two or more kinds of molecules are involved, as for example in the reaction of

hydrogen with chlorine: $:\overset{..}{Cl}:\overset{..}{Cl}: + H:H \rightarrow 2\ H:\overset{..}{Cl}:$ The two chlorines, kernel charge of $+7$, react with hydrogen to form hydrogen chloride, in which the chlorine is combined with hydrogen, $+1$. A similar type of change involves protons, i.e., the nuclei of hydrogen atoms. These are mobile and can leave one atom to attach themselves to another atom of lower kernel charge, provided, however, the latter has an unshared (lone or exposed) electron pair in its octet:

$$\overset{\textstyle H}{\underset{\textstyle H}{H:\overset{..}{Cl}: + :\overset{..}{N}:H \rightarrow :\overset{..}{Cl}: + H:\overset{..}{N}:H^+}}$$

$$\overset{}{\underset{\textstyle H}{H:\overset{..}{Br}: + H:\overset{..}{O}:H \rightarrow :\overset{..}{Br}:^- + H:\overset{..}{O}:H^+}}$$

When hydrogen chloride and ammonia come together, the proton leaves the chlorine, kernel charge $+7$, for the nitrogen, kernel charge $+5$. When hydrogen bromide dissolves in water, the proton leaves the bromine $(+7)$ for the oxygen $(+6)$.

Bond Energies:

When two neutral gaseous atoms combine to form a gaseous molecule, energy is evolved. The amount of energy evolved during the formation of one gram molecular weight of the substance is called the bond energy. It is, conversely, the amount of energy which is necessary to dissociate one gram molecular weight of the gaseous substance into neutral, gaseous atoms. The values of some common bonds, as given by Pauling, are shown in Table IV.

TABLE IV

Bond Energies of Some Bonds, in Kg. cal. per mole at 25° C

F–F	Cl–Cl	Br–Br	I–I
64.55	56.90	45.23	35.39
H–F	H–Cl	H–Br	H–I
147.3	101.0	86.2	70.8
H–H	H–C	H–N	H–O
102.36	97.70	83.35	109.94
C–C	C–N	C–O	C–S
83.87	66.4	81.8	73.5
C–F	C–Cl	C–Br	C–I
124.5	78.6	65.2	56.5
C=C	C–O	C≡C	C≡N
150.0	175.4	198.8	209.3

The methods by which bond energies are obtained usually involve calculation from thermochemical and spectroscopic data, because in general their direct determination is not possible. Although bond energies relate to gaseous reactions, they can be applied, generally with no great error, to reactions in the liquid state, since the molal heats of vaporization, which, strictly speaking, should be taken into account, are usually much smaller for organic substances than the bond energies. Moreover, since the values for the reactants and the products are of the same order of magnitude, the errors involved in neglecting them cancel each other to a considerable extent.

Knowing bond energy values, it is possible to calculate fairly accurately the heat effect in a given reaction without actually carrying out the measurement. Thus one can predict the heat of reaction when ethylene (C_2H_4) reacts with hydrogen to form ethane (C_2H_6). In doing this, it is permissible to write equations for reactions which do not take place under ordinary conditions. The hypothetical step reactions are shown below, and with them the energy changes, in Kg. cal. per mole, which are involved:

(a) $H–H \rightarrow H + H - 102.4$ Kg. cal.

(b) $H_2C=CH_2 \rightarrow H_2C–CH_2 - (150.0 - 83.0)$ Kg. cal.

(c) $H + H + H_2C–CH_2 \rightarrow H–CH_2–CH_2–H + (2 \times 100.0)$ Kg. cal.

(d) $H–H + H_2C=CH_2 \rightarrow H_3C–CH_3 + 30.6$ Kg. cal.

In (a), since energy must be supplied in order to break the H–H bond of the hydrogen molecule, the reaction is endothermic. In (b), energy must be supplied (150.0 Kg. cal.) in order to completely break the C=C double bond, but when the two carbons unite to form the C–C single bond, energy is evolved (83.0 Kg. cal.). The difference, 67.0 kg. cal., is the amount of energy which must be absorbed in order to convert a double bond into a single bond. This step, therefore, is also endothermic. In (c), since energy is evolved through the formation of two C–H single bonds, the reaction is exothermic. On adding (a), (b), and (c), it is evident, from (d), that the overall reaction is exothermic. This is a moderately large heat effect. The heat effect in many other reactions can be approximated by carrying out similar step reactions, or can be more quickly calculated by writing an equation in which only the bonds involved appear, and in which each bond is assigned its energy value. Thus, for the equation above we may write:

$$C-C + H-H \rightarrow 2\,C-H + C-C$$
$$150.0 \quad 102.4 \quad 2 \times 100 \quad 83.0$$

The sum of the left hand values is 252.4, and of the right hand values, 283.0 Kg. cal. Since the latter is the larger value, the bonds on the right hand side constitute the more stable combination, by 30.6 Kg. cal.

Usually, reactions in which large amounts of heat are evolved tend to take place spontaneously. Oftentimes a suitable catalyst must be present in order for the reaction to proceed with a detectable rate. The catalyst merely hastens the rate but does not, ordinarily, alter the energy effect. The reaction between ethylene and hydrogen, which is very slow without a catalyst, will proceed spontaneously under ordinary conditions when the proper catalyst is present (in this case platinum black), for the energy effect is moderately large. But the reverse change, the conversion of ethane to ethylene and hydrogen, does not proceed spontaneously under ordinary conditions, even in the presence of the catalyst. As a rule, reactions in which large amounts of energy are absorbed do not proceed spontaneously under ordinary conditions.

Ionic Character of Electron-Pair Bonds:

The electron-pair bond between two like atoms, viz., A:A, is a normal covalent bond. If the electron-pair bond between two unlike atoms, viz., A:B, is intermediate between the bonds A:A and B:B, it is a normal covalent bond also. However, if it is not intermediate between these two it is not a normal covalent bond because it has ionic character, in addition to the covalent character. This ionic character may be small, as for example in bromine chloride, or may be very large, as for example in hydrogen fluoride (see Table V). The determination of the amount of ionic character in an electron-pair bond is made by comparing the bond energy of A:B with the mean of the bond energies of A:A and B:B.

The bond energy of a bond between two unlike atoms, A:B, is either equal to or greater than one half the sum of A:A and B:B. If it is the former, then the following equation can be written:

$$\frac{1}{2} \text{ A:A} + \frac{1}{2} \text{ B:B} \rightarrow \text{A:B} + 0 \text{ Kg. cal.}$$

If it is the latter, that is, if there is an evolution of energy, then another equation, in which Δ is a positive number, can be written:

$$\frac{1}{2} \text{ A:A} + \frac{1}{2} \text{ B:B} \rightarrow \text{A:B} + \Delta \text{ Kg. cal.}$$

In the former case, the A:B bond, formed in the absence of any energy effect, resembles the A:A and B:B bonds in being a normal covalent bond. In the latter case, where the energy effect is positive (it is never negative), the A:B bond is more stable than it would be if it were a normal covalent bond. This excess of stability is ascribed to ionic character, that is, $A^{+-}B$ or $A^{-+}B$. The possibility of resonance, involving covalent and ionic structures, is the factor which increases the stability. The difference, Δ, between the observed bond energy and the value predicted from additivity, is therefore a measure of the ionic character of the bond, A:B. Thus in Table V, in which observed and predicted bond energies are shown for several halogen compounds, the deviations, Δ, are indications of the excess stability, and therefore of the degree of ionic character of the several bonds.

TABLE V.   IONIC CHARACTER OF SOME HALOGEN BONDS
BOND ENERGIES

	Actual	Predicted	Δ
HF	147.0	83.4	63.6
HCl	100.8	79.6	21.2
HBr	86.1	73.8	12.3
HI	70.6	68.8	1.8
ClF	87.9	60.6	27.3
BrCl	51.4	51.0	0.4
ICl	49.3	46.1	3.2
IBr	41.5	40.3	1.2

From Table V it is evident that in the hydrogen halides, HF has the greatest amount of ionic character and HI has the least. Also, of the interhalogen compounds, BrCl has practically zero ionic character. The BrCl bond therefore closely approximates a normal covalent bond.

Scale of Electronegativities:

The Δ values in Table V are a quantitative measure of the ionic character of some shared-electron single bonds. It is significant that a bond is purely covalent, i.e., lacks ionic character, when the two atoms in combination are alike in electronegativity, i.e., BrCl, while it possesses a large degree of ionic character when the atoms in combination are dissimilar in their degree of negativity, i.e., HF. Chemists have recognized for a great many years that fluorine is the most electronegative of all the elements, and that the electronegativity of the other halogens decreases in the order, fluorine, chlorine, bromine, and iodine. However, these Δ values give a quantitative measure of the degree of electronegativity. From them and from other values of a similar nature, it is possible to arrange the elements in order, starting with the most electropositive elements, the alkali metals, at one end and ending with the most electronegative element, fluorine, at the other end. A convenient plot of this type is due to Pauling, who sets the square root of Δ, expressed in volt electrons (1 v.e. = 23.05 Kg. cal.), equal to the difference in electronegativities of the elements. It is called the Electronegativity Map, Figure 1.

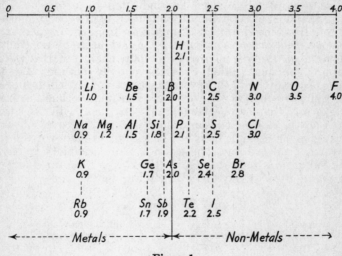

Figure 1

In Figure 1, the alkali metals, being the most electropositive, stand at one end and fluorine, the most electronegative, stands at the other. Thus the metals are on the left, the nonmetals on the right. Boron, at 2.0 on the scale, is taken as the point where neither metallic nor nonmetallic properties predominate. The difference between any two points, when squared and then multiplied by 23.05, gives the approximate value, in Kg. cal. per mole, corresponding to Δ, in Table V. It is thus possible from Figure 1 to calculate the approximate heat effect when any bond A:B, is formed from A:A and B:B. The greater the separation of any two elements on the map, the greater will be this value, and consequently the greater the ionic character of the bond.

On the basis of the scale of negativities, one can conclude that there is little ionic character in a C-C bond, some in the C-H bond, still more in the C-O and C-Cl bonds. Also, carbon is negative in comparison to hydrogen positive in comparison to nitrogen, oxygen, and the halogens.

In the double bonds the ionic character generally runs parallel to the ionic character of single bonds. Thus, the carbon-to-carbon double bond in ethylene is nonionic whereas the carbon to oxygen double bond in carbon dioxide has considerable ionic character, the carbon being positively, the oxygen negatively charged.

In the so-called *Semi-Polar Double Bond* (sometimes known as *a coordinate covalence*), which actually is a shared electron pair, one atom may be pictured as having contributed both electrons to the bond, as in sulfur dioxide: $\ddot{\text{:O:}}\overset{**}{*}\text{S:}\ddot{\text{O:}}$. Here the electrons supplied by the sulfur atom are indicated by asterisks. It is evident that in the right hand bond, both of the electrons came originally from the sulfur atom, and none from the oxygen atom. If now a *formal charge* is calculated for each atom by subtracting the charge on the kernel from the total charge of the valence electrons (splitting shared electrons equally between the two atoms which they join), these values are obtained in sulfur dioxide: left to right, $O = 0, S = +1, O = -1$. One bond is thus a covalent double bond (4 shared electrons) and the other, with the formal charges on the adjacent atoms, looks like both a covalent single bond and an ionic, single bond. The formula is often written with an arrow to indicate that the electron pair, originally belonging to the sulfur, has moved over to the oxygen. Sulfur dioxide is often represented by any one of the following:

O = S = O	:Ö::S:Ö:	O = S → O	$\overset{+}{O} = S - \overset{-}{O}$
classical formula	electronic formula	simplified electronic formulas	

Sometimes, in the electronic formula, the charges are indicated. In this particular case there is resonance between two equivalent electronic structures.

Some of the more common inorganic substances having the semipolar double bond are nitric acid, nitrate ion, and nitrogen pentoxide. It is probable that many compounds of second row and heavier elements, such as orthophosphoric acid and phosphate ion, sulfurous and sulfuric acids, sulfite and sulfate ions, chloric and perchloric acids, chlorate and perchlorate ions, and many others have semi-polar double bonds.

Some organic compounds with this type of bond are nitromethane, trimethylamine oxide, and dimethyl sulfone:

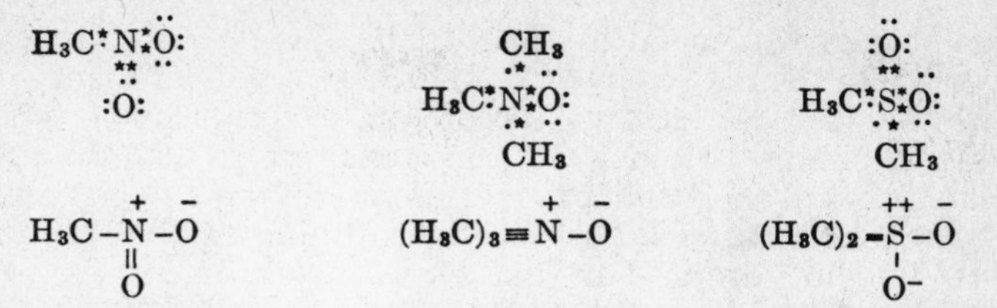

$$H_3C \overset{..}{\overset{*}{\underset{**}{N}}} \overset{..}{\overset{*}{O}}:$$
$$:\overset{..}{O}:$$

$$CH_3$$
$$H_3C \overset{.}{\overset{*}{\underset{.}{N}}} \overset{.}{\overset{*}{O}}:$$
$$CH_3$$

$$:\overset{..}{\overset{**}{O}}:$$
$$H_3C \overset{.}{\overset{*}{\underset{.}{S}}} \overset{..}{O}:$$
$$CH_3$$

$$H_3C - \overset{+}{N} - \overset{-}{O}$$
$$\overset{\|}{O}$$

$$(H_3C)_3 = \overset{+}{N} - \overset{-}{O}$$

$$(H_3C)_2 - \overset{++}{S} - \overset{-}{O}$$
$$O^-$$

nitromethane   trimethylamine oxide   dimethyl sulfone

A group of compounds in which one atom contributes both electrons to the bond is the coordination compounds of Werner; the copper ammonia salts, such as $Cu(NH_3)_4SO_4$; the ferrocyanides, $K_4Fe(CN)_6$; the metallic platinochlorides, K_2PtCl_6, etc. In these a central, heavy atom is surrounded by four or by six other atoms, the attachment being by means of the unshared electron pairs of the latter. The number of such atoms which are attached is known as the *coordination number*. Coordination compounds are common in the case of elements belonging to Group VIII, of the Periodic System, or elements near them, i.e., elements not close to the noble gases. The idea of a coordination number has been extended, within recent years, to include compounds in which the bonds are not coordinate covalent. For example, one speaks of the coordination number of carbon in carbon tetrachloride and of nitrogen in trimethylamine oxide as being 4. First row elements never have a coordination number greater than 4.

The Hydrogen Bond:

Hydrogen may, under certain conditions, have a coordination number of two. Ordinarily, the hydrogen nucleus, the proton, is associated closely with an electron pair, which holds it to some other atom. Typical examples are hydrogen, methane, ammonia, water, and hydrogen fluoride, in the non-associated form:

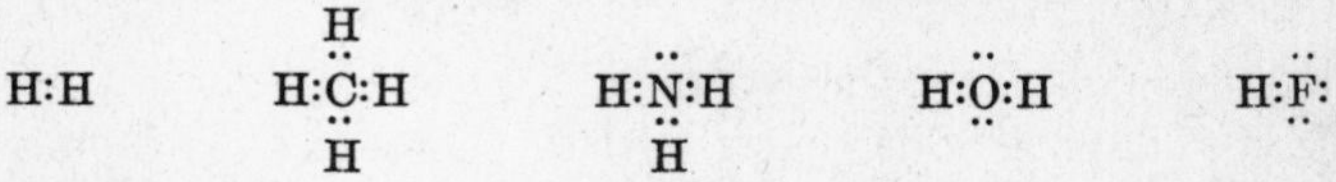

$$H:H \qquad H:\overset{H}{\underset{H}{\overset{..}{C}}}:H \qquad H:\overset{..}{\underset{H}{N}}:H \qquad H:\overset{..}{\underset{..}{O}}:H \qquad H:\overset{..}{\underset{..}{F}}:$$

In these molecules the coordination number of hydrogen is one. However, hydrogen fluoride also exists in the vapor state as the more complicated, polymeric form, H_6F_6. Water, in the liquid state, has a molecular weight greater than 18. Thus, both hydrogen fluoride and water are associated, that is, they do not have a structure corresponding to the simple formula above. It is believed that the hydrogen atom of one molecule is attracted by an unshared electron pair of another, thus forming a bond between the two atoms. A clear cut case is the bifluoride ion, HF_2^-, or $:\overset{..}{F}:H:\overset{..}{\underset{..}{F}}:$.

A qualitative explanation of this type of bond is based upon the assumption that, in a simple HF molecule, the electrons of the bond are so strongly attracted by the fluorine kernel, due to the small distance and the high kernel charge, they do not sufficiently neutralize the positive charge of the proton, which can then form an additional bond with a lone electron pair on another fluorine atom. On such a basis, one would expect this type of bond to be observed only with an atom of high kernel charge and small radius. Its stability should drop off as the kernel charge decreases, or the radius increases. Such is the case. Only in H-F, H-O, and H-N bonds does the hydrogen form a bond between atoms, and the tendency to do so decreases in the above order. Thus, liquid hydrogen fluoride is largely H_6F_6, water is a complicated system with $(H_2O)_2$, $(H_2O)_3$, etc., present, liquid ammonia is only slightly associated whereas liquid methane and liquid hydrogen are non-associative.

When hydrogen binds two atoms, its coordination number is two. It can never be more than two. This type of union has ionic character, because it is impossible for hydrogen to form more than one covalent bond.

Although the most stable hydrogen bond is the one associated with fluorine, the most important one is associated with oxygen. The O-H bond is common in inorganic and in organic chemistry; water and inorganic acids containing oxygen have this bond; also organic alcohols and acids. A common property of these compounds is the tendency of their molecules to associate. This is due to the formation of a hydrogen bond between the oxygen atoms of different molecules. For example, two molecules of acetic acid, CH_3COOH, are joined together in the form of a double molecule, i.e., dimer, through hydrogen atoms:

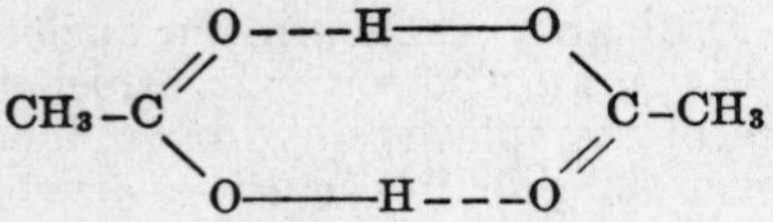

$$CH_3-C \begin{matrix} O---H-----O \\ \diagup \quad\quad\quad \diagdown \\ \diagdown \quad\quad\quad \diagup \\ O-----H---O \end{matrix} C-CH_3$$

Unsymmetrical Addition to Double Bonds:

When carbon dioxide reacts with water to form carbonic acid, the reaction may be pictured as taking place by the addition of the fragments of a water molecule to the $C = O$ double bond. From the scale of electronegativities, carbon is positively, oxygen negatively charged in carbon dioxide; also hydrogen is positively, oxygen negatively charged, in water:

$$O-\overset{+}{C}-\overset{-}{O} \rightleftarrows O-\overset{+}{C}-\overset{-}{O} + H^+ \rightleftarrows O-C-OH$$
$$\qquad\qquad \underset{OH^-}{\vdots \;\; \vdots} \qquad \underset{OH}{|}$$

Thus the positive hydrogen becomes attached to the negative oxygen in carbon dioxide, the negative hydroxyl to the positive carbon. Another hypothesization of the reaction is possible. Carbon dioxide resonates among three possible electronic structures:

$$:\ddot{O}::C:\ddot{O}: \qquad\qquad :\overset{+}{\ddot{O}}:::C:\overset{=}{\ddot{O}}: \qquad\qquad :\overset{=}{\ddot{O}}:C:::\overset{+}{\ddot{O}}:$$

$$\text{I} \qquad\qquad\qquad \text{II} \qquad\qquad\qquad \text{III}$$

If carbon dioxide reacts as if it were II or III, it is probable that the water molecule becomes attached, through the formation of a hydrogen bond, to the carbon dioxide molecule at the oxygen atom which carries a formal negative charge (IV). There could then be a rearrangement of the electronic structure, leaving the carbon temporarily deficient in electrons (V):

$$\overset{+}{:\ddot{O}}:::C:\overset{-}{\ddot{O}}:H:\ddot{O}:H \qquad :\ddot{O}::\overset{+}{C}:\ddot{O}:H + \overset{-}{:\ddot{O}}:H \qquad \begin{matrix} \overset{\displaystyle H}{} \\ :\ddot{O}: \\ :\ddot{O}::C:\ddot{O}:H \end{matrix}$$

$$\text{IV} \qquad\qquad\qquad \text{V} \qquad\qquad\qquad \text{VI}$$

The carbon atom, carrying a formal positive charge because of the electron deficiency, would then attract the negative hydroxyl ion, producing carbonic acid (VI).

The $C = O$ bond is present in many classes of organic compounds, such as aldehydes, ketones, acid chlorides, and acid anhydrides. These undergo many addition reactions in which a positively charged atom, often times hydrogen, becomes attached to the oxygen atom of the $C = O$ group, and the residual negatively charged part of the reacting molecule becomes attached to the carbon atom of the $C = O$ group. This is typified in the reaction of aldehyde with hydrogen cyanide:

$$\begin{matrix} \overset{\displaystyle H}{|} \\ CH_3 - \overset{+}{C} = \overset{-}{O} + \overset{+}{H} - \overset{-}{CN} \end{matrix} \rightarrow \begin{matrix} \overset{\displaystyle H}{|} \\ CH_3 - \overset{|}{C} - OH \\ | \\ CN \end{matrix}$$

$$\begin{matrix} \text{aldehyde} & \text{hydrogen} & \text{aldehyde} \\ & \text{cyanide} & \text{cyanohydrin} \end{matrix}$$

Another type of unsymmetrical addition involves those carbon-to-carbon double bonds which possess ionic character. This is discussed later.

Ionic Character in Carbon-to-Carbon Bonds:

Although in ethane, $H_3C:CH_3$, there is no ionic character in the C-C bond, ionic character is induced in this bond when one of the hydrogen atoms is replaced by chlorine. This is because Cl, by strongly attracting the electron pair which joins it to carbon, C-(1) (below), increases the effective kernel charge of C-(1). The latter now more strongly attracts its other six electrons, thus inducing some ionic character to the bond between it and carbon atom C-(2):

$$\begin{matrix} & H & H \\ :\ddot{Cl}: & C: & C: & H \\ & H & H \\ & (1) & (2) \\ & & I \end{matrix} \qquad \begin{matrix} H & & H \\ Cl\leftarrow\!\!-\!\!-C\leftarrow\!\!-\!\!-C\leftarrow\!\!-\!\!-H \\ H & & H \\ & II \end{matrix} \qquad \begin{matrix} H & & H \\ Cl^-\!\!\overset{+}{C}\!\!-\!\!\overset{+}{C}H \\ H & & H \\ & III \end{matrix}$$

The overall effect, as shown in I, is a displacement or shift of all of the electrons in the C_2H_5 part of the molecule towards the chlorine atom. This is represented in II by the dotted arrows, and in III by writing plus and minus charges. These last two indicate a definite, although small, amount of ionic character in these electron-pair bonds. The magnitude of this ionic character must be small because the original effect due to the ionic character in the C-Cl bond must be divided among the other three bonds of C-(1). If the carbon chain were longer, the effect would travel along it, but with a rapidly diminishing value.

This so-called *Inductive Effect* accounts for the influence of substituents, i.e., atoms or radicals replacing hydrogen, upon the ionization constants of organic acids. For example, this constant, K_A, for acetic acid, CH_3COOH, is 1.8×10^{-5}; for monochloroacetic acid, $ClCH_2COOH$, is 1.5×10^{-3}; for dichloroacetic acid, $Cl_2CHCOOH$, is 5.1×10^{-2}; and for trichloroacetic acid, Cl_3CCOOH, is a comparatively large value, for it is a strong acid. The tendency of the H of a carboxyl group, COOH, to drop off as a positively charged hydrogen ion is related to the repulsion exerted upon the proton by the oxygen kernel. If any influence tends to increase the effective kernel charge of the oxygen atom, the tendency towards ionization should likewise increase, i.e., the value for K_A should rise. It is believed that the chlorine induces an electronic shift thus:

$$\overset{\leftarrow}{Cl}-\overset{\leftarrow}{CH_2}-\overset{\leftarrow}{C}-\overset{\leftarrow}{O}-H$$
$$\underset{O}{\|}$$

This should increase the tendency of the proton to drop off from the oxygen atom, as discussed above. Other substituents produce different effects, due to the fact that some, for example nitro, NO_2, and cyano, CN, are more electronegative (attract the shared electron pair more) than chlorine, while others, for example methyl, CH_3, are more electropositive (attract the electron pair less).

Some of the more important univalent radicals commonly found in organic compounds, in the order of decreasing electronegativities (approximate order only) are nitro, NO_2; cyano, CN; sulfo, SO_2OH; nitroso, NO; formyl, CHO; benzoyl, COC_6H_5; acetyl, $COCH_3$; carboxyl, COOH; chloro, Cl, (bromo and iodo); hydroxyl, OH; methyl, CH_3; and ethyl, C_2H_5. The last two are more electropositive than hydrogen, whereas all the others are more electronegative than hydrogen.

Radicals often exert an influence upon the properties of organic compounds because of their electrochemical character. Some of the effects which they produce may be generalized as follows: when electronegative radicals are substituted for the hydrogen atoms in organic molecules, they generally (a) increase the K_A of acids (already mentioned); (b) increase the K_A of phenols; (c) decrease the K_B of organic bases; (d) increase the reactivity of aldehydes and ketones; (e) decrease the reactivity of a C=C double bond towards bromine and other similar reagents; (f) decrease the ease of substitution in the benzene ring; (g) exert a directing influence during benzene substitution.

Substituents induce ionic character in double bonds, in much the same way they do in single bonds, as discussed above, and in so doing they modify the chemical reactivity of double bonds. They also bring about, through conjugation and changes in polarizability, other effects of a more complicated nature which cannot be discussed here. In general, the replacement of one or more hydrogen atoms in ethylene by electronegative radicals decreases the tendency of the molecule to add bromine. Thus, ethylene (I), adds one mole of bromine rapidly; cinnamic acid, (II), in which two of the hydrogen atoms have been replaced by the negative radicals, C_6H_5 and COOH, adds bromine slowly; while tetrachloroethylene, in which all of the hydrogen atoms are replaced by negative chlorine atoms, does not add bromine:

$$\begin{array}{c} H \\ H \end{array} \!\!\! \diagdown C = C \diagup \!\!\! \begin{array}{c} H \\ H \end{array} + Br_2 \rightarrow \begin{array}{c} H \; H \\ | \; \; | \\ H-C-C-H \\ | \; \; | \\ Br \; Br \end{array}$$

I

$$\begin{array}{c} C_6H_5 \\ H \end{array} \!\!\! \diagdown C = C \diagup \!\!\! \begin{array}{c} H \\ COOH \end{array} + Br_2 \rightarrow \begin{array}{c} H \; H \\ | \; \; | \\ C_6H_5-C-C-COOH \\ | \; \; | \\ Br \; Br \end{array}$$

II

$$\begin{array}{c} Cl \\ Cl \end{array} \!\!\! \diagdown C = C \diagup \!\!\! \begin{array}{c} Cl \\ Cl \end{array} + Br_2 \rightarrow \text{no addition}$$

III

The possibility of resonance in II and III probably contributes to the stability of the olefinic bond. Many other addition reactions of the olefinic bond are retarded by the proximity of electronegative substituents.

The ionic character induced in a double bond by certain substituents has a directive effect upon the addition of an unsymmetrical reagent. This is shown by the nature of the products obtained when hydrogen bromide adds to propylene (IV) and to acrylic acid (V):

$$\begin{array}{c} H \\ H \end{array} \!\!\! \diagdown \overset{-}{C} = \overset{+}{C} \diagup \!\!\! \begin{array}{c} H \\ CH_3 \end{array} + HBr \rightarrow \begin{array}{c} H \; H \\ | \; \; | \\ H-C-C-CH_3 \\ | \; \; | \\ H \; Br \end{array}$$
$$\hspace{3cm} 1 \; 2$$

IV

$$\begin{array}{c} H \\ H \end{array} \!\!\! \diagdown \overset{+}{C} = \overset{-}{C} \diagup \!\!\! \begin{array}{c} H \\ COOH \end{array} + HBr \rightarrow \begin{array}{c} H \; H \\ | \; \; | \\ H-C-C-COOH \\ | \; \; | \\ Br \; H \end{array}$$

V

In IV, the electropositive character of methyl, CH_3, compared to H, allows carbon C-2 to appropriate more completely the electron pair which joins it to CH_3; the result is a decrease in the effective kernel charge of C-2.   Carbon C-1 therefore attracts the electron pair more than C-2 does; this gives rise to some ionic character in the double bond as indicated.   In V, the electronegative character of the carboxyl group, COOH, produces the opposite effect, leading to some ionic character in which the carbon atoms of the double bond carry charges opposite to those in propylene.   Although similar conclusions can be drawn from other addition reactions, the situation is complicated by the possibility of resonance, especially in the case of compounds of the vinyl chloride type (see Table III, P. 12.)

This chapter was written by Dr. H. J. Lucas, Professor of Organic Chemistry at the California Institute of Technology, Pasadena, California.

CHAPTER III

THE ALIPHATIC HYDROCARBONS

I. Alkanes, Saturated Hydrocarbons or Paraffins, Homologs of Methane

A. Introduction:

The hydrocarbons, as the name implies, contain only hydrogen and carbon. Hydrogen has a valence of one and the valence of carbon, with a few possible exceptions, is four. In the saturated hydrocarbon series, these valence relations may be illustrated by the following types:

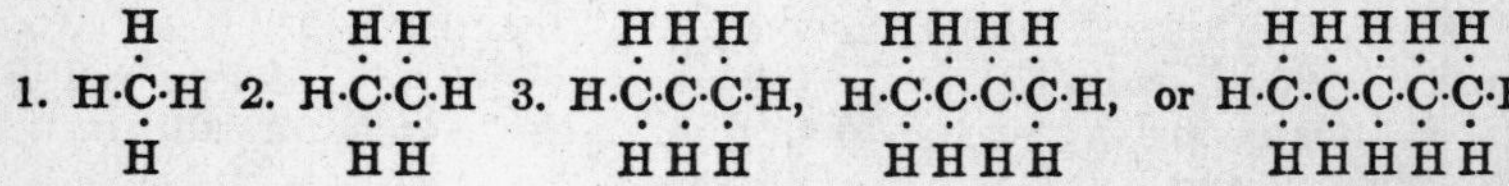

$$
\begin{array}{lllll}
& \text{H} & \text{H H} & \text{H H H} & \text{H H H H} & \text{H H H H H} \\
1. & \text{H·C·H} & 2.\ \text{H·C·C·H} & 3.\ \text{H·C·C·C·H,} & \text{H·C·C·C·C·H, or} & \text{H·C·C·C·C·C·H} \\
& \text{H} & \text{H H} & \text{H H H} & \text{H H H H} & \text{H H H H H} \\
& \text{methane} & \text{ethane} & \text{propane} & \text{butane} & \text{pentane}
\end{array}
$$

$$
\begin{array}{lll}
& & \text{H} \\
& & \text{H·C·H} \\
\text{H H H} & \text{H H H H} & \text{H | H} \\
4.\ \text{H·C · C · C·H or} & \text{H·C · C · C · C·H, and} & 5.\ \text{H·C—C—C·H} \\
\text{H | H} & \text{H | H H} & \text{H | H} \\
\text{H·C·H} & \text{H·C·H} & \text{H·C·H} \\
\text{H} & \text{H} & \text{H}
\end{array}
$$

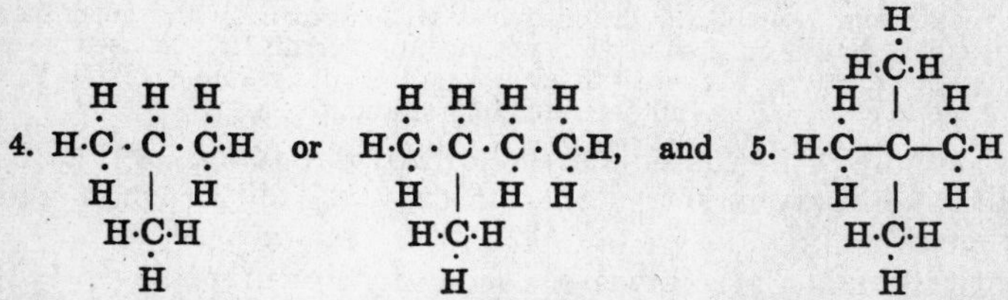

2-methylpropane (isobutane, trimethylmethane)　　2-methylbutane (isopentane, dimethylethylmethane)　　2,2-dimethylpropane (neopentane, tetramethylmethane)

The carbon atom in methane, as indicated above, has its four valences satisfied by four hydrogen atoms. In ethane, three of the valences of each carbon atom are satisfied by three hydrogen atoms and the fourth valence is satisfied by the other carbon atom. *A carbon atom that has only one of its valence bonds satisfied by another carbon atom is designated as a primary carbon atom.* In the above formulas, all but methane contain two or more *primary carbon atoms.* Since methane is the lone member in which the four valence bonds of a given carbon atom are taken up by four hydrogen atoms, one might expect methane and its derivatives to possess, as they actually do, certain specific chemical properties.

In the inner carbon groupings of propane, butane, and pentane, it is observed that only two of the valences are satisfied by hydrogen atoms and that the other two are satisfied by other carbon atoms. These illustrate *secondary carbon groupings.* The compounds 2-methylpropane (**isobutane**) and 2-methylbutane (isopentane) contain one so-called *tertiary carbon atom* each. A *quaternary carbon atom* is found in such compounds as 2, 2-dimethylpropane (neopentane).

Observe that butane and 2-methylpropane (isobutane) contain the same constituents by weight but contain different carbon chains. This illustrates *chain* isomerism. The same type of isomerism is exhibited by pentane, 2-methylbutane (isopentane), and 2, 2-dimethylpropane (neopentane).

In this hydrocarbon series, each member differs from the preceding member by a CH_2 group. Such a family of compounds, provided there is a progressive gradation in chemical reactivity and physical properties, is known as an *homologous series*.

Alkanes are represented by the type formulas C_nH_{2n+2} or $R \cdot H$, where "R" is always $C_nH_{2n+1} \cdot$ "R" is defined as the *alkyl* group and represents, as just indicated, any univalent aliphatic saturated hydrocarbon radical. This symbol, R, is used repeatedly throughout this outline, and the student should familiarize himself with its meaning and use. A few of the simple alkyl radicals are illustrated in the next division.

Alkanes occur in natural gas, marsh gas, petroleum, and in oil shale distillates.

The alkanes are employed as a source of heat and power, as solvents, and in organic syntheses. 2,2,4-Trimethylpentane (known by the misnomer isooctane, which has been assigned an octane number of 100) is used as a reference fuel in rating the anti-knock properties of gasoline. 2,2,3-Trimethylbutane (Triptane), has an octane rating of about 125.

The structure of methane is evidenced by the following:

1. Qualitative analysis shows that it contains only carbon and hydrogen.

2. Quantitative analysis gives a carbon-hydrogen ratio of 1/4, hence the empirical formula of methane is CH_4.

3. Molecular weight determinations also indicate that the molecular formula is CH_4.

4. Substitution reactions with a given reagent give mono-substituted derivatives which are all identical, hence the four hydrogen atoms in methane are identical and are, therefore, symmetrically distributed about the carbon atom.

5. The graphic formula of methane is, consequently, $H \cdot \overset{\displaystyle H}{\underset{\displaystyle H}{C}} \cdot H$

A similar line of reasoning is used to develop the structure of the other members of the series.

The alkanes exhibit *chain* isomerism. The graphic formulas for pentane, isopentane, and neopentane, shown on page 25, illustrate this type of isomerism.

B. Nomenclature:

The class name is alkanes, paraffins or saturated hydrocarbons. In the following tabulation, note that *ane* is the characteristic ending for the I.U.C. names of the members of the alkane series:

Structural Formulas	I.U.C Names	R. Alkyl Radicals	I.U.C. Names
CH_4..................	methane	CH_3..................	methyl
$CH_3 \cdot CH_3$..............	ethane	$CH_3 \cdot CH_2$..............	ethyl
$CH_3 \cdot CH_2 \cdot CH_3$..........	propane	$CH_3 \cdot CH_2 \cdot CH_2$..........	propyl
$CH_3 \cdot CH_2 \cdot CH_2 \cdot CH_3$......	butane	$CH_3 \cdot CH_2 \cdot CH_2 \cdot CH_2$......	butyl
$CH_3 \cdot CH_2 \cdot CH_2 \cdot CH_2 \cdot CH_3$..	pentane	$CH_3 \cdot CH_2 \cdot CH_2 \cdot CH_2 \cdot CH_2$.	pentyl, amyl

The next members of the series are, in order, hexane, heptane, octane, nonane, decane, hendecane or undecane, dodecane, etc.

C. Preparations:

THE ALKANES MAY BE PREPARED:

1. *By the substitution of a hydrogen atom for the $\cdot CO \cdot ONa$ group of the salt of a fatty acid when the salt is fused with soda-lime ($NaOH/CaO$):*

   $R \cdot CO \cdot ONa + NaOH/(CaO)$, fusion $\rightarrow R \cdot H + Na_2CO_3$, and
   $CH_3 \cdot CO \cdot ONa + NaOH/(CaO)$, fusion $\rightarrow CH_4 + Na_2CO_3$, or sodium ethanoate **(sodium acetate)** + soda-lime, heat $\rightarrow$ methane + sodium carbonate. In this preparation the calcium oxide slows down the reaction and tends to decrease the etching of the glassware. Since it does not enter into the reaction, the CaO is shown in parenthesis. *This convention is used throughout this outline.*

   For the preparation of the higher homologues, better yields are obtained by replacing the soda-lime with sodium methoxide, or $R \cdot CO \cdot ONa + Na \cdot O \cdot CH_3$, heat $\rightarrow R \cdot H + Na_2CO_3$ + other products.

2. *By the union of two alkyl residues when alkyl halides are treated with (a) metallic sodium in dry ether (Wurtz Reaction), or (b) the Grignard reagent in dry ether (Grignard Reaction):*

   a. $R \cdot X + 2 Na + X \cdot R$, dry ether $\rightarrow R \cdot R + 2 NaX$, and
   $CH_3 \cdot CH_2 \cdot I + 2 Na + I \cdot CH_2 \cdot CH_3$, dry ether $\rightarrow CH_3 \cdot CH_2 \cdot CH_2 \cdot CH_3 +$ 2 NaI, or iodoethane **(ethyl iodide)** + sodium in dry ether $\rightarrow$ butane + sodium iodide.

   b. $R \cdot X + R \cdot Mg \cdot X$, dry ether $\rightarrow R \cdot R + MgX_2$, and
   $CH_3 \cdot I + CH_3 \cdot CH_2 \cdot Mg \cdot I$, dry ether $\rightarrow CH_3 \cdot CH_2 \cdot CH_3 + MgI_2$, or iodomethane **(methyl iodide)** + ethylmagnesium iodide in dry ether $\rightarrow$ propane + magnesium iodide.

   The Grignard reagent is prepared by treating clean magnesium in dry ether with an alkyl halide:

   $R \cdot Br + Mg$, dry ether $\rightarrow R \cdot Mg \cdot Br$ (alkylmagnesium bromide).

   In laboratory practice, the Wurtz reaction is more satisfactory for condensations in which two identical alkyl groups are to be linked together; for the linking of different alkyl groups the Grignard reaction is more desirable.

3. *By the substitution of hydrogen (a) for halogen in alkyl halides when treated with concentrated hydriodic acid and red phosphorus (reduction), or (b) for the metallic groups in organometallic compounds when treated with water or dilute acids (hydrolysis), alcohols (alcoholysis), ammonia (ammonolysis), amines, etc.:*

   a. $3 R \cdot X + 6 HI/2 P$, heat $\rightarrow 3 R \cdot H + 3 HX + 2 PI_3$, and
   $3 CH_3 \cdot CH_2 \cdot Br + 6 HI/2 P$, heat $\rightarrow 3 CH_3 \cdot CH_3 + 3 HBr + 2 PI_3$, or bromoethane **(ethyl bromide)** + hydrogen iodide/red phosphorus, heat $\rightarrow$ ethane + hydrogen bromide + phosphorus triiodide.

b. 1. R·Mg·X (Grignard reagent) + HX, aq. → R·H + MgX₂, and

$CH_3 \cdot CH_2 \cdot Mg \cdot Cl + HCl$, aq. → $CH_3 \cdot CH_3 + MgCl_2$, or ethylmagnesium chloride + hydrochloric acid → ethane + m gnesium chloride.

2. R·Zn·X (or R₂Zn) + HX, aq. → R·H + ZnX₂ (or 2 R·H + ZnX₂).

A number of the organometallic compounds, such as R₂Zn, undergo spontaneous combustion in the presence of oxygen.

4. By the loss of alkenes from the cracking of higher alkanes in the liquid phase, through the application of heat (commercial):

$R \cdot CH_2 \cdot CH_2 \cdot R'$ + heat and pressure → $R \cdot H + H_2C:CHR'$, or petroleum residues + cracking → lower members of the saturated and unsaturated hydrocarbon series (where R′ is usually less than R).

Though the chemical nature of this reaction is not understood, recent studies have indicated that *free radicals* may play an important role in the mechanism involved. Cracking and polymerization have increased the yield of gasoline, obtained from crude, from 20 to over 70%. *n*-Butane, with AlCl₃ and 200 pounds pressure at 25° C., gives satisfactory yields of isobutane.

5. By the reductive addition of hydrogen to alkenes, CₙH₂ₙ (commercial):

R·HC:CH·R′ + H₂/(fine Pt, Pd or Ni) → R·CH₂·CH₂·R′, and

$CH_3 \cdot CH_2 \cdot HC:CH_2 + H_2$/(Pt, Pd or heated Ni) → $CH_3 \cdot CH_2 \cdot CH_2 \cdot CH_3$, or 1-butene (α-butylene) + hydrogen/ (Pt, Pd or Ni as a catalyst)→ butane. In laboratory practice, reducing agents capable of liberating hydrogen are frequently used (NaHg/H₂O, Na/EtOH, etc.).

6. By (a) the catalytic condensation of hydrogen with carbon monoxide to give mixtures of alkanes (Fischer-Tropsch Synthesis), or (b) the destructive distillation of coal (coking process):

a. x CO + y H₂/ (catalyst and pressure) → alkane mixtures resembling petroleum (commercial) + x H₂O.

The Fischer-Tropsch synthesis is now being used commercially in countries with a limited supply of petroleum. The reaction is carried out at 250-300° C., under about 200 pounds pressure, in the presence of a nickel-chromite catalyst. The reaction is readily controlled, and the products vary with the conditions.

b. Coal + proper coking → small yields of alkanes (commercial).

This is the so-called liquefaction process by which coal is heated to about 500° C. under a pressure of 4,000 to 5,000 pounds to yield hydrocarbon products. Four tons of coal yield about one ton of gasoline. The reaction is of value for countries with large coal but small petroleum supplies.

7. By the following specific methods for methane:

a. C + 2 H₂/(Ni catalyst at 475°, pressure) → CH₄, or carbon + hydrogen over nickel at 475° → methane.

b. CO + 3 H₂/(Ni, at 250°, pressure) → CH₄ + H₂O, or carbon monoxide + hydrogen over nickel at 250° → methane + water. Carbon dioxide may be reduced by a similar procedure.

c. Al₄C₃ + 12 H₂O → 3 CH₄ + 4 Al(O)OH·H₂O, or aluminum carbide + water → methane + aluminum hydroxide.

D. Physical Properties:

Of the normal (straight-chain) alkanes, C₁ to C₄ are gases, C₅ to C₁₅ are liquids, and C₁₆ and above are solids. In general, the boiling point, melting point, viscosity, and specific gravity, increase with an increase in the molecular weight. Branched-chain compounds, however, boil lower than the corresponding straight-chain compounds. This gradation of properties is shown by the boiling points of the following: methane, –161.66; ethane, –88.63; propane, –42.59; butane, –0.65; pentane, 36.0; isopentane, 28.0; and neopentane, 9.5. The alkanes are practically insoluble in water, colorless, and the higher members are usually odorless. Methane possesses a slightly alliaceous smell and some of the other members possess faint but characteristic odors.

E. Chemical Properties:

THE ALKANES REACT UNDER PROPER CONDITIONS:

1. By substitution with chlorine, bromine, nitric acid, and sulfuric acid, as indicated below:

Alkanes and chlorine	R	H + Cl	Cl	R is a primary, secondary, or tertiary saturated hydrocarbon radical.
Alkanes and bromine	R	H + Br	Br	
Alkanes and nitric acid	R	H + HO	NO_2	
Alkanes and sulfuric acid	R	H + HO	$SO_2{\cdot}OH$	

For example, $CH_4 + 4\ Cl_2$ (vapor phase) $\rightarrow CH_3Cl \rightarrow CH_2Cl_2 \rightarrow CHCl_3 \rightarrow CCl_4 + 4\ HCl$, or methane + chlorine → **methyl chloride, methylene chloride, chloroform, or carbon tetrachloride,** depending on the conditions and the relative amount of chlorine used in the reaction. A mixture of all of these products is usually obtained.

Nitration is effected by the use of concentrated nitric acid in the vapor phase and is a general reaction, whereas sulfonation is applicable to tertiary H-atoms only.

2. By cleavage when subjected to pyrolysis or severe oxidation:

a. $R{\cdot}CH_2{\cdot}CH_2{\cdot}R' + \text{heat and pressure} \rightarrow R{\cdot}H + H_2C{:}CH{\cdot}R'$, where R′ is usually lighter than R, or pyrolysis of the higher alkanes yields alkane and alkene residues.

Under certain conditions of pyrolysis, the alkanes yield carbon and hydrogen as the chief products. At high temperatures in the presence of moisture and appropriate catalysts, however, the reaction takes the following course: $CH_4 + H_2O/(\text{catalyst, heat}) \rightarrow CO + 3\ H_2$.

b. $2\ CH_3{\cdot}CH_3 + 7\ O_2\ (\text{burning}) \rightarrow 4\ CO_2 + 6\ H_2O + x$ calories, or alkanes + burning, excess oxygen → carbon dioxide + water + heat.

The vapor phase oxidation of benzene and naphthalene to maleic anhydride and phthalic anhydride respectively, represent important commercial processes [Ind. Eng. Chem., *32*, 1294 (1940)].

II. Alkenes and Alkynes, or Olefins and Acetylenes

A. Introduction:

Unsaturated hydrocarbons are similar to the alkanes, but they contain carbon atoms united by double or triple bonds as shown below:

```
 HHHH         HHHH      H  H        HH             H   H
H·C·C·C:C·H  H·C·C:C·C·H H·C·C:C·H  H·C·C·C:C·H   H·C·C:C·C·H
 HH           H   H       H CH3      HH             H   H
1-butene     2-butene  2-methylpropene 1-butyne      2-butyne
```

Unsaturated hydrocarbons containing a double bond are known as alkenes or **olefins,** and those containing triple bonds are known as alkynes or **acetylenes.** Owing to structural strains, double bonds and triple bonds may open and form additive products as illustrated below:

$$
\begin{array}{cccc}
\text{H} & \text{H} & \text{H} & \text{H} \\
\text{H·}\overset{\cdot}{\text{C}}\text{·} & \overset{\cdot}{\text{C}}\text{:} & \text{C·} & \overset{\cdot}{\text{C}}\text{·H} \\
\dot{\text{H}} & & \dot{\text{H}} &
\end{array}
+ \text{Br}_2 \rightarrow
\begin{array}{cccc}
\text{H} & \text{H} & \text{H} & \text{H} \\
\text{H·}\overset{\cdot}{\text{C}}\text{·} & \overset{\cdot}{\text{C}}\text{·} & \text{C·} & \overset{\cdot}{\text{C}}\text{·H} \\
\dot{\text{H}} & \text{Br} & \text{Br} & \dot{\text{H}}
\end{array}
\text{(2,3-dibromobutane,}
$$
β-butylene bromide), or

$$
\begin{array}{cccc}
\text{H} & \text{H} & & \\
\text{H·}\overset{\cdot}{\text{C}}\text{·} & \overset{\cdot}{\text{C}}\text{·} & \text{C:C·H} \\
\dot{\text{H}} & \dot{\text{H}} & &
\end{array}
+ \text{Br}_2 \rightarrow
\begin{array}{cccc}
\text{H} & \text{H} & \text{Br} & \text{Br} \\
\text{H·}\overset{\cdot}{\text{C}}\text{·} & \overset{\cdot}{\text{C}}\text{·} & \overset{\cdot}{\text{C}}\text{:} & \text{C·H} \\
\dot{\text{H}} & \dot{\text{H}} & &
\end{array}
\text{(1,2-dibromo-1-butene), then}
$$

$$
\begin{array}{cccc}
\text{H} & \text{H} & \text{Br} & \text{Br} \\
\text{H·}\overset{\cdot}{\text{C}}\text{·} & \overset{\cdot}{\text{C}}\text{·} & \overset{\cdot}{\text{C}}\text{:} & \text{C·H} \\
\dot{\text{H}} & \dot{\text{H}} & &
\end{array}
+ \text{Br}_2 \rightarrow
\begin{array}{cccc}
\text{H} & \text{H} & \text{Br} & \text{Br} \\
\text{H·}\overset{\cdot}{\text{C}}\text{·} & \overset{\cdot}{\text{C}}\text{·} & \text{C·} & \overset{\cdot}{\text{C}}\text{·H} \\
\dot{\text{H}} & \dot{\text{H}} & \text{Br} & \text{Br}
\end{array}
\text{(1,1,2,2-tetrabromobutane).}
$$

The alkenes and the alkynes form homologous series. *They are characterized, as indicated above, by addition reactions in preference to the typical substitution reactions of the alkanes.*

Alkenes are represented by the type formulas C_nH_{2n}, $R·HC:CH_2$, or $R·HC:CH·R$; the alkynes by C_nH_{2n-2}, $R·C:C·H$, or $R·C:C·R$.

Alkenes result from the destructive distillation of coal, or petroleum. Ethene is formed in what appears to be an anaerobic decomposition of carbohydrate material. An apple, for example, liberates about one cubic centimeter of ethene in thirty days. Alkynes are synthetic products. Traces of acetylene are formed when the flame of a Bunsen burner strikes back.

Alkenes and alkynes are employed as a source of heat and power and in organic synthesis. Ethene (**ethylene**) is used as an anesthetic and for accelerating the ripening of citrus fruit. Ethyne (**acetylene**) finds application in welding, in the production of **chloroprene,** and in the synthesis of glacial **acetic acid.**

The methods of preparation and the chemical properties will indicate the structures of the alkenes and the alkynes.

Different positions of the unsaturated linkages in identical carbon chains give rise to *position* isomerism, which is evidenced by both the alkenes and the alkynes.

B. Nomenclature:

The class names for compounds containing double and triple bonds are, respectively, alkenes or **olefins** and alkynes or **acetylenes.** The I.U.C. names of the alkenes are characterized by the ending *ene* and those of the alkynes by the ending *yne.*

Structural Formulas	Names
$H_2C:$ (unstable)	methene, or **methylene**
$H_2C:CH_2$	ethene, or **ethylene**
$CH_3·HC:CH_2$	propene, or **propylene**
$CH_3·CH_2·HC:CH_2$	1-butene, or α-**butylene**
$CH_3·HC:CH·CH_3$	2-butene, or β-**butylene**

CH₃·CH₂·CH₂·HC:CH₂ .... 1-pentene, α-amylene
CH₃·CH₂·HC:CH·CH₃ .... 2-pentene, β-amylene
(CH₃)₂C:C(CH₃)₂ 2,3-dimethyl-2-butene
H·C⫶ (unstable)...................... methyne
H·C⫶C·H........................... ethyne, or acetylene
CH₃·C⫶C·H......................... propyne, or methylacetylene
CH₃·CH₂·C⫶C·H.................... 1-butyne, or ethylacetylene
CH₃·C⫶C·CH₃...................... 2-butyne, or dimethylacetylene
CH₃·CH₂·CH₂·C⫶C·H............... 1-pentyne, or propylacetylene
CH₃·CH₂·C⫶C·CH₃................. 2-pentyne, or ethylmethylacetylene
CH₂·CH₂·CH₂·CH₂·C⫶C·H........... 1-hexyne, or butylacetylene

C. Preparations:

I. THE ALKENES MAY BE PREPARED:

1. *By the removal of* (a) *HX from alkyl halides,* (b) *2X from 1, 2-, or 2, 3-, etc., dihalogen compounds,* (c) *H₂O from alcohols, or* (d) *R·H from the higher hydrocarbons* (commercial):

   a. $R \cdot CH_2 \cdot CH_2 \cdot X + NaOH$, alc .→ $R \cdot HC:CH_2 + NaX + H_2O$, and $CH_3 \cdot CH_2 \cdot CH_2 \cdot Br + NaOH$, alc. → $CH_3 \cdot HC:CH_2 + NaBr + H_2O$, or 1-bromopropane (propyl bromide) + alcoholic sodium hydroxide → propene (propylene) + sodium bromide + water. *Dry powdered,* molten, or alcoholic potassium hydroxide may be used as the reagent in these reactions. The use of alcoholic alkalies tends toward the formation of ethers.

   b. $R \cdot CHX \cdot CH_2 \cdot X$ (or $R \cdot CHX \cdot CHX \cdot R$) + Zn → $R \cdot HC:CH_2$ (or $R \cdot HC:CH \cdot R$) + ZnX₂, and
   $CH_3 \cdot CHBr \cdot CH_2 \cdot Br + Zn → CH_3 \cdot HC:CH_2 + ZnBr_2$, or 1,2-dibromopropane (propylene bromide) + zinc dust → propene (propylene) + zinc bromide.

   c. $CH_3 \cdot CH_2OH + HO \cdot SO_2 \cdot OH ⇌ CH_3 \cdot CH_2 \cdot O \cdot SO_2 \cdot OH + H_2O$, then $CH_3 \cdot CH_2 \cdot O \cdot SO_2 \cdot OH +$ heat, 170° → $H_2C:CH_2 + H_2SO_4$, or ethanol (ethyl alcohol) + sulfuric acid ⇌ ethyl hydrogen sulfate and water. Then, ethyl hydrogen sulfate + heat, 170° → ethene (ethylene) + sulfuric acid. Ethanol is the only *primary* alcohol that yields olefins readily by this method, *hence a type reaction has not been given.* Secondary and tertiary alcohols, however, dehydrate more readily and give better yields of olefins under these conditions. P₂O₅, H₃PO₄, or Al₂O₃, with appropriate conditions, may be used as the dehydrating agent. Commercially, Al₂O₃ is used to prepare the ethylene which is used in the synthesis of ethylene bromide, which must be free of the more corrosive propylene bromide. The ease of dehydration of the alcohols decreases, according to the type of alcohol, in the order R₃C·OH, R₂CH·OH, R·CH₂·OH.

   d. $R' \cdot CH_2 \cdot CH_2 \cdot R +$ cracking → $R'HC:CH_2 + R \cdot H$, and
   $CH_3 \cdot CH_2 \cdot CH_2 \cdot CH_2 \cdot CH_2 \cdot CH_3 +$ cracking → $H_2C:CH_2 +$ $CH_3 \cdot CH_2 \cdot CH_2 \cdot CH_3$, or hexane + cracking (heat and pressure) → ethene (ethylene) + butane (commercial). Other possible products of this particular reaction are propene and propane, or pentene and methane.

2. By the reductive addition of two hydrogen atoms to yne compounds:
$R \cdot C : C \cdot H + H_2 /$ (Pt, Pd or heated Ni) $\rightarrow R \cdot HC : CH_2$, and
$CH_3 \cdot CH_2 \cdot C : C \cdot H + H_2/$(Pt, Pd or heated Ni) $\rightarrow CH_3 \cdot CH_2 \cdot HC : CH_2$,
or 1-butyne (**ethylacetylene**) + molecular hydrogen and a catalyst,
or appropriate reducing agents $\rightarrow$ 1-butene (α-**butylene**).

3 By the polymerization of simpler ene compounds:
Many olefins undergo polymerization, even at room temperature, in
consequence of the union of several molecules, when treated with aluminum
chloride, dilute sulfuric acid, zinc chloride, or boron fluoride: $2C_4H_8 \rightarrow$
C_8H_{16}; x $C_5H_{10} \rightarrow C_{10}H_{20}$, $C_{15}H_{30}$, etc.

II. THE ALKYNES MAY BE PREPARED:

1. By the removal of (a) 2 H X from 1,1-, 1,2-, 2,2-, or 2,3-, etc.,
dihalogen compounds, or (b) 4 X from 1,1,2,2-, or 2,2,3,3-, etc.,
tetrahalogen compounds:

    a. $R \cdot CH_2 \cdot CHX_2 + 2 NaOH$, alc. $\rightarrow R \cdot C : C \cdot H + 2 NaX + 2H_2O$, and
    $CH_3 \cdot CH_2 \cdot CHBr_2 + 2$ NaOH, alc. $\rightarrow CH_3 \cdot C : C \cdot H + 2$ NaBr $+ 2$ H_2O,
    or 1,1-dibromopropane (**propylidene bromide**) + alcoholic sodium
    hydroxide $\rightarrow$ propyne (**methylacetylene**) + sodium bromide + water.
    Dry powdered, molten, or alcoholic potassium hydroxide, or hot cal-
    cium oxide may be used as the reagent.

    b. $R \cdot CX_2 \cdot CHX_2 + 2$ Zn $\rightarrow R \cdot C : C \cdot H + 2$ ZnX_2, and
    $CH_3 \cdot CH_2 \cdot CI_2 \cdot CHI_2 + 2$ Zn $\rightarrow CH_3 \cdot CH_2 \cdot C : C \cdot H + 2$ ZnI_2, or
    1,1,2,2-tetraiodobutane + zinc dust $\rightarrow$ 1-butyne + zinc iodide.

2. By (a) the double decomposition of acetylides with alkyl halides,
or (b) the hydrolysis of acetylides:

    a. $R \cdot C : C \cdot M + X \cdot R \rightarrow R \cdot C : C \cdot R + MX$, and
    $CH_3 \cdot C : C \cdot Na + I \cdot CH_3 \rightarrow CH_3 \cdot C : C \cdot CH_3 + NaI$, or 1-propynylsodium
    (**sodium methylacetylide**) + iodomethane (**methyl iodide**) $\rightarrow$ 2-butyne
    (**dimethylacetylene**) + sodium iodide.

    b. $R \cdot C : C \cdot M + H_2O \rightarrow R \cdot C : C \cdot H + MOH$.

3. By the following specific reactions for ethyne (acetylene):

    a. $CaC_2 + 2 H_2O \rightarrow H \cdot C : C \cdot H + Ca(OH)_2$ (commercial), or
    calcium carbide + water $\rightarrow$ ethyne (**acetylene**) + calcium hydroxide.

    b. x C + y H_2, electric arc $\rightarrow H \cdot C : C \cdot H + CH_4 + CH_3 \cdot CH_3$, or
    carbon + hydrogen, electric arc $\rightarrow$ ethyne + methane + ethane.

    c. x CH_4, electric arc $\rightarrow H \cdot C : C \cdot H$ (commercial) + other
    products, or natural gas rich in methane, electric arc $\rightarrow$ ethyne
    (**acetylene**) + other products.

D. Physical Properties:
Of the normal alkenes, C_1 to C_4 are gases, C_5 to C_{15} are liquids, and C_{16}
and above are solids. In general, the boiling point, melting point, viscosity and
specific gravity increase with an increase in molecular weight. The boiling
points of the 1-alkenes are about the same as those of the corresponding alkanes. The alkenes are colorless, only slightly soluble in water, and the lowest
member ethene (**ethylene**), has a rather pleasant odor.
The physical properties of the alkynes are similar to those of the alkenes.
The following tabulation indicates gradation of properties:

	ethyne	propyne	butyne	pentyne	hexyne	heptyne	octyne
B.p.	−88.5	−23.	18.5	40	71.5	110.5	125.
F.p.	−81.8	−104.7	−130.	−95.	−150	−70	. . . .
D.	$.6208^{-8}_{4}{}^4$	$.6785^{-2}_{4}{}^7$	$.668^{2}_{4}$	$.6882^{15}_{4}$	$.7120^{15}_{4}$	$.7288^{15}_{4}$	.770

A graph of the freezing points of propyne, butyne, pentyne, hexyne, and heptyne, in which the carbon content is plotted as abscissas and the melting points as ordinates, illustrates what is known as the *saw tooth rule.*

E. Chemical Properties:

I. THE ALKENES REACT UNDER PROPER CONDITIONS:

1. *By (a) oxidative addition with cold dilute alkaline permanganate, chlorine, bromine, iodine in alcohol, and ozone, (b) reductive addition with hydrogen from appropriate reducing agents or molecular hydrogen in the presence of finely divided platinum, palladium, or heated nickel, and (c) simple addition with hydrochloric acid (in the presence of aluminum chloride, commercial), hydrobromic acid, hydriodic acid, sulfuric acid, nitrous anhydride, nitrogen tetroxide, and hypohalous acids to give the products indicated below:*

$$H_2C \cdot \quad OH \ Cl \ Br \ I \qquad H \quad H \quad H \quad H \quad H \qquad NO \quad NO \quad X$$
$$\mid \qquad\qquad\qquad\qquad\qquad O_3$$
$$R \cdot HC \cdot \quad OH, Cl, Br, I, \quad ; H; \ Cl, \ Br, \ I, \ OSO_3H, \ ONO, \ ONO_2, OH,$$

To illustrate, $R \cdot HC:CH_2 + HBr \rightarrow R \cdot CHBr \cdot CH_3$ (2-bromoalkane), or $3 \ R \cdot HC:CH_2 + 2 \ MnO_4^- + 4 \ H_2O \rightarrow 3 \ R \cdot CHOH \cdot CH_2OH + 2 \ MnO_2 + 2 \ OH^-$

In these reactions, when the reagent adds as hydrogen and a radical, the latter becomes attached, under proper conditions, to the carbon atom with the least number of hydrogen atoms, or in general the more negative group adds to the more positive carbon atom. This principle is known as *Markovnikov's* (**Markownikoff's**) *rule.* **Cf. Work of M. S Kharasch.**

Ethene and chlorine give small amounts of 1,1,2-trichloroethane. When ethene is heated with bromine and NaCl or NaNO₃, the products are $Br \cdot CH_2 \cdot CH_2 \cdot Cl$ and $Br \cdot CH_2 \cdot CH_2 \cdot ONO_2$.

Ethene and sulfuric acid give, depending on conditions, ethyl bisulfate, ethyl ether, ethylene, ethionic acid, and β-hydroxyethyl sulfonic acid.

Ethene, when treated with air in the presence of silver or gold, gives ethylene oxide (commercial), and olefins add alkanes (especially *t*-H) to give polygas (Houdry Process).

2. *By cleavage when subjected to vigorous oxidation:*

a. $CH_3 \cdot HC:CH_2 + (a + \frac{1}{2}b) \ O_2$ (burning) $\rightarrow a \ CO_2 + b \ H_2O + c \ C + x$ calories.

b. $3 \ R \cdot HC:CH \cdot R' + 8 \ MnO_4^-$, aq. $\rightarrow 3 \ R \cdot CO \cdot O^- + 3 \ R' CO \cdot O^- + 8 \ MnO_2 + 2 \ H_2O + 2 \ OH^-$

The isolation and identification of these products, or the products obtained from the hydrolysis of ozonides ($R_2C \overset{O_3}{\diagup\!\!-\!\!\diagdown} CH \cdot R' + H_2O/Zn \rightarrow R \cdot CO \cdot R + R' \cdot CHO + H_2O_2$), serve to locate the unsaturated group in the molecule.

3. *By polymerization to give indefinite mixtures of derivatives:*

Isobutylene (C_4H_8) + (catalyst) → di-isobutylene (C_8H_{16}), and iso-amylene (C_5H_{10}) + (catalyst) → di-isoamylene ($C_{10}H_{20}$), tri-isoamylene ($C_{15}H_{30}$), etc.

II. THE ALKYNES REACT UNDER PROPER CONDITIONS:

1. *By stepwise (a) oxidative addition with cold dilute alkaline permanganate, chlorine (in presence of $SbCl_5$), bromine, iodine in alcohol, and ozone (b) reductive addition with hydrogen from appropriate reducing agents or molecular hydrogen in the presence of finely divided platinum, palladium, or heated nickel, and (c) simple addition with hydrochloric acid (in the presence of aluminum trichloride), hydrobromic acid, hydriodic acid, sulfuric acid, nitrous anhydride, nitrogen tetroxide, hypohalous acids, and water in the presence of sulfuric acid and mercuric salts, to give the initial additive products indicated below:*

$$
\begin{array}{l}
\text{H·C· OH Cl Br I} \quad\quad \text{H H H H H} \quad\quad \text{NO NO X H}\\
\phantom{\text{H·C}}\|\phantom{\text{OH Cl Br I}}\;\;O_3\\
\text{R·C· OH, Cl, Br, I,} \quad ; \text{H; Cl, Br, I, OSO}_3\text{H, ONO, ONO}_2\text{, OH, OH}
\end{array}
$$

Only the first addition is indicated in this tabulation except perhaps for the reactions with alkaline permanganate and the catalytic addition of water.

In the reaction with alkaline permanganate solution, the initial intermediate may be of the alkene oxide type $(R·HC \overset{\diagdown O\diagup}{-} CH_2)$ which adds water to give the corresponding glycol, and in the catalytic addition of water the mechanism may involve the addition of sulfuric acid followed by hydrolysis to give a *final* product of the type $R·CO·CH_3$. (With ethyne the final product is ethanal.)

The second addition occurs, in general, in an analogous manner. For example, $R·C\!:\!C·H + 2\,HBr \rightarrow R·BrC\!:\!CH_2 \rightarrow R·CBr_2·CH_3$.

2. *By cleavage when subjected to vigorous oxidation:*

$$CH_3·C\!:\!C·H + (a + \tfrac{1}{2}b)\,O_2 \rightarrow a\,CO_2 + b\,H_2O + c\,C + x\text{ calories}$$

3. *By polymerization to give well defined derivatives:*
   $R·C\!:\!C·H$, passage through a hot tube → symmetrical trialkyl-benzenes, or $R·C\!:\!C·R$, a hot tube → hexalkylbenzene, and 3 $H·C\!:\!C·H$, passage through a hot tube → C_6H_6, benzene

4. *By substitution of metals or halogen for the hydrogen atom of the ynyl group ($·C\!:\!C·H$):*

   a. $R·C\!:\!C·H + Cu(NH_3)_2Cl$, aq. → $R·C\!:\!C·Cu + NH_4Cl + NH_3$, and $CH_3·C\!:\!C·H + Cu(NH_3)_2Cl$, aq. → $CH_3·C\!:\!C·Cu + NH_4Cl + NH_3$, or propyne (methylacetylene) + ammoniacal cuprous chloride → 1-propynylcopper (copper methylacetylide) + ammonium chloride + ammonia, and

   $R·C\!:\!C·H + Ag(NH_3)_2NO_3$, aq. → $R·C\!:\!C·Ag + NH_4NO_3 + NH_3$.

   These reactions are specific for the ynyl group ($·C\!:\!C·H$).

   b. $H·C\!:\!C·H + 2\,KOI \rightarrow I·C\!:\!C·I$ (1,2-diiodoethyne) + 2 KOH
      Both the mono and dihalo derivatives of acetylene have explosive properties, particularly in contact with air.

   c. $2\,H·C\!:\!C·H + H_2O$, superheated steam → $CH_3·CO·CH_2 + ?$

5. *By rearrangement (isomerization) to give isomeric forms:*
   a. $CH_3·CH_2·C\!:\!C·H + $ (ethanol, heat) → $CH_3·C\!:\!C·CH_3$, or
   b. $CH_3·CH_2·C\!:\!C·CH_3 + $ (Na, heat) → $CH_3·CH_2·CH_2·C\!:\!C·H$

III. Alkadienes or Diolefins

The alkadienes, as the name suggests, are hydrocarbons that contain two ene linkages per molecule, and are represented by the type formula C_nH_{2n-2}.

The naming of these compounds is similar to that for the ene-series except that the characteristic ending becomes *diene.*

The alkadienes are synthetic products. They may be prepared by the removal of 2 HX or 4 X from the proper halogen compounds by reactions analogous to those given for the preparation of ene compounds. Butadiene may be obtained by passing the vapors of cyclohexene over a red-hot platinum spiral, or by treating the 1,2-, 1,3- and 1,4-dichlorobutane mixture with soda lime at 700° C., or by pyrolysis of butane (50–60%, U. O. P.). Among the most important members of this group are 1,3-**butadiene** ($H_2C:CH \cdot HC:CH_2$) and the **substituted butadienes.**

A few reactions that are peculiar to the dienes are:

1. 1, 4-Additions to the 1, 3-alkadienes (α-γ-diolefins):

$$H_2C:CH \cdot HC:CH_2 + Br_2 \rightarrow H_2CBr \cdot HC:CH \cdot BrCH_2 \text{ (1, 4-dibromo-2-butene)}$$

This illustrates the first step in the addition to what are known as *conjugated double bonds.* The addition of halogen acids to unsaturated compounds of the general type of acrylic acid, may be interpreted as addition to a *conjugated system:*

$$H_2C:CH \cdot C(OH):O + H \cdot Br \rightarrow (H_2CBr \cdot HC:C[OH]OH) \rightarrow H_2CBr \cdot HCH \cdot CO \cdot OH$$
acrylic acid   unstable intermediate   3-bromopropanoic acid

2. Polymerizations to give synthetic rubber:

a. x $H_2C:C(CH_3) \cdot HC:CH_2$ (**isoprene**) + (catalyst) $\rightarrow$ rubber-like products

b. x $H_2C:C(Cl) \cdot HC:CH_2$ (**chloroprene**) + (polymerization) $\rightarrow$ **Neoprene,**

a synthetic rubber. Chloroprene is produced by the addition of hydrochloric acid to 1-buten-3-yne ($H_2C:CH \cdot C:C \cdot H$, **vinylacetylene**), which is obtained by the controlled polymerization of acetylene in the presence of cuprous salts.

3. Isomerization, of certain dienes, to give yne compounds:

$$(CH_3)_2C:C:CH_2 + (Na) \rightarrow (CH_3)_2CH \cdot C:C \cdot H$$

The reactions of maleic anhydride with conjugated alkadienes find technical use in the removal of dienes in the purification of benzene, the preparation of water-soluble salts from tung oil, and the preparation of resin intermediates. Butadiene, chloroprene, and isoprene are employed in the production of synthetic rubber.

IV. Commercial Applications

The most important applications of the aliphatic hydrocarbons are as (1) sources of energy and heat and (2) synthetic intermediates for the production of a variety of important derivatives.

There is an ever increasing demand for liquid petroleum gas (LPG) and liquid petroleum tank gas (essentially mixtures of propane and butane), which are used as a source of heat and/or power for cabs, buses, and cooking, as industrial fuels, and for the enrichment of natural gas.

The high-octane motor-fuels blends are rich in synthetics such as 2,2,4-trimethylpentane (*isooctane*) and 2,2,3-trimethylbutane (*triptane*), all obtained from petroleum in modern processing, and contain lesser amounts of unsaturated compounds.

Synthetic derivatives of various types have developed into major industries.

CHAPTER IV

ALIPHATIC HALOGEN DERIVATIVES

I. Monohaloalkanes or Alkyl Halides

A. Introduction:

The alkyl halides are compounds in which one hydrogen atom in an alkane has been replaced by a fluorine, chlorine, bromine, or iodine atom. They may be represented by the type formulas $R \cdot CH_2 \cdot X$, $R_2CH \cdot X$, and $R_3C \cdot X$ to designate respectively primary, secondary and tertiary compounds. The general formula may be written $R \cdot X$ or $C_nH_{2n+1}X$, where X represents F, Cl, Br, or I.

Position isomerism is illustrated by $CH_3 \cdot CH_2 \cdot CH_2 \cdot Cl$ (1-chloropropane, propyl chloride) and $CH_3 \cdot CHCl \cdot CH_3$ (2-chloropropane, isopropyl chloride).

Alkyl halides are very useful synthetic reagents. Chloroethane (ethyl chloride) finds a limited use as a local anesthetic and an extensive use in the production of tetraethyllead. The amyl chlorides, prepared from light petroleum fractions, are important synthetic reagents. Alkyl halides also find use as solvents. Methyl bromide is being used as an insecticide, particularly in the fruit orchards on the west coast.

B. Nomenclature:

Formulas	I.U.C. Names	Common Names
$CH_3 \cdot Cl$	chloromethane	methyl chloride
$CH_3 \cdot CH_2 \cdot Br$	bromoethane	ethyl bromide
$CH_3 \cdot CH_2 \cdot CH_2 \cdot I$	1-iodopropane	propyl iodide
$CH_3 \cdot CHI \cdot CH_3$	2-iodopropane	isopropyl iodide
$CH_3 \cdot CH_2 \cdot CH_2 \cdot CH_2 \cdot Cl$	1-chlorobutane	butyl chloride
$CH_3 \cdot CH_2 \cdot CHBr \cdot CH_3$	2-bromobutane	*sec*-butyl bromide
$CH_3 \cdot (CH_3)C(I) \cdot CH_3$	2-iodo-2-methylpropane	*ter*-butyl iodide
$CH_3 \cdot CH_2 \cdot CH_2 \cdot CH_2 \cdot CH_2 \cdot Cl$	1-chloropentane	amyl chloride
$CH_3 \cdot CH_2 \cdot CH_2 \cdot CHCl \cdot CH_3$ . .	2-chloropentane	*sec-act*-amyl chloride
$(CH_3)_2CH \cdot CH_2 \cdot CH_2 \cdot Cl$ . . .	3-chloro-2-methylbutane	isoamyl chloride
$(CH_3)_2CH \cdot CHI \cdot CH_3$	3-iodo-2-methylbutane	*sec*-isoamyl iodide
$CH_3 \cdot CH_2 \cdot CBr(CH_3)_2$	2-bromo-2-methylbutane	*ter*-amyl bromide
$(CH_3)_3C \cdot CH_2 \cdot Cl$	1-chloro-2, 2-dimethyl-propane	neopentyl chloride

C. Preparations:

THE ALKYL HALIDES MAY BE PREPARED:

1. By the replacement of the "OH" group in alcohols with an "X" group by means of (a) hydrogen chloride or bromide in the presence of sulfuric acid or zinc chloride, (b) dry hydrogen chloride or bromide, (c) phosphorus halides, or (d) thionyl chloride:

a. $R \cdot OH + HX / H_2SO_4 \rightarrow R \cdot X$ (any alkyl halide) $+ H_2SO_4 \cdot H_2O$,
and $CH_3 \cdot CH_2 \cdot CH_2 \cdot OH + HBr / H_2SO_4 \rightarrow CH_3 \cdot CH_2 \cdot CH_2 \cdot Br + H_2SO_4 \cdot H_2O$,

or 1-propanol (**propyl alcohol**) + hydrogen bromide in conc. sulfuric acid → 1-bromopropane (**propyl bromide**) + sulfuric acid hydrate. The HX for this reaction is usually produced from a salt and sulfuric acid ($NaBr/H_2SO_4$, etc.).

b. $R \cdot OH + HCl$, dry → $R \cdot Cl$ (any alkyl chloride) + H_2O, and $CH_3 \cdot CH_2 \cdot OH + HBr$, dry → $CH_3 \cdot CH_2 \cdot Br + H_2O$, or ethanol (**ethyl alcohol**) + dry hydrogen bromide → bromoethane (**ethyl bromide**) + water.

c. 1. $R \cdot OH + PX_5$ → $R \cdot X$ (any alkyl halide) + POX_3 + HX, and $CH_3 \cdot CH_2 \cdot OH + PCl_5$ → $CH_3 \cdot CH_2 \cdot Cl + POCl_3 + HCl$, or ethanol (**ethyl alcohol**) + phosphorus pentachloride → chloroethane (**ethyl chloride**) + phosphorus oxychloride + hydrogen chloride. In laboratory practice the use of iodine in the presence of red and yellow phosphorus is employed in the preparation of the alkyl iodides.

2. $3 R \cdot OH + PX_3$ → $3 R \cdot X$ (any alkyl halide) + H_3PO_3

d. $R \cdot OH + SOCl_2$ → $R \cdot Cl + SO_2 + HCl$

In each of these methods of preparation, alkyl halide formation decreases and olefin formation increases, according to the type of alcohol used, in the order $R \cdot CH_2 \cdot OH$, $R_2CH \cdot OH$, $R_3C \cdot OH$.

2. By the addition of halogen acids to unsaturated hydrocarbons:

1. $H_2C:CH_2 + HX$ → $CH_3 \cdot CH_2 \cdot X$ (a primary halide)
2. $R \cdot HC:CH_2 + HX$ → $R \cdot CHX \cdot CH_3$ (a secondary halide)
3. $R_2C:CH_2 + HX$ → $R_2CX \cdot CH_3$ (a tertiary halide)

The fluorides, likewise, may be obtained by this general reaction, as indicated by:
$CH_2 \cdot HC:CH_2 + H \cdot F/(dry)$ → $CH_2 \cdot CHF \cdot CH_3 + (CH_2 \cdot CH_2 \cdot CH_2F)$, and $CH_3 \cdot C:C \cdot H + 2 H \cdot F/$ (dry) → $CH_3 \cdot CF_2 \cdot CH_3 + (CH_3 \cdot CH_2 \cdot CHF_2)$. The ratio of products may be controlled to some extent by the experimental conditions.

3. By the halogenation of saturated hydrocarbons (commercial):

$R \cdot H + X_2$ → $R \cdot X$ (any alkyl halide) + HX, and

1. $CH_4 + Cl_2$ → $CH_3 \cdot Cl$ (chloromethane, **methyl chloride**) + HCl, or $CH_3 \cdot CH_3 + Cl_2$ → $CH_3 \cdot CH_2 \cdot Cl$ (chloroethane, **ethyl chloride**) + HCl

In the chlorination of the hydrocarbons, normally carried out in the vapor phase, mixtures result and the derivatives must be separated by fractional distillation.

2. $CH_3 \cdot CH_2 \cdot CH_3 + Cl_2$ → $CH_3 \cdot CHCl \cdot CH_3$ (and $CH_3 \cdot CH_2 \cdot CH_2 \cdot Cl$) + HCl

In this reaction, the ratio of isomers is dependent upon the relative activity and the relative number of the primary and secondary hydrogen atoms. Consequently, mixtures always result.

3. $(CH_3)_3C \cdot H + Cl_2$ → $(CH_3)_3C \cdot Cl$ (and $[CH_3]_2CH \cdot CH_2 \cdot Cl$) + HCl

The ratio of isomers is dependent upon the *relative activity* and the *relative number* of the *different types of hydrogen atoms* in the molecule. Mixtures always result.

Direct fluorination has been effected by carrying the reaction out in a copper spiral, which conducts the liberated heat away rapidly enough to prevent a violent reaction.

4. By the displacement of one halogen atom by another, as

   a. $R \cdot Cl + NaI/(acetone) \rightarrow R \cdot I + NaCl$.

      The progress of the reaction is dependent on the greater solubility of sodium iodide compared to sodium chloride in acetone.

   b. $R \cdot X + AgF$ (or HgF_2, or HgF) $\rightarrow R \cdot F + AgX$.

   c. $R \cdot CCl_3 + SbF_3$ (or SbF_3Cl_2) $\rightarrow R \cdot CF_3 + SbCl_3$.

D. Physical Properties:

    For any given alkyl group, the specific gravities and the boiling points decrease, according to the halogen present, in the order $R \cdot I$, $R \cdot Br$, $R \cdot Cl$, $R \cdot F$. Fluoromethane (methyl fluoride), chloromethane (methyl chloride), bromomethane (methyl bromide), fluoroethane (ethyl fluoride), and chloroethane (ethyl chloride) are gases but the rest of the lower members of the series are liquids. The first solid of the series is $C_{15}H_{31}I$. The alkyl halides are colorless when pure and have a pleasant odor. The chlorides are less dense and the bromides and iodides are denser than water. The alkyl halides are all insoluble in water but are soluble in alcohol, ether, and benzene.

E. Chemical Properties:

THE ALKYL HALIDES REACT UNDER PROPER CONDITIONS:

1. To give additive products with ammonia, amines, magnesium, zinc, etc.:

Reactants:	$R \cdot CH_2 \cdot X$,	$R_2CH \cdot X$, or	$R_3C \cdot X$,
+ NH_3, heat	$\rightarrow R \cdot CH_2 \cdot NH_2 \cdot HX$	$R_2CH \cdot NH_2 \cdot HX$	$R_3C \cdot NH_2 \cdot HX$
+ $R \cdot NH_2$, heat	$\rightarrow R \cdot CH_2 \cdot NH(HX)R$	$R_2CH \cdot NH(HX)R$	$R_3C \cdot NH(HX)R$
+ $R_2N \cdot H$, heat	$\rightarrow R \cdot CH_2 \cdot N(HX)R_2$	$R_2CH \cdot N(HX)R_2$	$R_3C \cdot N(HX)R_2$
+ R_3N, heat	$\rightarrow R \cdot CH_2 \cdot N(X)R_3$	$R_2CH \cdot N(X)R_3$	$R_3C \cdot N(X)R_3$
+ Mg, ether	$\rightarrow R \cdot CH_2 \cdot Mg \cdot X$	$R_2CH \cdot Mg \cdot X$	$R_3C \cdot Mg \cdot X$
+ Zn dust, heat	$\rightarrow R \cdot CH_2 \cdot Zn \cdot X$	$R_2CH \cdot Zn \cdot X$	$R_3C \cdot Zn \cdot X$
+ CO, cat.	$\rightarrow R \cdot CH_2 \cdot CO \cdot X$	$R_2CH \cdot CO \cdot X$	$R_3C \cdot CO \cdot X$

2. To give condensation products in the presence of sodium, the Grignard reagent, zinc alkyls, etc.:

  a. R $\boxed{X + Na}$ *b.* R $\boxed{X}$ *c.* R $\boxed{X}$ R $\boxed{ZnX}$

     R $\boxed{X + Na}$ R $\boxed{Mg \cdot X}$ R $\boxed{Zn \cdot X}$ or $R \cdot Zn$ $\boxed{X}$

    To illustrate, $R \cdot X + 2 Na$, ether $+ X \cdot R \rightarrow R \cdot R + 2 NaX$

    In these reactions, the yields of saturated hydrocarbons decrease and olefin formation increases according to the type of halide used, in the order $R \cdot CH_2 \cdot X$, $R_2CH \cdot X$, $R_3C \cdot X$. To date, the zinc alkyls have been obtained from the iodides only.

3. To give unsaturated products in the presence of alcoholic caustic soda or potash:

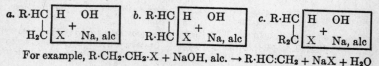

  a. $R \cdot HC$ $\boxed{\begin{matrix} H & OH \\ | & + \\ X & Na, alc \end{matrix}}$ *b.* $R \cdot HC$ $\boxed{\begin{matrix} H & OH \\ | & + \\ X & Na, alc \end{matrix}}$ *c.* $R \cdot HC$ $\boxed{\begin{matrix} H & OH \\ | & + \\ X & Na, alc \end{matrix}}$

with H_2C (a), $R \cdot HC$ (b), R_2C (c) below.

    For example, $R \cdot CH_2 \cdot CH_2 \cdot X + NaOH$, alc. $\rightarrow R \cdot HC \colon CH_2 + NaX + H_2O$

4. *To give double decomposition when treated with:*

a.	Aqueous sodium hydroxide	R	X + Na	OH, aq.
	Aqueous potassium hydroxide	R	X + K	OH, aq.
b.	Sodium alcoholate	R	X + Na	OR
	Potassium alcoholate	R	X + K	OR
c.	Sodium salts of fatty acids	R	X + Na	O·OC·R
	Potassium salts of fatty acids	R	X + K	O·OC·R
d.	Sodium cyanide	R	X + Na	CN (some R·NC)
	Potassium cyanide	R	X + K	CN (some R·NC)
e.	Silver cyanide	R	X + Ag	NC (some R·CN)
f.	Sodium nitrite	R	X + Na	O.NO (some R·NO₂)
	Potassium nitrite	R	X + K	O·NO (some R·NO₂)
g.	Silver nitrite	R	X + Ag	NO₂ (some R·O·NO)
h.	Silver oxide	R	X + Ag	O Ag + X R

As an example, R·X + NaOH, aq. → R·OH + NaX

If sulfur be substituted for oxygen in the first six reagents, the corresponding sulfur derivatives are obtained.

5. *To give hydrocarbons when treated with hydrogen from appropriate reducing agents, or concentrated hydriodic acid in the presence of red phosphorus:*

a. Activated hydrogen R X + H H/(Zn/ROH/CuSO₄, etc.)

b. Hydriodic acid R X + I H/(red phosphorus, heat), or

CH₃·CH₂·I + HI/(red phosphorus) → CH₃·CH₃ + I₂.

In general, the reactions indicated above are applicable to primary (R·CH₂X), secondary (R₂CH·X), and tertiary (R₃C·X) alkyl halides. The activity for a given halogen increases, for most reactions according to the type of radical involved, in the order R·CH₂·X, R₂CH·X, R₃C·X. For a given radical, the activity increases in the order R·F, R·Cl, R·Br, R·I. Bromides and iodides are commonly used, therefore, in laboratory syntheses, but since the chlorides are much cheaper they are more desirable for commercial processes. Neopentyl chloride is extremely stable to most reagents.

II. Polyhaloalkanes

A. Introduction:

Polyhaloalkanes, as the name implies, are compounds obtained by replacing two or more hydrogen atoms in a hydrocarbon by halogen atoms. The substitutions may occur on the same or on different carbon atoms. The properties of these compounds are similar to those of the alkyl halides except that the presence of additional halogen atoms in the molecule tends, in general, to alter the chemical activity of the halogen groups.

Where any halogen is represented by "X," the general type formula for polyhalogenated compounds may be given as $C_nH_{2n+2-y}X_y$, where y is two or more.

The polyhalogenated hydrocarbons are used in organic synthesis; the liquid members are important solvents. Carbon tetrachloride is employed as a dry cleaner, in fire extinguishers, and in the commercial production of

chloroform. Chloroform has anesthetic properties but, owing to its bad effect on the heart, it has a very limited clinical use, and is being replaced by ether, ethylene, ethylene chloride, and cyclopropane. Dichlorodifluoromethane and similar compounds are employed as refrigerants. Iodoform is an antiseptic. 1, 2-Dibromoethane (ethylene bromide) is used, along with tetraethyllead, in ethyl gasoline. Diiodomethane finds a limited use as a heavy liquid for the separation of minerals. Trichloroethylene is used as a degreasing agent, particularly before electroplating. Paraflow and Santopour are prepared respectively by condensing highly chlorinated paraffin wax with aromatic hydrocarbons and phenolic compounds. They are used to inhibit the separation of wax from lubricating oils.

Using pure hexachloroethane as a specific example, its structure would be determined about as follows:

1. Qualitative analysis shows that it contains only carbon and chlorine.
2. Quantitative analysis gives a carbon-chlorine ratio of 1:3, hence the empirical formula is CCl_3.
3. Molecular weight determinations give about 235 g. per mole, hence the molecular formula must be twice the empirical formula or $(CCl_3)_2$.
4. Since only one mono-chloroethane is known, the six hydrogen atoms in the parent compound must be very nearly equivalent or they must be symmetrically distributed with respect to the two carbon atoms.
5. Reduction of the chloroethanes give ethane. The two carbon atoms, therefore, must be linked together. Moreover, the graphic formula assigned to ethane is shown at the right, and replacement of two or more of the hydrogen atoms by halide atoms would give the formula for the corresponding halide.

$$\begin{array}{ccc} H & H \\ | & | \\ H\cdot C\cdot C\cdot H \\ | & | \\ H & H \end{array}$$

Position isomerism is illustrated by the following:

$CH_3\cdot CHCl\cdot CH_2Cl$	$CH_2Cl\cdot CH_2\cdot CH_2Cl$	$CH_3\cdot CH_2\cdot CHCl_2$
1, 2-dichloropropane	1, 3-dichloropropane	1, 1-dichloropropane
(propylene chloride)	(trimethylene chloride)	(propylidene chloride)

B. Nomenclature:

These compounds are named as substitution products of the corresponding hydrocarbons, using prefixed numerals to indicate the positions occupied by the halogen groups. In the following table, the compounds are named as *substitution products* (I.U.C. names) in the second column, and as *addition products* (common names) in the third column:

Structural Formulas	I.U.C. Names	Common Names
$CH_2Cl\cdot CH_2Cl$	1, 2-dichloroethane	ethylene chloride
$CH_3\cdot CHCl\cdot CH_2Cl$	1, 2-dichloropropane	propylene chloride
$CH_3\cdot CH_2\cdot CHCl\cdot CH_2Cl$	1, 2-dichlorobutane	α-butylene chloride
$CH_3\cdot CHBr\cdot CHBr\cdot CH_3$	2, 3-dibromobutane	β-butylene bromide
$CH_3\cdot CH_2\cdot CH_2\cdot CHI\cdot CH_2\cdot I$	1, 2-diiodopentane	α-amylene iodide

C. Preparations:

THE POLYHALOALKANES MAY BE PREPARED:

1. By the addition of halogen atoms to unsaturated hydrocarbons, the additions being on adjacent carbon atoms except in the case of the conjugated systems:

a. $R \cdot HC:CH_2$ (or $R \cdot HC:CH \cdot R'$) + X_2 → $R \cdot HCX \cdot XCH_2$ (or $R \cdot HCX \cdot XCH \cdot R'$), and

$CH_3 \cdot HC:CH_2 + Br_2$ → $CH_3 \cdot CHBr \cdot CH_2Br$ (1, 2-dibromopropane)

b. $R \cdot C:C \cdot H$ (or $R \cdot C:C \cdot R'$) + $2 X_2$ → $R \cdot CX_2 \cdot X_2CH$ (or $R \cdot CX_2 \cdot X_2C \cdot R'$)

c. $H_2C:CH \cdot HC:CH_2 + Br_2$ → $BrCH_2 \cdot HC:CH \cdot CH_2Br$ (1,4-dibromo-2-butene)

2. *By the substitution of carbonyl oxygen in aldehydes and ketones by chlorine when heated with phosphorus pentachloride:*

$R \cdot CHO$ (or $R \cdot CO \cdot R$) + PCl_5, heat → $R \cdot CHCl_2$ (or $R \cdot CCl_2 \cdot R$) + $POCl_3$, and

$CH_3 \cdot CO \cdot CH_3 + PCl_5$, heat → $CH_3 \cdot CCl_2 \cdot CH_3 + POCl_3$, but these dihalo-derivatives tend to lose HCl to give unsaturated substituted hydrocarbons:

$CH_3 \cdot CCl_2 \cdot CH_3$, heat → $CH_3 \cdot ClC:CH_2 + HCl$

3. *By the substitution of chlorine or bromine atoms for hydrogen atoms in saturated hydrocarbons under special conditions to give mixtures which must be separated by fractional distillation.*

Propane, for example, yields the following mono- and dichloropropanes:

x $CH_3 \cdot CH_2 \cdot CH_3$ + y Cl_2 → $CH_3 \cdot CH_2 \cdot CH_2 \cdot Cl$, $CH_3 \cdot CHCl \cdot CH_3$, $CH_2Cl \cdot CH_2 \cdot CH_2Cl$, $CH_3 \cdot CHCl \cdot CH_2Cl$, $CH_3 \cdot CH_2 \cdot CHCl_2$, and $CH_3 \cdot CCl_2 \cdot CH_3$ + y HCl

4. *By (a) the action of chlorine in alkaline solution on ethyl alcohol or acetone to give chloroform or (b) the controlled commercial reduction of carbon tetrachloride:*

a. 1. $CH_3 \cdot CH_2 \cdot OH + Cl_2/2 NaOH$ → $CH_3 \cdot CHO + 2 NaCl + 2 H_2O$, then
$CH_3 \cdot CHO + 3 Cl_2/3 NaOH$ → $CCl_3 \cdot CHO + 3 NaCl + 3 H_2O$, and
$CCl_3 \cdot CHO + NaOH$ → $CHCl_3$ (chloroform) + $H \cdot CO \cdot O \cdot Na$ (sodium formate)

2. $CH_3 \cdot CO \cdot CH_3 + 3 Cl_2/3 NaOH$ → $CCl_3 \cdot CO \cdot CH_3$ (1, 1, 1-trichloropropanone) + $3 NaCl + 3 H_2O$, then
$CCl_3 \cdot CO \cdot CH_3 + NaOH$ → $CHCl_3$ (trichloromethane, chloroform) + $CH_3 \cdot CO \cdot ONa$ (sodium ethanoate, sodium acetate). These are haloform reactions. Bromoform or iodoform may be made by similar reactions. The essential grouping for the preparation of CHX_3 by these reactions is $CH_3 \cdot CHOH \cdot$ or its first oxidation derivative, $CH_3 \cdot CO \cdot$.

b. $8 CCl_4 + 7 Fe + 4 H_2O$ → $8 CHCl_3 + Fe_3O_4 + 4 FeCl_2$

5. *By the substitution of chlorine atoms for sulfur atoms in carbon disulfide in the presence of a catalyst (antimony pentasulfide) to give carbon tetrachloride:*

$CS_2 + 3 Cl_2/(Sb_2S_5)$ → $CCl_4 + S_2Cl_2$ (commercial)

The sulfur monochloride in turn, in the presence of iron, converts carbon disulfide to carbon tetrachloride.

6. *By treating carbon tetrachloride with hydrogen fluoride in the presence of $SbCl_xF_y$ (where x + y = 5):*

$CCl_4 + 2 HF/(SbCl_xF_y)$ → CCl_2F_2, Freon 12 + 2 HCl.
Freon 11, CCl_3F, and Freon 13, $CClF_3$, are also obtained.

D. Physical Properties:

The polyhalogenated compounds, as a class, cannot be assigned common or general physical properties owing to the great difference in their composition. They occur as gases, liquids, and solids. The odors and physical properties vary widely. The specific gravity increases with an increase in the number of halogen atoms present in the molecule. Diiodomethane, CH_2I_2, is the densest organic liquid (D. 3.325). The members of this group are insoluble in water and are good solvents for fats.

E. Chemical Properties:

The polyhaloalkanes show the same reactions as the alkyl halides only when a single halogen atom is linked to a given carbon atom. When two or more halogen atoms are linked to a given carbon atom, the chemical activity of the halogen atoms is usually slightly abnormal.

A few specific reactions are given for chloroform:

1. $2\ CHCl_3 + O_2$ /(air and light) $\rightarrow 2\ Cl \cdot CO \cdot Cl$ **(phosgene)** $+ 2\ HCl$
2. $4\ CHCl_3 + 3\ O_2$ /(excess oxygen) $\rightarrow 4\ Cl \cdot CO \cdot Cl + 2\ Cl_2 + 2\ H_2O$
3. $CHCl_3 + HNO_3 \rightarrow CCl_3 \cdot NO_2$ **(chloropicrin,** poison, b.p.112) $+ H_2O$
   Chloropicrin is produced commercially by the action of bleaching powder and steam on calcium picrate.

III. Unsaturated Halo-Compounds

Although a relatively large number of unsaturated halo-compounds may be obtained experimentally, there are relatively few that are of much importance.

They are named as substitution products of the corresponding hydrocarbon, as addition products of dienes or alkynes, or by common names.

Two general methods are employed in their preparation: the removal of HX or 2 X from appropriate halo-compounds, or the addition of HX or X_2 to unsaturated hydrocarbons. For example: $CH_3 \cdot CCl_2 \cdot CH_3$, heat $\rightarrow$ $CH_3 \cdot ClC:CH_2$ (2-chloropropene) $+ HCl$, and $H \cdot C:C \cdot HC:CH_2$ **(vinylacetylene)** $+ HCl$, gas $\rightarrow H_2C:CCl \cdot HC:CH_2$ **(chloroprene).**

More recently (1938) propylene has been chlorinated at 400-600° to give allyl chloride, as:
$$CH_3 \cdot HC:CH_2 + Cl_2, 400\text{-}600° \rightarrow Cl \cdot CH_2 \cdot HC:CH_2\ (80\%) + HCl.$$

The production of this product is not surprising for an H-atom in the 4-position to an unsaturated linkage is replaced, and a negative-negative bond is formed in lieu of a negative-positive bond.

In their physical properties, these compounds lie intermediate between those of the unsaturated hydrocarbons and the saturated halo-compounds.

They show the typical chemical reactions of the unsaturated hydrocarbons and the alkyl halides except in so far as the presence of the one group may have a shading influence on the activity of the other group. One of the more important reactions is illustrated by the polymerization of **chloroprene**:
$H_2C:CCl \cdot HC:CH_2$, polymerization $\rightarrow$ a synthetic rubber

Vinyon is a copolymer of vinyl acetate with vinyl chloride (about 20/80 ratio). Vinylidine is a polymer of 1, 1-dichloroethylene, to give a string molecule of the general type:
$- (CH_2 \cdot CCl_2 \cdot CH_2 \cdot CCl_2 \cdot CH_2 \cdot CCl_2 \cdot CH_2 \cdot CCl_2)_n -$
The vinylidine chloride for the synthesis is prepared by:
$Cl \cdot CH_2 \cdot CH_2 \cdot Cl + Cl_2$, elevated temperature $\rightarrow H_2C:CCl_2 + 2\ HCl$.

The structure of the unsaturated halo-compounds is evidenced by the methods employed for their preparation and by their chemical properties. They show nuclear and position isomerism.

In general, the unsaturated halo-compounds are used as synthetic intermediates. **Chloroprene** is employed in the production of a synthetic rubber

CHAPTER V

ALIPHATIC OXYGEN DERIVATIVES:
ALCOHOLS, ETHERS, HALOHYDRINS, AND ALKENE OXIDES

I. Alcohols

A. Introduction:

Alcohols may be considered as hydrocarbon derivatives in which one or more hydrogen atoms have been replaced by the "OH" group.

The monohydroxy alcohols are classified as primary ($R \cdot CH_2 \cdot OH$), secondary ($R_2CH \cdot OH$), and tertiary ($R_3C \cdot OH$), and may be represented by the general formulas $R \cdot OH$ or $C_nH_{2n+1}OH$. Specific examples are:

H H H H	H H H H	CH_3
H·C·C·C·C·OH	H·C·C·C·C·H	$CH_3 \cdot C \cdot CH_3$
H H H H	H H O H	O
	H	H
1-butanol	2-butanol	2-methyl-2-propanol
(butyl alcohol)	(sec-butyl alcohol)	(tert-butyl alcohol)

Alcohols containing more than one hydroxyl group may be represented by the general formula $C_nH_{2n+2-y}(OH)_y$, where y is two or more. The common polyhydroxy alcohols usually have one hydroxyl group linked with each carbon atom in the molecule.

Methanol (**methyl alcohol**) occurs as the methyl esters of salicylic and pectinic acids in oil of wintergreen and in tobacco respectively. It is one of the products obtained from the destructive distillation of wood. Ethanol (**ethyl alcohol**) is formed during the fermentation of the sugars in fruit juices, or by the fermentation of starches. Some ethyl esters occur in nature. 1, 2-Ethanediol (**glycol**) is a synthetic product. 1, 2, 3-Propanetriol (**glycerol**) is an important constituent of both animal and vegetable fats and oils which are organic esters of glycerol. *i-Erythritol* occurs as the ester of orsellinic acid in numerous lichens and algae. *Arabitol* and *xylitol*, both pentahydroxy alcohols, are obtained by the reduction of the corresponding sugars, *arabinose* and *xylose*, which are widely distributed in nature in the form of their anhydrides, arabans and xylans. *Mannitol*, a hexahydroxy alcohol, is found in celery, in the larch, in manna, in rye bread, and in cane sugar. Other hexahydroxy alcohols are readily obtained by the reduction of the corresponding sugars.

The formulas for 1-pentanol, 2-pentanol, and 3-pentanol illustrate *position* isomerism among the alcohols. *Functional* isomerism of the alcohols and ethers of the same carbon content is shown by: $CH_3 \cdot CH_2 \cdot OH$ and $CH_3 \cdot O \cdot CH_3$, or $CH_3 \cdot CH_2 \cdot CH_2 \cdot OH$ and $CH_3 \cdot CH_2 \cdot O \cdot CH_3$.

Alcohols are used extensively in the preparation of derivatives, especially the esters, and as solvents. Methanol (**methyl alcohol**) is employed as an anti-freeze and in the production of methanal (**formaldehyde**) which is used in the synthesis of resins. Methanol is blended with benzene and gasoline and sold as Dynax for motor fuel for speed boats. Ethanol (**ethyl alcohol**) ranks first among organic chemicals in both quantity of production and value of product. It is used as a solvent, as a synthetic intermediate, as an anti-freeze, and as an ingredient of beverages. Isopropyl alcohol finds application as a solvent, as a synthetic intermediate, and as an anti-freeze. Butyl and **amyl** alcohols are employed extensively in the preparation of esters for use as solvents in the lacquer industry. Isoamyl acetate is sold under the name of banana oil. Ethanediol (**glycol**) is used extensively as an anti-freeze and in the formation of derivatives. 1,2,3-Propanetriol (**glycerol**) is employed in the synthesis of resins, in the production of **nitroglycerin,** in the synthesis of other derivatives, and as an anti-freeze.

B. Nomenclature:

The I.U.C. nomenclature for the alcohols replaces the final *e* of the hydrocarbon name by *ol* and prefixes a numeral to indicate the position of the hydroxyl group.

Structural Formulas	I.U.C. Names	Common Names
$CH_3 \cdot OH$	methanol	methyl alcohol, wood alcohol, carbinol
$CH_3 \cdot CH_2 \cdot OH$	ethanol	ethyl alcohol, grain alcohol, methylcarbinol
$CH_3 \cdot CH_2 \cdot CH_2 \cdot OH$	1-propanol	propyl alcohol, ethylcarbinol
$CH_3 \cdot CHOH \cdot CH_3$	2-propanol	isopropyl alcohol, *sec*-propyl alcohol, dimethylcarbinol
$CH_3 \cdot CH_2 \cdot CH_2 \cdot CH_2 \cdot OH$	1-butanol	butyl alcohol, propylcarbinol
$CH_3 \cdot CH_2 \cdot CHOH \cdot CH_3$	2-butanol	*sec*-butyl alcohol, ethyl methylcarbinol
$(CH_3)_2C(OH)CH_3$	2-methyl-2-propanol	*tert*-butyl alcohol, trimethylcarbinol
$CH_3 \cdot CH_2 \cdot CH_2 \cdot CH_2 \cdot CH_2 \cdot OH$	1-pentanol	*n*-amyl alcohol, butylcarbinol
$CH_3 \cdot CH_2 \cdot CH_2 \cdot CHOH \cdot CH_3$	2-pentanol	*sec-act*-amyl alcohol, methylpropylcarbinol
$CH_3 \cdot CH_2 \cdot CHOH \cdot CH_2 \cdot CH_3$	3-pentanol	diethylcarbinol
$(CH_3)_2CH \cdot CH_2 \cdot CH_2 \cdot OH$	3-methyl-1-butanol	isoamyl alcohol, isobutylcarbinol
$CH_2OH \cdot CH_2OH$	1, 2- ethanediol	glycol, ethylene glycol
$CH_2OH \cdot CHOH \cdot CH_2OH$	1, 2, 3-propanetriol	glycerol, glycerin
$i\text{-}CH_2OH \cdot CHOH \cdot CHOH \cdot CH_2OH$	*i*-1, 2, 3, 4-butanetetrol	*i*-erythritol, erythrol, erythrite, erythroglucin

C. Preparations:

I. MONOHYDROXY ALCOHOLS MAY BE PREPARED:

1. By the substitution of "O H" for "X" when alkyl halides are treated with aqueous sodium or potassium hydroxide:

$R \cdot X + NaOH$, aq.$\rightarrow R \cdot OH$ (any alkanol)$+ NaX$, and

$CH_3 \cdot CH_2 \cdot Cl + NaOH$, aq. $\rightarrow CH_3 \cdot CH_2 \cdot OH + NaCl$, or chloroethane (**ethyl chloride**) + sodium hydroxide, aqueous → ethanol (**ethyl alcohol**) + sodium chloride.

The yields of alcohols decrease and olefin formation increases, according to the type of radical involved, in the order $R \cdot CH_2 \cdot X$, $R_2CH \cdot X$, $R_3C \cdot X$.

2. *By the introduction of alkyl and hydrogen groups into a carbonyl or epoxy compound upon treatment with the Grignard reagent in dry ether and subsequent hydrolysis:*

a. 1. $H \cdot CHO + R \cdot Mg \cdot X \rightarrow R \cdot CH_2 \cdot O \cdot Mg \cdot X$, and

$R \cdot CH_2 \cdot O \cdot Mg \cdot X + HX$, aq. $\rightarrow R \cdot CH_2 \cdot OH$ (a primary alcohol) $+ MgX_2$, or

$$\overset{H}{\underset{H}{C}}:O + CH_3 \cdot Mg \cdot I \rightarrow CH_3 \cdot \overset{H}{\underset{H}{C}} \cdot O \cdot Mg \cdot I \text{ (ethoxymagnesium iodide,}$$
iodomagnesium ethoxide), and
$CH_3 \cdot CH_2 \cdot O \cdot Mg \cdot I + HCl$, aq. $\rightarrow CH_3 \cdot CH_2 \cdot OH + MgICl$, or

methanal (**formaldehyde**) + methylmagnesium iodide and subsequent hydrolysis $\rightarrow$ ethanol (**ethyl alcohol**) + chloromagnesium iodide.

2. $H_2C \overset{O}{\underset{\cdot \cdot}{\cdot}} CH_2 + R \cdot Mg \cdot X$, then HX, aq. $\rightarrow R \cdot CH_2 \cdot CH_2 \cdot OH + MgX_2$

b. 1. $R \cdot CHO + R' \cdot Mg \cdot X$, then HX, aq. $\rightarrow R \cdot CHOH \cdot R'$ (a *sec*-alcohol) $+ MgX_2$

2. $R \cdot HC \overset{O}{\underset{\cdot \cdot}{\cdot}} CH_2 + R' \cdot Mg \cdot X$, then HX, aq. $\rightarrow R \cdot CH_2 \cdot CHOH \cdot R' + MgX_2$

c. 1. $R_2C:O + R' \cdot MgX$, then 2 HX, aq. $\rightarrow R_2R'C \cdot OH$ (a *tert*-alcohol) $+ MgX_2$

2. $R_2C \overset{O}{\underset{\cdot \cdot}{\cdot}} CH_2 + R'_2Mg$, then HX, aq. $\rightarrow R_2C(OH) \cdot CH_2 \cdot R' + MgX_2 + R' \cdot H$

3. *By catalytic hydrogenation or by reduction of aldehydes or ketones in acid solution (reduction with metals in basic solution tends to yield pinacols):*

a. $R \cdot CHO + Zn + 2 H_2O$, acidic $\rightarrow R \cdot CH_2 \cdot OH$ (a primary alcohol) $+ Zn^{++} + 2 OH^-$, and
$CH_3 \cdot CH_2 \cdot CHO + Zn + 2 H_2O$, acidic $\rightarrow CH_3 \cdot CH_2 \cdot CH_2 \cdot OH$ (1-propanol, **propyl alcohol**) $+ Zn^{++} + 2 OH^-$

b. $R_2C:O + Zn + 2 H_2O$, acidic $\rightarrow R_2CH \cdot OH$ (a secondary alcohol) $+ Zn^{++} + 2 OH^-$

The alkene oxides, acids, esters, etc., may be reduced under proper conditions to yield alcohols.

4. *By the hydration of alkenes* (formed by cracking petroleum, commercial):

$$\overset{H\,H\,H}{\underset{H\,H\,H}{R' \cdot C \cdot C \cdot C \cdot R}} + \text{cracking} \rightarrow \overset{H\,H}{R' \cdot C:C \cdot H} + \overset{H}{H \cdot C \cdot R} \text{ (where } R' < R), \text{ and}$$

$$\overset{H\,H}{R' \cdot C:C \cdot H} + H_2SO_4 \rightarrow \overset{H\,H}{\underset{O\,H}{R' \cdot C \cdot C \cdot H}}, \text{ then } H_2O, \text{ excess} \rightarrow \overset{H\,H}{\underset{O\,H}{R' \cdot C \cdot C \cdot H}} + H_2SO_4$$
$$SO_2 \cdot OH \qquad\qquad H$$

The alcohol is separated by distillation.

5. *By the following specific methods:*

a. Methanol (**methyl alcohol**) is produced commercially by the use of heat and pressure on a mixture of carbon monoxide or carbon dioxide and hydrogen in the presence of a catalyst (Patart Process), or by the destructive distillation of wood:

   1. $CO + 2 H_2/$ (zinc chromite, 400–500° C., 200 atms.) → $CH_3 \cdot OH$

   By varying the catalyst and conditions, the Patart process becomes a general synthesis for the *n*-alkanols except ethanol, which is not readily obtained.

   2. $CO_2 + 3 H_2/$ (Fe, Cr, Zn oxides, 400° C., 2500 lbs.) → $CH_3 \cdot OH + H_2O$

b. Methanol (**methyl alcohol**) and ethanol (**ethyl alcohol**) may be obtained in poor yields by the action of nitrous acid on the corresponding amines:

   1. $CH_3 \cdot NH_2 + HO \cdot NO/$ ($Na \cdot NO_2/H^+$) → $CH_3 \cdot OH + N_2 + H_2O$
   2. $CH_3 \cdot CH_2 \cdot NH_2 + HO \cdot NO/(NaNO_2/H^+)$ → $CH_3 \cdot CH_2 \cdot OH + N_2 + H_2O$

c. Ethanol (**ethyl alcohol**) and butanol (**butyl alcohol**) are produced in tank car lots by the fermentation of the sugars in molasses or the starches in corn. Other fermentation products are 1-propanol, 2-methyl-1-propanol, 3-methyl-1-butanol (**isoamyl alcohol**), and 2-methyl-1-butanol (**active amyl alcohol**). Butanol may also be produced economically by the following reactions:

   $CH_3 \cdot CHO + CH_3 \cdot CHO /$ (NaOH, aq., dilute) → $CH_3 \cdot CHOH \cdot CH_2 \cdot CHO$ (**aldol**)

   $CH_3 \cdot CHOH \cdot CH_2 \cdot CHO +$ catalyst, heat → $CH_3 \cdot HC:CH \cdot CHO$ (**crotonaldehyde**)

   $CH_3 \cdot HC:CH \cdot CHO + 2 H_2/$ (catalyst) → $CH_3 \cdot CH_2 \cdot CH_2 \cdot CH_2 \cdot OH$ (1-butanol)

II. DIHYDROXY ALCOHOLS MAY BE PREPARED:

1. *By treating ethylene with* (a) *hypochlorous acid and subsequent hydrolysis, or* (b) *air in the presence of gold or silver and the addition of water, to give glycol:*

   a. $H_2C:CH_2 + HO \cdot Cl$, aq. → $HO \cdot CH_2 \cdot CH_2 \cdot Cl$, then $+ NaHCO_3$, aq. → $HO \cdot CH_2 \cdot CH_2 \cdot OH$ (**glycol,** commercial) $+ NaCl + CO_2$

   b. $H_2C:CH_2 + \frac{1}{2} O_2/$ (Au or Ag) → $H_2C \overset{O}{\overbrace{}} CH_2$, then $+ H_2O$ → $HO \cdot CH_2 \cdot CH_2 \cdot OH$ (**glycol,** commercial)

2. *By the hydrolysis of dihalogen derivatives of the hydrocarbons.*

3. *1, 3-Propanediol* (*trimethylene glycol*) *is a by-product in the soap industry.* The term **glycols** is applied, in general, to aliphatic dihydroxy alcohols.

III. TRIHYDROXY ALCOHOLS MAY BE PREPARED:

1. *By the saponification of the fats to give glycerol which is a by-product in the commercial production of soap.* The reaction for the saponification of tristearin is illustrative of this process:

$H_2C \cdot O \cdot OC(CH_2)_{16}CH_3$ $H_2C \cdot OH$

$H \cdot C \cdot O \cdot OC(CH_2)_{16}CH_3 + 3\ NaOH,\ aq. \rightarrow H \cdot C \cdot OH + 3\ Na \cdot O \cdot OC(CH_2)_{16}CH_3$

$H_2C \cdot O \cdot OC(CH_2)_{16}CH_3$ $H_2C \cdot OH$ (a soap)

2. By HOCl and H_2O on haloalkenes or H_2O on trihaloalkanes.

IV. POLYHYDROXY ALCOHOLS MAY BE PREPARED:

By isolation from natural products or by the reduction of the corresponding sugars.

D. Physical Properties:

Ethanol (ethyl alcohol, b.p. 78.5, m.p. -117.3, D. 0.7928_4^{20}) is infinitely soluble in water, chloroform, methanol, and ether. Refractive indices, densities, and boiling points of alcohols increase with an increase in carbon content, but the solubility in water decreases. For a given carbon content, however, the refractive indices, densities, and boiling points decrease as the chin becomes more branched but the solubility in water increases. The introduction of additional "OH" groups into a molecule tends to intensify the characteristic properties of the hydroxyl group. In general, for a given number of carbon atoms, the boiling point, the sweetness, and the solubility of an alcohol in water, increase with the number of hydroxyl groups.

E. Chemical Properties:

I. THE MONOHYDROXY ALCOHOLS REACT UNDER PROPER CONDITIONS:

1. By replacement of (a) the hydrogen atom of the hydroxyl group, and (b) the hydroxyl group, as indicated below when treated with:

a. Active metals $R \cdot O$ | H | + | | Na
Acid halides[1] $R \cdot O$ | H | + | X | $OC \cdot R$
Organic acids/(H_2SO_4, P_2O_5, etc.) $R \cdot O$ | H | + | HO | $OC \cdot R$
Organic acid anhydrides $R \cdot O$ | H | + $R \cdot CO \cdot O$ | $OC \cdot R$
Alkyl hydrogen sulfates $R \cdot O$ | H | + $HOSO_2O$ | R
Grignard reagent $R \cdot O$ | H | + | R | $Mg \cdot X$

To illustrate, $2\ R \cdot OH + 2\ Na \rightarrow 2\ R \cdot ONa + H_2$

b. Hydriodic acid/(red phosphorus)   R | OH | + | H | I
Hydrobromic acid (or HCl), dry   R | OH | + | H | Br
Hydrochloric acid (or HBr)/c.H_2SO_4 R | OH | + | H | Cl
Sulfuric acid   R | OH | + | H | $O \cdot SO_2 \cdot OH$
Nitric acid   R | OH | + | H | $O \cdot NO_2$
Phosphorus trihalides   3 R | OH | + | P | X_3
Phosphorus pentahalides   R | O | $H + Cl$ | PCl_3 | Cl

For Example, $R \cdot OH + HI/(red\ phosphorus) \rightarrow R \cdot I + H_2O$, but $3\ R \cdot I + 3\ HI/2\ P$, heat $\rightarrow 3\ R \cdot H$ (an alkane) $+ 2\ PI_3$

2. By dehydration, to give unsaturated derivatives:

The ease of dehydration increases, according to the type, in the order $R \cdot CH_2 \cdot OH$, $R_2CH \cdot OH$, $R_3C \cdot OH$. An example is: $:C$ | H | c.H_2SO_4, $:C$ | OH | P_2O_5, etc.

$R \cdot CHOH \cdot CH_2 \cdot R + P_2O_5 \rightarrow R \cdot HC:CH \cdot R$ (an alkene) $+ 2\ HPO_3$

---

[1] The esters, formed in this reaction from *t*-alcohols, undergo cleavage by the by-product HX to give the corresponding *t*-alkyl halide. If the HX is removed as formed by aniline, this side reaction is inhibited.

3. *By oxidation, to give derivatives, when treated with:*

a. One mole of dichromic acid per three moles of

R·CH(OH)　|　H $+ 1$ OH$^-$　|　1 OH$^-$

b. Two moles of dichromic acid per three moles of

R·C(OH)　|　H$_2$ $+ 2$ OH$^-$　|　2 OH$^-$

c. One mole of dichromic acid per three moles of

R$_2$C(OH)　|　H $+ 1$ OH$^-$　|　1 OH$^-$

Primary alcohols give, in order, aldehydes, acids, oxidized derivatives, and, finally, cleavage of the carbon chain. Secondary alcohols give ketones, oxidized derivatives, and cleavage, whereas tertiary alcohols give oxidized derivatives and then cleavage. The particular nature of the oxidation process is dependent upon both the oxidizing agent and the temperature, but more upon the temperature than the nature of the reagent. By the use of copper and similar catalysts, however, the primary alcohols may be oxidized by air to the corresponding aldehydes. The activity of the different types of · alcohols toward most reagents increases in the order R·CH$_2$·OH, R$_2$CH·OH, R$_3$C·OH, but their ease of oxidation increases in the reverse order, R$_3$C·OH, R$_2$CH·OH, R·CH$_2$·OH.

a. R·CH$_2$·OH + oxidation $\rightarrow$ R·CHO $\rightarrow$ R·CO·OH $\rightarrow$ oxidized derivative $\rightarrow$ cleavage, and

3 CH$_3$·CH$_2$·OH + Na$_2$Cr$_2$O$_7$/4 H$_2$SO$_4$ $\rightarrow$ 3 CH$_3$·CHO + Na$_2$SO$_4$ + Cr$_2$(SO$_4$)$_3$ + 7 H$_2$O, or ethanol (**ethyl alcohol**) + a given amount of sodium dichromate in sulfuric acid $\rightarrow$ ethanal (**acetaldehyde**) + sodium sulfate + chromic sulfate + water, and

3 CH$_3$·CH$_2$·OH + 2 Na$_2$Cr$_2$O$_7$/8 H$_2$SO$_4$ $\rightarrow$ 3 CH$_3$·CO·OH + 2 Na$_2$SO$_4$ + 2 Cr$_2$(SO$_4$)$_3$ + 11 H$_2$O, or ethanol (**ethyl alcohol**) + a slight excess of sodium dichromate in sulfuric acid $\rightarrow$ ethanoic acid (**acetic acid**) + sodium sulfate + chromic sulfate + water, or with nitric acid, by varying the conditions, the following *oxidized derivatives* may be obtained:

$$\begin{array}{llllll} \text{CH}_2\text{·OH} \rightarrow \text{CHO} & \text{CO·OH} \rightarrow \text{CO·OH} \rightarrow & \text{CO·OH} \rightarrow \text{CO·OH} \\ \text{CH}_3 & \text{CHO} & \text{CH}_3 & \text{CH}_2\text{·OH} & \text{CHO} & \text{CO·OH,} \end{array}$$

or **ethyl alcohol**, **glyoxal**, **acetic acid**, **glycolic acid**, **glyoxylic acid**, and **oxalic acid**. Oxalic acid, in turn, undergoes cleavage to give carbon dioxide and water when treated with sodium or potassium permanganate: 5 HOOC·COOH + 2 KMnO$_4$/3 H$_2$SO$_4$ $\rightarrow$ 10 CO$_2$ + 2 MnSO$_4$ + K$_2$SO$_4$ + 8 H$_2$O

b. R$_2$CH·OH + oxidation $\rightarrow$ R$_2$C:O $\rightarrow$ oxidized derivative $\rightarrow$ cleavage

c. R$_3$C·OH + oxidation $\rightarrow$ oxidized derivative $\rightarrow$ cleavage

II. THE DIHYDROXY ALCOHOLS:

These are illustrated by **glycol,** and contain two OH groups and react accordingly. The equations for the oxidation and nitration of ethanediol (glycol) are:

1. CH$_2$·OH + oxi. $\rightarrow$ CHO $\rightarrow$ CO·OH or CHO $\rightarrow$ CO·OH $\rightarrow$ CO·OH

　　CH$_2$·OH　　　　CH$_2$·OH CH$_2$·OH CHO　　CHO　　CO·OH, **or**

ethanediol (glycol) + oxi. → hydroxyethanal (giycolic aldehyde) → hydroxyethanoic acid (glycolic acid) or ethanedial (glyoxal) → oxoethanoic acid (glyoxylic acid) → ethanedioic acid (oxalic acid).

2. $HO \cdot CH_2 \cdot CH_2 \cdot OH + 2 HNO_3 / (H_2SO_4) \rightarrow O_2NO \cdot CH_2 \cdot CH_2 \cdot ONO_2 + 2 H_2O$, or ethanediol (glycol) + nitric acid/(sulfuric acid) → 1,2-ethanediol dinitrate (glycol dinitrate, an explosive) + water.

III. THE TRIHYDROXY ALCOHOLS

These are illustrated by 1, 2, 3-propanetriol (**glycerol**), and they react about as one who is familiar with the chemistry of the OH group would predict. Three representative reactions are given for glycerol:

1. $CH_2OH \cdot CHOH \cdot CH_2OH + HCl$ (100-110° C) → $CH_2OH \cdot CHOH \cdot CH_2 \cdot Cl + H_2O$, or one mole of 1,2,3-propanetriol (**glycerol**) + one mole of dry HCl, 100-110° C. → 3-chloro-1,2-propanediol (**glycerol chlorohydrin**) + water.

2. $CH_2OH \cdot CHOH \cdot CH_2OH + 2 HCl$ (100-110° C.) → $CH_2Cl \cdot CHOH \cdot CH_2Cl + 2 H_2O$

3.
$$\begin{array}{lll} H_2C \cdot OH & HO \cdot NO_2 & H_2C \cdot O \cdot NO_2 \\ H \cdot \overset{.}{C} \cdot OH + HO \cdot NO_2 / (H_2SO_4) \rightarrow & H \cdot \overset{.}{C} \cdot O \cdot NO_2 + 3 H_2O, \text{ or} \\ H_2C \cdot OH & HO \cdot NO_2 & H_2C \cdot O \cdot NO_2 \end{array}$$

1,2,3-propanetriol trinitrate (**nitroglycerin**, glycerol trinitrate, glyceryl nitrate, trinitrin, glonoin) is prepared by slowly spraying the glycerin into the acid mixture (40-44% nitric acid, rest sulfuric acid) in a nitrator at about 5° C.

IV. THE POLYHYDROXY ALCOHOLS:

Their reactions are, in general, an extension of the type reactions of the simpler alcohols.

II. Ethers

A. Introduction:

Ethers may be considered as derivatives of the hydrocarbons in which two alkyl groups are attached to an oxygen atom, $R \cdot O \cdot R$, or as derivatives of the alcohols in which the hydrogen atom of the hydroxyl group has been replaced by an alkyl group. In simple ethers the alkyl groups are identical, as in $CH_3 \cdot O \cdot CH_3$ (methoxymethane, **methyl ether**); but in mixed ethers the alkyl groups are different, as in $CH_3 \cdot CH_2 \cdot O \cdot CH_3$ (methoxyethane, **ethyl methyl ether**). Ethoxyethane ($CH_3 \cdot CH_2 \cdot O \cdot CH_2 \cdot CH_3$, **ethyl ether**) is the most important member of the group but 2-isopropoxypropane (**isopropyl ether**) is gradually displacing ethoxyethane (**ethyl ether**) for certain uses. Complex and unsaturated ethers have been prepared but the study of their preparations and properties is beyond the scope of this outline.

The type formula for ethers is $R \cdot O \cdot R'$, where R' may or may not be identical with R.

Ethers are used extensively as solvents. Methyl ether is used to some extent as a refrigerant, and ethyl ether, when mixed with gasoline, is marketed as a cold weather fuel. Ethoxyethane (ethyl ether) has a limited use as a general anesthetic. Diethylene glycol ($O[CH_2 \cdot CH_2OH]_2$), which has both alcohol and ether properties, is a very good organic solvent.

The ethers are *metameric* among themselves, illustrations being $CH_3 \cdot O \cdot CH_2 \cdot CH_2 \cdot CH_3$, $CH_3 \cdot CH_2 \cdot O \cdot CH_2 \cdot CH_3$, and $CH_3 \cdot O \cdot CH(CH_3)_2$. They are *functional* isomers of the alcohols with the same carbon content:

$$CH_3 \cdot O \cdot CH_3 \text{ and } CH_3 \cdot CH_2 \cdot OH.$$

B. Nomenclature:

Structural Formulas	I.U.C. Names	Common Names
$CH_3 \cdot O \cdot CH_3$	methoxymethane	methyl ether, dimethyl ether
$CH_3 \cdot O \cdot CH_2 \cdot CH_3$	methoxyethane	ethyl methyl ether
$CH_3 \cdot CH_2 \cdot O \cdot CH_2 \cdot CH_3$	ethoxyethane	ethyl ether, diethyl ether
$CH_3 \cdot CH_2 \cdot O \cdot CH_2 \cdot CH_2 \cdot CH_3$	1-ethoxypropane	ethyl propyl ether
$CH_3 \cdot CH_2 \cdot O \cdot CH(CH_3)_2$	2-ethoxypropane	ethyl isopropyl ether
$CH_3 \cdot CH_2 \cdot CH_2 \cdot O \cdot CH_2 \cdot CH_2 \cdot CH_3$	1-propoxypropane	propyl ether, dipropyl ether
$(CH_3)_2HC \cdot O \cdot CH(CH_3)_2$	2-isopropoxypropane	isopropyl ether, di-isopropyl ether

C. Preparations:

THE ETHERS MAY BE PREPARED:

1. *By the reaction of an alcoholate or silver oxide on an alkyl halide:*

    a. $R \cdot X + Na \cdot O \cdot R \rightarrow R \cdot O \cdot R + Na \cdot X$ (Williamson Reaction), and $CH_3 \cdot CH_2 \cdot I + Na \cdot O \cdot CH_2 \cdot CH_3 \rightarrow CH_3 \cdot CH_2 \cdot O \cdot CH_2 \cdot CH_3$ (ethyl ether) + NaI

    b. $R \cdot X + Ag \cdot O \cdot Ag + X \cdot R$, heat $\rightarrow R \cdot O \cdot R + 2 AgX$, and $2 CH_3 \cdot I + Ag_2O$, heat $\rightarrow CH_3 \cdot O \cdot CH_3$ (methyl ether) + 2 AgI

2. *By the dehydration of the alcohols:*

    a. $R \cdot OH + HO \cdot SO_2 \cdot OH$, cold $\rightarrow R \cdot O \cdot SO_2 \cdot OH + H_2O$, and $R \cdot O \cdot SO_2 \cdot OH + HO \cdot R$, heat $\rightarrow R \cdot O \cdot R + H_2SO_4$, or $CH_3 \cdot CH_2 \cdot OH + HO \cdot SO_2 \cdot OH$, cold $\rightarrow CH_3 \cdot CH_2 \cdot O \cdot SO_2 \cdot OH$ (ethyl hydrogen sulfate) + H_2O, and $CH_3 \cdot CH_2 \cdot O \cdot SO_2 \cdot OH + HO \cdot CH_2 \cdot CH_3$, 140°C. $\rightarrow$ $CH_3 \cdot CH_2 \cdot O \cdot CH_2 \cdot CH_3 + H_2SO_4$

    b. $2 R \cdot OH + (Al_2O_3 \text{ catalyst})$ 240-260°C. $\rightarrow R \cdot O \cdot R + H_2O$

    Ether formation decreases and olefin formation increases according to the type of alcohol used, in the order $R \cdot CH_2 \cdot OH$, $R_2CH \cdot OH$, $R_3C \cdot OH$.

D. Physical Properties:

Ethers are much more volatile than the isomeric alcohols. Ethoxyethane (ethyl ether) is a colorless, volatile, inflammable liquid, lighter than water, slightly soluble in water, and has a characteristic odor. Volatility, inflammability, and solubility in water decrease with an increase in carbon content, but the densities show a gradual increase.

E. Chemical Properties:

THE ETHERS, THOUGH QUITE INERT TO MOST CHEMICAL REAGENTS, REACT:

1. By single cleavage at the oxygen linkage, when treated with:

a. Hydriodic acid, conc., cold   R | O·R + H | I
b. Sulfuric acid, conc., heat   R | O·R + H | O·SO$_2$·OH
c. Steam (H$_2$O, under pressure, 150°C)   R | O·R + H | OH
d. Dilute acids, heat and pressure   R | O·R + H | OH

To illustrate, R·O·R + HI, cold → R·OH + R·I

2. By double cleavage when treated with:

	R	O	R
a. Phosphorus pentachloride, heat[1] | Cl | PCl$_3$ | Cl
b. Hydriodic acid, conc., hot | I | H   H | I
c. Sulfuric acid, two moles, heat | HO·SO$_2$·O | H   H | O·SO$_2$·OH

As an example, R·O·R + PCl$_5$ → 2 R·Cl + POCl$_3$

3. By substitution on the hydrocarbon chain:

a. Chlorine or bromine, proper conditions   R·HC | H + Cl | Cl
       |
       R·O

b. 1. CH$_3$·CH$_2$·O·CH$_2$·CH$_3$ + Cl$_2$, dark → CH$_3$·CHCl·O·CH$_2$·CH$_3$ + HCl
   2. CH$_3$·CH$_2$·O·CH$_2$·CH$_3$ + 10 Cl$_2$, light → CCl$_3$·CCl$_2$·O·CCl$_2$·CCl$_3$ + 10 HCl

4. By the formation of unstable peroxides:

Ethers react with ozone to give peroxides which form explosive mixtures with air. Peroxide formation has been recorded in ether that has stood for sometime in contact with air, and serious explosions have resulted. If peroxides are suspected of being present, they should be decomposed by treatment with zinc dust and a small amount of sulfuric acid in acetic acid.

III. Halohydrins and Alkene Oxides

A. Introduction:

The halohydrins may be considered as substituted hydrocarbons or as substituted alcohols. The alkene oxides may be regarded as derivatives of the halohydrins.

The **halohydrins**, especially 2-chloroethanol (**ethylene chlorohydrin**), are used in organic synthesis. The alkene oxides are important synthetic reagents, and epoxyethane (**ethylene oxide**) is used as an insecticide and as a fumigant. The hydroxy ethers, prepared from the alkene oxides and the alcohols, have the combined solvent properties of both the alcohols and the ethers, and the very lowest members are miscible with water. Certain hydroxy ethers are marketed under the trade name of **Cellosolve**.

Both the **halohydrins** and the **alkene oxides** exhibit *position* isomerism.

---

[1] The PCl$_5$, when heated, gives PCl$_3$ and Cl$_2$, and the latter effects chlorination so that the cleavage products are usually polychlorides.

B. Nomenclature:

Structural Formulas	I.U.C. Names	Common Names
$Br \cdot CH_2 \cdot CH_2 \cdot OH$	2-bromoethanol	ethylene bromohydrin
$Cl \cdot CH_2 \cdot CH_2 \cdot OH$	2-chloroethanol	ethylene chlorohydrin
$CH_3 \cdot CHOH \cdot CH_2 \cdot Cl$	1-chloro-2-propanol	propylene chlorohydrin
$Br \cdot CH_2 \cdot CH_2 \cdot CH_2 \cdot OH$	3-bromo-1-propanol	trimethylene bromohydrin
$CH_2Cl \cdot CHOH \cdot CH_2 \cdot OH$	3-chloro-1,2-propanediol	α-chlorohydrin
$CH_2Br \cdot CHBr \cdot CH_2 \cdot OH$	2,3-dibromo-1-propanol	β-dibromohydrin
$H_2C \overset{.O.}{\cdot} CH_2$	epoxyethane	ethylene oxide, oxirane
$CH_3 \cdot HC \overset{.O.}{\cdot} CH_2$	1,2-epoxypropane	propylene oxide, propene oxide
$CH_3 \cdot CH_2 \cdot HC \overset{.O.}{\cdot} CH_2$	1,2-epoxybutane	α-butylene oxide
$CH_3 \cdot HC \overset{.O.}{\cdot} CH \cdot CH_3$	2,3-epoxybutane	β-butylene oxide

C. Preparations:

I. THE HALOHYDRINS MAY BE PREPARED:

1. *By the addition of hypochlorous or hypobromous acids to alkenes:*

$$R \cdot HC{:}CH_2 + HO \cdot X \rightarrow R \cdot CHOH \cdot CH_2 \cdot X, \text{ and}$$

$$CH_3 \cdot HC{:}CH_2 + HO \cdot Cl \rightarrow CH_3 \cdot CHOH \cdot CH_2Cl \text{ (propylene chlorohydrin)}$$

2. *By the addition of halogen acids to alkene oxides:*

$$R \cdot HC \overset{.O.}{\cdot} CH_2 + HX \rightarrow R \cdot HCOH \cdot CH_2 \cdot X \, (+ \, R \cdot HCX \cdot CH_2 \cdot OH),$$
and

$$H_2C \overset{.O.}{\cdot} CH_2 + HCl \rightarrow CH_2OH \cdot CH_2 \cdot Cl, \text{ or epoxyethane (ethylene oxide)}$$
+ hydrochloric acid → 2-chloroethanol (**ethylene chlorohydrin**)

3. *By the partial hydrolysis of dihaloalkanes:*

$$R \cdot CHX \cdot CH_2 \cdot X + NaOH, \text{ aq.} \rightarrow R \cdot CHOH \cdot CH_2 \cdot X + NaX, \text{ and}$$
$$R_2C(X) \cdot CHX \cdot R + NaOH, \text{ aq.} \rightarrow R_2C(OH) \cdot CHX \cdot R + NaX, \text{ or}$$
the ease of hydrolysis increases, according to the type of halogen atom involved, in the order $R \cdot CH_2 \cdot X$, $R_2CH \cdot X$, $R_3C \cdot X$.
$CH_3 \cdot CHCl \cdot CH_2 \cdot Cl + NaOH$, aq. → $CH_3 \cdot CHOH \cdot CH_2 \cdot Cl + NaCl$, or 1,2-dichloropropane (**propylene chloride**) + aqueous sodium hydroxide → 1-chloro-2-propanol (**propylene chlorohydrin**) + sodium chloride.

4. *By the substitution of "X" for "OH" in glycols with HCl or HBr under controlled conditions:*

$$R \cdot CHOH \cdot CH_2 \cdot OH + HCl \rightarrow R \cdot CHCl \cdot CH_2 \cdot OH + H_2O, \text{ and}$$

$CH_2OH \cdot CH_2 \cdot OH + HCl \rightarrow CH_2OH \cdot CH_2 \cdot Cl + H_2O$, or 1,2-ethanediol (**glycol**) + hydrochloric acid → 2-chloroethanol (**ethylene chlorohydrin**) + water.

II. THE ALKENE OXIDES MAY BE PREPARED:

By the loss of "HX" from alkene halohydrins with solid NaOH or by catalytic addition of oxygen to alkenes:

$$R \cdot CHOH \cdot CH_2 \cdot X + NaOH, \rightarrow R \cdot HC \overset{.O.}{\cdot} CH_2 + NaX + H_2O, \text{ and}$$

$$CH_3 \cdot CHOH \cdot CH_2 \cdot Cl + NaOH, \text{ solid} \rightarrow CH_3 \cdot HC \overset{.O.}{\cdot} CH_2 + NaCl + H_2O, \text{ or}$$

1-chloro-2-propanol (**propylene chlorohydrin**) + solid sodium hydroxide → 1,2-epoxypropane (**propylene oxide**) + sodium chloride + water.

D. Physical Properties:

The lower members of the **halohydrins** are liquids which are denser than water. The hydroxyl group, of course, tends to increase their solubility in water but the presence of the halide group tends to counteract this effect. Some members of the series are, therefore, quite soluble in water while others are only slightly soluble. Both the hydroxyl and halide groups tend to decrease volatility, as observed from the following: Ethane boils at −88.6, chloroethane at 12.2, ethanol at 78.5, 2-chloroethanol at 128.8, 2-bromoethanol at 150.3, and 2-iodoethanol at 85/25mm.

The **alkene oxides** have boiling points intermediate between those of the corresponding hydrocarbon and the monohydroxy alcohol of the same carbon content. Epoxyethane (**ethylene oxide**, b.p. 10.7) is a pleasant smelling gas, and 1-2-epoxypropane (**propylene oxide**, b.p. 35) is a colorless liquid.

E. Chemical Properties:

I. THE HALOHYDRINS, *being mixed compounds, should show the properties both of alkyl halides and alcohols except in so far as one group may influence or interfere with the activity of the other group.*

Among the more important reactions are (1) reduction, (2) esterification, (3) oxidation, and (4) alkene oxide formation, when treated with:

1. Reducing agents (Zn/H⁺, NaHg/H₂O, etc.)
2. Salts of organic acids
3. Oxidizing agents
4. Solid sodium or potassium hydroxides

R·CHOH·CH₂	X + H	H
R·CHOH·CH₂	X + Na	O·OC·R
R(CH₂X)C(OH)	H + OH⁻	OH⁻
H₂C −	X Na⁺	

Or, as an example,   R·HC·O   H⁺ OH⁻

$$R \cdot CHOH \cdot CH_2 \cdot X + 2\,H^+ / Zn \rightarrow R \cdot CHOH \cdot CH_3 + HX + Zn^{++}$$

II. THE ALKENE OXIDES, *having only three atoms in the ring, are quite reactive chemically.*

Their principal reactions are of an additive nature. Some typical products are indicated below:

H₂C·O− H   H H H H H   H   H   H   MgX

H₂C−   OH, H, Cl, Br, X, O·CH₂·CH₂·OH, O·OC·R, NH₂, O·R, R,

or, to illustrate, $H_2C \overset{.O.}{\cdot} CH_2 + H \cdot OH \rightarrow CH_2OH \cdot CH_2OH$ (**glycol**)

CHAPTER VI

ALIPHATIC OXYGEN DERIVATIVES

I. Saturated Aldehydes and Ketones

A. Introduction:

Aldehydes and ketones may be considered as derivatives of hydrocarbons in which two hydrogen atoms on the same carbon atom have been replaced by an oxygen atom. In aldehydes the carbonyl group (C:O) is linked to one carbon atom and one hydrogen atom (R·CO·H), but in ketones it is linked to two carbon atoms (R·CO·R). This similarity of structure tends toward similar methods of preparation and similar chemical properties of the aldehydes and ketones, but the difference in the groups linked to the carbonyl group gives to each series certain individual characteristics.

The type formula for the aldehydes is R·CHO and that for the ketones is R·CO·R.

Normally, aliphatic aldehydes occur only in traces in nature. **Citronellal** ($H_2C:C[CH_3]·CH_2·CH_2·CH_2·CH [CH_3]·CH_2·CHO$) is found in citronella oil and in lemon grass oil. Citral ($[CH_3]_2C:CH·CH_2·CH_2·C[CH_3]:CH·CHO$, **geranial**) occurs in the oil of lemons and oranges, and the corresponding alcohol is found in lemon grass oil (70–80%). Propanone ($CH_3·CO·CH_3$, **acetone**) occurs as a metabolic product in the urine, especially that of diabetics. 2-Hendecanone ($CH_3·CO·C_9H_{19}$, **methyl nonyl ketone**) occurs in the oil of rue and, with homologs, in cocoanut oil. Ketones occur in perfumes, as violet, and in some terpenes as odoriferous substances.

Owing to its exceptional chemical properties, **formaldehyde** is used as a protein preservative, as a germicide, as a fungicide, as a deodorant, in embalming fluids, in the manufacture of artificial resins, in silvering mirrors, and in tanning. The aldehydes and ketones are important synthetic intermediates, and acetone is an important commercial solvent.

Among themselves, the aldehydes and ketones exhibit chain isomerism as illustrated by butanal ($CH_3·CH_2·CH_2·CHO$) and 2-methylpropanal ($[CH_3]_2CH·CHO$), or 2-pentanone ($CH_3·CH_2·CH_2·CO·CH_3$) and 3-methyl-2-butanone ($[CH_3]_2CH·CO·CH_3$). They are functional isomers of each other, of the unsaturated alcohols, and of the alkene oxides of the same carbon content. Compare, for example, butanal ($CH_3·CH_2·CH_2·CHO$), butanone ($CH_3·CH_2·CO·CH_3$), 3-buten-1-ol ($H_2C:CH·CH_2·CH_2·OH$), and 2-3-epoxybutane ($CH_3·HC \overset{O}{\cdot\cdot} CH·CH_3$).

B. Nomenclature:

The I.U.C. nomenclature replaces the final *e* of the corresponding hydrocarbon name by *al* for the aldehydes and by *one* for the ketones. The ketones usually require a prefixed numeral to indicate the relative position of the carbonyl group, but it is understood that the aldehyde group occupies position 1. The common names of the aldehydes are derived from the names of the corresponding acids by replacing *ic acid* by *aldehyde.*

Structural Formulas	Names	Structural Formulas Names
$H \cdot CHO$	methanal, **formaldehyde**	....
$CH_3 \cdot CHO$	ethanal, **acetaldehyde**	....
$CH_3 \cdot CH_2 \cdot CHO$	propanal, **propionaldehyde**	$CH_3 \cdot CO \cdot CH_3,$
		propanone, **acetone**
$CH_3 \cdot CH_2 \cdot CH_2 \cdot CHO$	butanal, **butyraldehyde**	$CH_3 \cdot CH_2 \cdot CO \cdot CH_3,$
		butanone, **ethyl**
		methyl ketone

C. Preparations:

I. THE ALDEHYDES MAY BE PREPARED:

1. By the removal of two hydrogen atoms from a primary alcohol group linked to the desired radical through (a) oxidation, or (b) catalytic dehydrogenation:

   a. $R \cdot CH_2 \cdot OH + $ mild oxidation $\rightarrow R \cdot CHO + H_2O,$ and

     $3\, CH_3 \cdot CH_2 \cdot OH + [K_2Cr_2O_7/4\, H_2SO_4,\, aq.] \rightarrow 3\, CH_3 \cdot CHO + Cr_2(SO_4)_3 + K_2SO_4 + 7\, H_2O,$ or ethanol (**ethyl alcohol**) $+$ [dichromic acid, aq.] $\rightarrow$ ethanal (**acetaldehyde**) $+$ chromic sulfate $+$ potassium sulfate $+$ water. This method is restricted to the volatile aldehydes.

   b. $R \cdot CH_2 \cdot OH + [(Cu,\, 300°C)] \rightarrow R \cdot CHO$ (an aldehyde) $+ H_2$

2. By (a) heating calcium salts of fatty acids containing the desired radical with calcium formate, or (b) by passing the vapors of the acid mixed with formic acid vapors over manganous oxide at 300°C:

   a. $(R \cdot CO \cdot O)_2Ca + (H \cdot CO \cdot O)_2Ca,$ fuse $\rightarrow 2\, R \cdot CHO + 2\, CaCO_3,$ and

$$CH_3 \cdot CO \begin{array}{|c|} \hline O \cdot Ca \cdot O \\ \hline CO \cdot O \cdot Ca \cdot O \cdot OC \\ \hline \end{array} \begin{array}{c} OC \cdot CH_3 \\ H \end{array} \rightarrow 2\, CH_3 \cdot CHO \text{ (ethanal, } \textbf{acetaldehyde)} + 2\, CaCO_3$$

   b. $R \cdot CO \cdot OH + H \cdot CO \cdot OH / (MnO,\, 300°C.) \rightarrow R \cdot CHO + CO_2 + H_2O$

     This is essentially a simultaneous reduction and oxidation respectively of the carbonyl carbon atoms of the two acids and is similar to a Cannizzaro reaction.

3. By the action of the Grignard reagent with (a) ethyl formate or ethyl orthoformate, or (b) hydrogen cyanide in ether solution, followed by hydrolysis:

   a. $R \cdot Mg \cdot X + H \cdot CO \cdot O \cdot CH_2 \cdot CH_3 \rightarrow R \cdot CHO + CH_3 \cdot CH_2 \cdot O \cdot Mg \cdot X$

   b. $R \cdot Mg \cdot X + H \cdot CN,$ ether $\rightarrow R \cdot C(:N \cdot Mg \cdot X)H,$ then
     $R \cdot C(:N \cdot Mg \cdot X)H + H_2O / 2\, HX \rightarrow R \cdot CHO + MgX_2 + NH_4X$

4. By the hydrolysis of the corresponding dihaloalkanes:
   $R \cdot CHCl_2 + H_2O / (PbO)$ boil $\rightarrow R \cdot CHO + 2\, HCl$

5. By the following commercial methods for (a) methanal (formaldehyde), and (b) ethanal (acetaldehyde):

   a. Formalin (a 40% aqueous solution of **formaldehyde**) is produced by passing the vapors of **methyl alcohol,** mixed with air, over heated copper gauze. The heated copper catalyzes the dehydrogenation of the alcohol, and the combustion of the hydrogen provides the heat that is required for the continuation of the reaction. The copper may be replaced by silver.

$$CH_3 \cdot OH + CuO/(Cu/air) \rightarrow H \cdot CHO \text{ (methanal, } \textbf{formaldehyde)} + Cu + H_2O$$

Formaldehyde may occur in plants as an unstable intermediate which polymerizes to carbohydrate material:

$$CO_2 + H_2O, \text{ photosynthesis} \rightarrow H \cdot CHO + O_2$$

b. **Acetaldehyde** is produced (1) by absorption of **acetylene** in sulfuric acid containing mercuric salts and subsequent hydrolysis, (2) by the passing of vapors of **ethyl alcohol**, mixed with air, over heated copper gauze, or (3) as a by-product in the preparation of acetic anhydride:

1. $H \cdot C \vdots C \cdot H + 2 H_2SO_4/(Hg^{++}) \rightarrow [CH_3 \cdot CH(O \cdot SO_2 \cdot OH)_2]$, then $[CH_3 \cdot CH(O \cdot SO_2 \cdot OH)_2] + H_2O \rightarrow CH_3 \cdot CHO$ (ethanal, commercial) $+ 2 H_2SO_4$

2. $CH_3 \cdot CH_2 \cdot OH + CuO/(Cu/air) \rightarrow CH_3 \cdot CHO + Cu + H_2O$

3. $H \cdot C \vdots C \cdot H + 2 CH_3 \cdot CO \cdot OH/(\text{catalyst}) \rightarrow CH_3 \cdot CH(O \cdot OC \cdot CH_3)^2$ (**ethylidene diacetate**), then heat $\rightarrow CH_3 \cdot CO \cdot O \cdot OC \cdot CH_3$ (**acetic anhydride**, commercial) $+ CH_3 \cdot CHO$

II. THE KETONES MAY BE PREPARED:

1. *By the removal of two hydrogen atoms from a secondary alcohol group linked to the desired radicals through* (a) *oxidation, or* (b) *catalytic dehydrogenation:*

   a. $R_2CH \cdot OH + \text{mild oxidation} \rightarrow R \cdot CO \cdot R + H_2O$, and

   $3 CH_3 \cdot CHOH \cdot CH_3 + Cr_2O_7^{=} + 8 H^+ \rightarrow 3 CH_3 \cdot CO \cdot CH_3$ (**propanone, acetone**) $+ 2 Cr^{+++} + 7 H_2O$

   b. $R_2CH \cdot OH + (Cu, 300°C.) \rightarrow R \cdot CO \cdot R$ (a ketone) $+ H_2$

2. *By* (a) *heating calcium salts of fatty acids containing the desired radicals, or* (b) *by passing the vapors of the acids over manganous oxide at 300°C:*

   a. $(R \cdot CO \cdot O)_2Ca + \text{fusion} \rightarrow R \cdot CO \cdot R + CaCO_3$, and

   $CH_3 \cdot CO \boxed{O \cdot Ca \cdot O \cdot OC} CH_3, \text{ fuse} \rightarrow CH_3 \cdot CO \cdot CH_3$ (**acetone**) $+ CaCO_3$

   b. $2 R \cdot CO \cdot OH, \text{ vapors} + (MnO, 300°C.) \rightarrow R \cdot CO \cdot R + CO_2 + H_2O$

   Either of these reactions may be employed for the production of mixed ketones as $R \cdot CO \cdot R'$, but the results are less satisfactory.

3. *By the action of the Grignard reagent with* (a) *esters other than those of formic acid or acyl halides, or* (b) *alkyl nitriles or amides in ether solution, followed by hydrolysis:*

   a. $R \cdot Mg \cdot X + R \cdot CO \cdot O \cdot Et, \text{ ether} \rightarrow R \cdot CO \cdot R + Et \cdot O \cdot Mg \cdot X$, and

   $CH_3 \cdot Mg \cdot X + Et \cdot O \cdot OC \cdot CH_2 \cdot CH_3, \text{ ether} \rightarrow CH_3 \cdot CO \cdot CH_2 \cdot CH_3 + Et \cdot O \cdot Mg \cdot X$

   b. $R \cdot Mg \cdot X + R \cdot CN \rightarrow R \cdot C(:N \cdot Mg \cdot X)R$, then

   $R \cdot C(:N \cdot Mg \cdot X)R + H_2O/2 HX \rightarrow R \cdot CO \cdot R + MgX_2 + NH_4X$

4. *By the hydrolysis of the corresponding dihaloalkanes:*

$$R_2CCl_2 + H_2O/(PbO), \text{ boil} \rightarrow R_2C:O + 2 \text{ HCl}$$

Ketonic hydrolysis of alkyl acetoacetates as a source of synthetic ketones will be considered later.

5. *By the following commercial methods for acetone:*

a. The fermentation of starch by *Clostridium acetobutylicum* (Weizmann) bacteria yields a solvent mixture containing about 60% butanol, 30% acetone, and 10% ethanol.

b. The controlled oxidation or catalytic dehydrogenation of **isopropyl alcohol** (cf. II.1., above). The isopropyl alcohol is produced by absorption of propene in sulfuric acid and subsequent hydrolysis.

c. The dehydrogenation and decarboxylation of ethyl alcohol (U. S. Patent No. 1,665,350):

$$2 \text{ } CH_3 \cdot CH_2 \cdot OH + H_2O, \text{ catalyst} \rightarrow CH_3 \cdot CO \cdot CH_3 + CO_2 + 4 \text{ } H_2$$

d. Vapor phase dehydration and decarboxylation of acetic acid (cf. II.2.b., above). Fusion of calcium acetate is the old commercial method (cf. II.2.a., above).

D. Physical Properties:

Formaldehyde and **acetaldehyde** (b.p. 21) are colorless gases. Dodecanal (**lauraldehyde**, m.p. 44.5), is the first solid of the straight chain series. The boiling points of the aldehydes are much lower than those of the corresponding alcohols, and their solubility in water decreases with an increase in the carbon content.

Acetone (b.p. 56.1) is a colorless liquid, lighter than water, completely miscible with water, and has a sweetish odor. The first solid member of the series is 6-hendecanone (**diamyl ketone,** m.p. 15–16).

E. Chemical Properties:

The most active constituent of both the aldehydes and ketones is the carbonyl group (C:O) which characterizes most of their typical reactions. The hydrogen atom linked to this group in the aldehydes gives them, however, certain specific properties which serve to differentiate them from ketones.

THE ALDEHYDES AND KETONES REACT UNDER PROPER CONDITIONS:

1. *By oxidation when treated with chromic acid or other appropriate oxidizing agents* (cf. V.I, E.I.3):

R·CHO + oxidation → R·CO·OH → oxidized derivatives → cleavage, and $R_2C:O$ + oxidation → oxidized derivatives → cleavage, or

$3 \text{ } CH_3 \cdot CHO + Cr_2O_7^= + 8 \text{ } H^+ \rightarrow 3 \text{ } CH_3 \cdot CO \cdot OH + 2 \text{ } Cr^{+++} + 4 \text{ } H_2O$, and $CH_3 \cdot CO \cdot CH_2 \cdot CH_2 \cdot CH_3 + Cr_2O_7^= + 8H^+ \rightarrow CH_3 \cdot CO \cdot OH + CH_3 \cdot CH_2 \cdot CO \cdot OH + 2 \text{ } Cr^{+++} + 4 \text{ } H_2O$, or this last reaction may yield methanoic (**formic**) acid (or its oxidized derivative) and butanoic (**butyric**) acid.

The aldehydes are very susceptible to oxidation, but the ketones are attacked less readily. The lower aldehydes are oxidized to the corresponding acids by air in the presence of catalysts such as manganese acetate, and all the members of the series are partially oxidized by Fehling

solution, Tollens reagent, and similar substances, or by permanganate
solution, or chromic acid. Schiff reagent gives a red coloration with
aldehydes in the cold, but not with ketones. Simple ketones give negative
results with Fehling solution and with Tollens reagent. To illustrate:
$R \cdot CHO + 2\,Cu^{++} + 2\,OH^-/(\text{Rochelle salt}/NaOH,\ \text{Fehling}) \rightarrow R \cdot CO \cdot OH$
$+ 2\,Cu^+ + H_2O$, and $R \cdot CHO + 2\,Ag(NH_3)_2^+/H_2O$ (Tollens) $\rightarrow R \cdot CO \cdot OH +$
$2\,Ag^\circ + 2\,NH_4^+ + 2\,NH_3$.

The cuprous ion is precipitated as red cuprous oxide and the silver is
normally deposited as a silver mirror. The aliphatic aldehydes show a
slight tendency toward autooxidation and reduction (cf. Cannizzaro
reaction), especially so if the *CHO* group is linked to a hydrogen atom or a
tertiary carbon atom, or if $Al(OEt)_3$ is used as a catalyst (Tischtschenko
reaction).

2. *By reduction when treated with appropriate reducing agents*
(Zn/H^+, Na/ROH, $NaHg/H_2O$), or by passage with hydrogen
over catalysts such as copper, silver, or nickel:
$R \cdot CHO + 2\,Na/2\,EtOH \rightarrow R \cdot CH_2 \cdot OH$ (primary) $+ 2\,EtONa$,
and $R_2C{:}O + 2\,Na/2\,EtOH \rightarrow R_2CH \cdot OH$ (secondary) $+ 2\,EtONa$

Reduction of ketones, especially with amalgamated magnesium, results
in the formation of pinacols. Through the loss of water, the pinacols
rearrange to give pinacolones.

3. *By addition when treated with the Grignard reagent, hydrogen*
cyanide, sodium hydrogen sulfite, or ammonia, as indicated below:

Reagent	R·CHO	$R_2C{:}O$
+ R′·MgX	→ R·CH(O·MgX)·R′	$R_2C(O \cdot MgX) \cdot R'$ (cf. Ch. V.I, C.I.2)
+ H·CN	→ R·CH(O·H)·CN	$R_2C(O \cdot H) \cdot CN$
+ H·OSO₂Na	→ R·CH(O·H)·SO₃Na	$R_2C(O \cdot H) \cdot SO_3Na$
+ H·NH₂	→ R·CH(O·H)·NH₂	complex derivatives

Formaldehyde with the Grignard reagent and subsequent hydrolysis
gives primary alcohols, other aldehydes give secondary alcohols, and
ketones give tertiary alcohols. This is, in reality, a reductive addition,
the carbonyl group being reduced to a carbinol group. The sodium
hydrogen sulfite addition products are usually crystalline if the aldehyde
is of high molecular weight, or if the ketone has a methyl group attached
to the carbonyl group. Formaldehyde reacts with ammonia to give
hexamethylenetetramine, which illustrates the abnormal activity somewhat
common to the first member of a homologous series.

4. *By substitution when treated with the following:*

a. Hydroxylamine   RCH │O + H₂│ N·OH   R₂C │O + H₂│ N·OH
   Hydrazine   RCH │O + H₂│ N·NH₂   R₂C │O + H₂│ N·NH₂
   Phenylhydrazine RCH │O + H₂│ N·NH·C₆H₅   R₂C │O + H₂│ N·NH·C₆H₅
   Substituted hydrazine RCH │O + H₂│ N·NH·CO·NH₂ R₂C │O + H₂│ N·NH·CO·NH₂
   Alcohols   RCH │O + 2H│ OR¹   no reaction with ketones²

b. Phosphorus
   pentahalides   RCH │O + PCl₃│ Cl₂   R₂C │O + PCl₃│ Cl₂

c. Halogen in alkali   HC:O   RC:O (cf. IV.II,C.4.a)
                    R·HC │H + X│ X   R·HC │H + X│ X

d. Halogen gas   R·CO │H + X│ X   R·CO·HRC │H + X│ X

---

[1] Effective catalysts, in order of increasing importance, are $ZnCl_2$, $FeCl_3$, $CaCl_2$, and BF_3.
Propanal is several hundred times as reactive as is butanal, with respect to this reaction.
[2] Acetone gives a 70% yield of acetal with ethanol.

The first four reactions under *a* may be considered as additions followed by the loss of a molecule of water, or they may be considered as condensation reactions. The products under *c* represent the first step in the haloform reaction. Trichloroacetaldehyde ($CCl_3 \cdot CHO$, chloral) is obtained when **acetaldehyde** is treated with chlorine water or by the action of chlorine on anhydrous alcohol. Chloral is of interest because it combines with water to give a crystalline solid ($CCl_3 \cdot CH[OH_2]$, chloral hydrate) which has two hydroxyl groups attached to the same carbon atom, as evidenced by the fact that it gives both semi- and true acetals with alcohols. Reaction *d* is of no practical value in the aliphatic series.

5. *By (a) condensation in the presence of calcium hydroxide, barium hydroxide, dilute sodium hydroxide, or zinc chloride, (b) polymerization in the presence of catalysts, and (c) resin formation in the presence of proper condensing agents:*

   a. $R \cdot CHO + R' \cdot CH_2 \cdot CHO / (Ca[OH]_2, aq.) \rightarrow R \cdot CHOH \cdot CHR' \cdot CHO$, then $R \cdot CHOH \cdot CHR' \cdot CHO$, by loss of $H_2O \rightarrow RHC:CR' \cdot CHO$

   When two moles of acetaldehyde condense in the manner indicated in the first equation, the product is known as aldol. This type of reaction is known, therefore, as *aldol condensation*. It may be considered as an addition reaction in which *an alpha hydrogen atom* becomes attached to the oxygen atom of the other molecule. The reaction may take place between aldehydes, aldehydes and ketones, or ketones.

   b. 1. $x \, H \cdot CHO$, aq., evaporate $\rightarrow (H \cdot CHO)_x$ (paraformaldehyde)

   2. $4 \, CH_3 \cdot CHO + $ trace H_2SO_4, 0°C. $\rightarrow (CH_3 \cdot CHO)_4$ (metaldehyde, a solid)

   3. $3 \, CH_3 \cdot CHO + $ trace $H_2SO_4 \rightarrow (CH_3 \cdot CHO)_3$ (paraldehyde, a liquid)

   None of these polymers have aldehyde properties. When **paraformaldehyde** is heated, formaldehyde is regenerated, and **acetaldehyde** is obtained by treating its polymers with mineral acids. Paraldehyde is a ring structure containing alternate carbon and oxygen atoms.

   c. 1. $H \cdot CHO + C_6H_5 \cdot OH$ (phenol) $\rightarrow$ a synthetic resin, as "Bakelite"

   2. $R \cdot CHO + $ conc. aq. NaOH, heat $\rightarrow$ resin formation

II. Unsaturated Aldehydes and Ketones

A. Introduction:

The unsaturated aldehydes and ketones may be considered as derivatives of the unsaturated hydrocarbons in which two hydrogen atoms on the same carbon atom have been replaced by an oxygen atom, or as the first oxidation products respectively of the unsaturated primary and secondary alcohols.

The general type formula for the *enal* and *enone* series is $C_nH_{2n-2}O$.

Propenal (**acrolein**) is produced in small amounts when fats are heated or burned.

Due to their rather exceptional chemical properties, these compounds are useful synthetic intermediates.

The ene linkage and the carbonyl group both give rise to the possibility of position isomerism, and the unsaturated aldehydes are functional isomers of the unsaturated ketones.

B. Nomenclature:

Type Formulas	I.U.C. Names	Common Names
$H_2C:CH\cdot CHO$	propenal	acrolein, acrylaldehyde
$CH_3\cdot HC:CH\cdot CHO$	trans-2-butenal	crotonaldehyde, crotonic aldehyde
$CH_3\cdot HC:C(CH_3)\cdot CHO$	2-methyl-2-butenal	tiglaldehyde, $\alpha,\ \beta$-dimethylacrolein
$H_2C:CH\cdot CO\cdot CH_3$	butenone	methyl vinyl ketone
$CH_3\cdot HC:CH\cdot CO\cdot CH_3$	3-penten-2-one	ethylidene acetone
$H_2C:CH\cdot CH_2\cdot CH_2\cdot CO\cdot CH_3$	5-hexen-2-one	allyl acetone
$(CH_3)_2\cdot C:CH\cdot CO\cdot CH_3$	4-methyl-3-penten-2-one	mesityl oxide
$(CH_3)_2C:CH\cdot CO\cdot HC:C(CH_3)_2$	2,6-dimethyl-2,5-hepta-dien-4-one	phorone

C. Preparations:

I. THE UNSATURATED ALDEHYDES MAY BE PREPARED:

1. By (a) the careful oxidation of allyl alcohol, or (b) by the dehydration of glycerol, to give acrolein:

   a. $H_2C:CH\cdot CH_2\cdot OH$ + mild oxidation $\rightarrow H_2C:CH\cdot CHO$ (propenal)

   b. $CH_2OH\cdot CHOH\cdot CH_2OH + (KHSO_4)$, heat, 200°C. $\rightarrow H_2C:CH\cdot CHO +$ 2 H_2O

2. By the aldol condensation of acetaldehyde followed by dehydration to give crotonaldehyde:

   $CH_3\cdot CHO + CH_3\cdot CHO$/(dilute alkali) $\rightarrow CH_3\cdot CHOH\cdot CH_2\cdot CHO$ (aldol),
   then $CH_3\cdot CHOH\cdot CH_2\cdot CHO$ + (dilute acid), heat $\rightarrow CH_3\cdot HC:CH\cdot CHO$

II. THE UNSATURATED KETONES MAY BE PREPARED:

1. By the aldol condensation of ketones, followed by dehydration:

   2 $CH_3\cdot CO\cdot CH_3 + (Ba[OH]_2$, dilute alkali) $\rightarrow (CH_3)_2C(OH)\cdot CH_2\cdot CO\cdot CH_3$,
   then $(CH_3)_2C(OH)\cdot CH_2\cdot CO\cdot CH_3 + (H^+)$, 100°C. $\rightarrow (CH_3)_2C:CH\cdot CO\cdot CH_3 + H_2O$

   The product is mesityl oxide. Iodine may also be used to effect dehydration of the diacetone.

2. By the action of acids on acetone to give both mesityl oxide and phorone (but the yields are very poor):

   x $CH_3\cdot CO\cdot CH_3 + (H^+)$, heat $\rightarrow (CH_3)_2C:CH\cdot CO\cdot CH_3 + (CH_3)_2C:CH\cdot CO\cdot HC:C(CH_3)_2$

D. Physical Properties:

Propenal (acrolein, b.p. 57) and 2-butenal (crotonaldehyde, b.p. 105) are both colorless liquids. Acrolein has a pungent, irritating odor. Mesityl oxide (b.p. 128.7) is a colorless liquid with a characteristic peppermint odor, and phorone (m.p. 28) is a solid.

E. Chemical Properties:

THE UNSATURATED ALDEHYDES AND KETONES REACT UNDER PROPER CONDITIONS:

1. *To give derivatives that are characteristic* (a) *of the carbonyl group* (cf. Ch. VI, Div. I, Sec. E), *and* (b) *of the unsaturated carbon-to-carbon linkage* (cf. Ch. III, Div. II, Sec. E), except in so far as the presence of the one group might effect the normal activity of the other group.

2. *To give the following more or less characteristic reactions:*

a. $H_2C{:}CH{\cdot}CHO + HBr \rightarrow CH_2Br{\cdot}CH_2{\cdot}CHO$ (3-bromopropanal)

b. $H_2C{:}CH{\cdot}CHO + 2\ NaHSO_3$, aq. $\rightarrow CH_2(SO_3Na){\cdot}CH_2{\cdot}CH(SO_3Na){\cdot}OH$

c. $(CH_3)_2C{:}CH{\cdot}CO{\cdot}CH_3 + HO{\cdot}NH_2 \Big\langle$
$\qquad\qquad (CH_3)_2C{\cdot}CH_2{\cdot}CO{\cdot}CH_3$, or
$\qquad\qquad\quad H{\cdot}\overset{\cdot\cdot}{N}{\cdot}OH \quad N{\cdot}OH$
$\qquad\qquad (CH_3)_2{\cdot}HC{\cdot}CH{\cdot}\overset{\cdot\cdot}{C}{\cdot}CH_3$

d. $CH_3{\cdot}HC{:}CH{\cdot}CO{\cdot}CH_3 + C_2H_5{\cdot}Mg{\cdot}Br$, then hyd. $\Big\langle$
$\qquad\qquad CH_3{\cdot}CH{\cdot}CH_2{\cdot}CO{\cdot}CH_3$, or
$\qquad\qquad\qquad \overset{\cdot\cdot}{C}_2H_5$
$\qquad\qquad\qquad\qquad\qquad C_2H_5$
$\qquad\qquad CH_3{\cdot}HC{:}CH{\cdot}\overset{\cdot\cdot}{C}(OH){\cdot}CH_3$

Such reagents as hydrogen cyanide, sodium hydrogen sulfite, hydroxylamine, and the Grignard reagent do not add, normally, to *ene* linkages. The activating effect of the carbonyl group on the *ene* linkage is interpreted as being due to a conjugated system of double bonds (cf. Ch. III, Div. III).

III. Ketenes

Ketene ($H_2C{:}C{:}O$) and its alkyl derivatives ($RHC{:}C{:}O$, aldoketenes and $R_2C{:}C{:}O$, ketoketenes) make up the members of the ketene series. They are, in a sense, unsaturated ketones, but are usually known as *ketenes.*

Ketene may be formed (a) by the loss of methane from **acetone** when passed through a hot tube, or (b) by the removal of two bromine atoms from bromoacetyl bromide upon treatment with zinc dust: $CH_3{\cdot}CO{\cdot}CH_3$ + passage through hot tube, $700°C. \rightarrow H_2C{:}C{:}O + CH_4$, or $CH_2Br{\cdot}CO{\cdot}Br$ + Zn dust, heat $\rightarrow H_2C{:}C{:}O + ZnBr_2$.

Ketene is a poisonous, colorless gas with an extremely unpleasant odor (b.p. -56).

Ketene is extremely reactive, most of its reactions being of an additive character, as illustrated below:

$H_2C{:}C{:}O + H_2O \rightarrow CH_3{\cdot}CO{\cdot}OH$ (ethanoic acid, **acetic acid**)

$H_2C{:}C{:}O + CH_3{\cdot}CH_2{\cdot}OH \rightarrow CH_3{\cdot}CO{\cdot}O{\cdot}CH_2{\cdot}CH_3$ (ethyl ethanoate)

$H_2C{:}C{:}O + CH_3{\cdot}CO{\cdot}OH \rightarrow CH_3{\cdot}CO{\cdot}O{\cdot}OC{\cdot}CH_3$ (**acetic anhydride,** commercial)

$H_2C{:}C{:}O + NH_3 \rightarrow CH_3{\cdot}CO{\cdot}NH_2$ (ethanamide, **acetamide**)

$H_2C{:}C{:}O + Br_2 \rightarrow CH_2Br{\cdot}CO{\cdot}Br$ (bromoethanoyl bromide)

$H_2C{:}C{:}O + R{\cdot}CH_2{\cdot}CO{\cdot}R/(cat.) \rightarrow R{\cdot}HC{:}O\ (O{\cdot}OC{\cdot}CH_3){\cdot}R$ (an enol acetate; Degering-Gwynn reaction)

As evidenced by the preceding reactions, the ketenes are important synthetic intermediates. The aldoketenes polymerize quite readily. Carbon suboxide ($O{:}C{:}C{:}C{:}O$), obtained by the dehydration of malonic acid, may be regarded as a diketene.

CHAPTER VII

ALIPHATIC OXYGEN DERIVATIVES: SATURATED AND UNSATURATED MONO- AND DIACIDS

I. Saturated Monoacids

A. Introduction:

Acids may be considered as derivatives of the hydrocarbons in which one or more of the hydrogen atoms have been replaced by a carboxyl group ($\cdot CO \cdot OH$), or as the third oxidation product of the $\cdot CH_3$ group of a hydrocarbon. The monoacids (**monocarboxylic, monobasic**) contain only one carboxyl group per molecule, whereas the diacids (**dicarboxylic, dibasic**) contain two carboxyl groups per molecule. The only triacid of importance is 2-hydroxy-1, 2, 3-propanetricarboxylic (**citric**) acid. Methanoic (**formic**) acid is abnormal in that the carboxyl group is linked to a hydrogen atom ($H \cdot CO \cdot OH$) and it has, therefore, certain properties that are common to the aldehydes.

The type formulas for the saturated monoacids are $R \cdot CO \cdot OH$ or $C_nH_{2n+1} CO \cdot OH$.

Methanoic (**formic**) acid occurs in ants and nettles. Ethanoic (**acetic**) acid is found in the form of its esters in odoriferous oils of many plants and in fruit juices. Butanoic (**butyric**) acid occurs in rancid butter, in Limburger cheese, in perspiration, and as glycerol esters in butter and fats. Pentanoic ($C_4H_9 \cdot COOH$, **valeric**), hexanoic ($C_5H_{11} \cdot COOH$, **caproic**), octanoic ($C_7H_{15} \cdot COOH$, **caprylic**), decanoic ($C_9H_{19} \cdot COOH$, **capric**), hexadecanoic ($C_{15}H_{31} \cdot COOH$, **palmitic**), and octadecanoic ($C_{17}H_{35} \cdot COOH$, **stearic**) acids are widely distributed in the form of their glycerol esters in fats and oils of plant and animal origin. The aliphatic acids occurring in nature usually have an even number of carbon atoms.

Methanoic (**formic**) acid, because of its slightly germicidal properties, finds use in disinfecting hides and in the brewing industries. It is used as a coagulant for rubber latex. Ethanoic (**acetic**) acid is used in the dyeing of textiles, in the tanning of leather, in the preparation of important solvents as ethyl, amyl. and butyl ethanoates, and in the form of its anhydride in the manufacture of cellulose acetate for photographic films and for rayon. Ethanoic acid is an excellent solvent for most organic and even some inorganic compounds. Propanoic (**propionic**) acid, which is produced as a by-product in the Patart (methanol) synthesis, is used in the production of cellulose acetate-propionate, which is less soluble than is cellulose acetate. A mixture of hexadecanoic (**palmitic**) and octadecanoic (**stearic**) acids is used, along with paraffin, in the manufacture of candles. The sodium and potassium salts of these acids are water soluble and are used extensively as soaps. Octadecanoic (**stearic**) acid is used in making shoe and metal polishes, and in the production of cosmetic creams. It is also used as a softener in rubber.

Chain isomerism in the acid series is illustrated by pentanoic acid ($CH_3 \cdot CH_2 \cdot CH_2 \cdot CH_2 CO \cdot OH$) and 3-methylbutanoic acid ($[CH_3]_2 CH \cdot CH_2 \cdot CO \cdot OH$).

Examples of position isomerism are afforded by butanoic acid ($CH_3 \cdot CH_2 \cdot CH_2 \cdot CO \cdot OH$) and 2-methylpropanoic acid ($CH_3 \cdot CH[CH_3] \cdot CO \cdot OH$). They are functional isomers of the esters of the same carbon content, examples being propanoic acid ($CH_3 \cdot CH_2 \cdot CO \cdot OH$) and methyl ethanoate ($CH_3 \cdot CO \cdot O \cdot CH_3$, methyl acetate).

B. Nomenclature:

The I.U.C. names for the acids are derived from the name of the corresponding hydrocarbon by replacing the final *e* with *oic* and adding the word *acid*, as shown below:

Structural Formulas	I.U.C. Names	Common Names
$H \cdot CO \cdot OH$	methanoic acid	formic acid
$CH_3 \cdot CO \cdot OH$	ethanoic acid	acetic acid
$CH_3 \cdot CH_2 \cdot CO \cdot OH$	propanoic acid	propionic acid
$CH_3 \cdot CH_2 \cdot CH_2 \cdot CO \cdot OH$	butanoic acid	butyric acid
$CH_3 \cdot CH_2 \cdot CH_2 \cdot CH_2 \cdot CO \cdot OH$	pentanoic acid	valeric acid

In the I.U.C. system of nomenclature the carboxyl group occupies position 1. In naming derivatives of the acids by their common names, the letters of the Greek alphabet are used to designate relative position , the α-carbon atom being the one linked to the carboxyl group, β-, γ-, and δ- designating respectively the carbon atom next in order from the carboxyl group.

C. Preparations:

THE SATURATED MONOACIDS MAY BE PREPARED:

1. *By the oxidation (a) of the corresponding primary alcohols or aldehydes, (b) of the proper secondary alcohols or ketones, or (c) of the appropriate tertiary alcohols:*

   a. $3 \ R \cdot CH_2 \cdot OH + 2 \ Cr_2O_7^- /16 \ H^+$ aq., heat $\rightarrow 3 \ R \cdot CO \cdot OH + 4 \ Cr^{+++} + 11 \ H_2O$, and

   $3 \ CH_3 \cdot CH_2 \cdot OH + 2 \ Cr_2O_7^-/16 \ H^+$, heat $\rightarrow 3 \ CH_3 \cdot CO \cdot OH$ (ethanoic acid) $+ 4 \ Cr^{+++} + 11 \ H_2O$.

   b. $3 \ CH_3 \cdot CH_2 \cdot CH_2 \cdot CHOH \cdot CH_3 + Cr_2O_7^- /8H^+$ aq., $\rightarrow 3 \ CH_3 \cdot CH_2 \cdot CH_2 \cdot CO \cdot CH_3$ (2-pentanone) $+ 2 \ Cr^{+++} + 7 \ H_2O$, or

   $CH_3 \cdot CH_2 \cdot CH_2 \cdot CO \cdot CH_3 + Cr_2O_7^- /8 \ H^+$ aq., $\rightarrow CH_3 \cdot CH_2 \cdot CO \cdot OH$ (propanoic acid) $+ CH_3 \cdot CO \cdot OH$ (ethanoic acid) $+ 2 \ Cr^{+++} + 4 \ H_2O$.

   Methanoic acid (or its oxidized derivative) and butanoic acid are also formed in this reaction but the carbonyl group tends to remain with the smaller primary radical. If a secondary alkyl radical is attached to the carbon atom bearing the secondary hydroxyl group, the oxidation proceeds as follows:

   $(CH_3)_2CH \cdot CHOH \cdot CH_2 \cdot CH_3 + Cr_2O_7^- /8 \ H^+$ aq., $\rightarrow (CH_3)_2C{:}O$ (propanone) $+ CH_3 \cdot CH_2 \cdot CO \cdot OH$ (propanoic acid) $+ 2 \ Cr^{+++} + 5 \ H_2O$.

   The acetone may be oxidized, in turn, to give formic acid (or its oxidized derivative) and acetic acid. If a tertiary alkyl radical is attached to the carbon atom bearing the secondary hydroxyl group, the oxidation takes the following course:

3 $(CH_3)_3C \cdot CHOH \cdot CH_2 \cdot CH_3$ + 4 $Cr_2O_7^-$/32 H$^+$, aq., → 3 $(CH_3)_3C \cdot CO \cdot OH$ (2, 2-dimethylpropanoic acid) + 3 $CH_3 \cdot CO \cdot OH$ (ethanoic acid) + 8 Cr^{+++} + 19 H_2O.

c. $(CH_3)_2COH \cdot CH_2 \cdot CH_3$ + $Cr_2O_7^-$/8H$^+$, aq., → $CH_3 \cdot CH_2$ $CO \cdot CH_3$ (butanone) + $H \cdot CO \cdot OH$ (methanoic acid)+ 2 Cr^{+++} + 5 H_2O

This equation gives one set of initial cleavage products, but acetone and acetic acid may also be formed in this reaction as the first products of the oxidation.

2. By use of the Grignard reagent with carbon dioxide in dry ether, followed by hydrolysis:

$R \cdot MgX$, dry ether + CO_2 → $R \cdot CO \cdot OMgX$, then

$R \cdot CO \cdot OMgX$ + HX, aq., → $R \cdot CO \cdot OH$ + MgX_2, and

$CH_3 \cdot MgX$, dry ether + CO_2 → $CH_3 \cdot CO \cdot OMgX$ (**halomagnesium acetate**), then

$CH_3 \cdot CO \cdot OMgX$ + HX, aq. → $CH_3 \cdot CO \cdot OH$ (ethanoic acid) + MgX_2

The first step in this reaction involves reductive addition. By the use of a recent development in procedure, the reaction with carbon dioxide is effected by pouring the Grignard reagent upon dry ice. An analogous synthesis is $R \cdot Na$ + CO_2 → $R \cdot CO \cdot ONa$ (Wanklyn reaction).

3. By (a) the loss of CO_2 upon heating malonic or alkylmalonic acids, or (b) the acid hydrolysis of acetoacetic esters (see Ch. VIII, Div. IV, E. *2.b.* 2.c)

a. $HO \cdot OC \cdot CH_2 \cdot CO \cdot OH$ + heat, above m.p. → $CH_3 \cdot CO \cdot OH$ + CO_2

4. By the hydrolysis of acid derivatives:

a. $R \cdot CO \cdot ONa$ (a salt) + H$^+$, aq. → $R \cdot CO \cdot OH$ + Na$^+$

b. $(R \cdot CO)_2O$ (an acid anhydride) + H_2O/ (H$^+$) → 2 $R \cdot CO \cdot OH$

c. $R \cdot CN$ (a nitrile) + H_2SO_4, then 2 H_2O, heat → $R \cdot CO \cdot OH$ + NH_4HSO_4

d. $R \cdot CO \cdot NH_2$ (an amide) + H$^+$/H_2O, heat → $R \cdot CO \cdot OH$ + NH_4^+

e. $R \cdot CO \cdot Cl$ (an acid chloride) + OH$^-$, aq. → $R \cdot CO \cdot OH$ + Cl$^-$

f. $R \cdot CO \cdot O \cdot R$ (an ester) + OH$^-$, aq., → $R \cdot CO \cdot O^-$ + $R \cdot OH$ (cf. a)

g. $R \cdot CCl_3$ (a 1, 1, 1-trihalide) + 3 OH$^-$, aq. → $R \cdot CO \cdot OH$+ 3 Cl$^-$ + H_2O

5. By the following methods for (a) methanoic acid, and (b) ethanoic acid:

a. 1. $HOOC \cdot COOH$ + (glycerol), 90-140°C. → $H \cdot CO \cdot OH$ + CO_2 (Lab.)

2. CO + NaOH, conc., 200°C., 6-10 atms. → $H \cdot CO \cdot ONa$, then $H \cdot CO \cdot ONa$ + H_2SO_4 → $H \cdot CO \cdot OH$ + $NaHSO_4$ (commercial)

b. 1. $H \cdot C\!:\!C \cdot H$ + H_2O/ (Hg^{++}/H_2SO_4) → $CH_3 \cdot CHO$, then 2 $CH_3 \cdot CHO$ + O_2/ (air, $[CH_3 \cdot CO \cdot O]_2Mn$) → 2$CH_3 \cdot CO \cdot OH$

The manganese acetate prevents the formation of peroxides. If cobalt is used as the catalyst, commercial yields of cobaltic acetate are obtained.

2. $H \cdot C\!:\!C \cdot H$ + 2 $CH_3 \cdot CO \cdot OH$/ (catalyst) → $CH_3CH(O \cdot OC \cdot CH_3)_2$ (**ethylidene diacetate**), then

$CH_3 \cdot CH (O \cdot OC \cdot CH_3)_2$ + heat → $(CH_3 \cdot CO)_2O + CH_3 \cdot CHO$, then
2 $CH_3 \cdot CHO + O_2 /$ (air, $[CH_3 \cdot CO \cdot O]_2Mn$) → $2CH_3 \cdot CO \cdot OH$ (a new commercial process). The acetic acid used in the initial step of this reaction is recovered as acetic anhydride.

3. $CH_3 \cdot OH + CO$, catalyst → $CH_3 \cdot CO \cdot OH$ (commercial)

4. $CH_3 \cdot CH_2 \cdot OH + O_2 /$ (bacterial) → $CH_3 \cdot CO \cdot OH + H_2O$ (vinegar process)

5. Wood + distillation → $CH_3 \cdot CO \cdot OH + CH_3 \cdot OH + CH_3 \cdot CO \cdot CH_3$

D. Physical Properties:

Methanoic (formic, m.p. 8.5, b.p. 100.5) acid is a colorless, hygroscopic liquid with a pungent, irritating odor like that of SO_2. It forms a constant boiling mixture with water (77.5% formic acid, b.p. 107.1). Formic acid is very corrosive and irritates the skin, causing painful blisters. Ethanoic (acetic, m.p. 16.67°, b.p. 118) acid is a colorless liquid with a penetrating odor and a sour taste, and is soluble in water, alcohol, and ether in all proportions. Butanoic (butyric) acid is a thick, sour liquid, with a disagreeable odor. Hexadecanoic (palmitic, m.p. 64) and octadecanoic (stearic, m.p. 69.3) acids are colorless, waxy solids. The boiling points increase and the solubility in water rapidly decreases for the straight chain members as the carbon content increases.

The strength of an acid is usually expressed in terms of a dissociation constant K_A which is defined as follows:

K_A = conc. of hydrogen ion × (conc. of anion/conc. of undissoc. acid)

The K_A for a few acids are: **dichloroacetic acid**, 5×10^{-2}; **chloroacetic acid**, 1.5×10^{-3}; **formic acid**, 2.1×10^{-4}; **acetic acid**, 1.8×10^{-5}; **carbonic acid**, first hydrogen, 1×10^{-7}; and **carbonic acid**, second hydrogen, 5×10^{-11}, whereas for water it is about 1×10^{-16} at room temperature.

E. Chemical Properties:

THE MONOACIDS REACT UNDER PROPER CONDITIONS:

1. *By substitution (a) for the hydrogen atom of the hydroxyl group, (b) for the hydroxyl group, (c) for the two oxygen atoms of the carboxyl group, (d) for the carboxyl group, and (e) for hydrogen atoms linked to the alpha carbon atom, when treated with:*

a.
1. Active metals — $R \cdot CO \cdot O$ | H | Na
2. Ammonium hydroxide — $R \cdot CO \cdot O$ | H + HO | NH_4
3. Metallic hydroxides, etc. — $R \cdot CO \cdot O$ | H + HO | Na
4. Organic acids/(dehyd. agent) — $R \cdot CO \cdot O$ | H + HO | $OC \cdot R$
5. Acid chlorides — $R \cdot CO \cdot O$ | H + Cl | $OC.R$

b.
1. Alcohols/(H^+, dehyd. agent) — $R \cdot CO$ | OH + H | OR
2. Organic acids/(dehyd. agent) — $R \cdot CO$ | OH + H | $O \cdot OC \cdot R$
3. a. Phosphorus trihalides — 3 $R \cdot CO$ | OH + P | Cl_3
   b. Phosphorus pentahalides — $R \cdot CO$ | O | H + Cl | PCl_3 | Cl
4. Hydrogen/(catalysts) — $R \cdot CO$ | OH + H | H

c.
1. Phosphorus pentachloride, heat — $R \cdot C$ | O + PCl_3 | Cl_2
   L | O | H + Cl | PCl_3 | Cl
2. Ammonia, then P_2O_5, then heat — $R \cdot C$ | $O \cdot OH$ + H_3 | N

d. Soda lime (NaOH/CaO), heat   R $\boxed{\text{CO·O } \boxed{\text{H + HO}} \text{ Na NaO}}$ H

e. Halogen/(catalyst; S or P, sunlight)   RHC $\boxed{\text{H + Cl}}$ Cl
                                                              |
                                                          CO·OH

2. By dehydration and decarboxylation when the vapors of the acids are passed over manganous oxide at 300°C.:

R·CO $\boxed{\text{OH + H}}$ | O·OC R

The acids, in general, resist oxidation. Formic acid is very readily oxidized, however, to carbon dioxide and water. Acetic acid is very stable toward oxidizing agents, but with an increase in the length of the carbon chain the organic acids become less resistant to oxidation. The higher fatty acids are broken down into smaller units by the loss of two carbon atoms at a time for utilization as fuel in the animal organism. Acids also resist reduction, but they may be reduced catalytically (cf. b. 4, above) or more readily in the form of some of their derivatives.

Decarboxylation is also effected by the electrolysis of the sodium salts to give hydrocarbons.

Organic acids are usually identified by the synthesis and study of derivatives such as the amides, the anilides, the *p*-halophenacyl esters (R·CO·O·CH₂·CO·C₆H₄·X), the *p*-phenylphenacyl esters, the *p*-nitrobenzyl esters, and the semicarbazones of the esters.

II. Saturated Diacids

A. Introduction:

The saturated diacids may be thought of as derivatives of the alkanoic acids (saturated fatty acids) in which a hydrogen atom of the hydrocarbon chain has been replaced by a carboxyl group (·CO·OH).

The type formula for the saturated diacids is C_nH_{2n} (CO·OH)₂. Carbonic acid (HO·CO·OH) is sometimes considered as the first member of this series.

Ethanedioic (**oxalic**) acid occurs as the acid potassium salt, known as salt of sorrel, in tobacco, in rhubarb, and in other plants. Propanedioic (**malonic**) acid has not been isolated from natural products, but butanedioic (**succinic**) acid occurs in fossil gums, in lettuce, and in numerous other plants.

Ethanedioic (**oxalic**) acid is used in the dye and textile industries, in cleaning to remove iron rust and iron-ink stains, in quantitative analysis for the separation of calcium from magnesium, and as a reagent in the laboratory preparation of allyl alcohol and formic acid from glycerol (cf. Ch. VII, Div. I, Sec. C. 5. a. 1). Propanedioic (**malonic**) acid is used in the form of its ester in the *malonic ester synthesis* for the preparation of monoacids and barbitals, the latter being important hypnotics. Butanedioic (**succinic**) acid may be used as a primary standard in alkalimetry, iodimetry, and as its salt in acidimetry. The rare earths of the yttrium group are separated by the fractional crystallization of their butanedioates (**succinates**). Pentanedioic (**glutaric**) acid has a limited use, but hexanedioic (**adipic**) acid is used as a substitute for tartaric acid in baking powders, beverages, and various medicinal preparations. Its esters are used in the manufacture of perfumes, lacquers, plastics, and artificial leather.

Chain isomerism in this series is illustrated by 2, 2-dimethylbutanedioic acid and 2-ethylbutanedioic acid, whereas examples of position isomerism are afforded by butanedioic acid and 2-methylpropanedioic acid.

B. Nomenclature:

The diacids may be named (1) by adding the suffix "dioic acid" to the name of the corresponding hydrocarbon, (2) by adding the words "dicarboxylic acid" to the name of the hydrocarbon nucleus to which the carboxyl groups are linked and prefixing numerals to designate the relative positions of the acid groups, or (3) by the use of common names.

Structural Formulas	I.U.C. Names	Common Names
HOOC·COOH	ethanedioic acid, bicarboxylic acid	oxalic acid
HOOC·CH$_2$·COOH	propanedioic acid, methanedicarboxylic acid	malonic acid
HOOC·CH$_2$·CH$_2$·COOH	butanedioic acid, 1, 2-ethanedicarboxylic acid	succinic acid
HOOC(CH$_2$)$_3$COOH	pentanedioic acid, 1, 3-propanedicarboxylic acid	glutaric acid

C. Preparations:

THE SATURATED DIACIDS MAY BE PREPARED:

1. *By the same methods, in general, given in the preceding division for the introduction of carboxyl groups* (cf. Ch. VII, Div. I, Sec. C.):

2. *By the following special methods for* (a) *ethanedioic acid,* (b) *propanedioic acid, and* (c) *butanedioic acid:*

   a. 1. Sucrose + x HNO$_3$, conc./ (V$_2$O$_5$) → HOOC·COOH (**oxalic acid**) + oxides of N (Lab)

   2. Sawdust + NaOH, aq., conc., 240-50°C. → NaO·OC·CO·ONa, then NaO·OC·CO·ONa + 2 H$_2$SO$_4$, aq. → HO·OC·CO·OH + 2 NaHSO$_4$

   3. NaOH, conc. aq. + CO (pressure) → H·CO·ONa, then 2 H·CO·ONa + heat, 200°C. → NaO·OC·CO·ONa + H$_2$, then NaO·OC·CO·ONa + 2 H$_2$SO$_4$, aq. → HO·OC·CO·OH + 2 NaHSO$_4$ (new com'l)

   b. NaO·OC·CH$_2$·Cl + Na·CN → NaO·OC·CH$_2$·CN + NaCl, and NaO·OC·CH$_2$·CN + H$_2$SO$_4$, then 2 H$_2$O → HO·OC·CH$_2$·CO·OH + Na(NH$_4$)SO$_4$

   c. X·CH$_2$·CH$_2$·X + 2 KCN → NC·CH$_2$·CH$_2$·CN (butanedinitrile) + 2 KX, and NC·CH$_2$·CH$_2$·CN + 2 H$_2$SO$_4$, then 4 H$_2$O → HO·OC·CH$_2$·CH$_2$·CO·OH (butanedioic acid, **succinic acid**) + Na$_2$SO$_4$ + (NH$_4$)$_2$SO$_4$

D. Physical Properties:

Ethanedioic (**oxalic**, m.p. 189), propanedioic (**malonic**, m.p. 135.6), butanedioic (**succinic**, m.p. 185), pentanedioic (**glutaric**, m.p. 97.5), and hexanedioic (**adipic**, m.p. 151.1) acids are all colorless, crystalline solids. A graphical plot of the melting points against the carbon content of these acids, illustrates what is known as the *saw tooth* rule. These five acids are soluble in water.

E. Chemical Properties:

THE SATURATED DIACIDS REACT UNDER PROPER CONDITIONS:

1. *To give, in general, normal carboxyl derivatives* (cf. Ch. VII, Div. I, Sec. E).

   There are, of course, two carboxyl groups to react, hence it should be noticed that the presence of a second substituent group causes a limited shading of the properties of the first group and vice versa.

2. *To give derivatives peculiar to (a) ethanedioic acid when treated with concentrated sulfuric acid, permanganate solution, or glycerol and heat, (b) propanedioic, butanedioic, and pentanedioic acids when subjected to heat, (c) the calcium salts of the dioic acids when fused, and (d) malonic acid esters when treated with metallic sodium followed by treatment with alkyl iodides (Malonic Ester Synthesis); and the dialkylated product treated with urea (Fischer-Dilthey Synthesis):*

   a. 1. $HO \cdot OC \cdot CO \cdot OH + H_2SO_4$, conc., heat $\rightarrow CO_2 + CO + H_2SO_4 \cdot H_2O$

   2. $5\ HO \cdot OC \cdot CO \cdot OH + 2\ MnO_4^-/6\ H^+$, aq. $\rightarrow 10\ CO_2 + 8\ H_2O + 2\ Mn^{++}$

   The ease of oxidation is the distinguishing characteristic.

   3. *a.* $CH_2 \cdot OH + H \cdot O \cdot OC \cdot CO \cdot OH \rightarrow CH_2 \cdot O \cdot CO \cdot H$, by loss $\rightarrow CH_2$
      | ‖

      CHOH   (by loss of H_2O)   CHOH   of CO_2 and H_2O CH

      CH_2OH and CO_2) CH_2OH at $120°C.$ CH_2OH

      glycerol   monoformin   allyl alcohol

   b. If the monoformin produced in the above reaction is treated with an excess of oxalic acid and the mixture heated to the distillation point, a cycle results which gives formic acid as the distillate:
      $CH_2 \cdot O \cdot CO \cdot H + HO \cdot OC \cdot CO \cdot OH$, distill $\rightarrow CH_2 \cdot O \cdot OC \cdot CO \cdot OH$

      CHOH   CHOH $+ H \cdot CO \cdot OH$

      $CH_2OH \leftarrow$ then by the loss of CO_2, CH_2OH

b. 1. $HO \cdot OC \cdot CO \cdot OH + heat \rightarrow$ sublimes unchanged

   2. $HO \cdot OC \cdot CH_2 \cdot CO \cdot OH + heat$, 130-135°C. $\rightarrow CH_3 \cdot CO \cdot OH + CO_2$, also $HO \cdot OC \cdot CH_2 \cdot CO \cdot OH + heat/ (P_2O_5) \rightarrow C_3O_2 + 2\ H_2O$ (Diels-Wolf reaction)

   3. $H_2C \cdot CO \cdot OH + heat$ (or Ac_2O or $\rightarrow H_2C \cdot C{:}O + H_2O$ (butanedioic, or
      | $POCl_3$ and heat) | $>O$ succinic anhy-

      $H_2C \cdot CO \cdot OH$ $H_2C \cdot C{:}O$ dride)

   4. $H_2C \cdot CO \cdot OH + heat$ (or Ac_2O or $\rightarrow H_2C \cdot C{:}O + H_2O$ (pentanedioic
      | or glutaric

      $H \cdot C \cdot H$ $POCl_3$ and heat) H_2C O   anhydride)

      $H_2C \cdot CO \cdot OH$ $H_2C \cdot C{:}O$

   5. $HO \cdot OC\ (CH_2)_4 CO \cdot OH + heat \rightarrow$ sublimes, mostly unchanged

c. $CH_2 \cdot CH_2 \cdot CO \cdot O$ $CH_2 \cdot CH_2$
 $|$ $ > Ca + fusion \rightarrow$ $|$ $ > C{:}O$ (cyclopentanone) $+ CaCO_3$
 $CH_2 \cdot CH_2 \cdot CO \cdot O$ $CH_2 \cdot CH_2$

Owing to structural strains, cyclic compounds with less than five or more than six atoms in the ring are relatively unstable.

d. When ethyl malonate is treated with sodium ethylate, sodium malonic ester ($Et \cdot O \cdot OC \cdot HC{:}C(ONa) \cdot O \cdot Et$) is obtained which yields alkyl malonic esters upon treatment with alkyl halides:

$EtO \cdot OC \cdot CH_2 \cdot CO \cdot OEt + Na \cdot OEt \rightarrow EtO \cdot OC \cdot HC{:}C(ONa) \cdot OEt +$
$EtOH$, then

$EtO \cdot OC \cdot HC{:}C(ONa) \cdot OEt + R \cdot I \rightarrow EtO \cdot OC \cdot HC(R) \cdot CO \cdot OEt + NaI$,
then

$ EtO \cdot OC \cdot HC(R) \cdot CO \cdot OEt + 2 NaOH$, aq. $\rightarrow$
$ NaO \cdot OC \cdot HC(R) \cdot CO \cdot ONa + 2 EtOH$, then

$ NaO \cdot OC \cdot HC(R) \cdot CO \cdot ONa + 2 H_2SO_4$, aq. $\rightarrow$
$ HO \cdot OC \cdot HC(R) \cdot CO \cdot OH + 2 NaHSO_4$, then

$ HO \cdot OC \cdot HC(R) \cdot CO \cdot OH +$ heat above m.p. $\rightarrow$
$ R \cdot CH_2 \cdot CO \cdot OH + CO_2$

The second hydrogen atom in malonic ester may be replaced in a similar manner by an alkyl group before the last three steps indicated above are carried out. This series of reactions gives rise to an almost unlimited number of normal and α-substituted aliphatic monoacids.

When the dialkyl malonic esters are treated with urea in the presence of sodium ethoxide, they give dialkyl barbituric acid (Fischer-Dilthey Synthesis):

$CO \cdot OR + H \cdot N \cdot H / (C_2H_5 \cdot OH/Na) \rightarrow$ $O{:}C - N \cdot H + 2 R \cdot OH$
$|$ $|$ $|$ $|$
$R' \cdot C \cdot R''$ $C{:}O$ $R' \cdot C \cdot R''$ $C{:}O$
$|$ $|$ $|$ $|$
$CO \cdot OR + H \cdot N \cdot H$ $O{:}C - N \cdot H$

If the two alkyl groups are ethyl (C_2H_5), the product is veronal; if one of the alkyl groups is ethyl (C_2H_5) and the other is phenyl (C_6H_5), the product is luminal or phenobarbital; if one of the groups is ethyl and the other isoamyl, the product is amytal; and if one of the groups is ethyl and the other *s*-butyl, the product is nimytal.

III. Unsaturated Monoacids

A. Introduction:

The unsaturated monoacids may be considered as derivatives of the alkenes or the alkynes in which a hydrogen atom has been replaced by a carboxyl group.

The type formula for the alkenoic acids (oleic acid series) is $C_nH_{2n-1}CO \cdot OH$ and that for the alkynoic acids is $C_nH_{2n-3}CO \cdot OH$. The latter are uncommon and will not be considered in the subsequent discussion.

trans-2-Butenoic (**crotonic**) acid occurs as the glyceryl ester in croton oil. 9-Octadecenoic (**oleic**) acid is widely distributed as the glyceryl ester in fats and oils of vegetable and animal origin.

Glyceryl esters containing unsaturated acid residues are *partially* hydrogenated to give such products as Crisco, Trido, and Snowdrift. Oleic acid is used in the manufacture of soap. Linseed and tung oils, esters of unsaturated acids, are employed as drying oils in the paint industry. The esters of

acrylic acid undergo polymerization to give resins. Most of the other unsaturated monoacids have little or no commercial importance.

Nuclear isomerism is illustrated by 2-butenoic acid ($CH_3 \cdot HC:CH \cdot CO \cdot OH$) and 3-butenoic acid ($H_2C:CH \cdot CH_2 \cdot CO \cdot OH$). Examples of *position isomerism* are 3-butenoic acid ($H_2C:CH \cdot CH_2 \cdot CO \cdot OH$) and 2-methylpropenoic acid ($H_2C:C(CH_3) \cdot CO \cdot OH$). *Geometric, cis-trans,* or *olefin isomerism,* is shown by

trans-2-butenoic (**crotonic**) acid and

cis-2-butenoic (**isocrotonic**) acid.

$$
\begin{array}{cc}
H \cdot C \cdot CO \cdot OH & H \cdot C \cdot CO \cdot OH \\
\parallel & \parallel \\
CH_3 \cdot C \cdot H & H \cdot C \cdot CH_3 \\
\textit{trans}\text{-2-butenoic acid} & \textit{cis}\text{-2-butenoic acid}
\end{array}
$$

B. Nomenclature:

The I.U.C. names of the alkenoic acids are derived from the names of the corresponding hydrocarbons by replacing the final *e* with *oic*, adding the word *acid*, and prefixing the proper numerals:

Structural Formulas	I.U.C. Names	Common Names
$H_2C:CH \cdot CO \cdot OH$	propenoic acid	acrylic acid
$CH_3 \cdot HC:CH \cdot CO \cdot OH$	*trans*-2-butenoic acid	crotonic acid
$H_2C:CH \cdot CH_2 \cdot CO \cdot OH$	3-butenoic acid	vinylacetic acid

C. Preparations:

THE UNSATURATED MONOACIDS MAY BE PREPARED:

1. *By the mild oxidation of the corresponding alcohols or aldehydes:*

   a. $R \cdot HC:CH \cdot CH_2 \cdot OH + 2\ Ag_2O,\ aq. \rightarrow R \cdot HC:CH \cdot CO \cdot OH + 4\ Ag^\circ + H_2O$

   b. $R \cdot HC:CH \cdot CHO + Ag_2O,\ aq. \rightarrow R \cdot HC:CH \cdot CO \cdot OH + 2\ Ag^\circ$

2. *By the removal of hydrogen halide from adjacent carbon atoms of the appropriate halogen-substituted acid:*

   $R \cdot CH_2 \cdot CHX \cdot CO \cdot OH + 2\ NaOH,\ alc. \rightarrow R \cdot HC:CH \cdot CO \cdot ONa + NaX + 2\ H_2O,$ then

   $R \cdot HC:CH \cdot CO \cdot ONa + HCl,\ aq. \rightarrow R \cdot HC:CH \cdot CO \cdot OH + NaCl$

3. *By the removal of water or ammonia from adjacent carbon atoms of 3-hydroxy- or 3-aminoalkanoic acids:*

   a. $R \cdot CHOH \cdot CH_2 \cdot CO \cdot OH + heat \rightarrow R \cdot HC:CH \cdot CO \cdot OH + H_2O$

   b. $R \cdot CHNH_2 \cdot CH_2 \cdot CO \cdot OH + heat \rightarrow R \cdot HC:CH \cdot CO \cdot ONH_4,$ then
   $R \cdot HC:CH \cdot CO \cdot ONH_4 + HCl,\ aq. \rightarrow R \cdot HC:CH \cdot CO \cdot OH + NH_4Cl$

4. *By the Knoevenagel condensation for 2, 3-acids above acrylic acid:*
   $R \cdot CHO + H_2C(COOH)_2,\ 100°C. \rightarrow RHC:C(COOH)_2 + H_2O$
   then

$$RHC:C(COOH)_2, \triangle, \text{ above m.p. } \rightarrow R \cdot HC:CH \cdot CO \cdot OH + CO_2$$

The 2, 3-unsaturated acids may be obtained also by:

$$H_2C:CH \cdot CH_2 \cdot I + Na \cdot CN \rightarrow CH_3 \cdot HC:CH \cdot CN + NaI, \text{ then}$$

$$CH_3 \cdot HC:CH \cdot CN + 2 H_2O/H^+ \rightarrow CH_3 \cdot HC:CH \cdot CO \cdot OH + NH_4^+$$

D. Physical Properties:

Propenoic (**acrylic**) acid is a colorless liquid, soluble in water, and boils at 141.9. *Trans*-2-butenoic (**crotonic**, m.p. 72) acid is a colorless, crystalline solid, and is somewhat soluble in water. 9-Octadecenoic (**oleic**, m.p. 14) acid is a liquid at room temperatures, and is insoluble in water.

E. Chemical Properties:

THE UNSATURATED MONOACIDS REACT UNDER PROPER CONDITIONS:

1. *To give, in general, normal carboxyl derivatives* (cf. Ch. VII, Div. I, Sec. E).

The presence of the *ene* group will have, in many cases, a shading influence on the normal activity of the acid group.

An unsaturated carbon to carbon linkage usually increases the strength of an acid, the effect being most pronounced when the unsaturated linkage occupies the beta-gamma position. An unsaturated linkage in the alpha position tends to retard esterification, but unsaturation increases the rate of oxidation due to cleavage at the unsaturated linkage.

2. *To give additive products characteristic of the ene compounds* (cf. Ch. III, Div. II, Sec. E).

Some of the more important reactions are given below:

   a. $R \cdot HC:CH \cdot CO \cdot OH + H_2/$ (Pt, Pd or heated Ni) $\rightarrow R \cdot CH_2 \cdot CH_2 \cdot COOH$, and $CH_3(CH_2)_7 \cdot HC:CH(CH_2)_7 \cdot CO \cdot OH$ (oleic acid) $+ H_2/$ (Ni, 325°C.) $\rightarrow CH_3(CH_2)_7 \cdot CH_2 \cdot CH_2(CH_2)_7 \cdot CO \cdot OH$ (stearic acid). The more highly unsaturated monoacids are *partially* hydrogenated in this manner on a commercial basis.

   b. $R \cdot HC:CH \cdot CO \cdot OH + HX,$ aq. $\rightarrow R \cdot CHX \cdot CH_2 \cdot CO \cdot OH$
     In the 2, 3-unsaturated acids, the halogen atom takes up the position remote from the carboxyl group if the addition is carried out in aqueous solution.

   c. $R \cdot HC:CH \cdot CO \cdot OH + O_3 \rightarrow R \cdot HC \overset{\cdot O_3 \cdot}{:} CH \cdot CO \cdot OH,$ then

$$R \cdot HC \overset{\cdot O_3 \cdot}{:} CH \cdot CO \cdot OH + H_2O \rightarrow R \cdot CHO + OHC \cdot CO \cdot OH + H_2O_2$$

3. *To give isomeric forms in dilute alkali due to a tendency for the ene linkage to shift toward the carboxyl group:*

$$CH_3 \cdot CH_2 \cdot HC:CH \cdot CH_2 \cdot CO \cdot OH + OH^- \rightarrow CH_3(CH_2)_2 \cdot HC:CH \cdot CO \cdot O^- + H_2O$$

IV. Unsaturated Diacids

A. Introduction:

The unsaturated diacids may be regarded as derivatives of unsaturated hydrocarbons in which two hydrogen atoms have been replaced by carboxyl groups. Only *cis*-butenedioic (**maleic**) acid and *trans*-butenedioic (**fumaric**) acid will be considered.

The type formula for the alkenedioic acids is $C_nH_{2n-2}(COOH)_2$.

trans-Butenedioic (fumaric) acid occurs in many plants, in Iceland moss, and in some fungi.

Maleic and **fumaric acids** are used in the production of alkyd resins. **Maleic acid** is used in *diene* synthesis, and has commercial possibilities as an antioxidant.

The chemical properties of these acids will indicate that **maleic acid is** *cis*-butenedioic acid and that **fumaric acid is** *trans*-butenedioic acid.   The type of isomerism exhibited by these acids is known as *geometrical, cis-trans, or olefin isomerism*.

B. Nomenclature:

The diacids may be named (1) by adding the suffix *dioic acid* to the name of the corresponding hydrocarbon, (2) by adding *dicarboxylic acid* to the name of the hydrocarbon nucleus to which the carboxyl groups are attached, or (3) by common names.   In the first two systems, numerals are prefixed, when necessary, and in the latter system Greek letters are used to designate the position of the unsaturated group.   Where geometric isomers exist, the names are preceded by *cis-* or *trans-* to identify the compound specifically.

C. Preparations:

THE BUTENEDIOIC ACIDS MAY BE PREPARED:

1. *By (a) the removal of water from 2-hydroxybutanedioic (malic) acid, (b) the oxidation of benzene vapors and subsequent hydrolysis, or (c) the isomerization of trans-butenedioic acid and subsequent hydrolysis, to give in each case cis-butenedioic (maleic) acid:*

   a. (HO)HC·CO·OH + heat → H·C·CO·OH (**maleic acid.**   Some **fumaric**

       | ‖

     (H)HC·CO·OH   H·C·CO·OH **acid is also formed**) + H_2O

   b. $2\ C_6H_6 +\ \underline{9\ O_2\ }$ → 2 H·C·CO

            $(\overline{V_2O_5,\ 450°C})$ ‖ >O, then +

                              H·C·CO

            2 H·C·CO·OH + 4 CO_2 + 2 H_2O

     2 H_2O → ‖

                H·C·CO·OH

   c. HO·OC·C·H + heat (or PCl₅, → H·C·CO   H·C·CO·OH

       ‖   ‖ >O, then H_2O →   ‖

     H·C·CO·OH PCl₃ or P₂O₅) H·C·CO   H·C·CO·OH

2. *By (a) heating cis-butenedioic acid at 200°C, in a sealed tube to give trans-butenedioic acid, or (b) by varying the conditions for the oxidation of benzene (cf. maleic acid):*

   a. H·C·CO·OH + heat at 200°C., in a →   H·C·CO·OH (**fumaric acid**)

     ‖   ‖

     H·C·CO·OH   sealed tube   HO·OC·C·H

D. Physical Properties:

cis-Butenedioic (**maleic,** m.p. 130.5) acid is a colorless crystalline solid, and is quite soluble in water. *trans*-Butenedioic (**fumaric,** m.p. 287) acid is also a colorless crystalline solid, but is only slightly soluble in water. Of the two, **maleic acid** is more highly dissociated in water solution.

E. Chemical Properties:

THE BUTENEDIOIC ACIDS REACT UNDER PROPER CONDITIONS:

1. *To give typical carboxyl and ene derivatives in so far as the shading influence of the one group does not interfere with the normal chemical reactivity of the other group.*

2. *To give (a) identical products, or (b) different products, depending on the reagent employed:*

   a. 1. Both acids may be reduced by the use of appropriate reducing agents, or by molecular hydrogen in the presence of a catalyst, to give succinic acid $(CH_2 \cdot CO \cdot OH)_2$.

   2. Both acids react with aqueous sodium hydroxide at 100°C. to give an inactive mixture of sodium 2-hydroxybutanedioates (**sodium malates**):

   $$H \cdot C \cdot CO \cdot OH \text{ or } \quad H \cdot C \cdot CO \cdot OH + H_2O/2 \text{ NaOH,}$$
   $$\overset{\|}{H \cdot C \cdot CO \cdot OH} \quad \overset{\|}{HO \cdot OC \cdot C \cdot H} \quad aq., \quad 100°C. \rightarrow CH_2 \cdot CO \cdot ONa$$
   $$| \\ HO \cdot CH \cdot CO \cdot ONa \\ + 2 \ H_2O$$

   The **sodium malate** obtained in this reaction has a molecular structure which contains an asymmetric grouping. Asymmetric groupings suggest optical activity, but since the two possible isomers are formed in equivalent quantities (external compensation) the resulting product is optically inactive.

   b. 1. Heat converts both acids into the **maleic anhydride,** but fumaric anhydride is unknown.

   2. When oxidized with dilute permanganate solution, **maleic acid** yields *meso*-**tartaric acid,** whereas fumaric acid yields *dl*-**tartaric acid.**

   c. The octyl ester of maleic acid is used in the synthesis of surface active agents such as the Aerosols (cf. p. 281, B. 3), and maleic anhydride is used extensively in the production of synthetic resins of various types by use of the Diels-Alder reaction (cf. p. 341, item 18).

CHAPTER VIII

ALIPHATIC OXYGEN DERIVATIVES: THE SUBSTITUTED ACIDS

I. Halogen-Substituted Acids

A. Introduction:

The halogen-substituted acids may be considered as derivatives of the hydrocarbons, or as derivatives of the aliphatic acids in which one or more hydrogen atoms in the hydrocarbon radical have been replaced by halogen atoms. The general type formula for the halo-monoacids is $C_nH_{2n+1-y}X_y COOH$, where y is one or more.

Halogen-substituted acids are used principally as intermediates in the preparation of other acid derivatives.

Position isomerism is illustrated by 2-chloropropanoic acid ($CH_3 \cdot CHCl \cdot CO \cdot OH$, α-chloropropionic acid) and 3-chloropropanoic acid ($CH_2Cl \cdot CH_2 \cdot CO \cdot OH$, β-chloropropionic acid), and *chain isomerism* by 2-chloropentanoic acid ($CH_3 \cdot CH_2 \cdot CH_2 \cdot CHCl \cdot CO \cdot OH$, α-chlorovaleric acid) and 2-chloro-3-methylbutanoic acid ($[CH_3]_2CH \cdot CHCl \cdot CO \cdot OH$,

α-chloro-isovaleric acid). *Optical isomerism*, illustrated at the right,

$$\overset{H}{\underset{X}{CH_3 \cdot C \cdot CO \cdot OH}} \quad \text{and} \quad \overset{H}{\underset{X}{HO \cdot OC \cdot C \cdot CH_3}}$$

exists if the molecule is *asymmetric* in which case one structure is the mirror image of the other but not superposable (cf. Ch. XXVIII).

B. Nomenclature:

Structural Formulas	I.U.C. Names	Common Names
$Cl \cdot CH_2 \cdot CO \cdot OH$	chloroethanoic acid	chloroacetic acid
$Cl_2CH \cdot CO \cdot OH$	dichloroethanoic acid	dichloroacetic acid
$Cl_3C \cdot CO \cdot OH$	trichloroethanoic acid	trichloroacetic acid
$CH_3 \cdot CHBr \cdot CO \cdot OH$	2-bromopropanoic acid	α-bromopropionic acid
$I \cdot CH_2 \cdot CH_2 \cdot CO \cdot OH$	3-iodopropanoic acid	β-iodopropionic acid
$CH_3 \cdot CH_2 \cdot CBr_2 \cdot CO \cdot OH$	2,2-dibromobutanoic acid	α, α-dibromobutyric acid
$CH_3 \cdot CHBr \cdot CHBr \cdot CO \cdot OH$	2, 3-dibromobutanoic acid	α, β-dibromobutyric acid
$CH_3 \cdot CH_2 \cdot CH_2 \cdot CHCl \cdot CO \cdot OH$	2-chloropentanoic acid	α-chlorovaleric acid

C. Preparations:

THE HALOGEN-SUBSTITUTED ACIDS MAY BE PREPARED:

1. By the direct halogenation of acids or acid halides to give 2-halo acids:

    a. $R \cdot CH_2 \cdot CO \cdot OH + Cl_2 /$ (I_2, S, or P/sunlight) →
        $R \cdot CHCl \cdot CO \cdot OH + HCl$

b. $R \cdot CH_2 \cdot CO \cdot Br + Br_2 \rightarrow R \cdot CHBr \cdot CO \cdot Br + HBr$, and

$R \cdot CHBr \cdot CO \cdot Br + H_2O$ (excess) $\rightarrow R \cdot CHBr \cdot CO \cdot OH + HBr$

A good modification of this method is the bromination of malonic acid, and subsequent loss of carbon dioxide:

$R \cdot CH(COOH)_2 + Br_2 \rightarrow R \cdot CBr(COOH)_2 + HBr$, then

$R \cdot CBr(COOH)_2$, by loss of CO_2 on heating $\rightarrow R \cdot CHBr \cdot CO \cdot OH$

The alpha iodo acids may be prepared as follows:

$R \cdot CHBr \cdot CO \cdot OH + KI \rightarrow R \cdot CHI \cdot CO \cdot OH + KBr$

2. *By the addition of halogen acids or halogen to alkenoic acids:*

   a. $R \cdot HC{:}CH \cdot CO \cdot OH + HX$, aq. $\rightarrow R \cdot CHX \cdot CH_2 \cdot CO \cdot OH$ (3-haloalkanoic acid)

   If the HX addition to the 2-alkenoic acids is carried out in water solution, the halogen takes up the position remote from the carboxyl group.

   b. $R \cdot HC{:}CH \cdot CO \cdot OH + Cl_2$ (or Br_2 or $I_2/Et \cdot OH$) $\rightarrow$
   $R \cdot CHCl \cdot CHCl \cdot CO \cdot OH$

3. *By the replacement of the hydroxyl group of hydroxy acids by treatment with (a) PCl_5 or PCl_3, or (b) HCl or HBr:*

   a. $R \cdot CHOH \cdot CO \cdot OH + 2 PCl_5 \rightarrow R \cdot CHCl \cdot CO \cdot Cl + 2 POCl_3 + 2 HCl$, and

   $R \cdot CHCl \cdot CO \cdot Cl + H_2O$ (excess) $\rightarrow R \cdot CHCl \cdot CO \cdot OH + HCl$

   b. $R \cdot CHOH \cdot CO \cdot OH + HBr/H_2SO_4 \rightarrow R \cdot CHBr \cdot CO \cdot CH + H_2SO_4 \cdot H_2O$

D. Physical Properties:

Chloroethanoic (**chloroacetic**, m.p. 61.2, b.p. 189.5) acid is a colorless crystalline solid, and is very soluble in water, but dichloroethanoic (**dichloroacetic**, b.p. 193.5) acid is a water-soluble colorless liquid, whereas trichloroethanoic (**trichloroacetic**, m.p. 57.5, b.p. 195.3) acid is a water-soluble colorless crystalline solid. The ionization of the acids tends to increase with the introduction of halogen groups.

E. Chemical Properties:

THE HALOGEN SUBSTITUTED ACIDS REACT UNDER PROPER CONDITIONS:

1. *To give normal derivatives of the halogen and carboxyl groups in so far as the shading influence of the one group does not interfere with the normal activity of the other group* (cf. Ch. IV, Div. I, Sec. E, and Ch. VII, Div. I, Sec. E).

By masking the properties of the carboxyl group by derivative formation, a number of the reactions of the halogen group may be carried out more satisfactorily. To illustrate:

$R \cdot CHX \cdot CO \cdot ONa + NaCN \rightarrow R \cdot CH(CN) \cdot CO \cdot ONa + NaX$,

$R \cdot CHX \cdot CO \cdot ONa + Na_2SO_3 \rightarrow R \cdot CH(SO_3Na) \cdot CO \cdot ONa + NaX$

$R \cdot CHX \cdot CO \cdot ONa + 2 NH_3 \rightarrow R \cdot CH(NH_2) \cdot CO \cdot ONa + NH_4X$, and

$R \cdot CHX \cdot CO \cdot ONa + AgNO_2 \rightarrow R \cdot CH(NO_2) \cdot CO \cdot ONa + AgX$

$R \cdot CHX \cdot CO \cdot OEt + Na \cdot OR$, alc. $\rightarrow R \cdot CH(OR) \cdot CC \cdot OEt + NaX$

2. *To give characteristic derivatives when the 2-, 3-, or 4-halo acids are subjected to heat in the presence of dilute alkali or water:*

    a. $R \cdot CHX \cdot CO \cdot OH + 2\ NaOH$, aq. $\rightarrow R \cdot CHOH \cdot CO \cdot ONa + NaX + H_2O$

    b. 1. $R \cdot CHX \cdot CH_2 \cdot CO \cdot OH + (H_2O$ or $KOH/Et \cdot OH)$, heat $\rightarrow$
       $R \cdot HC{:}CH \cdot CO \cdot OH + HX$

       2. $R \cdot CHX \cdot CH_2 \cdot CO \cdot OH + Na_2CO_3$, aq., heat $\rightarrow R \cdot HC{:}CH_2 + NaX + CO_2 + NaHCO_3$

    c. $R \cdot CHX \cdot CH_2 \cdot CH_2 \cdot CO \cdot OH + (H_2O)$, heat $\rightarrow R \cdot \overline{CH \cdot CH_2 CH_2 CO \cdot O} + HX$

II. Hydroxy-Substituted Monoacids

A. Introduction:

The hydroxy-substituted acids may be considered as derivatives of the hydrocarbons, or as derivatives of the aliphatic acids in which one or more hydrogen atoms in the hydrocarbon radical have been replaced by a hydroxyl group (OH).

The general type formula for the hydroxy-monoacids is $C_nH_{2n+1-y}(OH)_y \cdot CO \cdot OH$, where y is one or more.

Hydroxyethanoic (**hydroxyacetic**, glycolic) acid occurs in unripe grapes. *d*-2-Hydroxypropanoic (*d*-**lactic**, sarcolactic) acid is found in muscle tissue and, at times, in the blood and urine. *dl*-2-Hydroxypropanoic (*dl*-**lactic**) acid occurs in sauerkraut, sour milk, and in the human digestive tract.

Chain isomerism is illustrated by 2-hydroxybutanoic ($CH_3 \cdot CH_2 \cdot CHOH \cdot CO \cdot OH$, **α-hydroxybutyric**) acid and 2-hydroxy-2-methylpropanoic ($[CH_3]_2 \cdot COH \cdot CO \cdot OH$, **α-hydroxyisobutyric**) acid. 2-Hydroxypropanoic ($CH_3 \cdot CHOH \cdot CO \cdot OH$, **lactic**) acid and 3-hydroxypropanoic ($CH_2OH \cdot CH_2 \cdot CO \cdot OH$, **hydracrylic**) acid are examples of *position isomerism.* *Optical isomerism*, which is illustrated by the lactic acids shown here, exists if the molecule is

$$\overset{H}{CH_3 \cdot \underset{OH}{C} \cdot CO \cdot OH} \text{ and } \overset{H}{HO \cdot OC \cdot \underset{HO}{C} \cdot CH_3}$$

asymmetric in which case one structure is the mirror image of the other but the structures are not superposable.

2-Hydroxypropanoic (**lactic**) acid is used to remove lime from dehaired hides, in chrome mordanting, in the acid dyeing of wools, in infant foods, in soft drinks, in candies, in baking powders, in jams and jellies, in preserving olives, in poultry and stock foods, and in the production of **ethyl lactate** and **antimony lactate**. Ethyl lactate is used as a solvent for pyroxylin lacquers, and antimony lactate is employed as a mordant in dyeing.

B. Nomenclature:

The hydroxy acids are named as substituted acids.

C. Preparations:

THE HYDROXY-SUBSTITUTED MONOACIDS MAY BE PREPARED:

1. *By the hydrolysis of halogen-substituted acids:*

$R \cdot CHX \cdot CO \cdot OH + 2\ NaOH$, aq., heat $\rightarrow R \cdot CHOH \cdot CO \cdot ONa + NaX + H_2O$, then

$R \cdot CHOH \cdot CO \cdot ONa + H_2SO_4$, aq. $\rightarrow R \cdot CHOH \cdot CO \cdot OH + NaHSO_4$

2. *By the addition of hydrogen cyanide to aldehydes or ketones, followed by hydrolysis, to give 2-hydroxy acids:*

   a. $R \cdot CHO + HCN \rightarrow R \cdot CHOH \cdot CN$, then $2 \ H_2O/H^+ \rightarrow$
      $R \cdot CHOH \cdot CO \cdot OH + NH_4^+$

   b. $R_2C{:}O + HCN \rightarrow R_2C(OH)CN$, then $2 \ H_2O/H^+ \rightarrow$
      $R_2C(OH) \cdot CO \cdot OH + NH_4^+$

3. *By the action of nitrous acid on the corresponding amino acid:*

   $R \cdot CH(NH_2) \cdot CO \cdot OH + HONO/ (NaNO_2/H^+) \rightarrow$
   $R \cdot CH(OH) \cdot CO \cdot OH$ (low yields) $+ N_2 + H_2O$

4. *By the controlled (a) reduction of aldehydic or ketonic acids, or*

   (b) *oxidation of glycols, or acids containing a tertiary α-carbon atom:*

   a. 1. $H \cdot OC(CH_2)_xCO \cdot OH + 2 \ Na/2 \ EtOH$ (or $NaHg/H_2O$) $\rightarrow$
         $CH_2OH(CH_2)_xCO \cdot OH + 2 \ EtONa$

      2. $R \cdot CO \cdot CO \cdot OH + 2 \ Na/2 \ EtOH$ (or $NaHg/H_2O$) $\rightarrow$
         $R \cdot CHOH \cdot CO \cdot OH + 2 \ EtONa$

   b. 1. $CH_2OH (CH_2)_xCH_2OH + O_2/(air/Pt) \rightarrow CH_2OH (CH_2)_x CO \cdot OH + H_2O$

      2. $3 \ (CH_3)_2CH \cdot CO \cdot OH + 2 \ MnO_4^-/H_2O \rightarrow 3(CH_3)_2C(OH) \cdot CO \cdot OH +$
         $2 \ MnO_2 + 2 \ OH^-$

D. Physical Properties:

Hydroxyethanoic (**hydroxyacetic**, m.p. 79) acid is a solid, soluble in water, and intermediate between methanoic (**formic**) and ethanoic (**acetic**) acids in its ionization. *dl*-2-Hydroxypropanoic (*dl*-lactic, m.p. 18, b.p. 122/15mm.) acid is a water-soluble, colorless, odorless, viscous liquid.

E. Chemical Properties:

THE HYDROXY-SUBSTITUTED MONOACIDS REACT UNDER PROPER CONDITIONS:

1. *To give normal derivatives of the hydroxyl and carboxyl groups in so far as the shading influence of the one group does not interfere with the normal activity of the other group (cf. Ch. V, Div. I, Sec. E, and Ch. VII, Div. 1, Sec. E).*

2. *To give characteristic derivatives when the 2-, 3-, 4-, or 5- hydroxy-substituted acids are subjected to heat:*

   a. $R \cdot CH \cdot CO \boxed{OH \ H}$ $R \cdot CH \cdot C{:}O$
      $\quad \ \ O \qquad \quad \ \ O \ + heat \rightarrow \ \ O \ \ O$ (2-hydroxyalkanoic lactide)
      $\boxed{H \ HO} \ OC \cdot HC \cdot R \qquad O{:}C \cdot HC \cdot R$ $+ \ 2 \ H_2O$

   b. $R \cdot CHOH \cdot CH_2 \cdot CO \cdot OH + heat \rightarrow R \cdot HC{:}CH \cdot CO \cdot OH + H_2O$

   c. 1. $R \cdot CHOH \cdot CH_2 \cdot CH_2 \cdot CO \cdot OH + heat \rightarrow R \cdot CH \cdot CH_2 \cdot CH_2 \cdot CO \cdot O + H_2O$

2. $R \cdot CHOH \cdot CH_2 \cdot CH_2 \cdot CH_2 \cdot CO \cdot OH + heat \rightarrow R \cdot \underset{|}{CH} \cdot CH_2 \cdot CH_2 \cdot CH_2 \cdot CO \cdot O$

$+ H_2O$, or the 4- and 5-hydroxy acids give five- and six-membered ring structures which are known as **lactones.**

$CH_3 \cdot CH \cdot CH_2 \cdot CH_2 \cdot CO \cdot O$, for example, is 4-hydroxypentanoic lactone

(γ-valerolactone).

In alkaline solution, the lactones give the salts of the acids.

III. Hydroxy-Substituted Di- and Triacids

A. Introduction:

These compounds may be considered as hydroxyl and carboxyl derivatives of the corresponding hydrocarbons, or as hydroxyl derivatives of the corresponding acids. The most important members of this group are the *tartaric acids* and *citric acid.*

2-Hydroxybutanedioic **(malic)** acid is widely distributed in berries, fruits (first isolated from unripe apples), carrots, tobacco, and other plant products. *d*-2,3-Dihydroxybutanedioic (*d*-**tartaric**) acid occurs along with the racemic (*dl*-**tartaric**) acid, both free and in the form of their potassium and calcium salts. Potassium acid tartrate is a constituent of grape juice. 2-Hydroxy-1, 2, 3-propanetricarboxylic **(citric)** acid is found in sugar beets, in legume seeds, and in most fruit juices except from plums.

2, 3-Dihydroxybutanedioic **(tartaric)** acid is used principally in the form of its salts. The acid potassium salt **(cream of tartar)** is an ingredient of tartrate baking powders, and potassium sodium tartrate (Rochelle salt) is used as a laxative and as an ingredient of Fehling solution. Antimony potassium tartrate **(tartar emetic)** is used as an emetic and in mordant dyeing. 2-Hydroxy-1, 2, 3-propanetricarboxylic **(citric)** acid is employed in the preparation of synthetic lemonade, in calico printing, and in such salts as ferric ammonium citrate, which is used in making blue-prints, and magnesium citrate, which is employed as a laxative.

Optical isomerism is illustrated by the **tartaric** acids of which four forms are known: *d*-tartaric acid, *meso*-tartaric acid, *l*-tartaric acid, and *dl*-tartaric acid. The *d*- and *l*-forms are mirror images of each other, and an equal mixture of these forms gives an inactive product, which is externally compensated, known as *dl*- or *racemic* tartaric acid. The *meso*-form, which is internally compensated, is also optically inactive. Pasteur made a classical contribution to the phenomena of optical activity when he isolated the sodium ammonium salts of *d*- and *l*-tartaric acids in crystalline form and showed that the one type of crystal was the mirror image of the other. Optical activity exists whenever the molecule is asymmetric (cf. Ch. XXVII).

Configuration is preferably designated by *D*- and *L*-, and optical activity by *d*- (dextro) and *l*- (levo).

B. Nomenclature:

CO·OH	CO·OH	HO·OC	CH₂·CO·OH
H·C·OH	H·C·OH	HO·C·H	|
HO·C·H	H·C·OH	H·C·OH	HO·C·CO·OH
CO·OH	CO·OH	HO·OC	|
			CH₂·CO·OH

L^+-2,3-dihydroxy-	*i*-2,3-dihydroxy-	D^--2,3-dihydroxy-	2-hydroxy-1,2,3-pro-
butanedioic|acid,	butanedioic acid,	butanedioic acid,	panetricarboxylic acid,
L^+-**tartaric acid**	*meso*-**tartaric acid**	D^--**tartaric acid**	**citric acid**

C. Preparations:

THE HYDROXY-SUBSTITUTED DI- AND TRIACIDS MAY BE PREPARED:

1. By use of the normal methods for the introduction of hydroxyl and carboxyl groups (cf. Ch. V, Div. I, Sec. C, and Ch. VII, Div. I, Sec. C).

a. 1.

$$\begin{array}{ccc} & \text{CN} & \text{CO·OH} \\ \text{CHO} + 2\,\text{HCN} \rightarrow & \text{CHOH, then } 4\,\text{H}_2\text{O}/2\,\text{H}^+ \rightarrow & \text{CHOH (tartaric acid)} + 2\,\text{NH}_4{}^+ \\ \text{CHO} & \text{CHOH} & \text{CHOH} \\ & \text{CN} & \text{CO·OH} \end{array}$$

2. $\text{HO·OC·CHBr·CHBr·CO·OH} + 2\,\text{AgOH, aq.} \rightarrow \text{HO·OC·CHOH·CHOH·CO·OH} + 2\,\text{AgBr}$

b.

$$\begin{array}{cccc} \text{CH}_3 & \text{CH}_2\text{Cl} & \text{CH}_2\text{CN} & \text{CH}_2\text{·CO·OH} \\ \text{C:O} + \text{Cl}_2 \rightarrow & \text{C:O} + \text{KCN} \rightarrow & \text{C:O} + 2\,\text{H}_2\text{O}/(\text{H}^+) \rightarrow & \text{C:O} + \text{Et OH}/(\text{H}^+) \rightarrow \\ \text{CH}_2\text{·CO·OEt} & \text{CH}_2\text{·CO·OEt} & \text{CH}_2\text{·CO·OEt} & \text{CH}_2\text{·CO·OEt} \end{array}$$

| eth l 3-oxobu-tanoate | ethyl 4-chloro-3-oxobutanoate | ethyl 4-c ano-3-oxobutanoate | 3-oxopentanedioic acid mono-ethyl ester |
| acetoacetic ester | chloroaceto-acetic ester | cyanoaceto-acetic ester | mono-ethyl ester of acetone dicarboxylic acid |

$$\begin{array}{cc} \text{CH}_2\text{·CO·OEt} & \text{CH}_2\text{·CO·OEt} & \text{CH}_2\text{·CO·OH} \\ \text{C:O} + \text{H·CN} \rightarrow & \text{C(OH)CN} + 4\,\text{H}_2\text{O}/2\,\text{H}^+ \rightarrow \text{HO·C·CO·OH} \\ \text{CH}_2\text{·CO·OEt} & \text{CH}_2\text{·CO·OEt} & \text{CH}_2\text{·CO·OH} \end{array}$$

| diethyl ester of acetonedicar-boxylic acid | addition product | | citric acid |

2. By the oxidation of maleic or fumaric acid with permanganate solution (cf. Ch. VII, Div. IV, Sec. E).

3. By isolation from natural products (commercial).

D. Physical Properties:

2-Hydroxybutanedioic (malic, m.p. 100) acid is a water-soluble, colorless solid. The natural product is levorotatory but undergoes *muta-rotation* (cf. Ch. X, Sec. A) upon dilution with water. d-2,3-Dihydroxybutanedioic (d-tartaric, m.p. 170) acid occurs as colorless, water-soluble monoclinic crystals. Upon heating with water at 165°C. for about forty eight hours, with or without the presence of hydrochloric acid, d-tartaric acid is partly converted into a mixture of racemic and meso acids. 2-Hydroxy-1,2,3-propane-tricarboxylic (citric, m.p. 100) acid crystallizes with one molecule of water to give a white solid. The anhydrous acid melts at 153°C.

E. Chemical Properties:

THE HYDROXY-SUBSTITUTED DI- AND TRIACIDS REACT UNDER PROPER CONDITIONS:

1. To give normal hydroxyl and carboxyl derivatives in so far as the shading influence of the one group does not interfere with the normal activity of the other group (cf. Ch. V, Div. I, Sec. E, and Ch. VII, Div. I, Sec. E).

2. To give characteristic reactions when treated with (a) cupric hydroxide, or (b) with sodium acid carbonate and heat:

a. $\begin{array}{l}\text{CO·OH}\\\text{ĊHOH}\\\text{ĊHOH}\\\text{ĊO·OH}\end{array}$ + CuSO₄/4 NaOH, aq. → $\begin{array}{l}\text{CO·ONa}\\\text{H·Ċ·O·}\\\text{H·Ċ·O·}\\\text{CO·ONa}\end{array}$Cu + Na₂SO₄ + 4 H₂O

$\qquad\qquad\qquad\qquad\qquad\qquad\qquad\qquad$ (probable structure of the copper-tartrate complex)

This product is the active constituent in Fehling solution. It is soluble in water and permits, consequently, a rather large concentration of the bivalent copper in alkaline solution. When treated with a reducing agent, the cupric form is reduced to the cuprous form and precipitated as red cuprous oxide (Cu_2O).

b. $\begin{array}{l}\text{CO·OH}\\\text{ĊHOH}\\\text{ĊHOH}\\\text{ĊO·OK}\end{array}$ + NaHCO₃, heat → $\begin{array}{l}\text{CO·ONa}\\\text{ĊHOH}\\\text{ĊHOH}\\\text{ĊO·OK}\end{array}$ + CO₂ + H₂O

$\qquad\qquad\qquad\qquad\qquad\qquad\qquad\qquad$ (Rochelle salt)

This equation represents the reaction of the acid potassium tartrate as a constituent of baking powder. The potassium salt is used in addition to the free acid in order to slow down the reaction and give a more gradual production of carbon dioxide.

IV. Oxo-Substituted Acids

A. Introduction:

The oxo-substituted acids may be considered as carbonyl and carboxyl derivatives of the corresponding hydrocarbons, or as carbonyl (:C:O) derivatives of the corresponding acids.

The type formula for the aldehydic acids may be represented by $H\cdot OC(CH_2)_x CO\cdot OH$, and that for the ketonic acids by $R\cdot CO(CH_2)_x CO\cdot OH$.

Oxoethanoic (glyoxylic) acid is found in gooseberries, in grapes, and in the fruits and leaves of some plants. 2-Oxopropanoic (pyruvic) acid may occur as an intermediate in the metabolism of the sugars. 3-Oxobutanoic (acetoacetic) acid occurs, along with **acetone** and β-hydroxybutyric acid, in the urine of diabetics.

Oxoethanoic (glyoxylic) acid is used as a reagent in the *glyoxylic acid test* for proteins. The ketonic acids, in general, have very little intrinsic value, but acetoacetic ester is an important starting material for the preparation of ketones, and straight and branched chain aliphatic acids.

Keto-enol isomerism, a form of *tautomerism,* is illustrated by the following: $CH_3\cdot C(:O)\cdot CH_2\cdot CO\cdot OC_2H_5$ (keto, 93%) $\rightleftarrows$ $CH_3\cdot C(OH):CH\cdot CO\cdot OC_2H_5$ (enol, 7%). The enol form appears to be the active member of the system. As it is used up in a reaction, the equilibrium shifts to the right. The rate and extent of this shift may be measured by the addition of bromine. By lowering the temperature of the ester to -78°C., it is possible to separate the keto form by crystallization. The pure keto forms melt at -39°C. When the tautomers may be separated, the phenomenon is referred to as desmotropism.

B. Nomenclature:

Structural Formulas	I.U.C. Names	Common Names
H·OC·CO·OH	oxoethanoic acid	glyoxylic (glyoxalic) acid
H·OC·CH₂·CO·OH	3-oxopropanoic acid	formylacetic acid
CH₃·CO·CO·OH	2-oxopropanoic acid	pyruvic (α-ketopropionic) acid
CH₃·CO·CH₂·CO·OH	3-oxobutanoic acid	acetoacetic (diacetic) acid

C. Preparations:

THE OXOSUBSTITUTED ACIDS MAY BE PREPARED:

1. By the methods normally employed for the introduction of the desired groups (cf. Ch. VI, Div. I, Sec. C, and Ch. VII, Div. I, Sec. C), *as, for example:*

   a. 1. $CCl_3 \cdot CHO$ (chloral) $+ 2 H_2O \rightarrow HO \cdot OC \cdot CHO + 3 HCl$

      2. $CHBr_2 \cdot CO \cdot OH + H_2O$ (hyd.) $\rightarrow OHC \cdot CO \cdot OH + 2 HBr$

      3. $2 CH_2OH \cdot CO \cdot OH + O_2$ (oxi.) $\rightarrow 2 OHC \cdot CO \cdot OH + 2 H_2O$

   b. 1.a. $CH_3 \cdot CBr_2 \cdot CO \cdot OH + Ag_2O \rightarrow CH_3 \cdot CO \cdot CO \cdot OH + 2 AgBr$

      b. $2 CH_3 \cdot CHOH \cdot CO \cdot OH + O_2$ (oxi.) $\rightarrow 2 CH_3 \cdot CO \cdot CO \cdot OH + 2 H_2O$

      c. $HO \cdot OC \cdot CHOH \cdot CHOH \cdot CO \cdot OH + $ heat $\rightarrow CH_3 \cdot CO \cdot CO \cdot OH + CO_2 + H_2O$

      d. $CH_3 \cdot CO \cdot CN + H_2SO_4/2H_2O \rightarrow CH_3 \cdot CO \cdot CO \cdot OH + NH_4HSO_4$

    2. $2 CH_3 \cdot CHOH \cdot CH_2 \cdot CO \cdot OH + O_2$ (mild oxidation) $\rightarrow 2 CH_3 \cdot CO \cdot CH_2 \cdot CO \cdot OH$ (**acetoacetic acid**) $+ 2 H_2O$

2. By a special synthesis for (a) *acetoacetic ethyl ester* (*Claisen Reaction*) *or* (b) *alkylacetoacetic ethyl ester* (Acetoacetic Ester Synthesis):

   a. $CH_3 \cdot CO \cdot OEt + H \cdot CH_2 \cdot CO \cdot OEt/2$ Na$/(Et \cdot OH$, or EtONa alone$) \rightarrow$
     $CH_3 \cdot C(ONa){:}CH \cdot COO \cdot Et + EtONa + H_2$, then
     $CH_3 \cdot C(ONa){:}CH \cdot CO \cdot OEt + $ cold, dilute HCl $\rightarrow CH_3 \cdot CO \cdot CH_2 \cdot CO \cdot OEt + NaCl$

     The ester separates as an oil which is isolated, dried, and purified by fractional distillation under reduced pressure. By hydrolysis with barium hydroxide, the barium salt of the acid is obtained. The ester is much more stable than the free acid, and is normally employed in synthesis as shown below.

   b. $CH_3 \cdot CO \cdot CH_2 \cdot CO \cdot OEt + EtONa \rightarrow CH_3 \cdot C(ONa){:}CH \cdot CO \cdot OEt + EtOH$,
     then $CH_3 \cdot C(ONa){:}CH \cdot CO \cdot OEt + R \cdot I \rightarrow CH_3 \cdot CO \cdot CHR \cdot CO \cdot OEt + NaI$

     Another alkyl group may be introduced by subsequent treatment with sodium ethoxide, followed by treatment with an alkyl iodide. In these syntheses, the ester separates as an oil which is isolated, dried, and purified by fractionation under diminished pressure.

D. Physical Properties:

   Oxoethanoic (**glyoxylic**) acid is a colorless, crystalline, slightly water-soluble solid, which is readily decomposed by heat. 2-Oxopropanoic (**pyruvic**, m.p. 14, b.p. 165) acid is a water-soluble liquid with an irritating odor. Ethyl 3-oxobutanoate (**acetoacetic ester**, b.p. 180) is a colorless, mobile liquid. The free acid is unstable.

E. Chemical Properties:

THE OXO-SUBSTITUTED ACIDS REACT UNDER PROPER CONDITIONS:

1. To give normal carbonyl and carboxyl derivatives except in so far as the presence of the one group may modify the normal activity of the other group (cf. Ch. VI, Div. I, Sec. E, and Ch. VII, Div. I, Sec. E):

2. To give characteristic reactions, as follows:

   a. 1. $CHO \cdot CO \cdot OH + H_2O \rightarrow CH(OH)_2 \cdot CO \cdot OH$ (glyoxylic acid hydrate)

      This is one of the few cases where two hydroxyl groups on the same carbon atom appear to be comparatively stable.

    2. $2\ CHO \cdot CO \cdot OH + 3\ NaOH$, aq. $\rightarrow NaO \cdot OC \cdot CO \cdot ONa +$ $CH_2OH \cdot CO \cdot ONa + 2\ H_2O$

      This is another illustration of the Cannizzaro reaction, an example of autooxidation and reduction. Normal carbonyl reactions are obtained with $NaHSO_3$, NH_2OH, $Ag(NH_3)_2OH$, etc.

   b. 1. a. $CH_3 \cdot CO \boxed{CO \cdot O} H + $ (dilute H_2SO_4), 150°C. $\rightarrow CH_3 \cdot CHO + CO_2$

      b. $CH_3 \cdot CO \cdot CO \cdot OH + 2\ Ag(NH_3)_2OH$, aq. $\rightarrow CH_3 \cdot CO \cdot ONH_4$ $+ (NH_4)_2CO_3 + 2\ Ag° + NH_3$

    2. a. $CH_3 \cdot CO \cdot CH_2 \cdot CO \cdot OH + $ mild heating $\rightarrow CH_3 \cdot CO \cdot CH_3 + CO_2$

      b. $CH_3 \cdot CO \cdot CH_2 \cdot CO \cdot OC_2H_5 + EtONa \rightarrow CH_3 \cdot C(ONa){:}CH \cdot CO \cdot OC_2H_5$ $+ EtOH$, then $CH_3 \cdot C(ONa){:}CH \cdot CO \cdot OC_2H_5 + R \cdot I \rightarrow CH_3 \cdot CO \cdot$ $CH(R) \cdot CO \cdot OC_2H_5 + NaI$

      This ester may be hydrolyzed as indicated below to give a mixed ketone or a straight-chain acid. Moreover, the whole series of reactions may be repeated, then hydrolyzed to give branched ketones or branched acids. Since any alkyl group may be introduced, the synthesis is very versatile.

      c. The members of the acetoacetic ester series undergo (1) *ketonic cleavage* in dilute acid or alkali, and (2) *acid cleavage* in concentrated alkali:

      (1). $CH_3 \cdot CO \cdot CH_2 \boxed{\begin{array}{c} CO \\ H \quad O \end{array}} O \cdot C_2H_5$ (2). $CH_3 \cdot CO \boxed{\begin{array}{c} CH_2 \cdot CO \quad O \\ NaO \quad H \qquad Na \end{array}} C_2H_5$ OH

V. Other Substituted Acids

The amino-substituted acids will be considered along with the other nitrogen derivatives (Ch. XIII). The cyano acids and the sulfo acids find their principal use as synthetic intermediates. They may be prepared by the methods normally employed for the introduction of the cyano and sulfo groups.

CHAPTER IX

ALIPHATIC OXYGEN DERIVATIVES: THE ACID DERIVATIVES

Acid derivatives are compounds which have been obtained by substitution in the carboxyl group of an acid. The following partial list is illustrative: anhydrides ($R \cdot CO \cdot O \cdot OC \cdot R$), acid halides ($R \cdot CO \cdot X$), esters ($R \cdot CO \cdot O \cdot R$), salts ($R \cdot CO \cdot OM$), amides ($R \cdot CO \cdot NH_2$), and acid nitriles ($R \cdot CN$). The amides and acid nitriles will be considered in Chapter XIII.

I. Acid Anhydrides

A. Introduction:

Acid anhydrides, as the name implies, result from an inter-molecular reaction between organic acids involving the loss of water, or the anhydrides may be regarded as acid derivatives in which the hydrogen atom of the carboxyl group has been replaced by an acyl ($R \cdot CO \cdot$) group.

The type formula for the monoacid anhydrides is $R \cdot CO \cdot O \cdot OC \cdot R$, and that for the diacid anhydrides is $(CH_2)_x (CO)_2O$, where "x" is normally two or more.

Ethanoic (acetic) anhydride is used extensively in the production of cellulose acetate for use in the film, rayon, and allied industries. The anhydrides are used somewhat as synthetic intermediates. The reaction with acetic anhydride is frequently employed in the chemical laboratory to determine the number of hydroxyl groups per unit structure of a given compound.

The anhydrides exhibit chain and position isomerism.

B. Nomenclature:

The acid anhydrides are named from the corresponding acids:

Structural Formulas	I.U.C. Names	Common Names
$(H \cdot CO)_2O$ (unstable)	methanoic anhydride	formic anhydride
$(CH_3 \cdot CO)_2O$	ethanoic anhydride	acetic anhydride
$(CH_3 \cdot CH_2 \cdot CO)_2O$	propanoic anhydride	propionic anhydride
$(CH_3 \cdot CH_2 \cdot CH_2 \cdot CO)_2O$	butanoic anhydride	butyric anhydride
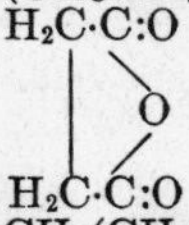 $H_2C \cdot C{:}O$ $\diagdown$ O $\diagup$ $H_2C \cdot C{:}O$	butanedioic anhydride	succinic anhydride
$CH_2(CH_2 \cdot CO)_2O$	pentanedioic anhydride	glutaric anhydride

C. Preparations:

THE MONOACID ANHYDRIDES MAY BE PREPARED:

1. By the treatment of the salts of the acids with (a) alkanoyl (acid) chlorides, (b) phosphorus oxychloride, (c) sulfuryl chloride, or (d) sulfur dichloride:

a. $R \cdot CO \cdot ONa + Cl \cdot OC \cdot R \rightarrow R \cdot CO \cdot O \cdot OC \cdot R + NaCl$, and

$CH_3 \cdot CO \cdot O \cdot Na + Cl \cdot OC \cdot CH_3 \rightarrow CH_3 \cdot CO \cdot O \cdot OC \cdot CH_3$ (**acetic anhydride**) + NaCl

b. $6 R \cdot CO \cdot O \cdot Na + POCl_3$, heat $\rightarrow 3 (R \cdot CO)_2 O + 3 NaCl + Na_3PO_4$

c. $4 R \cdot CO \cdot O \cdot Na + SO_2Cl_2 /(SO_2 + Cl_2) \rightarrow 2 (R \cdot CO)_2 O + 2 NaCl + Na_2SO_4$

d. $8 R \cdot CO \cdot O \cdot Na + 3 SCl_2 \rightarrow 4 (R \cdot CO)_2 O + 6 NaCl + Na_2SO_4 + 2 S$

The first reaction is especially useful in the preparation of mixed anhydrides, and the second reaction is applicable to the preparation of the higher anhydrides.

2. *By the treatment of acids with (a) acid chlorides, or (b) phosphorus pentoxide:*

a. $R \cdot CO \cdot OH + Cl \cdot OC \cdot R \rightleftarrows R \cdot CO \cdot O \cdot OC \cdot R + HCl$

b. $2 R \cdot CO \cdot OH + P_2O_5 \rightarrow R \cdot CO \cdot O \cdot OC \cdot R + 2 HPO_3$

Neither of these reactions gives good yields.

3. *By the commercial treatment of (a) acetic acid with ketene, (b) sodium acetate with chlorine in the presence of sulfur dichloride, or (c) acetic acid with acetylene in the presence of a catalyst, followed by heat, to give in each case acetic anhydride:*

a. $CH_3 \cdot CO \cdot OH + O{:}C{:}CH_2$ (**ketene**) $\rightarrow CH_3 \cdot CO \cdot O \cdot OC \cdot CH_3$

b. $8 CH_3 \cdot CO \cdot ONa + 2 Cl_2 /SCl_2 \rightarrow 4 (CH_3 \cdot CO)_2 O + 6 NaCl + Na_2SO_4$

c. $2 CH_3 \cdot CO \cdot OH + H \cdot C{:}C \cdot H /(catalyst) \rightarrow (CH_3 \cdot CO \cdot O)_2 CH \cdot CH_3$, then $(CH_3 \cdot CO \cdot O)_2 CH \cdot CH_3$ (**1,1-ethanediol diethanoate, ethylidene diacetate**) + heat $\rightarrow (CH_3 \cdot CO)_2 O + CH_3 \cdot CHO$

Besides the ordinary (or intermolecular) anhydride, **acetic acid forms an intramolecular or pseudo anhydride:**

$CH_3 \cdot CO \cdot OH +$ (Pt catalyst), heat $\rightleftarrows H_2C{:}C{:}O$ (**ketene**, about 10%) $+ H_2O$

The first two diacids, **oxalic** and **malonic,** do not form normal anhydrides, but C_3O_2 may be regarded as the pseudo anhydride of malonic acid. The anhydrides of the other diacids are normally formed by heating (cf. Ch. VII, Div. II, Sec. E).

D. Physical Properties:

Ethanoic (acetic, b.p. 139) anhydride, the most important member of the series, is a liquid with a sharp, pungent odor, and is soluble in about ten volumes of water. There is a decrease in solubility in water and an increase in the boiling point with an increase in the carbon content. The anhydrides boil somewhat higher than the acids from which they are derived and are generally soluble in alcohol, ether, and other organic solvents.

E. Chemical Properties:

THE ANHYDRIDES OF THE MONOACIDS REACT UNDER PROPER CONDITIONS:

1. *To give double decomposition when treated by:* $R \cdot CO \cdot O \mid OC \cdot R$

a.	1.	Acid hydrolysis /(H₂O /H⁺)		H	OH
	2.	Alkaline hydrolysis /(NaOH /H₂O)		Na	OH
b.		Alcoholysis		H	OR
c.		Ammonolysis		H	NH₂
d.	1.	Aminolysis, with primary amines		H	NHR
	2.	Aminolysis, with secondary amines		H	NR₂

2. *To give triple decomposition products when treated with phosphorus pentachloride:*

$$R \cdot CO \boxed{\begin{array}{c} \cdot \; O \\ \hline PCl_3 \end{array}} \cdot OC \cdot R$$
$$Cl \qquad \qquad Cl$$

3. *To give alpha halogen-substitution products when treated with chlorine or bromine in the presence of a catalyst:*

$$R \cdot CO \cdot O \cdot OC \cdot \overset{\cdot}{\underset{}{HC}} \boxed{H + X} X \quad \overset{R}{}$$

The halogen acid formed in this reaction may react to split the halogen-anhydride to give an acid halide and a halogen-substituted acid:

$$R \cdot CO \cdot O \cdot OC \cdot CHX \cdot R + HX, \text{ gas} \rightarrow R \cdot CO \cdot X + HO \cdot OC \cdot CHX \cdot R$$

4. *To give cellulose acetate when cotton, cotton linters, or other waste forms of pure cellulose are treated with a mixture of acetic acid, acetic anhydride, and sulfuric acid.*

II. Acid Halides

A. Introduction:

The acid halides may be considered as derivatives of the acids in which the OH of the carboxyl group has been replaced by a halogen atom. The type formula for the monoacid halides is $R \cdot CO \cdot X$.

Acid chlorides are used as acylating agents to introduce the acyl ($R \cdot CO \cdot$) group into alcohols, amines, and similar compounds. Ethanoyl (acetyl) chloride is used in quantitative analytical work to determine the number of hydroxyl groups in a given compound. It is also a useful reagent for differentiating between the various classes of amines.

The acid halides may be considered as functional isomers of the halogen-substituted aldehydes or ketones as, for example,

$$CH_3 \cdot CH_2 \cdot CO \cdot Cl, \quad CH_3 \cdot CHCl \cdot CHO, \quad \text{and} \quad CH_3 \cdot CO \cdot CH_2Cl.$$

B. Nomenclature:

The acid halides are named from the corresponding acids by changing the suffix *ic* to *yl*, and adding the name of the halide concerned:

Structural Formulas	I.U.C. Names	Common Names
$H \cdot CO \cdot Cl$ (unstable)	methanoyl chloride	formyl chloride
$CH_3 \cdot CO \cdot Cl$	ethanoyl chloride	acetyl chloride
$CH_3 \cdot CH_2 \cdot CO \cdot Br$	propanoyl bromide	propionyl bromide
$CH_3 \cdot CH_2 \cdot CH_2 \cdot CO \cdot Cl$	butanoyl chloride	butyryl chloride

C. Preparations:

THE ACID HALIDES MAY BE PREPARED:

1. *By the treatment of the corresponding acid with* (a) *phosphorus pentachloride,* (b) *phosphorus trichloride, or* (c) *thionyl chloride:*

   a. $R \cdot CO \cdot OH + PCl_5 \rightarrow R \cdot CO \cdot Cl + POCl_3 + HCl$, and

   $CH_3 \cdot CO \cdot OH + PCl_5 \rightarrow CH_3 \cdot CO \cdot Cl$ (acetyl chloride) $+ POCl_3 + HCl$

   b. $3 R \cdot CO \cdot OH + PCl_3 \rightarrow 3 R \cdot CO \cdot Cl + H_3PO_3$

   c. $R \cdot CO \cdot OH + SOCl_2 \rightarrow R \cdot CO \cdot Cl + SO_2$ (gas) $+ HCl$ (gas)

   The particular reagent used in these preparations is determined somewhat by the volatility of the acid chloride obtained, that is, the product and by-products should be capable of separation by distillation.

2. *By the treatment of salts of the corresponding acids with* (a) *phosphorus pentachloride,* (b) *phosphorus trichloride,* (c) *phosphorus oxychloride, or* (d) *sulfuryl chloride:*

   a. $R \cdot CO \cdot ONa + PCl_5 \rightarrow R \cdot CO \cdot Cl + POCl_3 + NaCl$, and

   $CH_3 \cdot CO \cdot ONa + PCl_5 \rightarrow CH_3 \cdot CO \cdot Cl$ (ethanoyl chloride) $+ POCl_3 + NaCl$

   b. $3 R \cdot CO \cdot ONa + PCl_3 \rightarrow 3 R \cdot CO \cdot Cl + Na_3PO_3$

   c. $2 R \cdot CO \cdot ONa + POCl_3 \rightarrow 2 R \cdot CO \cdot Cl + NaPO_3 + NaCl$

   d. $2 R \cdot CO \cdot ONa + SO_2Cl_2/(SO_2 + Cl_2) \rightarrow 2 R \cdot CO \cdot Cl + Na_2SO_4$ (commercial process)

D. Physical Properties:

The acid halides are pungent, sharp-smelling, colorless liquids which are denser than water. Ethanoyl (acetyl) chloride boils at 52.0°C.

E. Chemical Properties:

THE ACID HALIDES REACT UNDER PROPER CONDITIONS:

1. *By replacement of the halide atom when treated with:*

a. Water (hydrolysis)	$R \cdot CO$	$X +$	H	OH
b. Alcohols (alcoholysis)	$R \cdot CO$	$X +$	H	OR
c. Ammonia (ammonolysis)	$R \cdot CO$	$X +$	H	NH_2
d. 1. Primary amines (aminolysis)	$R \cdot CO$	$X +$	H	NHR
2. Secondary amines (aminolysis)	$R \cdot CO$	$X +$	H	NR_2
e. 1. Alkoxides	$R \cdot CO$	$X +$	K	OR
2. Salts of organic acids	$R \cdot CO$	$X +$	K	$O \cdot OC \cdot R$
f. Sodium or potassium cyanide	$R \cdot CO$	$X +$	K	CN
g. Sodium or potassium hydrosulfide	$R \cdot CO$	$X +$	K	SH
h. Appropriate reducing agents	$R \cdot CO$	$X +$	H	H
i. Grignard reagent (poor yields)	$R \cdot CO$	$X +$	XMg	R

2. *By replacement of the carbonyl oxygen atom with chlorine atoms when heated with phosphorus pentachloride in a sealed tube:*

$R \cdot CO \cdot Cl + PCl_5$, heat in sealed tube $\rightarrow R \cdot CCl_3 + POCl_3$

3. *By replacement of the α-hydrogen atom by treatment with halogen.*

III. Esters

A. Introduction:

Esters are derivatives of inorganic or organic acids. Examples of inorganic esters are **methyl chloride** ($CH_3 \cdot Cl$, an alkyl halide), methyl nitrate ($CH_3 \cdot O \cdot NO_2$), methyl nitrite ($CH_3 \cdot O \cdot NO$), methyl hydrogen sulfate ($CH_3 \cdot O \cdot SO_2 \cdot OH$), methyl sulfate ($[CH_3]_2SO_4$), tributyl phosphate ($[C_4H_9]_3PO_4$), triceryl phosphate ($[C_{26}H_{53}]_3PO_4$), and triethyl phosphate ($[C_2H_5]_3PO_4$). These will not be considered in the following discussion.

Organic esters are derivatives of organic acids in which the OH of the carboxyl group has been replaced by an alkoxy group. To illustrate their formation:

$CH_3 \cdot CO \cdot OH + H \cdot O \cdot CH_2 \cdot CH_3 /(H^+) \rightleftarrows CH_3 \cdot CO \cdot O \cdot CH_2 \cdot CH_3$ (an ester) $+ H_2O$

Most of the natural fats are mixtures of esters of **stearic** ($C_{17}H_{35} \cdot CO \cdot OH$), **palmitic** ($C_{15}H_{31} \cdot CO \cdot OH$), and **oleic** ($C_{17}H_{33} \cdot CO \cdot OH$) acids with glycerol.

$H_2C \cdot O \cdot OC \cdot C_{17}H_{35}$	$H_2C \cdot O \cdot OC \cdot C_{15}H_{31}$	$H_2C \cdot O \cdot OC \cdot C_{17}H_{33}$
$H \cdot C \cdot O \cdot OC \cdot C_{17}H_{35}$	$H \cdot C \cdot O \cdot OC \cdot C_{15}H_{31}$	$H \cdot C \cdot O \cdot OC \cdot C_{17}H_{33}$
$H_2C \cdot O \cdot OC \cdot C_{17}H_{35}$	$H_2C \cdot O \cdot OC \cdot C_{15}H_{31}$	$H_2C \cdot O \cdot OC \cdot C_{17}H_{33}$
1,2,3-propanetriol trioctadecanoate, **glyceryl tristearate**, or **stearin**, m.p. 54.5	1, 2, 3-propanetriol trihexadecanoate, **glyceryl tripalmitate**, or **palmitin**, m.p. 65.1	1,2,3-propanetriol *cis*-9-trioctadecenoate, **glyceryl trioleate**, or **olein**, m.p. 17

In most cases the glyceryl radical is combined with two or three different organic acid radicals to give a mixed ester.

Fatty oils are glyceryl esters, but the hydrocarbon radicals are sufficiently unsaturated to impart liquid properties. Examples are **olein** and **linseed** oil. The fatty oils may be converted to solids by catalytic hydrogenation (reductive addition of hydrogen):

$H_2C \cdot O \cdot OC(CH_2)_7HC:CH(CH_2)_7CH_3$ $H_2C \cdot O \cdot OC(CH_2)_7HCH \cdot HCH(CH_2)_7CH_3$
$H \cdot C \cdot O \cdot OC(CH_2)_7HC:CH(CH_2)_7CH_3 + 3H_2 \rightarrow H \cdot C \cdot O \cdot OC(CH_2)_7HCH \cdot HCH(CH_2)_7CH_3$
$H_2C \cdot O \cdot OC(CH_2)_7HC:CH(CH_2)_7CH_3$ (Ni) $H_2C \cdot O \cdot OC(CH_2)_7HCH \cdot HCH(CH_2)_7CH_3$

    glyceryl oleate (a liquid)   **glyceryl stearate (a solid)**

This type of reaction is employed commercially to prepare lard substitutes from vegetable oils. The hydrogenation is stopped when the desired consistency is obtained. The radicals less saturated than oleate are converted to oleate or saturated radicals as completely as possible with the minimum hydrogenation of the oleate radicals.

Waxes, in the chemical sense, are esters of monohydroxy alcohols and monoacids of high molecular weight. Examples of these higher esters are **myricyl palmitate** ($C_{15}H_{31} \cdot CO \cdot O \cdot C_{31}H_{63}$), which occurs in beeswax, and hexacosyl hexacosanoate ($C_{25}H_{51} \cdot CO \cdot O \cdot C_{26}H_{53}$, **ceryl cerotate**), which is the chief component of

Chinese wax. The natural waxes also contain free acids, free alcohols, and hydrocarbons.

Lipides, some of which are esters, are a group of substances which are soluble in fat solvents and occur particularly in the brain tissues. The general formula for the lecithins is shown at the right.

$$H_2C \cdot O \cdot OC \cdot R$$
$$H \cdot C \cdot O \cdot OC \cdot R'$$
$$H_2C \cdot O \cdot P(\rightarrow O) \cdot O \cdot CH_2 \cdot CH_2 \cdot N(CH_3)_3$$
$$OH \qquad\qquad OH$$

Simple organic esters are widely distributed in the essential oils of plants and give fruits their characteristic odors. Methyl ethanoate (methyl acetate), for example, occurs in peppermint, and linalyl acetate is an essential ingredient of lavender flavors. 3-Methylbutyl ethanoate (isoamyl acetate) is found in pear oil, and octyl ethanoate (octyl acetate) occurs in oranges. Fats are found both in plants and animals. The honeycomb of bees and the head of the sperm whale are among the sources of the waxes. Lipides occur quite commonly in animal tissues, especially in the brain.

Esters of low molecular weight are used extensively as solvents, in the lacquer industry, and as artificial flavors and perfumes.

The fats are one of the most important components of the human diet. They serve as a source of heat and energy, regulate the rate of digestion, aid in restricting the diet, and act as carriers in supplying the animal organism with vitamins A and D. The natural glycerides are the raw materials for the commercial production of glycerol and soap. Glycerides, blended with hydrocarbons, are also used somewhat extensively as lubricants in steam cylinder oils, and soap is an important ingredient of a number of greases.

The fatty oils are likewise valuable as food. Some unsaturated fatty oils absorb oxygen at the unsaturated groups to give hard, resinous materials. The use of linseed and similar fatty oils in paints and varnishes is dependent upon this property. Cottonseed and similar fatty oils are partially hydrogenated to give cooking fats such as Crisco and Snowdrift.

Carnauba wax is used in varnish, in candles, in commercial "water waxes" for floors, and to adulterate beeswax. Beeswax finds use in candle manufacture and in pharmacy. Lanolin (from wool grease) is employed as a base for ointments, salves, and emulsions. Chinese wax (secreted by an insect) is used in medicine, in candles, and as a furniture polish.

Lipides doubtless serve important functions in the animal body, and are used in candy making.

Among themselves, the esters exhibit chain and position isomerism, and they are functional isomers of the acids of the same carbon content.

B. Nomenclature:

The system of naming the esters is similar to that for naming salts.

Structural Formulas	I.U.C. Names	Common Names
$H \cdot CO \cdot O \cdot CH_3$	methyl methanoate	methyl formate
$H \cdot CO \cdot O \cdot CH_2 \cdot CH_3$	ethyl methanoate	ethyl formate
$CH_3 \cdot CO \cdot O \cdot CH_2 \cdot CH_3$	ethyl ethanoate	ethyl acetate
$CH_3 \cdot CH_2 \cdot CO \cdot O \cdot CH_3$	methyl propanoate	methyl propionate
$CH_3 \cdot CH_2 \cdot CH_2 \cdot CO \cdot O \cdot CH_2 \cdot CH_3$	ethyl butanoate	ethyl butyrate
$CH_3 \cdot CH_2 \cdot CH_2 \cdot CH_2 \cdot CO \cdot O \cdot CH_3$	methyl pentanoate	methyl valerate

C. Preparations:

THE ESTERS MAY BE PREPARED:

1. *By the action of an acid on an alcohol in the presence of hydrogen ion and some means for the removal of water:*

$R \cdot CO \cdot OH + HO \cdot R'/(H^+ \text{ as a catalyst}) \rightleftarrows R \cdot CO \cdot O \cdot R' + H_2O$, and

$CH_3 \cdot CO \cdot OH + HO \cdot CH_2 \cdot CH_3/(H^+) \rightleftarrows CH_3 \cdot CO \cdot O \cdot CH_2 \cdot CH_3$ (ethyl acetate) + H_2O

By the removal of water as it is formed, this reaction may be driven to completion to the right. In laboratory practice a dehydrating agent such as sulfuric acid is used, but in commercial production the water is removed by azeotropic distillation (cf. Groggins, P.H., Unit Processes in Organic Synthesis, p. 498–506, 1935 ed.)

2. *By the action of salts of very active or heavy metals on alkyl halides:*

$R \cdot CO \cdot OM + X \cdot R' \rightarrow R \cdot CO \cdot O \cdot R' + MX$, and

$CH_3 \cdot CO \cdot O \cdot Ag + I \cdot CH_2 \cdot CH_3 \rightarrow CH_3 \cdot CO \cdot O \cdot CH_2 \cdot CH_3$ (ethyl acetate) + AgI

3. *By the action of acid chlorides on (a) alcohols, or (b) alkoxides:*

   a. $R \cdot CO \cdot Cl + H \cdot O \cdot R' \rightarrow R \cdot CO \cdot O \cdot R' + HCl$, and

     $CH_3 \cdot CO \cdot Cl + H \cdot O \cdot CH_3 \rightarrow CH_3 \cdot CO \cdot O \cdot CH_3$ (methyl acetate) + HCl

   b. $R \cdot CO \cdot Cl + Na \cdot O \cdot R' \rightarrow R \cdot CO \cdot O \cdot R' + NaCl$

4. *By the action of acid anhydrides on alcohols:*

$R \cdot CO \cdot O \cdot OC \cdot R + HO \cdot R' \rightarrow R \cdot CO \cdot O \cdot R' + R \cdot CO \cdot OH$

5. *By the (a) catalytic condensation of aldehydes, or (b) dehydrogenation of alcohols:*

   a. $2 \ CH_3 \cdot CHO + [Al(O \cdot Et)_3] \rightarrow CH_3 \cdot CO \cdot O \cdot CH_2 \cdot CH_3$ (Tishchenko Reaction)

     The yields from this reaction are as high as 99%. Other effective catalysts are $AlCl_3 \cdot 3Al(O \cdot Et)_3$ and $HgCl_2 \cdot 2Al(O \cdot Et)_3$.

   b. $2 \ R \cdot CH_2 \cdot OH + \text{loss of } 2 \ H_2 \rightarrow R \cdot CO \cdot O \cdot CH_2 \cdot R + 2 \ H_2$ (U.S.P. 1,875,540)

6. *By the isolation of fats, fatty oils, waxes, and lipides from natural products (commercial).*

7. *By the treatment of chloroform with sodium ethylate to give orthoformic ester:*

$CHCl_3 + 3 \ Na \cdot O \cdot CH_2 \cdot CH_3 \rightarrow CH(O \cdot CH_2 \cdot CH_3)_3 + 3 \ NaCl$

Ethyl orthoformate is a useful reagent in many syntheses.

D. Physical Properties:

The lower esters are pleasant smelling liquids, but the higher members are solids. In odor, 3-methylbutyl ethanoate (isoamyl acetate) resembles banana oil; octyl ethanoate (octyl acetate), oranges; and pentyl butanoate (amyl butyrate), apricots. The properties of the fats are largely dependent upon the nature of the hydrocarbon radicals involved. In general, melting points increase with increase in the size and saturation of the hydrocarbon radicals. The melting points of the fatty oils decrease with decrease in the size and with an increase in the degree of unsaturation of the hydrocarbon radicals. The solid properties of the waxes increase with increase in the size of the hydrocarbon radicals. The esters, fats, fatty oils, waxes, and lipides are all more or less soluble in ether, in alcohol, and, in general, in organic solvents.

E. Chemical Properties:

THE ESTERS REACT UNDER PROPER CONDITIONS:

1. *By replacement of (a) the alkyl group by metallic atoms, hydrogen atoms, or a higher alkyl radical, (b) the alkoxy group with an amine or hydrogen group, or (c) the alkoxy oxygen atom by chlorine atoms, when treated with:*

a. 1. Aqueous alkali hydroxides (saponification)   R·CO·O $\boxed{\text{R} + \text{HO}}$ Na
   2. Hydriodic acid, conc., (splitting)   R·CO·O $\boxed{\text{R} + \text{I}}$ H
   3. Higher alcohols/(catalyst) (alcoholysis)   R·CO·O $\boxed{\text{R} + \text{HO}}$ R
   4. Water/H$^+$, heat, pressure   R·CO·O $\boxed{\text{R} + \text{HO}}$ H

b. 1. Dry ammonia (ammonolysis)   R·CO $\boxed{\text{O R} + \text{H}}$ NH$_2$
   2. Appropriate reducing agents as Na/alc., R·CO $\boxed{\text{O R} + \text{H}}$ H
      or catalytic reduction.

c. Phosphorus pentachloride (or trichloride)   R·CO $\boxed{\text{O} \boxed{\text{R} + \text{Cl}}\,\text{PCl}_3}$ Cl

More specifically, saponification may be illustrated by the alkaline hydrolysis of **palmitin** (a fat):

H$_2$C·O·OC·C$_{15}$H$_{31}$ H$_2$C·OH
  | |
H·C·O·OC·C$_{15}$H$_{31}$ + 3 NaOH, aq. → H·C·OH + 3 C$_{15}$H$_{31}$·CO·ONa (so-
  | |
H$_2$C·O·OC·C$_{15}$H$_{31}$ H$_2$C·OH   dium palmitate, *a soap*)

An analogous reaction is effected by enzymatic hydrolysis during the digestion of the fats and fatty oils.

Ammonolysis of the lower esters is effected in laboratory practice by the use of concentrated ammonium hydroxide. Ethanamide (**acetamide**) is readily prepared by such a procedure.

In the reduction of the esters, an aldehyde is presumably an intermediate and, if so, is reduced in turn to the alcohol. The catalytic hydrogenation of the fatty oils, illustrated in section A, is an important commercial reaction. The degree of unsaturation of a given fatty oil may be determined by a quantitative measurement of the amount of iodine that it will add at the unsaturated linkages. The results are expressed in terms of the "iodine number," which is defined as *the number of grams of iodine which combine with 100 g. of a fat or fatty oil.*

2. *By addition of Grignard reagent, in excess, to give tertiary alcohols upon subsequent hydrolysis:*

R·CO·O·R′ + R″·MgX, ether → R·CO·R″ (a ketone) + R′·O·MgX, then

     R·CO·R″ + R″·MgX, ether → R(R″)$_2$C·O·MgX, then

         R(R″)$_2$C·O·MgX + HX, aq. → R(R″)$_2$C·OH (tertiary) + MgX$_2$

The simple esters, fats, fatty oils, waxes, and lipides, enter, in general, into the same type of reactions, but the degree of reactivity is determined by the complexity of the structure of the compound involved.

IV. Salts

A. Introduction:

Salts are compounds in which the hydrogen atom of the carboxyl group has been replaced by a metallic atom.

Salts of the monoacids may be represented by the type formula $R \cdot CO \cdot OM$ where "M" is a univalent metallic group. In case "M" is a bivalent metallic atom the formula becomes $(R \cdot CO \cdot O)_2M$ for the monoacids, or $(CH_2)_x(CO \cdot O)_2M$ for the diacids.

Acid potassium ethanedioate (potassium **oxalate**) occurs in rhubarb, in tobacco, and in other plants. Acid potassium *d*-2,3-dihydroxybutanedioate (potassium *d*-**tartrate**) is a constituent of grape juice, and acid and normal salts of 2-hydroxy-1,2,3-propanetricarboxylic (**citric**) acid are found in the citrus fruits.

As indicated by their reactions under Div. E, the salts find extensive use in organic synthesis. The sodium and potassium salts of the higher acids are used extensively as soaps. **Ammonium acetate** is used as a sudorific, and aluminum acetate is taken in small doses as an astringent for diarrhea. Ferrous acetate, ferric acetate, and aluminum acetate, are all used as mordants in dyeing. Paris green, a mixture of **copper acetate** and copper arsenate, is used as an insecticide.

The salts exhibit chain and position isomerism.

B. Nomenclature:

In naming salts, the name of the metallic group is followed by the name of the corresponding acid in which the *ic* ending has been replaced by *ate* as shown below:

Structural Formulas	I.U.C. Names	Common Names
$H \cdot CO \cdot ONa$	sodium methanoate	sodium formate
$CH_3 \cdot CO \cdot OK$	potassium ethanoate	potassium acetate
$(CH_3 \cdot CH_2 \cdot CO \cdot O)_2Ca$	calcium propanoate	calcium propionate
$CH_3 \cdot CH_2 \cdot CH_2 \cdot CO \cdot ONa$	sodium butanoate	sodium butyrate
$(CH_2 \cdot CO \cdot O)_2Ca$	calcium butanedioate	calcium succinate

C. Preparations:

THE SALTS OF ORGANIC ACIDS MAY BE PREPARED:

1. *By the same general methods available for the preparation of salts of inorganic acids, as treatment of acids with* (a) *active metals,* (b) *metallic hydroxides,* (c) *metallic oxides,* (d) *metallic carbonates, and* (e) *ammonia:*

   a. $2 \ R \cdot CO \cdot OH + 2 \ M \rightarrow 2 \ R \cdot CO \cdot OM + H_2$ (where M is univalent), and

   $2 \ CH_3 \cdot CO \cdot OH + Zn \rightarrow (CH_3 \cdot CO \cdot O)_2Zn$ (zinc acetate) $+ H_2$

   b. $R \cdot CO \cdot OH + M \cdot OH \rightarrow R \cdot CO \cdot OM + H_2O$

   c. $2 \ R \cdot CO \cdot OH + MO \rightarrow (R \cdot CO \cdot O)_2M$ (where M is bivalent) $+ H_2O$

   d. $2 \ R \cdot CO \cdot OH + M_2CO_3 \rightarrow 2 \ R \cdot CO \cdot OM + CO_2 + H_2O$

   e. $R \cdot CO \cdot OH + NH_3 \rightarrow R \cdot CO \cdot ONH_4$

2. *By the action of alkali hydroxides on esters (saponification):*

    a. $R \cdot CO \cdot OR + NaOH$, aq. $\rightarrow R \cdot CO \cdot ONa + R \cdot OH$

    b. Fats and oils (glyceryl esters) are saponified by alkali hydroxides to give **glycerol** and salts of the higher fatty acids (*soaps*). The potassium salts are soft soaps, whereas the sodium salts are hard soaps. Saponification is illustrated by:

$$H_2C \cdot O \cdot OC \cdot C_{15}H_{31}$$
$$|$$
$$H \cdot C \cdot O \cdot OC \cdot C_{15}H_{31} + 3\ NaOH, \text{aq.} \rightarrow$$
$$|$$
$$H_2C \cdot O \cdot OC \cdot C_{15}H_{31}$$

1,2,3-propanetriol trihexadecanoate (palmitin, a fat)

$$CH_2OH$$
$$|$$
$$CHOH + 3\ NaO \cdot OC \cdot C_{15}H_{31} \text{ (sodium}$$
$$|$$
$$CH_2OH \quad \text{hexadecanoate, sodium}$$

1,2,3-propanetriol (glycerol)   palmitate)

3. *By heating sodium hydroxide with carbon monoxide to 150–170°C. under six to seven atmospheres pressure to give sodium formate:*

    $NaOH + CO$, 150–170°C., 6 to 7 atmospheres $\rightarrow H \cdot CO \cdot ONa$

D. Physical Properties:

The organic salts of the alkali metals are quite soluble in water, slightly soluble in alcohol, and insoluble in ether. Salts of the heavier metals are usually relatively insoluble. Metallic salts of the organic acids, as a rule, possess rather high and indefinite melting points. They are usually decomposed by heating.

E. Chemical Properties:

THE SALTS OF THE ORGANIC ACIDS REACT UNDER PROPER CONDITIONS:

1. *To give (a) acids, (b) esters, or (c) anhydrides, when treated with:*

a. Strong mineral acids	$R \cdot CO \cdot O$	Na + X	H
b. Alkyl halides	$R \cdot CO \cdot O$	Na + X	R
c. Acyl halides	$R \cdot CO \cdot O$	Na + X	$OC \cdot R$

In the preparation of non-volatile acids from their salts, barium compounds and sulfuric acid are commonly used, the barium sulfate being removed by filtration.

2. *To give acyl chlorides when treated with (a) phosphorus pentachloride, (b) phosphorus oxychloride, or (c) sulfuryl chloride:*

    a. $R \cdot CO \cdot ONa + PCl_5 \rightarrow R \cdot CO \cdot Cl + PO \cdot Cl_3 + NaCl$

    b. $2\ R \cdot CO \cdot ONa + POCl_3 \rightarrow 2\ R \cdot CO \cdot Cl + NaPO_3 + NaCl$

    c. $2\ R \cdot CO \cdot ONa + SO_2Cl_2/(SO_2 + Cl_2) \rightarrow 2\ R \cdot CO \cdot Cl + Na_2SO_4$

    Note that when phosphorus pentachloride is used, only one fifth of the available chlorine atoms are utilized, but that phosphorus oxychloride (b.p. 107.23°C), a liquid, is a by-product. This may react as indicated in the second equation to utilize two more chlorine atoms and give inorganic solids as the by-products. In the third equation, all of the chlorine is utilized and solid sodium sulfate is the only by-product.

3. *To give decomposition products which vary with the salt and the nature of the treatment:*

   a. $(H \cdot CO \cdot O)_2 Hg$ (mercuric formate), aq. + heat → $H \cdot CO \cdot OH + CO_2 + Hg°$

   b. $2 \, H \cdot CO \cdot ONa$ + heat, 200°C, → $(CO \cdot ONa)_2$ (sodium oxalate) + H_2

   c. $H \cdot CO \cdot O \cdot Na$ + $R \cdot CO \cdot O \cdot Na$, fuse → $R \cdot CHO$ (an aldehyde) + Na_2CO_3

   d. $2 \, R \cdot CO \cdot O \cdot Na$ + fusion → $R \cdot CO \cdot R$ (a ketone) + Na_2CO_3

   e. $R \cdot CO \cdot ONa$ + $NaOH/(CaO)$, fuse → $R \cdot H$ (an alkane) + Na_2CO_3

   f. $R \cdot CO \cdot ONH_4$ + heat → $R \cdot CO \cdot NH_2$ (an amide) + H_2O

In *c* and *d*, calcium salts are frequently employed. Traces of ketones may be formed in reaction *c*.

V. Other Acid Derivatives

The 1,1,1-trihalogen compounds have been considered previously (Ch. IV Div. II) and the amides and nitriles will be considered later (Ch. XIII Div's. V, VIII). The rest of the acid derivatives are relatively unimportant A few examples are illustrated below:

Thiolic Acids	Thionic Acids	Hydroxamic Acids	Hydroximic Acids
$R \cdot CO \cdot SH$	$R \cdot CS \cdot OH$	$R \cdot CO \cdot NH \cdot OH$	⇌ $R \cdot C(:N \cdot OH) \cdot OH$
$CH_3 \cdot CO \cdot SH$	$CH_3 \cdot CS \cdot OH$	$CH_3 \cdot CO \cdot NH \cdot OH$	⇌ $CH_3 \cdot C(:N \cdot OH) \cdot OH$
ethanethiolic (thioacetic) acid	ethanethionic acid	ethanehydroxamic (acetohydroxamic) acid	ethanehydroximic (acetohydroximic) acid

Hydroximic Acid Chlorides	Imide Chlorides	Amidines
$R \cdot C(:N \cdot OH) \cdot Cl$	$R \cdot C(:N \cdot H) \cdot Cl$	$R \cdot C(:N \cdot H) \cdot NH_2$
$CH_3 \cdot C(:N \cdot OH) \cdot Cl$	$CH_3 \cdot C(:N \cdot H) \cdot Cl$	$CH_3 \cdot C(:N \cdot H) \cdot NH_2$
acetohydroximyl chloride	ethanimidyl (acetimidyl) chloride	ethanamidine, **acetamidine**

Thioimido Esters	Thiamides	Amide Chlorides
$R \cdot C(:N \cdot H) \cdot S \cdot R$	$R \cdot CS \cdot NH_2$	$R \cdot CCl_2 \cdot NH_2$
$CH_3 \cdot C(:N \cdot H) \cdot S \cdot C_6H_5$	$CH_3 \cdot CS \cdot NH_2$	$CH_3 \cdot CCl_2 \cdot NH_2$
phenyl ethanethiolimidate, **phenyl thioacetimidate**	ethanethiamide, **thioacetamide**	1,1-dichloroethylamine, **acetamide chloride**

CHAPTER X

CLASSIFIED REPLACEMENT REACTIONS

I. Replacement of Hydrogen Atoms:

1. When linked to a saturated carbon atom by treatment with:

a.	Bromine	$\equiv$C	H +	Br	Br
b.	Chlorine	$\equiv$C	H +	Cl	Cl
c.	Nitric acid	$\equiv$C	H +	HO	NO_2
d.	Sulfuric acid	$\equiv$C	H +	HO	$SO_2 \cdot OH$

2. When linked to an yne carbon atom by treatment with:

a.	Ammonical cuprous chloride	–C:C	H +	$Cl(NH_3)_2$	Cu
b.	Ammonical silver nitrate	–C:C	H +	$NO_3(NH_3)_2$	Ag
c.	Potassium hypoiodite	–C:C	H +	KO	I

*3. When linked to an alcoholic **hydroxyl** group by treatment with:*

a.	Active metals	$\equiv$C·O	H +		2 Na
b.	Acid halides	$\equiv$C·O	H +	X	OC·R
c.	Acid anhydrides	$\equiv$C·O	H +	$R \cdot CO \cdot O$	OC·R
d.	Grignard reagent	$\equiv$C·O	H +	R	MgX
e.	Dehydrogenation catalysts	$\overset{O}{=}\dot{C}$	$\overset{H}{H}$ +	Cu, 300°	
f.	Oxidizing reagents	$\overset{O}{=}\dot{C}$	$\overset{H}{H}$ +	oxidation	

In the last two examples, the hydroxyl group is linked to a carbon atom which is linked in turn to at least one hydrogen atom.

4. When linked to an acid hydroxyl group by treatment with:

a.	Active metals	2 –CO·O	H +		2 Na
b.	Alkalies	–CO·O	H +	HO	Na
c.	Carbonates	–CO·O	H +	$Na \cdot O \cdot CO \cdot O$	Na
d.	Oxides	2 –CO·O	H +	O	M_2
e.	Alcohols	–CO·O	H +	HO	R/H_2SO_4
f.	Acid chlorides	–CO·O	H +	X	OC·R
g.	Organic acids	–CO·O	H +	HO	$OC \cdot R/P_2O_5$

II. Replacement of Halogen Atoms:

1. By double decomposition of alkyl halides when treated with:

a.	Hydrogen/catalyst	≡C	X +	H	H
	Hydrogen iodide/(P, red)	≡C	X +	I	H
b.	Sodium/(dry ether)	≡C	X + 2Na + X	C≡
	Grignard reagent/(ether)	≡C	X +	X·Mg	R
	Frankland reagent/(benzene)	≡C	X +	X·Zn	R
c.	Alkalies, aq.	≡C	X +	Na	OH
d.	Alcoholates	≡C	X +	Na	O·R
	Silver oxide	≡C	X + AgOAg + X	C≡
e.	Salts of organic acids	≡C	X +	Na	O OC R
f.	Metallic cyanides, NaCN, KCN,	≡C	X +	Na	CN
	but AgCN reacts as AgNC	≡C	X +	Ag	NC
g.	Metallic nitrites, NaNO₂, KNO₂,	≡C	X +	Na	O·NO
	but AgONO reacts as AgNO₂	≡C	X +	Ag	NO₂

2. By addition of alkyl halides when treated with:

a.	Mg/(dry ether)	$\equiv$C·X + Mg, ether → $\equiv$C·Mg·X
b.	Zn (dust, heat)	$\equiv$C·X + Zn, heat → $\equiv$C·Zn·X
c.	Ammonia	$\equiv$C·X + NH₃ → $\equiv$C·NH₃X
d.	*prim.*-Amines	$\equiv$C·X + R·NH₂ → $\equiv$C·NH₂RX
	sec.-Amines	$\equiv$C·X + R₂NH → $\equiv$C·NHR₂X
	tert.-Amines	$\equiv$C·X + R₃N → $\equiv$C·NR₃X

3. By the loss of (a) a hydrogen and a halogen atom, or (b) two halogen atoms, when treated with:

a. NaOH/ROH, or KOH/ROH
$$\begin{array}{l} =\text{C} \\ =\dot{\text{C}} \end{array} \begin{array}{l} \text{X} + \text{KOH} \\ \text{H} \quad \overline{\text{ROH}} \end{array}$$

b. Zn
$$\begin{array}{l} =\text{C} \\ =\dot{\text{C}} \end{array} \begin{array}{l} \text{X} \\ \text{X} \end{array} + \text{Zn}$$

If twice as many halogen atoms are present in the molecule, two HX will be lost in the first reaction and two X₂ in the second. All of the reactions indicated above are applicable, with certain restrictions, to the polyhalogenated compounds.

4. By double decomposition of acyl halides when treated with:

a.	Hydrogen or reducing agents	−CO	X +	H	H
b.	Water	−CO	X +	H	OH
c.	Alcohols	−CO	X +	H	O·R
d.	Ammonia	−CO	X +	H	NH₂
e.	Primary amines	−CO	X +	H	NHR
f.	Secondary amines	−CO	X +	H	NR₂
g.	Organic acids	−CO	X +	H	O·OC·R
h.	Alkalies	−CO	X +	Na	OH
i.	Alcoholates	−CO	X +	Na	O·R
j.	Salts of organic acids	−CO	X +	Na	O·OC·R
k.	Metallic cyanides	−CO	X +	Na	CN
l.	Thioalcoholates	−CO	X +	Na	S·R

III. Replacement of the Alcoholic Hydroxyl Group:

1. By double decomposition when treated with:

a.	HBr, aq/H_2SO_4, or HCl, aq/H_2SO_4	$\equiv$C	OH +	H	Br
b.	Dry HBr or dry HCl	$\equiv$C	OH +	H	Br
c.	HI/(P, red, heat)	$\equiv$C	OH +	H	I
d.	Nitric acid	$\equiv$C	OH +	H	O·NO_2
e.	Sulfuric acid	$\equiv$C	OH +	H	O·SO_2·OH
f.	Organic acids/(H_2SO_4, etc.)	$\equiv$C	OH +	H	O·OC·R
g.	Phosphorus pentahalides	$\equiv$C	O $\boxed{H + X}$ PX_3	X
h.	Phosphorus trihalides	3 $\equiv$C	OH +	P	X_3
i.	Thionyl chloride	2 $\equiv$C	OH +	SO	Cl_2
j.	Alkyl acid sulfates	$\equiv$C	OH +	HSO_3	OR
k.	Acid anhydrides	$\equiv$C	OH +	R·CO	O·OC·R
l.	Hydrogen/(catalyst)	$\equiv$C	OH +	H	H

2. By the loss of water when treated with:

Dehydrating agents, heat, etc.

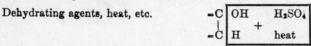

Corresponding reactions with the hydroxy acids give:

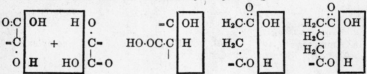

IV. Replacement of the Acid Hydroxyl Group:

1. By double decomposition when treated with:

a.	Hydrogen/(catalyst)	–CO	OH +	H	H
b.	Phosphorus pentahalides	–CO	O $\boxed{H + X}$ PX_3	X
c.	Phosphorus trihalides	3 –CO	OH +	P	X_3
d.	Thionyl chloride	2 –CO	OH +	SO	Cl_2
e.	Organic acids/(H_2SO_4, etc.)	–CO	OH +	H	O·OC·R

2. By the inter- or intramolecular loss of water:
This has been illustrated by the hydroxy acids.

V. Replacement of the Alkoxy Group:

1. In ethers when treated with:

a.	Hydrogen iodide, conc., cold	$\equiv$C	OR +	H	I
b.	Sulfuric acid, conc.	$\equiv$C	OR +	H	O·SO_2·OH
c.	Water, heat and pressure	$\equiv$C	OR +	H	OH

2. *In esters when treated with:*

a. Hydrogen/(catalyst)	$-CO$	OR	+	H	H	
Appropriate reducing agents	$-CO$	OR	+	H	H	
b. Ammonia	$-CO$	OR	+	H	NH_2	
c. Ethyl acetate/Na/ROH	$-CO$	OR	+	H	$CH_2 \cdot CO \cdot O \cdot Et$	
d. Phosphorus pentahalides	$-CO$	O	$\boxed{R + X}$	PX_3	X	

VI. Replacement of the Carbonyl Oxygen:

1. *By double decomposition when aldehydes and ketones are treated with:*

a. Hydroxylamine	$-C$	O	+	H_2	$N \cdot OH$
b. Hydrazine	$-C$	O	+	H_2	$N \cdot NH_2$
c. Phenylhydrazine	$-C$	O	+	H_2	$N \cdot NH \cdot C_6H_5$
d. Semicarbazide	$-C$	O	+	H_2	$N \cdot HN \cdot CO \cdot NH_2$

2. *By addition when aldehydes and ketones are treated with:*

	$R \cdot C(H)$	O	R_2C	O
	↓	↓	↓	↓
a. Hydrogen	H	H	H	H
b. Hydrogen cyanide	CN	H	CN	H
c. Sodium bisulfite	SO_3Na	H	SO_3Na	H
d. Grignard Reagent	R	MgX	R	MgX
e. Ammonia	NH_2	H	complexes	

3. *By double decomposition when aldehydes, ketones or acid chlorides are treated with:*

a. Phosphorus pentahalides $\qquad -C \boxed{O \qquad + PX_3}\ X_2$

VII. Replacement of an Acyl Group:

By double decomposition when treated with:

Water/(Sulfuric acid) $\qquad -CO \cdot O \boxed{OC \cdot R + \quad HO}\ H$

VIII. Replacement of the Carboxyl Group:

1. *By the loss of carbon dioxide when malonic acid, or alkyl malonic acids are treated with:*

Heat, at or above the m.p. $\qquad HO \cdot OC \cdot \overset{|}{\underset{|}{C}} \boxed{(C:O) \cdot O}\ H$

2. *By the loss of (a) carbon dioxide and water, or (b) carbon dioxide as carbonate, when:*

a. Acids/(MnO), are heated	R	$\boxed{CO \cdot O \cdot H\ +\ HO}$	$OC \cdot R$
b. Salts are fused with NaOH	R	$CO \cdot ONa\ +\ NaO$	H
c. Salts are fused	R	$CO \cdot ONa\ +\ NaO$	$OC \cdot R$

IX. Replacement of the Alkyl Group:

1. *In esters when treated with:*

a. Alkalies	−CO·O	R +	HO Na
b. Alcohols	−CO.O	R +	HO R′
c. Halogen acids, dry	−CO.O	R +	I H

2. *In ethers when treated with:*

	R	· O ·	R
a. Hydriodic acid, conc., hot	I	H H	I
b. Phosphorus pentahalides	X	PX$_3$	X
c. Sulfuric acid, conc.	HO·SO$_2$O	H H	O·SO$_2$·OH

CHAPTER XI

ALIPHATIC OXYGEN DERIVATIVES

THE CARBOHYDRATES

A. Introduction:

The word "carbohydrates" implies "hydrates of carbon." They occur in large quantities as the result of photosyntheses, and contain, for the most part, hydrogen and oxygen in the same ratio as in water. There are, however, carbohydrates that do not conform to the "hydrate rule," and there are compounds that conform to the rule that are not carbohydrates. To the former class belong such compounds as rhamnose ($C_6H_2[H_2O]_5$) and rhamnoheptose ($C_7H_2[H_2O]_6$), and in the latter group are methanal (H·CHO, formaldehyde), ethanoic acid ($CH_3·CO·OH$, acetic acid), and 2-hydroxypropanoic acid ($CH_3·CHOH·CO·OH$, lactic acid). The term hydrate, moreover, carries the concept of loosely bound water in contradistinction to water of constitution. A typical hydrate can be represented by a reversible system such as the following: $CuSO_4·5 H_2O + heat \rightleftarrows CuSO_4 + 5 H_2O$. The so-called "hydrates of carbon" cannot be represented by such a system. The term carbohydrate appears, therefore, to be a misnomer and it may ultimately be replaced by "saccharide." "Glucide" has also been proposed.

The carbohydrates are, actually or potentially, hydroxy or polyhydroxy oxo derivatives of the hydrocarbons; that is to say, they may be considered as polyhydroxy aldehydes, or polyhydroxy ketones, or as compounds which yield hydrolytic products that are either polyhydroxy aldehydes, or polyhydroxy ketones, or a mixture of both.

The carbohydrates may be classified, according to composition and structure, as follows:

I. Monosaccharides (monosaccharoses)
    a. Diose (two oxygen atoms)
    b. Trioses (three oxygen atoms)
        1. Aldotrioses
        2. Ketotrioses
    c. Tetroses (four oxygen atoms)
        1. Aldotetroses
        2. Ketotetroses
        3. Methyl aldotetroses
    d. Pentoses (five oxygen atoms)
        1. Aldopentoses

        2. Ketopentoses
        3. Methyl aldopentoses
    e. Hexoses (six oxygen atoms)
        1. Aldohexoses
        2. Ketohexoses
        3. Methyl aldohexoses
    f. Heptoses (seven oxygen atoms)
    g. Octoses (eight oxygen atoms)
    h. Nonoses (nine oxygen atoms)
    i. Glucosides, fructosides, etc.

II. Disaccharides (disaccharoses)

III. Trisaccharides (trisaccharoses)

IV. Tetrasaccharides (tetrasaccharoses)

V. Polysaccharides (polysaccharoses)
   a. Pentosans (anhydrides of pentoses)
      1. Arabinosans (anhydrides of arabinose)
      2. Lyxosans (anhydrides of lyxose)
      3. Ribosans (anhydrides of ribose)
      4. Xylosans (anhydrides of xylose)
   b. Hexosans (anhydrides of hexoses)
      1. Allosans (anhydrides of allose)

2. Altrosans (anhydrides of altrose)

3. Dextrans (anhydrides of dextrose)
   a. Dextrins
   b. Glycogen
   c. Starch
   d. Cellulose

4. Galactosans (anhydrides of galactose)

5. Gulosans (anhydrides of gulose)

6. Idosans (anhydrides of idose)

7. Levulosans (anhydrides of levulose)

8. Mannosans (anhydrides of mannose)

9. Talosans (anhydrides of talose)

It should be noted that the terms arabans, xylans, galactans, and mannans, in which the "os" is omitted, represent common usage. The allosans, altrosans, gulosans, idosans, lyxosans, ribosans, and talosans are rare, if they occur at all, in nature.

The carbohydrates may be represented, with a few exceptions, by the type formula $n(C_x[H_2O]_x) - (n-1)H_2O$, where x is the number of carbon atoms in a building unit and n is the number of building units per molecule.

Carbohydrates are one of the most important sources of food and raw materials, as indicated by the following:

Glycolic aldehyde and d-glyceraldehyde may occur as metabolic intermediates; the former in the conversion of glycine into D-glucose and the latter in the biochemical utilization of the simple sugars. L-Arabinose is the building unit of the arabinosans (arabans) which are found in cherry gum and in gum arabic. D-Xylose is widely distributed in the form of its anhydride the xylosans (xylans) in bran, corn cobs, cotton seed hulls, oat hulls, peanut shells, straw, and most woods. Xylose is, accordingly, sometimes referred to as wood sugar. D-Glucose occurs in ripe fruits, flowers, honey, leaves, roots, sap, urine of diabetics, and in small amounts in normal urine and in the blood (about 0.1%). D-Galactose is the essential constituent of agar-agar, it plays an important role in the physiology of the cerebrosides of the brain, and is one of the building units of lactose and raffinose. D⁻-Fructose is an ingredient of honey, is the essential constituent of inulin, and is one of the building units of sucrose and raffinose. Sucrose occurs in almonds, coffee, flowers, honey, maple sap, sugar beet, sorghum cane, sugar cane, walnuts, and in small amounts, associated with D-glucose or D⁻-fructose or both, in most plants. Maltose is found in beer, in corn syrup, and in malt, and as an intermediate in the hydrolysis of starch. Lactose occurs in milk (3-7%) and is, consequently, a by-product in the manufacture of cheese. Dextrins occur in honey, in vegetable juices, in some plants, and as intermediates in the hydrolysis of starch. Glycogen is found in the liver (about 10%), in muscle tissue, and in small amounts in the blood, brain, kidneys, pancreas, spleen, and in a large number of fungi. Starch is present in practically all green plants, but is most abundant in the storage cells such as the seeds and the tubers. Cellulose is the structural material of the cell walls of almost all plants. Inulin occurs in certain lichens, chicory root, dahlia bulbs, dandelion roots, sweet potatoes, and Jerusalem artichokes.

The economical importance of carbohydrate material is shown by the commercial uses listed below; concerning which C. M. A. Stine of du Pont has said: "there are ten thousand individual products made from cellulose, and the industry is still in its infancy."

1. OF THE MONOSACCHARIDES:

The dioses, trioses, and tetroses are of scientific value only. The pentoses serve as food for herbivorous animals and find use in the production of furfural. D-Glucose, D^--fructose, and D-galactose serve as a source of energy for the human organism. Commercial glucose is marketed in two grades: products designed specifically for food, such as Karo, corn syrup, and dextrose, and products designed for commercial uses such as the silvering of mirrors, the manufacture of ethyl alcohol, the production of vinegar, and other commercial products.

2. OF SUCROSE, MALTOSE AND LACTOSE:

The importance of sucrose is best illustrated by its historical development. The use of honey, milk, and fruits doubtless represents the beginning of sugar consumption by the human race. Sugar cane was known in ancient China, India, and Egypt. As early as the seventh century sugar had already become a commercial product. In the fifteenth century the culture of sugar cane was introduced into Brazil and the West Indies. At present it is produced in Australia, Brazil, China, Cuba, Egypt, India, Jamaica, Japan, Peru, Philippine Islands, other tropical countries, and some of the southern States.

The extraction of sugar from beets was first accomplished at the beginning of the nineteenth century. The process then introduced was about 3% efficient. In 1828 in France and in 1836 in Germany, however, it was successfully produced from beets. In U. S. the beet sugar industry has assumed commercial proportions in the states of California, Colorado, Indiana, Kansas, Michigan, Montana, Nebraska, South Dakota, Utah, and Wisconsin, and is produced on a smaller scale in some of the other states. Raw sugar is imported into this country, under quota, and refined locally.

Lactose is used extensively in infant foods. The galactose unit in lactose seems to be essential to the proper development of the nerve sheaths. For economic reasons, purified maltose has not assumed commercial production.

3. OF STARCH, DEXTRIN, GLYCOGEN, AND INULIN:

Starch serves as a very important source of food for both human beings and animals. It is used mostly in its naturally occurring forms, such as potatoes, carrots, beets, parsnips, wheat, oats, rice, corn, and similar products. Because of its adhesive properties, wheat starch is used in the production of paste. Arrowroot starch is used principally for starching cloth and for sizing cloth and paper. Rice starch finds special uses in the preparation of toilet powders and in the finishing of cotton cloth. Both potato and corn starch find use in the manufacture of corn syrup and ethyl alcohol. Sago starch is used principally as a food. Starch may be nitrated to give the explosive "nitrostarch."

Dextrin is used for thickening colors in calico printing, in making mucilage, in brewing, for thickening tanning extracts, and in the confectionery industry.

Glycogen in the liver serves as a source of reserve energy for the animal organism. It has been called animal starch because glycogen is the form in which the animal organisms store carbohydrates.

The inulin industry appears to be in its infancy but looks to a future as a source of levulose.

4. OF CELLULOSE PRODUCTS:

   a. Those in which the cellulose is not changed chemically:

   1. Hemicellulose, which is composed largely of pentoses combined with a uronic acid, is found in young shoots and leaves, and serves as an important source of food for herbivorous animals and as a source of vitamins for vitamin-conscious man.

   2. Natural cellulose products are represented by all kinds of cotton, linen, hemp, and jute goods in the form of string, thread, rope, cloth, or paper. Cotton goods are derived from the "boll" of the cotton plant; linen goods, from the flax plant. The leaves and stalks of the hemp plant are used in the production of twine and rope. Sacking, burlap, and carpets are manufactured from the stalks of the jute plant.

   3. In the paper industry the raw materials consist of cotton rags, linen rags, wood, straw, and hemp. Bond papers, originally, were made wholly from linen. They now contain other materials. Newspaper stock and similar grades of paper are made almost wholly from such woods as beech, birch, fir, larch, pine, poplar, and spruce.

   4. Mercerized cotton was introduced on a commercial basis about 1896. In this process, the cotton is passed through a 30% NaOH bath in the cold to give a sodium cellulose and then washed well with hot and cold water to give a hydrocellulose. The surface fibers are changed slightly in composition resulting in a product with a greater sheen. By this treatment the flat cotton fibers become cylindrical and thicker and increase in weight from 8-10%. They also become smooth and translucent and possess a luster similar to that of silk. On drying, the mercerized fibers shrink from 20-25% of their original length but increase in tensile strength by as much as 68%. If the fibers are dried under tension, shrinkage is prevented and a 35% increase in tensile strength is still obtained. The silky character of the product may be increased by successive treatment with calcium acetate, soap, and dilute acetic acid. Mercerized cotton dyes more readily than cotton. The process was named after its inventor, Mercer, who discovered it in 1844.

   5. Cellulose is also used extensively in the lumber industry, in the production of lumber substitutes, and as a fuel.

   b. Those in which the cellulose is changed chemically:

   1. Rayon is the trade name applied to fibers which are produced (1) by dissolving cellulose in a suitable solvent and then reprecipitating the cellulose in the form of threads which may possess a high degree of sheen, or (2) by the acylation of cellulose to give the cellulose acetate products. In 1930, U. S. produced about 125,000,000 pounds of rayon, and by 1940 the dollar value of the annual production of synthetic fibers was over five times that of natural silk. The industry dates back to Chardonnet in 1885. Previously it had been found that fine fibers of collodion could be used in the production of carbon filaments for incandescent electric lights. This led Chardonnet to try spinning collodion filaments. As a result he obtained the first artificial silk.

      Most rayon, as produced in the past, has been inferior to natural silk both in strength, especially when wet, and in fineness of fiber. Recent developments, however, have robbed the silk worm of the last of its mystic arts and industry is now producing rayon which compares in fineness with natural silk, and, in tensile strength, is even superior. Lusterless rayon, also, is being produced.

      For the production of rayon, three distinct processes are now in use: acetate, cuprammonium, and viscose. The nitrocellulose process, which represents the first rayon development, has been discontinued as a commercial process for the production of rayon.

2. Cellulose nitrate (nitrocellulose) products.

 a. Explosives.

 1. Smokeless powder is usually a mixture of cellulose nitrates and nitroglycerin (glyceryl trinitrate).

 2. Cordite is a colloidal mixture of nitrocellulose with nitroglycerin (20-40%). It possesses greater propelling and less shattering properties than ordinary smokeless powder. The after gases pit the bore of the gun, hence its use is being abandoned.

 3. Military guncotton (M.G.C.) contains 13-13.3% of nitrogen and is used in small weapons. Its decomposition products are CO, CO_2, H_2O, H_2 and N_2, hence smokeless.

 4. Pyro (pyrocollodion), used for cannon, contains 12.5-12.7% nitrogen.

 5. In the production of the cellulose nitrate for the above products, purified or dried cotton (usually waste material) is treated with a mixture of HNO_3/H_2SO_4. The concentration of the acids and the time of heating varies according to the nature of the desired product. Most linters are nitrated in stainless steel pots and the contact with the nitrating acid is of short duration. The nitrated product is then dropped into a centrifugal or hydroextractor.

 The particular product that is used for guncotton is soluble in nitrobenzene, ethyl acetate, and acetone but is insoluble in water, alcohol, ether, acetic acid, and nitroglycerin. One kilo of guncotton yields 741 liters of gas at ordinary temperatures. It is used almost universally as a propellant, whereas T.N.T. (trinitrotoluene) is employed in the bursting charge of the so-called high-explosive shells. Tetryl (tetranitromethylaniline) is used as a booster and in anti-aircraft shells.

 b. Cellulose plastics.

 1. Celluloid is made from cellulose nitrate containing from 10.8-11.6% nitrogen. It is obtained by mixing dried "pyroxylin" (a mixture of the lower nitrates of cellulose) with the proper amount of camphor in the presence of alcohol under heat and pressure. Coloring matter is added as desired and the pasty mass is pressed into sheets or blocks which are readily machined into the desired products. Celluloid is plastic while hot, but becomes brittle and hard on cooling. It is transparent in thin sheets, but opaque in thick pieces. Opaqueness may be increased by the addition of ZnO. The products are used in the production of imitation ivory, tortoise shell, and onyx. Mineral salts decrease the flammability.

 2. Pyralin is made by pressing thin sheets of celluloid together to give the stratified appearance of natural products. About a quarter million pounds are produced each day for various industrial uses.

 c. Lacquers are produced by dissolving pyroxylin in proper solvents such as butyl acetate or other low-boiling solvents, and then adding the proper resins and appropriate pigments. Dip and spraying lacquers are thin solutions of such a composition that the excess lacquer will drip off or dry rapidly and leave a smooth surface. The non-volatile or film-forming portion of a lacquer is composed of nitrocellulose, a gum or resin, and plasticizer or softener. Brushing lacquers must contain a solvent that will not dry too quickly.

The reason that nitrocellulose can be used in lacquers is that it has been subjected to heat and pressure in water suspension for some twenty minutes. With the lowering of not more than 0.05% in nitrogen content there is a change in the material so that it forms low viscosity solutions. Otherwise, it would have to be applied in such dilute solutions that many coats would be required for a commercial finish.

The use of Duco and similar products have reduced painting time from about twenty days to twenty hours or less.

d. Bronzing liquids consist of a 3% solution of pyroxylin in amyl acetate or other solvents to which have been added aluminum powder and resins.

e. Dopes consist of 6-10% cellulose nitrate in proper solvents. Such products are used for finishing leather and for coating airplane wings.

f. Collodion is a 3-5% solution of pyroxylin in a 3/1 ether-ethyl alcohol mixture. It is used as a coating for sensitive photographic plates. Sometimes the properties of collodion are altered slightly by adding turpentine or castor oil.

g. Artificial leather is made by coating the surface of a suitable fabric with pyroxylin and a plasticizer (castor oil) and then stamping the product with a suitable design.

h. Photographic films are produced principally by spreading a solution of pyroxylin and camphor in amyl acetate or similar solvent onto a clean, polished, heated drum and continuously taking off a sheet of film. After the solvent evaporates, the film is stripped off and cut into proper sizes. Since films of this composition are highly flammable and give off noxious fumes when burning, cellulose nitrate films are being displaced by cellulose acetate films.

3. Cellulose acetate solutions of the proper consistency are passed through a slit to give thin sheets that are used in the film industry. Cellulose acetate is also used as an insulating coating for copper wire.

4. Cellophane contains about 17% glycerol. It is a viscose product. Viscose is poured on a suitable surface and the cellulose hydrate precipitated out. A better process forces the viscose through a very narrow slot into a regenerating bath. By coating with a transparent lacquer the product becomes moistureproof. Cellophane is used extensively as a wrapping material.

5. Vulcanized fiber is produced by treating unloaded, unsized rag paper with a warm 60-70% solution of $ZnCl_2$. The treated paper is passed between rollers where a number of sheets may be fused together. It is then thoroughly washed. Three to four weeks are required to wash 1/4 inch material and six to eight months for the proper washing of two-inch material.

6. Parchment paper is prepared by immersing a good grade of paper in cold, concentrated sulfuric acid and then washing thoroughly. The surface of the paper is converted into a modified cellulose which is tough and resistant to water.

Structural relationships in the carbohydrate group are of primary importance, and a few representative examples will be considered.

I. ESTABLISHMENT OF THE STRUCTURE OF D-GLUCOSE:

In any studies on structure, the purity of the compound under consideration is of the utmost importance. The compound must be carefully purified by some of the methods available for the purification of organic compounds.

1. The empirical formula of glucose is shown to be CH_2O:

Qualitative analysis indicates that glucose contains only carbon, hydrogen, and oxygen. Quantitative analysis gives 40% C, 6.7% H, and 53.3% O (by difference). From such data, the empirical formula is obtained as follows:

$40.0 \div 12.0 = 3.33$, then $\div 3.33 = 1$, or C_1
$6.7 \div 1.008 = 6.67$ $\div 3.33 = 2$, or H_2
$53.3 \div 16.0 = 3.33$ $\div 3.33 = 1$, or O_1, or CH_2O.

2. The molecular formula is shown to be $(CH_2O)_6$ or $C_6H_{12}O_6$:

Molecular weight determinations on D-glucose by the freezing point method, give a value of about 180. Since this is six times the weight of the empirical formula, the molecular formula must be represented by six empirical units, that is $(CH_2O)_6$ or $C_6H_{12}O_6$.

3. The skeleton formula is shown to be $C \cdot C \cdot C \cdot C \cdot C \cdot C$:

Since the glucose molecule contains six carbon atoms, as just established, the determination of the carbon skeleton becomes the next logical point of attack. Although a number of reactions may be employed to evidence the relative arrangement of the carbon atoms in D-glucose, the following reactions will be sufficiently informative:

D-Glucose + HI/red Phos., 100°C., → 2-iodohexane or 3-iodohexane, and 2- or 3-iodohexane + HI/red Phos., heat → n-hexane.

Since n-hexane is known to be $CH_3 \cdot CH_2 \cdot CH_2 \cdot CH_2 \cdot CH_2 \cdot CH_3$, the carbon atoms in the glucose molecule must form a continuous chain and the skeleton formula becomes $C \cdot C \cdot C \cdot C \cdot C \cdot C$.

4. The semiconstitutional formula is shown to be $C_6H_7O(OH)_5$:

Since the carbon atoms in the glucose molecule are linked to form a continuous chain, and since there is one oxygen atom and two hydrogen atoms for each carbon atom, the presence of hydroxyl groups is to be expected. The number of hydroxyl groups may be determined quantitatively by acylation.

Upon acetylation of D-glucose, an acetate derivative is obtained which is found to have a molecular weight of about 390. This is an increase of 210 over the molecular weight of D-glucose. Each acetyl group $(CH_3 \cdot CO \cdot)$ introduced replaces one hydrogen atom and represents, therefore, an increase of 42 in the molecular weight of the compound. Since $210 \div 42 = 5$, it appears that five acetyl groups have been introduced into the molecule to give a penta-acetyl-D-glucose. Inasmuch as this reaction (in the absence of water, amines, phenols, or other interfering substances), is specific for the alcoholic OH group, five hydroxyl groups must exist in the D-glucose molecule, and the semiconstitutional formula becomes $C_6H_7O(OH)_5$.

5. The constitutional formula is shown to be $CH_2OH(CHOH)_4CHO$:

Up to this point, one oxygen atom in the D-glucose molecule has not been identified in terms of its chemical properties. Numerous reactions of D-glucose indicate, however, that it contains either an aldehyde or a ketone group. The question "which one?" was answered by Kiliani. He treated D-glucose with hydrogen cyanide, hydrolyzed to the corresponding acid, reduced with hydrogen iodide in the presence of red phosphorus, and obtained n-heptanoic acid $(CH_3 \cdot CH_2 \cdot CH_2 \cdot CH_2 \cdot CH_2 \cdot CH_2 \cdot CO \cdot OH)$. Any ketose would have given a branched-chain acid. D-Glucose, therefore, must be a polyhydroxy aldehyde. Since five hydroxyl groups are present, and since the Kiliani synthesis contradicts the possibility of two or more of the OH groups being attached to the same carbon atom, they must be distributed so as to link one with each of the five remaining carbon atoms. Filling in with the remaining six hydrogen atoms, the constitutional formula is found to be $CH_2OH \cdot CHOH \cdot CHOH \cdot CHOH \cdot CHOH \cdot CHO$ or CH_2OH $(CHOH)_4CHO$.

6. *The semistructural formula is shown to be*

$CH_2OH \cdot HCOH \cdot HCOH \cdot HOCH \cdot HCOH \cdot CHO:$

Emil Fischer, reasoning from the synthesis of sugars by the use of the Kiliani synthesis and from the van't Hoff-LeBel hypothesis which calls for isomeric forms of asymmetric carbon groupings (the number of isomers being 2^n, where n represents the number of asymmetric carbon atoms when the terminal groups are different), predicted the possibilities indicated in the following diagrammatic representation:

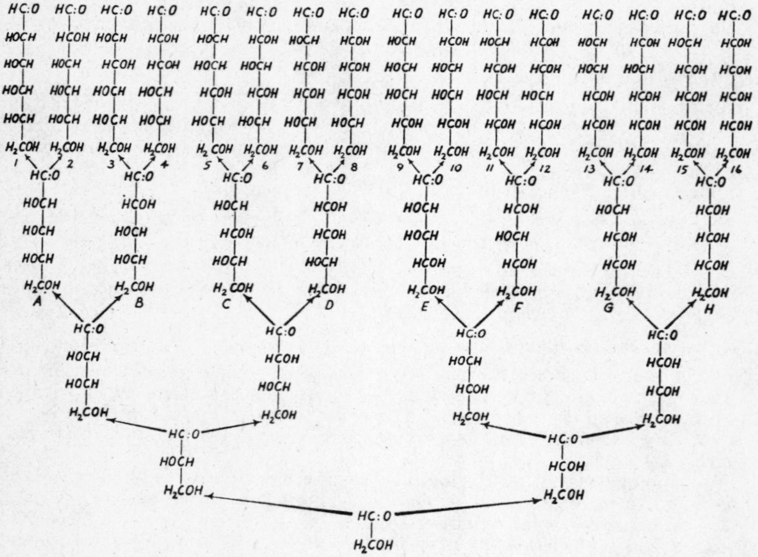

This diagram indicates the existence of sixteen stereoisomeric aldohexoses, which have all been synthesized subsequently in the laboratory. Formulas 1 to 8 have the OH of the lower asymmetric grouping to the left and were arbitrarily assigned by Rosanoff (1906) to the *L*-series since they were derived from *l*-glyceraldehyde. Formulas 9 to 16 were assigned, accordingly, to the *D*-series. Fischer was confronted, therefore, with the problem of selecting one of the formulas of the *D*-series to represent *D*-glucose. Its mirror image in the *L*-series would automatically identify the formula for *L*-glucose. This problem was solved by Emil Fischer by reasoning from experimental facts. For the purpose of the present discussion, the following facts are available:

a. 1. *D*- or *L*-galactose + HNO_3 (sp.g. 1.15) → an inactive diacid (mucic acid)

2. *D*- or *L*-galactose + HCN + hyd. + HNO_3 (sp.g. 1.15) → two active diacids

3. D- and L-galactose + epimeric change → D- and L-talose, respectively

4. D- or L-allose + HNO_3 (sp.g. 1.15) → an inactive diacid (allomucic acid)

5. D- and L-allose + epimeric change → D- and L-altrose, respectively

b. 1. D- and L-glucose + inversion → L- and D-gulose, respectively

2. D-glucose + H_2NOH + $Ac_2O/AcONa$ + $NH_3/AgNO_3/EtOH$ + HCl, aq. → D-arabinose

L-gulose + H_2NOH + $Ac_2O/AcONa$ + $NH_3/AgNO_3/EtOH$ + HCl, aq. → L-xylose

3. D-arabinose + HNO_3 (sp.g. 1.15) → an active diacid (d-trihydroxyglutaric acid)

L-xylose + HNO_3 (sp.g. 1.15) → an inactive diacid (i-trihydroxyglutaric acid)

4. D- and L-glucose + epimeric change → D- and L-mannose, respectively

5. D- and L-gulose + epimeric change → D- and L-idose, respectively.

Interpretation of these facts, considered in the order listed, lead to the following conclusions:

a. 1. A compound, even though it contains asymmetric groupings, is optically inactive if it is symmetrical with respect to a point, a line, or a plane. Formula 1, by conversion to the formula of the corresponding diacid, takes the form shown at the right which is symmetrical with respect to a line. That is, the upper half of the formula is the mirror image of the lower half of the formula, and the corresponding compound is optically inactive. Applying this test to the corresponding formulas for the diacids of the sixteen possible aldohexoses, it becomes apparent that the pairs 1, 16 and 7, 10 are the only ones that will satisfy the conditions of fact *a*. 1. The galactose pair, consequently, must be 1, 16 or 7, 10.

$$\begin{array}{c} CO \cdot OH \\ HO \cdot C \cdot H \\ HO \cdot C \cdot H \\ \hline HO \cdot C \cdot H \\ HO \cdot C \cdot H \\ CO \cdot OH \end{array}$$

2. Carrying out the reactions indicated in fact *a*. 2, on the compound represented by formula 1, should yield isomeric compounds with the formulas shown at the right. An examination of these formulas reveals the fact that only one of them represents an optically active compound. Hence formula 1 cannot represent the structure of L-galactose. By use of the same line of reasoning, it is found that the pair 7, 10 represent respectively L- and D-galactose.

$$\begin{array}{cc} CO \cdot OH & CO \cdot OH \\ HO \cdot C \cdot H & H \cdot C \cdot OH \\ HO \cdot C \cdot H & HO \cdot C \cdot H \\ HO \cdot C \cdot H & HO \cdot C \cdot H \\ HO \cdot C \cdot H & HO \cdot C \cdot H \\ HO \cdot C \cdot H & HO \cdot C \cdot H \\ CO \cdot OH & CO \cdot OH \end{array}$$

3. Epimers (so named by Votocek in 1911) are shown at the right. Epimeric change involves the transposition of the hydrogen atom and the hydroxyl group on the alpha carbon atom ($HO \cdot C \cdot H \rightleftarrows H \cdot C \cdot OH$). By such a transposition on formula 7, one obtains formula 8, and the corresponding change on formula 10 gives formula 9. The 8, 9 pair, therefore, represent respectively the L- and D-taloses.

$$\begin{array}{cc} CHO & CHO \\ HO \cdot C \cdot H & H \cdot C \cdot OH \\ H \cdot C \cdot OH & H \cdot C \cdot OH \\ H \cdot C \cdot OH & H \cdot C \cdot OH \\ HO \cdot C \cdot H & HO \cdot C \cdot H \\ CH_2OH & CH_2OH \end{array}$$

4. As previously shown, the only formulas that represent compounds which would give inactive diacids are 1, 7, 10, and 16. The 7, 10 pair, however, has been assigned to L- and D-galactose. Hence the structures represented by formulas 1 and 16 must be assigned to L- and D-allose.

5. Epimeric change on the compound represented by formula 1 would give a product with a structure corresponding to that represented by formula 2. Formula 2 must be assigned, therefore, to L-altrose and D-altrose must be represented by formula 15.

   Up to this point, formulas 1, 2, 7, 8, 9, 10, 15, and 16 have been assigned to definite sugars other than L- or D-glucose. This leaves formulas 3, 4, 5, 6, 11, 12, 13, and 14 for further consideration.

b. 1. *Inversion* involves a transposition of the terminal groups of an aldose ($CHO \rightarrow CH_2OH$, and $CH_2OH \rightarrow CHO$). The same effect is obtained in so far as the formula is concerned if the asymmetric nucleus is rotated 180° in the plane of the paper. Rotation of the asymmetric nucleus of formula 3, for example, gives formula 12, formula 4 gives itself, formula 5 gives 14, and formula 6 gives itself. These relationships, in conformity with fact *b.* 1, mean that if formula 3 represents L-glucose, formula 12 must represent D-gulose. Likewise, if formula 5 represents L-glucose, formula 14 must represent D-gulose.

2. Relationship with the pentose sugars must now be established. D-glucose, when subjected to the Wohl degradation, gives D-arabinose, whereas L-gulose is degraded to L-xylose. The degradation product of the structure represented by formula 3 is indicated by B, and that from 5 by C. Hence if formula 3 represents the structure of L-glucose, formula B must represent that of L-arabinose. Likewise, if formula 5 represents L-glucose, then the structure of L-arabinose must be represented by formula C. It becomes necessary, therefore, to determine whether formula B or formula C must be assigned to L-arabinose.

3. Since L-arabinose and L-xylose must be represented by these two formulas, and since L-arabinose gives an active diacid whereas L-xylose gives an inactive diacid, it is apparent that formula B must represent L-arabinose and formula C must represent L-xylose. If, however, L-arabinose is represented by formula B, L-glucose must be represented by formula 3, and formula 14 must be assigned to D-glucose. Likewise, if formula C is assigned to L-xylose, the structure of L-gulose is correctly represented by formula 5, and that of D-gulose by formula 12.

4. The formula which represents the epimer of L-glucose is 4, and must be assigned, therefore, to L-mannose and D-mannose must be represented by formula 13.

5. The epimer of L-gulose must be represented by formula 6. Consequently, L-idose is represented by formula 6, and formula 11 must be assigned to D-idose.

   It was by some such line of reasoning that Fischer assigned the sixteen possible aldohexose structures to specific sugars, and established the fact, since confirmed by independent proofs, that the semistructural formula for D-glucose is correctly represented by $CH_2OH \cdot HCOH \cdot HCOH \cdot HOCH \cdot HCOH \cdot CHO$.

7. *The structure of D-glucose in aqueous solution is best represented by the following equilibrium:*

$$CH_2OH \cdot HCOH \cdot HCOH \cdot HOCH \cdot HCOH \cdot CHO \text{ (aldehyde formula)}$$

$$+ H_2O \downarrow\uparrow \quad -H_2O$$

$$\begin{array}{ccc}
\text{alpha-D-Glucose, } [\alpha]_D^{20} = & \text{(Aldehydrol} & \text{beta-D-Glucose, } [\alpha]_D^{20} = +17.5 \\
+113.4 & \text{formula)} & \\
\text{(pyranoid ring formula)} & & \text{(pyranoid ring formula)}
\end{array}$$

Some of the experimental facts which demand the representation of two distinct forms of D-glucose, are as follows:

a. D-Glucose is known in two distinct forms.

b. The comparative ease of glucoside formation indicates the presence of one active hydrogen atom in the molecule.

c. By changing the conditions of alkylation, distinctly different alkyl α-D-glucosides or alkyl β-D-glucosides may be obtained.

d. D-Glucose and most of the reducing sugars, show *mutarotation.*

Obviously enough, these facts call for an extension of the semistructural formula assigned to D-glucose by Emil Fischer. Consideration of these facts in the order listed, lead to the following conclusions:

a. By the formation of an oxide ring, the carbonyl carbon atom has become asymmetric and has given rise to two distinct forms which have been designated as α-D-glucose and β-D-glucose.

b. It is observed that the *carbonylic hydroxyl group* (the OH group linked to the original carbonyl carbon atom) is neither of the primary, secondary nor tertiary types. Inasmuch as an oxygen atom is linked to the same carbon atom, the carbonylic hydroxyl group might be expected to be more reactive than the rest of the hydroxyl groups. Hence the ease with which the monoalkyl glucosides are formed can be satisfactorily explained.

c. By changing the conditions of alkylation, it is quite probable that a shift in the oxide ring might be effected. Theoretically, the oxide ring might be formed between the terminal group and any one of the other carbon atoms of the chain. Such possibilities account for the formation of abnormal alkyl *alpha-* and *beta-*glucosides.

d. By postulating the equilibrium shown above, mutarotation can be satisfactorily explained. Freshly prepared solutions of α-D-glucose show a specific rotation of +113.4, but upon standing this value drops to a constant at + 52.5. Freshly prepared solutions of β-D-glucose, on the other hand, give an initial value of +17.5 and a final value of +52.5. It appears, therefore, that +52.5 represents the specific rotation of an equilibrium mixture. This change in specific rotation, which continues until equilibrium is established is caused by a change in the relative proportions of the optically active isomers, and is defined as *mutarotation.*

It appears, therefore, that D-glucose in aqueous solution is best represented by the equilibrium system postulated above which contains a minimum concentration of the free aldehyde form. Crystalline D-glucose may be α-D-glucose hydrate, α-D-glucose, β-D-glucose, or β-D-glucose hydrate depending on the conditions of crystallization.

8. *Ordinary α- or β-D-glucose is shown to contain a pyranoid ring:*

A number of independent proofs have shown that the pyranoid ring represents the stable form of *D*-glucose. Starting with normal methyl α-*D*-glucoside, Helferich demonstrated this by the following series of reactions:

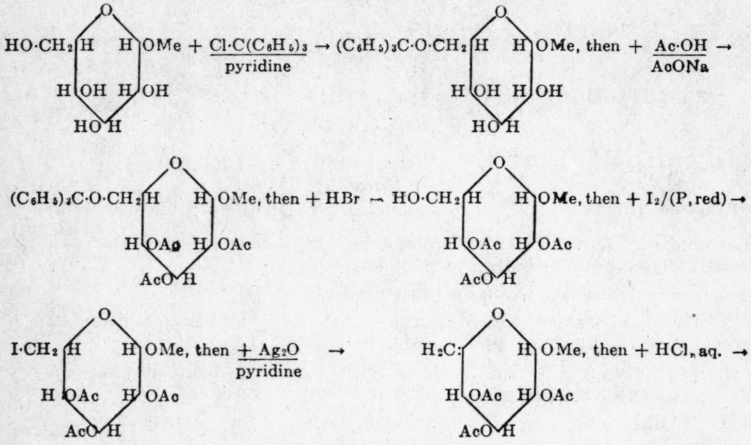

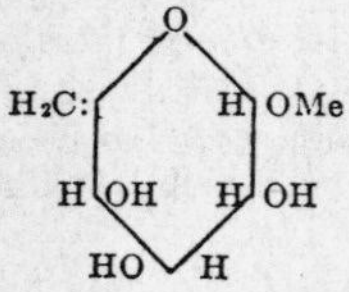

$HO \cdot CH_2$|H   H|OMe + $\underset{pyridine}{\underline{Cl \cdot C(C_6H_5)_3}}$ → $(C_6H_5)_3C \cdot O \cdot CH_2$|H   H|OMe, then + $\underset{AcONa}{\underline{Ac \cdot OH}}$ →
H|OH H|OH   H|OH H|OH
HO|H   HO|H

$(C_6H_5)_3C \cdot O \cdot CH_2$|H   H|OMe, then + HBr → $HO \cdot CH_2$|H   H|OMe, then + $I_2/(P, red)$ →
H|OAc H|OAc   H|OAc H|OAc
AcO|H   AcO|H

$I \cdot CH_2$|H   H|OMe, then + $\underset{pyridine}{\underline{+ Ag_2O}}$ → $H_2C{:}$ H|OMe, then + HCl_n aq. →
H|OAc H|OAc   H|OAc H|OAc
AcO|H   AcO|H

$H_2C{:}$ H|OMe
H|OH H|OH
HO|H

The final product, shown to the left, is non-enolic and non-reducing. Therefore *D*-glucose must, under these conditions, contain a pyranoid ring structure. If a furanoid ring were present, the enolic, reducing product shown at the right would be one of the end products of these reactions.

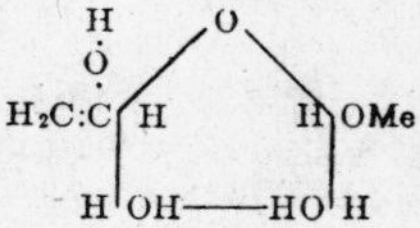

H
O
$H_2C{:}C$|H   H|OMe
H|OH——HO|H

II. ESTABLISHMENT OF THE STRUCTURE OF D⁻-FRUCTOSE:

The structure of D^--fructose may be developed (a) stepwise as in the case of *D*-glucose, or (b) by comparison with closely related sugars. Employing the latter method, it is known that D^--fructose, upon reduction, yields *D*-sorbitol and *D*-mannitol. Furthermore, D^--fructose, *D*-glucose, and *D*-mannose all yield the same ozazone. This means that the lower four carbon groupings in D^--fructose are identical with the corresponding groupings in *D*-glucose. It has been shown that D^--fructose, as a constituent of sucrose, exists in the *furanoid ring* but in the un-

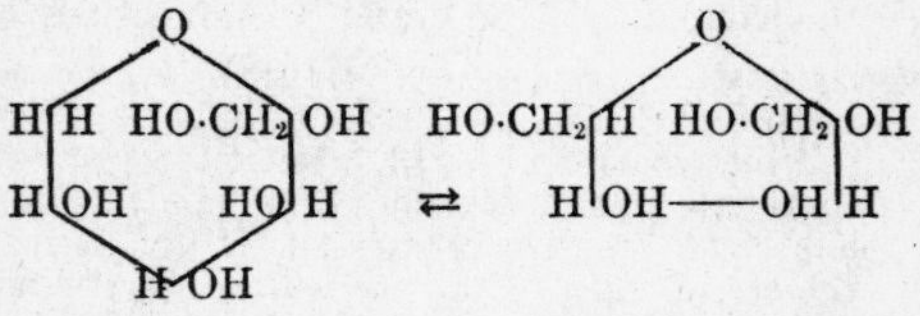

H|H $HO \cdot CH_2$|OH $HO \cdot CH_2$|H $HO \cdot CH_2$|OH
H|OH HO|H   ⇌   H|OH——OH|H
H|OH

β-δ-D^--fructose   β-γ-D^--fructose
(pyranoid ring)   (furanoid ring)

combined form it probably exists mostly in the form of the *pyranoid ring*. An equilibrium system, analogous to the one postulated for *D*-glucose, may be set up for *D⁻*-fructose.

III. ESTABLISHMENT OF THE STRUCTURE OF SOME OF THE DISACCHARIDES:

In order to assign a structure to a disaccharide, one must know (1) the component monosaccharide units, (2) the specific oxygen "bridge" in each building unit, (3) the particular carbon atoms involved in the glucosidic linkage, and (4) the specific form (alpha or beta) of one or both units.

Considerable work has been done in establishing the structures of the disaccharides. The evidence at present indicates that the best formulations of maltose (4-α-*D⁺*-glucosyl-*D⁺*-glucose), cellobiose (4-β-*D⁺*-glucosyl-*D⁺*-glucose), lactose (4-β-*D⁺*-galactosyl-*D⁺*-glucose), and sucrose (α-*D⁺*-glucosyl β-γ-*D⁻*-fructoside) are those indicated below:

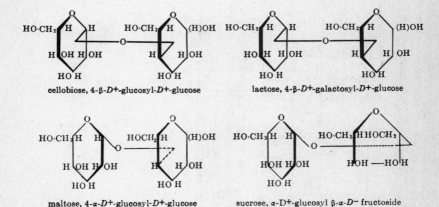

cellobiose, 4-β-*D⁺*-glucosyl-*D⁺*-glucose

lactose, 4-β-*D⁺*-galactosyl-*D⁺*-glucose

maltose, 4-α-*D⁺*-glucosyl-*D⁺*-glucose

sucrose, α-*D⁺*-glucosyl β-α-*D⁻* fructoside

As an illustration of the methods employed for establishing the structure of a disaccharide, the following specific example is given in which the facts lead to the conclusion that maltose is 4-α-D-glucosyl-D-glucose.

1. Maltose is a reducing sugar. Gram for gram, it is about one-half as efficient a reducing agent as is *D*-glucose.

2. Maltose + $CH_3 \cdot OH$/(dry HCl, reflux) → α- and β-methylmaltosides.

3. α-Methylmaltoside + complete methylation → heptamethyl-α-methylmaltoside.

4. Heptamethyl-α-methylmaltoside + H_2O/(5% HCl) → 2, 3, 6-trimethylglucose + 2, 3, 4, 6-tetramethylglucose. If both of these cleavage products exist in the form of the pyranoid ring (and there is no evidence to the contrary), the only possible linkage is the one shown above, for enzyme hydrolysis indicates that maltose contains an α-glucosidic linkage. Other independent proofs confirm this structure.

IV. ESTABLISHMENT OF THE STRUCTURE OF THE POLYSACCHARIDES:

Despite a large amount of competent research on the structures of the polysaccharides, the results have been rather inconclusive. The provisional structural units in starch and cellulose are shown below, these units being taken x times to make up the molecule:

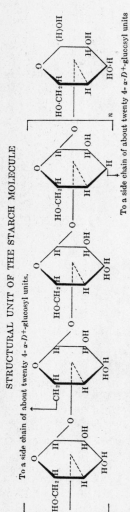

STRUCTURAL UNIT OF THE STARCH MOLECULE

This is a diagramatic representation of the structural units of starch. It is now (1941) believed that the parent chain in the starch molecule is made up of about twenty 4-α-D+-glucosyl units, and that alternate units of this parent structure have side chains of about twenty 4-α-D+-glucosyl units each. These side chains, in turn, may have additional side chains.

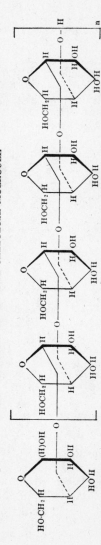

STRUCTURAL UNIT OF THE CELLULOSE MOLECULE

This is a diagramatic representation of the structural unit of cellulose, which seems to be built up of 4-β-D+-glucosyl units to give a long chain or string molecule, whereas the starch molecule has a tree-like structure.

B. Nomenclature:

Most of the members of the carbohydrate group are known by their common names. These names have been considered in connection with the development of the corresponding structures. Systematic names, however, may be used. D-Glucose ($CH_2OH\cdot HCOH\cdot HCOH\cdot HCOH\cdot HCOH\cdot CHO$), for example, may be named as pentahydroxyhexanal $+-++$, and L-glucose ($CH_2 OH\cdot HOCH\cdot HOCH\cdot HCOH\cdot HOCH\cdot CHO$) as pentahydroxyhexanal $-+--$. All of the monosaccharides may be named by use of this system.

Wohl's work on glyceraldehyde enables tracing of family relationships back to either d-glyceraldehyde or l-glyceraldehyde. The family relationship of a sugar is determined by the spatial arrangement of the CHOH group next to the terminal CH_2OH group. In the D-family of sugars, this CHOH group is written $H\cdot C\cdot OH$, and in the L-family it becomes $HO\cdot C\cdot H$. Classification on this basis simplifies structural relationships. The letters d- and l- have been used, unfortunately, to indicate both structural relationship and the

$$
\begin{array}{cc}
CHO & CHO \\
H\cdot C\cdot OH & HO\cdot C\cdot H \\
CH_2OH & CH_2OH \\
\textit{d}\text{-glyceral-} & \textit{l}\text{-glyceral-} \\
\text{dehyde} & \text{dehyde}
\end{array}
$$

sign of optical activity. If a sugar with a d-structure is levorotatory, both facts may be indicated by the symbol d'-. d'-Fructose, for example, is related structurally to the d- series but its optical activity places it in the levorotatory group. A *better practice* is to use D- and L- to show structural relationship, and to reserve d- and l- for the sign of optical activity. When it is desirable to indicate both structural relationship and the opposite sign of optical activity, the symbols D $(^+)$- and L $(^-)$- or D^-- and L^+- should be used.

Polyhydroxy aldehydes are termed *aldoses*, and polyhydroxy ketones are known as *ketoses*. The general structure of the simple sugars is indicated by such names as *aldotetrose* (aldo, signifying aldehyde and tetrose, indicating four oxygen atoms in the molecule), *ketopentose*, *aldohexose*, *ketoheptose*, and *methyl aldoheptose* (methyl, indicating the presence of a methyl group: aldo, signifying an aldehyde group; and heptose, to indicate the presence of seven oxygen atoms in the aldose radical).

The ending "ose" denotes carbohydrate material. All carbohydrates contain, actually or potentially, the "ose group" which is shown just below. The endings "itol" (formerly "ite") and "onic" are used to designate respectively the corresponding polyhydroxy alcohols and the polyhydroxy monoacids (arabitol from arabinose, and arabonic acid from arabinose). The loss of a molecule of water between

$$
\begin{array}{cc}
HO & O \\
\cdot C\cdot & C\cdot
\end{array}
$$

the carboxyl group of the polyhydroxy acid and the γ- or δ-hydroxyl group results in the formation of an oxide ring structure which is known as a *lactone*. A *lactal* (semiacetal, pyranoid or furanoid ring) contains a similar oxide ring structure which is formed, supposedly, by the migration of a hydrogen atom from the δ- or γ-hydroxy group of an aldose or ketose to the carbonyl group, or by hydration and subsequent dehydration. This results in the linkage of the carbonyl carbon atom through an oxygen atom to another carbon atom in the same carbon chain.

In order to condense formulas, as much as clarity will permit, the use of the radicals shown at the right will be employed in writing equations which involve only the activity of the carbonyl group.

Saccharin, which is the imide of o-sulfobenzoic acid and was discovered by Ira Remsen, is several hundred times as sweet as sugar, but is not related structurally to the carbohydrates.

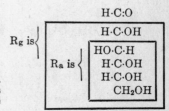

$$
R_g \text{ is} \left\{
\begin{array}{l}
H\cdot C\text{:}O \\
H\cdot C\cdot OH
\end{array}
\right.
$$

$$
R_a \text{ is} \left\{
\begin{array}{l}
HO\cdot C\cdot H \\
H\cdot C\cdot OH \\
H\cdot C\cdot OH \\
CH_2OH
\end{array}
\right.
$$

C. Preparations:

I. THE MONOSACCHARIDES MAY BE PREPARED:

1. *By the aldol condensation* (cf. Ch. VI, Div. I, Sec. E.5.a)

   a. x H·CHO/(lime water) → CH_2OH·CHO, CH_2OH·CHOH·CHO, CH_2OH·CHOH·CHOH·CHO, etc.

   b. 2 CH_2OH·CHO/ (lime water) → CH_2OH·CHOH·CHOH·CHO

   c. x CH_2OH·HCOH·CHO ⇄ y CH_2OH·CO·CH_2OH/ (lime water) → z CH_2OH·HCOH·HCOH·HOCH·CO·CH_2OH

   This reaction is of historical importance as Emil Fischer, by the condensation of *dl*-glyceraldehyde in dilute alkali, obtained *dl*-fructose as the first synthetic crystalline sugar.

2. *By the Kiliani synthesis* (R_a means CH_2OH·HCOH·HCOH·HOCH·):

$$2\,\begin{array}{c}\text{H·C:O}\\ \dot{R}_a\end{array} + \frac{3\%\ \text{HCN}}{\text{trace NH}_3} \rightarrow \begin{array}{c}\text{CN}\\ \text{H·}\dot{\text{C}}\text{·OH}\\ \dot{R}_a\end{array} + \begin{array}{c}\text{CN}\\ \text{HO·}\dot{\text{C}}\text{·H}\\ \dot{R}_a\end{array}, \frac{\text{then hydrolysis}}{\text{d·H}_2\text{SO}_4\text{, boil}} \rightarrow$$

$$\begin{array}{c}\text{CO·OH}\\ \text{H·}\dot{\text{C}}\text{·OH}\\ \dot{R}_a\end{array} + \begin{array}{c}\text{CO·OH}\\ \text{HO·}\dot{\text{C}}\text{·H}\\ \dot{R}_a\end{array}$$

The acid mixture is then treated with active brucine to give the corresponding salts. These are separated by fractional crystallization, the brucine is removed by precipitation with a base and subsequent extraction of the solution with ether, then the solution is acidified with sulfuric acid to liberate the free acids, and the polyhydroxy monoacids isolated in the form of their lactones. Selecting one of the isolated lactones, the next step in the reaction is:

$$\left.\begin{array}{l}\text{O:C}\\ \text{H·}\dot{\text{C}}\text{·OH}\\ \text{HO·}\dot{\text{C}}\text{·H}\\ \text{H·}\dot{\text{C}}\text{·O}\end{array}\right] \begin{array}{l}\\ \\ \\ \\ \text{H·}\dot{\text{C}}\text{·OH}\\ \dot{\text{C}}\text{H}_2\text{OH (}D\text{-gluconolactone)}\end{array} + 2\,\text{Na/(Hg)/2 H}_2\text{SO}_4\text{, aq.} \rightarrow \begin{array}{c}\text{O:C·H}\\ \text{H·}\dot{\text{C}}\text{·OH}\\ \text{HO·}\dot{\text{C}}\text{·H}\\ \text{H·}\dot{\text{C}}\text{·OH}\\ \text{H·}\dot{\text{C}}\text{·OH}\\ \dot{\text{C}}\text{H}_2\text{OH}\\ (D\text{-glucose})\end{array} + 2\,\text{NaHSO}_4$$

It was the use of this reaction that enabled Kiliani to prove that *D*-glucose is a polyhydroxy aldehyde.

3. *By mild oxidation, or more vigorous oxidation followed by reduction, of the polyhydroxy alcohols:*

a. Low yields of the aldoses may be obtained by direct oxidation of the corresponding alcohols:

$$CH_2OH + Br_2, aq./2 \, C_6H_5CO \cdot ONa \rightarrow H \cdot C{:}O \, (arabinose) +$$
$$\dot{R}_a \qquad\qquad\qquad\qquad\qquad\qquad \dot{R}_a$$

$$2 \, NaBr + 2 \, C_6H_5 \cdot CO \cdot OH$$

b. Much better yields of the aldoses are obtained by the oxidation of a polyhydroxy alcohol to the corresponding monoacid, and then reducing the monoacid lactone to the corresponding aldose:

$$CH_2OH + 2 \, Br_2/H_2O/4 \, C_6H_5 \cdot CO \cdot ONa, \, aq. \rightarrow O{:}C \cdot OH$$
$$\dot{R}_g \qquad\qquad\qquad\qquad\qquad\qquad\qquad\qquad \dot{R}_g$$

(gluconic acid) + 4 NaBr + 4 C₆H₅·CO·OH

The gluconic acid, through the loss of a molecule of water, changes over into the lactone which is then reduced to the corresponding aldose by the use of sodium amalgam in the presence of dilute sulfuric acid (cf. Reduction of *D*-gluconolactone, under Kiliani synthesis).

4. By use of the following intraconversion reactions:

a. Conversion of a *D*- or *L*-aldose to an *L*- or *D*-aldose (changing ends, inversion):

1. Progressive oxidation of *D*-glucose, for example, to the corresponding diacid, followed by progressive reduction, gives *L*-gulose. The consecutive steps in the reaction are: (1) oxidation with nitric acid (sp.g. 1.15) to *D*-gluconic acid, (2) loss of water to give *D*-gluconolactone, (3) oxidation with nitric acid (sp.g. 1.15) to *L*-guluronolactone, (4) oxidation with nitric acid (sp.g. 1.15) to *D*-saccharolactone, (5) reduction with sodium amalgam in dilute sulfuric acid to *D*-glucuronic acid, (6) alkaline reduction with sodium amalgam in water to sodium *L*-gulonate, (7) loss of water when acidified to give *L*-gulonolactone, and (8) reduction with sodium amalgam in dilute sulfuric acid to give *L*-gulose. The steps are indicated below by number:

Changing ends may or may not yield the same series, depending on the configuration of the sugar.

2. Inversion of aldoses may be effected by the following series of reactions in which, for example, D-glucose is changed to L-gulose. The successive steps in the reaction are (1) acylation with acetic anhydride in the presence of sodium acetate to give a mixture of the α- and β-forms of pentacetyl-D-glucose, (2) treatment with dry hydrogen bromide in acetic anhydride to give a bromo (tetracetyl-D-glucose) mixture of the α- and β-forms, (3) reaction with l-menthol in quinoline to give a mixture of the tetracetyl menthyl α- and β-D-glucosides which are separated by fractional crystallization, (4) hydrolysis of the β-form with barium hydroxide in an alcoholic-water mixture at 60°C., or of the α-form by dilution with two thousand parts of water, to give the corresponding menthyl D-glucoside, (5) oxidation with bromine in a pyridine solution of sodium hydroxide to give menthyl α-D- glucuronic acid (if the α-form is selected), (6) hydrolysis with dilute hydrochloric acid to give D-glucuronic acid, (7) alkaline reduction with sodium amalgam in water to give sodium gulonate, (8) loss of water when acidified to give L-gulonolactone, (9) reduction with sodium amalgam in dilute sulfuric acid to give L-gulose, and (10) a change to the lactal structure. The successive steps are indicated below by number:

b. Epimeric change on isomeric aldoses is effected by (1) oxidation to the corresponding monoacid (cf. C. 3. b), (2) heating in pyridine for a number of hours to give an equilibrium mixture of the isomeric acids, (3) fractional crystallization of the brucine salts of the acids (cf. C. 2), (4) liberating and isolating the acid lactone (cf. C. 2), and (5) reduction of the lactone (cf. C. 2).

c. Conversion of an aldose to the corresponding α-ketose is effected by (1) formation of the osazone by treatment with three moles of phenylhydrazine hydrochloride in the presence of sodium acetate, (2) hydrolysis of the osazone with fuming hydrochloric acid to give the osone, and (3) the reduction of the osone by the use of zinc and acetic acid to give the corresponding α-ketose. D-Glucose, for example, may be converted to D^--fructose by this series of reactions, the steps being indicated by number:

H·C $\boxed{O + H_2}$ NNH$\phi \rightarrow$ H·CNNHϕ H·C:NNHϕ H·C:NNHϕ

H·C·OH   (1)   C (O $\boxed{H)H + H·N}$ H$\rightarrow$C $\boxed{O + H_2}$ NNH$\phi \rightarrow$ C:NNHϕ

R$_a$ R$_a$ H·N·ϕ R$_a$ R$_a$

H·C $\boxed{NNH\phi + 2H}$ OH   H·C:O   CH$_2$OH

C $\boxed{NNH\phi + 2H}$ OH$\rightarrow$ C:O, then + Zn/2 CH$_3$·CO·OH, aq. $\rightarrow$C:O +

R$_a$ (2)   R$_a$ (3)   R$_a$

NH$_3$ + ϕNH$_2$ (aniline) + 4 H$_2$O + (CH$_3$·CO·O)$_2$Zn

d. Conversion of a ketose to the corresponding aldoses may be effected by (1) reduction with sodium amalgam in water to the corresponding polyhydroxy alcohols (two isomers result), (2) oxidation with bromine in the presence of sodium carbonate to the corresponding monoacids (cf. C. *3. b*), (3) fractional crystallization, liberation, and isolation of the acids in the form of their lactones (cf. C. *2*), and (4) reduction of the isolated lactone (in this case, *D*-gluconolactone) to the corresponding aldose (cf. C. *2*). The steps in the conversion of *D⁻*-fructose to *D*-glucose are indicated below by number:

CH$_2$OH   CH$_2$OH   CH$_2$OH   O:C·OH   O:C·OH

C:O   (1)   H·C·OH or HO·C·H   (2)   H·C·OH or HO·C H

R$_a$ $\rightarrow$ R$_a$ R$_a$ $\rightarrow$ R$_a$ R$_a$

O:C$\boxed{}$ O:C·H

(3)   H·C·OH   (4)   H·C·OH

$\rightarrow$ R$_{a-H}$ $\rightarrow$ R$_a$

e. Intraconversion may be effected by the use of dilute alkalies such as the alkaline earths, sodium acetate, lead oxide, guanidine, and similar substances. Under such conditions of alkalinity, Lobry de Bruyn and van Ekenstein found that the reducing sugars are all slowly transformed into mixtures of a number of isomers. For example, *D*-glucose $\rightleftarrows$ *D⁻*-fructose $\rightleftarrows$ *D*-mannose $\rightleftarrows$ other isomers.

5. *By degradation reactions, such as that of* Wohl, Ruff, Weerman, Löb, Neuberg, *or* Guebert. In that of Wohl (1) the aldose is treated with hydroxylamine (H$_2$NOH) to give the corresponding oxime, then with (2) acetic anhydride in the presence of sodium acetate (Ac$_2$O /AcONa) to remove water and acetylate the nucleus, then (3) alcoholic ammoniacal silver nitrate treatment results in the ammonolysis of the acetyl groups and the subsequent loss of hydrogen cyanide to give an aldose containing one less carbon atom, which (4) reacts with the acetamide present to give a derivative that (5) is hydrolyzed with dilute hydrochloric acid to give the aldose. *D*-Glucose, for example, may be degraded to *D*-arabinose as indicated below:

H·C:O   H·C:N·OH   C:N   C:N

H·C·OH   (1)   H·C·OH   (2)   H·C·OAc   (3) $\overline{\text{H·C·O}}$ H

HO·C·H $\rightarrow$ HO·C·H $\rightarrow$ AcO·C·H $\rightarrow$ HO·C·H

H·C·OH   H·C·OH   H·C·OAc   H·C·OH

H·C·OH   H·C·OH   H·C·OAc   H·C·OH

CH$_2$OH   CH$_2$OH   CH$_2$OAc   CH$_2$OH

$$(3) \quad \text{H·C:O}$$
$$\rightarrow \text{HO·C·H}$$
$$\text{H·C·OH}$$
$$\text{H·C·OH}$$
$$\text{CH}_2\text{OH}$$

$$(4) \quad \text{H·C (NHAc)}_2$$
$$\rightarrow \text{HO·C·H}$$
$$\text{H·C·OH}$$
$$\text{H·C·OH}$$
$$\text{CH}_2\text{OH}$$

$$\text{H·C:O}$$
$$(5) \quad \text{HO·C·H}$$
$$\rightarrow \text{H·C·OH}$$
$$\text{H·C·OH}$$
$$\text{CH}_2\text{OH}$$

6. *By hydrolysis of polysaccharides (commercial):*

   a. By hydrolytic processes, the arabinosans yield arabinose, xylosans yield xylose, dextrosans yield glucose, levulosans yield levulose (D-fructose), galactosans yield galactose, and mannosans yield mannose.

   b. Starch is hydrolyzed commercially by heating it in the presence of dilute acid. The product desired determines the time of heating. The industrial products, corn syrup (commercially known as *glucose*, which is a mixture of dextrin, maltose and D-glucose), grape sugar, and dextrin are all produced from starch by merely controlling the extent of the hydrolysis.

7. *By the condensation of aldoses with nitromethane, conversion to the salt with alkali, and subsequent hydrolysis with acid.*

II. THE DISACCHARIDES MAY BE PREPARED:

1. *By isolation from natural products:*

   a. Sucrose is obtained from the sugar cane or the sugar beet. The steps in the commercial preparation of sucrose are maceration, extraction, clarification, concentration, crystallization, separation, drying and refining.

   b. Maltose is obtained from sprouted barley or from the partial hydrolysis of starch.

   c. Lactose is obtained as a by-product in the cheese industry. The casein and fat constituents of the milk are precipitated with rennet and separated by filter presses. The filtrate (whey) is neutralized with calcium carbonate, filtered, and processed to recover the lactose. The yield, on the basis of the whole milk, is about 2.3%.

2. *By special synthesis:*

   Sucrose, maltose, and lactose, according to claims, have all been synthesized in the laboratory, but laboratory methods may never compete with the photosynthetic processes of nature.

III. THE POLYSACCHARIDES MAY BE PREPARED:

By the isolation from natural products which are, of course, the materials produced by plants through photosynthesis:

1. Dextrin is obtained in the controlled hydrolysis of starch.

2. Glycogen is obtained by extraction of liver tissue.

3. Starch is produced on a commercial scale from corn, wheat, potatoes, rice, arrowroot, and the sago palm. The process usually involves steeping in water to burst the cells, maceration in roller mills, mixing with water, sieving through screens and bolting cloth, and fractional settling to separate the gluten from the starch.

D. Physical Properties:

Monosaccharides, in general, are readily soluble in water, have a sweet taste which is intensified by increasing the number of OH groups, and are usually crystalline though they do not all crystallize readily.

Glycolic aldehyde is sweet, soluble in cold water and alcohol, but insoluble in ether. It can be distilled in vacuum and is somewhat volatile with steam. It can be obtained as a crystalline product, m.p. 95-97.

Glyceraldehyde (m.p. 138) is a white powder, not very sweet, soluble in water, slightly soluble in alcohol, and insoluble in ether.

Dihydroxyacetone is sweet, very soluble in cold water, slightly soluble in ether, almost insoluble in hot acetone, and insoluble in ligroin. Dried over sulfuric acid, it shows an indefinite melting point (68-75), which may be due to polymerization.

Arabinose is precipitated by alcohol, and yields an orange-yellow osazone. It is best distinguished from xylose by the formation and identification of its *p*-bromophenylhydrazone. Xylose is distinguished from arabinose by conversion into the double salt, cadmium bromide and cadmium xylonate (the double salt of xylonic acid).

D^--Fructose forms anhydrous crystals (from alcohol) which are somewhat hygroscopic. It forms an insoluble methylphenylhydrazone whereas the corresponding derivative of *D*-glucose is soluble. α-*D*-Glucose (m.p. 146) has a specific rotation of +113.4; β-*D*-glucose, +17.5; α, β-*D*-glucose, +52.2; methyl α-*D*-glucoside, +157; methyl β-*D*-glucoside, −33. *D*-Galactose (m.p. 118-20) crystallizes with one molecule of water of hydration. *D*-Mannose (m.p. 132) can be precipitated by alcohol.

Most of the di- and trisaccharides are soluble in water, and crystallize quite readily. Lactose loses water of hydration at about 130, turns yellow at about 160-180 due to the formation of the so-called lactocaramel which melts at about 200. Its osazone crystallizes in tufts of fine needles. Maltose loses its water of hydration at 100. It can be precipitated in wart-like anhydrous aggregates from alcohol. Its osazone forms tufts of needles somewhat blunter than those of phenyl glucosazone. Sucrose melts at 186 and then, if allowed to cool, solidifies to a white glassy mass. If heated for some time just above its melting point, it solidifies on cooling to a light brown product known as barley sugar. If heated at 210, sucrose loses water and forms caramel which is used for flavoring and coloring confections. Raffinose (m.p. 119) loses water of hydration at 100-10.

Taking sucrose as a standard at 100, the relative sweetness of the more common sugars is indicated in the following tabulation:

D^--Fructose	173.3	Rhamnose	32.5
D-Glucose	74.3	D-Galactose	32.1
D-Xylose	40.0	Raffinose	22.6
Maltose	32.5	Lactose	16.0

Polysaccharides, as a rule, are not sweet, are not soluble in water, and are not readily obtained in a crystalline form. Dextrin has an insipid mucilaginous taste, and gives a red to brown coloration with iodine in the presence of potassium iodide (unless the test is obscured by the presence of starch). Glycogen tends to give an opalescent solution which may be clarified by treating with acetic acid. With iodine in potassium iodide, glycogen gives a reddish brown coloration. Ordinary air-dried starch is tasteless. It contains about 18% of water but becomes anhydrous on heating to 110. On boiling starch with water, the granules break to give a colloidal solution which gives a blue coloration with iodine in potassium iodide. The coloration disappears on heating and reappears on cooling. Cellulose may be fibrous, cellular, or woody. There are many indications that it may be colloidal in nature. Inulin is practically tasteless and gives a negative test with iodine in the presence of potassium iodide.

E. Chemical Properties:

I. THE REDUCING SUGARS REACT UNDER PROPER CONDITIONS:

1. *To give* (a) *polyhydroxy alcohols by catalytic hydrogenation (commercial) or by treatment with sodium amalgam and water, or* (b) *hydrocarbons when treated with concentrated hydriodic acid in the presence of red phosphorus:*

    a. 1. $R_x CHO + 2 NaHg + 2 H_2O \rightarrow R_x CH_2OH + 2 NaOH + Hg$, where R_x is $CH_2OH(CHOH)_n$.

        2. $2 R_x \cdot CO \cdot CH_2OH + 4 NaHg + 4 H_2O \rightarrow R_x \cdot HCOH \cdot CH_2 OH + R_x \cdot HOCH \cdot CH_2OH$ (the two isomers are always produced) $+ 4 NaOH + Hg$

    b. 1. $R_x CHO + 3y HI / y P \rightarrow R \cdot CH_3$ (as final product) $+ y PI_3 + z H_2O$

        2. $R_x \cdot CO \cdot CH_2OH + 3y HI / y P \rightarrow R \cdot CH_2 \cdot CH_3$ (final product) $+ y PI_3 + z H_2O$

    The polyhydroxy acids are reduced under similar circumstances to the corresponding fatty acid. To reduce the osones, zinc and acetic acid are employed (cf. Sec. C. 4. c).

2. *To give oxidation products which vary with the reagent:*

    a. Aldoses are oxidized with bromine water in the presence of sodium benzoate or carbonate to give the salt of the corresponding monoacid. Benedict solution, Fehling solution, and Tollens reagent react in a similar manner and are used, due to an observable change in the metallic group, in testing for the reducing sugars. The products formed depend on the sugar, the reagent, and the treatment. Schiff reagent, though positive for the aldehydes, gives negative results with the aldoses.

    $R_x \cdot CHO + Br_2 / 2 Na_2CO_3 / H_2O \rightarrow R_x \cdot CO \cdot OH + 2 NaBr + 2 NaHCO_3$

    b. Aldoses, by protection of the carbonylic hydroxyl group through glucoside formation (usually of the menthyl type), may be oxidized by bromine in a pyridine solution of sodium hydroxide to the corresponding "uronic" acid (cf. Sec. C. I. *4. a.* 2, steps 5 and 6).

c. Aldoses, when oxidized with dilute nitric acid (sp.g. 1.15), give the lactone of the corresponding diacid (cf. Sec. C. I. *4. a.* 1, steps 1, 2, 3 and 4).

d. 1. Aldoses, when oxidized with dilute nitric acid in the absence of air, yield the corresponding α-oxo acids:

$R_x \cdot CHOH \cdot CHO + 4 \ HNO_3$, dilute, absence of air $\rightarrow$
$R_x \cdot CO \cdot CO \cdot OH + 3 \ H_2O + 4 \ NO_2$

2. Ketoses, when treated with (a) dilute nitric acid in the presence of air, or (b) mercuric oxide in the presence of barium hydroxide, yield, respectively, a mono- and a diacid, and two monoacids:

a. $CH_2OH \cdot HCOH \cdot HCOH \cdot HOCH \cdot CO \cdot CH_2OH + 8 \ HNO_3$, aq. $\rightarrow$ $HO \cdot OC \cdot HCOH \cdot HCOH \cdot CO \cdot OH$ (*meso***tartaric acid**) $+ \ HO \cdot OC \cdot CH_2OH$ (glycolic acid) $+ 8 \ NO_2 + 5 \ H_2O$, or

b. $R_x \cdot CHOH \cdot CO \cdot CH_2OH + 2 \ HgO / (OH^-, aq.) \rightarrow$ $R_x \cdot CO \cdot OH + HO \cdot OC \cdot CH_2OH + 2 \ Hg$

These two equations indicate that the cleavage products depend on the nature of the oxidant and the experimental conditions. Numerous patents have been granted on claims for definite cleavage products by the use of a given oxidizing reagent.

Fermentation reactions may be regarded as oxidation reactions that are catalyzed by enzymes or ferments.

e. The use of vigorous oxidizing agents, such as hot chromic acid, results, as a rule, in the quantitative production of carbon dioxide and water.

3. *To give addition products (a) with hydrogen cyanide, or (b) with the salts of barium, calcium, sodium, zinc, etc.*

a. The addition of hydrogen cyanide is the first step in the Kiliani synthesis (cf. Sec. C. *2*). This addition is analogous to the corresponding reaction of the aldehydes and ketones, but the aldoses and ketoses do not form additive products with the Grignard reagent, sodium hydrogen sulfite, or ammonia.

b. Pseudo addition products with salts, as illustrated by $C_6H_{12}O_6 \cdot NaCl$, are quite common among the simple sugars.

4. *To give substitution products:*

a. Of the carbonylic hydroxyl group to give glucosides (cf. C. I. *4. a.* 2)

1.

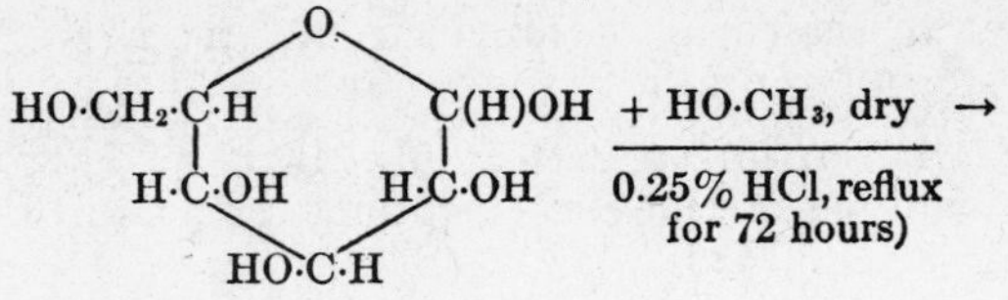

$$HO \cdot CH_2 \cdot \underset{|}{C} \cdot H \qquad \underset{|}{C}(H)OH \; + \; HO \cdot CH_3, \text{ dry } \rightarrow$$
$$H \cdot \underset{|}{C} \cdot OH \qquad H \cdot \underset{|}{C} \cdot OH \qquad \overline{0.25\% \text{ HCl, reflux}}$$
$$HO \cdot C \cdot H \qquad\qquad\qquad \text{for 72 hours)}$$

D^+-glucose

$$HO \cdot CH_2 \cdot \underset{|}{C} \cdot H \qquad H \cdot \underset{|}{C} \cdot OCH_3 \text{ and } HO \cdot CH_2 \cdot \underset{|}{C} \cdot H \quad CH_3O \cdot \underset{|}{C} \cdot H$$
$$H \cdot \underset{|}{C} \cdot OH \qquad H \cdot \underset{|}{C} \cdot OH \qquad\qquad H \cdot \underset{|}{C} \cdot OH \qquad H \cdot \underset{|}{C} \cdot OH$$
$$HO \cdot C \cdot H \qquad\qquad\qquad\qquad HO \cdot C \cdot H$$

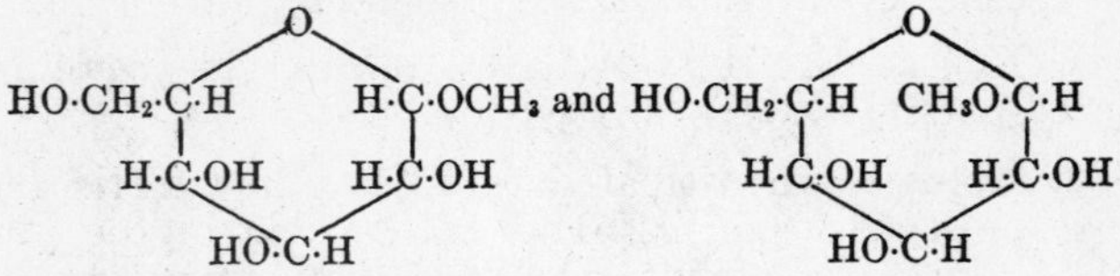

methyl α-D^+-glucoside   methyl β-D^+-glucoside

2. The phosphoric acid esters which are of considerable biochemical importance belong to this group of derivatives.

b. Of the carbonyl group when treated with (1) hydrazine or substituted hydrazines, (2) hydroxylamine, or (3) thioalcohols:

1. $R_x \cdot CHO + H_2N \cdot NH_2/(Ac \cdot OH) \rightarrow R_x \cdot C(H) : N \cdot NH_2$ (a hydrazone) + H_2O, or

$(R_x)_2C:O + H_2N \cdot NH\phi/(Ac \cdot OH) \rightarrow (R_x)_2C:N \cdot NH\phi$ (a phenylhydrazone) + H_2O

These equations are similar to the corresponding reactions of the aldehydes and ketones, but aldoses and ketoses react with three moles of phenylhydrazine to give osazones (cf. C. I. 4. c).

2. $R_x \cdot CHO + H_2NOH \rightarrow R_x \cdot C(H):NOH$ (an oxime) + H_2O, or

$(R_x)_2C:O + H_2NOH \rightarrow (R_x)_2C:NOH$ (an oxime) + H_2O

These equations should be compared with the corresponding reactions of the aldehydes and ketones. The reaction of hydroxylamine with an aldose is the first step in the Wohl degradation (cf. C. I. 5).

3. $R_x \cdot CHO + 2 \, HS \cdot CH_2 \cdot CH_3/(\text{dry HCl}) \rightarrow$
$R_x \cdot CH(S \cdot CH_2 \cdot CH_3)_2 + H_2O$

c. Of all the hydroxyl groups when treated with (1) acetic anhydride in the presence of sodium acetate, or (2) a carbon tetrachloride solution of methyl sulfate in the presence of sodium hydroxide:

1.

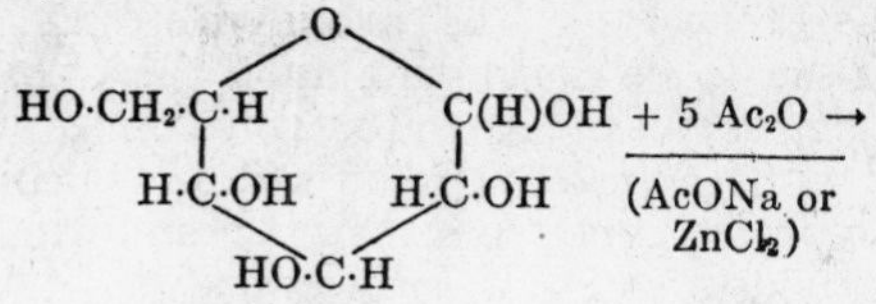

$$HO \cdot CH_2 \cdot C \cdot H \qquad C(H)OH + 5\ Ac_2O \rightarrow$$
$$H \cdot C \cdot OH \qquad H \cdot C \cdot OH \qquad \overline{(AcONa\ or}$$
$$HO \cdot C \cdot H \qquad ZnCl_2)$$

D^+-glucose

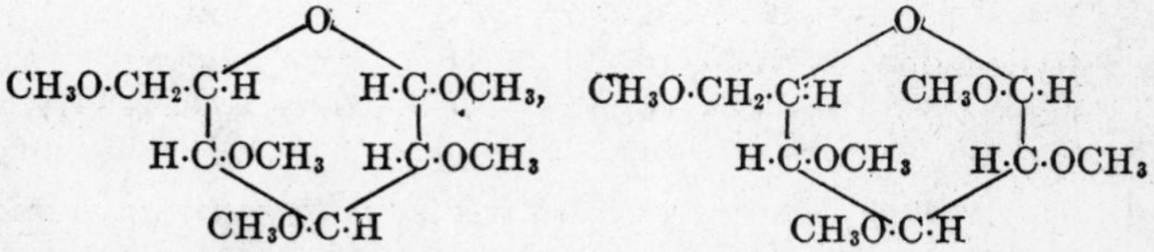

$$AcO \cdot CH_2 \cdot C \cdot H \qquad H \cdot C \cdot OAc \quad \text{and} \quad AcO \cdot CH_2 \cdot C \cdot H \qquad AcO \cdot C \cdot H$$
$$H \cdot C \cdot OAc \qquad H \cdot C \cdot OAc \qquad H \cdot C \cdot OAc \qquad H \cdot C \cdot OAc$$
$$AcO \cdot C \cdot H \qquad\qquad AcO \cdot C \cdot H$$

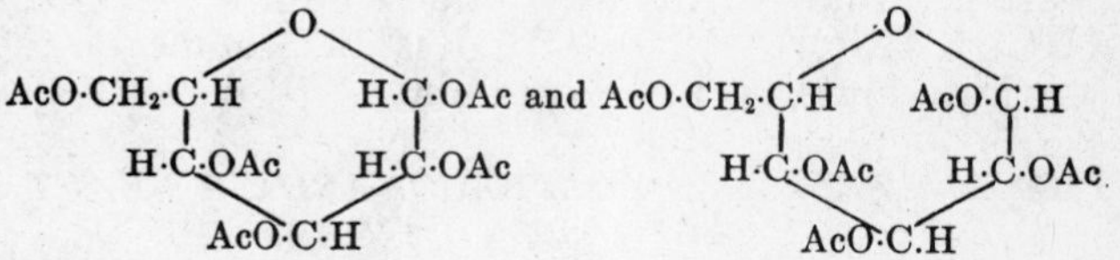

pentacetyl-α-D^+-glucose   pentacetyl-β-D^+-glucose

2.

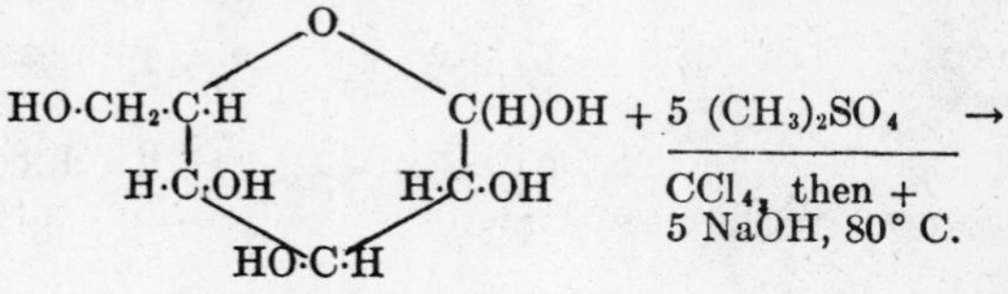

$$HO \cdot CH_2 \cdot C \cdot H \qquad C(H)OH + 5\ (CH_3)_2SO_4 \rightarrow$$
$$H \cdot C \cdot OH \qquad H \cdot C \cdot OH \qquad \overline{CCl_4,\ \text{then} +}$$
$$HO \cdot C \cdot H \qquad 5\ NaOH,\ 80°\ C.$$

D^+-glucose

$$CH_3O \cdot CH_2 \cdot C \cdot H \qquad H \cdot C \cdot OCH_3, \quad CH_3O \cdot CH_2 \cdot C \cdot H \qquad CH_3O \cdot C \cdot H$$
$$H \cdot C \cdot OCH_3 \qquad H \cdot C \cdot OCH_3 \qquad H \cdot C \cdot OCH_3 \qquad H \cdot C \cdot OCH_3$$
$$CH_3O \cdot C \cdot H \qquad\qquad CH_3O \cdot C \cdot H$$

methyl tetramethyl-α-D^+-   methyl tetramethyl-β-D^+-
glucoside   glucoside

5. *To give degradation products of the same series* when treated by the method of Wohl, Ruff, Weerman, Löb, Neuberg, or Guebert (cf. C. I. 5).

6. *To give isomeric products through intraconversion* (cf. C. I. 4).

7. *To give* (a) *furfural from the pentoses,* or (b) *(hydroxymethyl)-furfural from the hexoses, when distilled with strong mineral acids* as, for example, 12% hydrochloric acid:

a. $C_5H_{10}O_5 + \dfrac{\text{loss of 3 H}_2O}{(12\%\ HCl,\ distil)} \rightarrow$

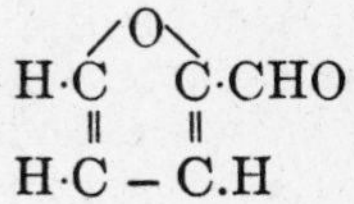

$$H \cdot C \quad C \cdot CHO \ (furfural)$$
$$H \cdot C - C \cdot H$$

b. $C_6H_{12}O_6 + \dfrac{\text{loss of 3 H}_2O}{(12\%\ HCl,\ distil)} \rightarrow$

$$HO \cdot CH_2 \cdot C \quad C \cdot CHO$$
$$H \cdot C - C \cdot H$$
(hydroxymethyl)furfural

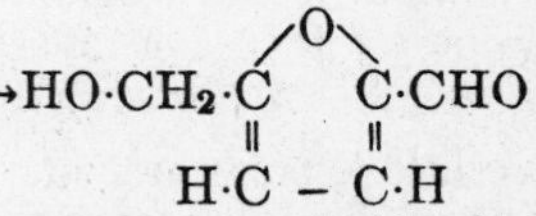

The Molisch test for carbohydrate and protein material is supposed to be dependent upon the formation of furfural or its derivatives. A water solution of the unknown is treated with a few drops of a dilute alcoholic solution of *alpha*-naphthol and concentrated sulfuric acid introduced with a pipette so as to form two distinct layers. A violet ring at the juncture of the liquids indicates the presence of carbohydrate material. Glucosamine ($CH_2OH \cdot HCOH \cdot HCOH \cdot HOCH \cdot HCNH_2 \cdot CHO$) is the source of the carbohydrate material in the protein.

II. THE DI-, TRI-, AND POLYSACCHARIDES REACT UNDER PROPER CONDITIONS:

1. *To give monosaccharides as the ultimate hydrolytic products:*

    a. Sucrose + (sucrase, or dilute acid) → *D*-glucose + *D$^-$*-fructose

     Maltose + (maltase, or dilute acid) → 2 *D*-glucose

     Lactose + (lactase, or dilute acid) → *D*-glucose + *D*-galactose

    b. Dextrin + (diastase, or dilute acid)   → x maltose
     Starch + (diastase, or acid) → dextrins → x maltose
     Glycogen + (glycogenase, or acid)   → x glucose
     Cellulose + (cellulase at 67°C., or acid) → x cellobiose, and
     Cellobiose + (cellobiase at 46°C., or acid) → 2 glucose.

2. *To give acetylated derivatives when treated with acetic anhydride:*

    Cellulose, for example, reacts with acetic anhydride in the presence of acetic acid and sulfuric acid to give cellulose acetate (and some cellobiose octa-acetate).

3. *To give the following derivatives of cellulose:*

    a. Cellulose reacts with nitric acid in the presence of sulfuric acid to give nitrocellulose, or with concentrated nitric acid (sp.g. 1.25) to give oxycellulose.

    b. Cellulose, upon prolonged heating with concentrated nitric acid, yields oxalic acid, carbon dioxide, and water. Similar results are obtained upon prolonged heating with concentrated sodium hydroxide, whereas dilute alkali gives no appreciable reaction. Ten to 28% sodium hydroxide gives a product known as sodium cellulose.

    c. Cellulose is disintegrated by Schweitzer reagent (a specially prepared solution of cupric ammonium hydroxide), from which it may be precipitated by acids.

    d. Cellulose is dissolved, due to hydrolysis, by concentrated sulfuric acid, phosphoric acid, hydrofluoric acid, and similar reagents. Vigorous oxidizing agents effect decomposition.

CHAPTER XII

ALIPHATIC SULFUR COMPOUNDS

I. Alkanethiols or Thioalcohols

A. Introduction:

The alkanethiols (**thioalcohols, mercaptans**) may be considered as hydrocarbon derivatives of hydrogen sulfide, or as derivatives of the alcohols in which the oxygen atom has been replaced by a sulfur atom. The general type formulas are $C_nH_{2n+1}SH$ or $R \cdot SH$. Like the corresponding alcohols, these derivatives may be of three types; primary ($R \cdot CH_2 \cdot SH$), secondary ($R_2CH \cdot SH$), and tertiary ($R_3C \cdot SH$).

Ethanethiol is a normal metabolic product of yeast and some of the other microörganisms and is produced, therefore, by the decaying of organic material which contains sulfur. It is also a constituent of coal gas. Organic sulfur compounds occur in petroleum, as a result of which special refining procedures have been introduced.

Due to the fact that very small traces of the **mercaptans** can be detected by odor, small traces are added to otherwise odorless but toxic refrigerants. They are also used in detecting leaks in gas mains, and find a limited use in organic synthesis.

The alkanethiols exhibit chain and position isomerism among themselves and are functional isomers of the alkylthioalkanes (**thioethers**) containing the same number of carbon atoms.

B. Nomenclature:

In naming these derivatives by the I.U.C. system, the suffix "thiol" is added to the name of the corresponding hydrocarbon. Examples are:

Structural Formulas	I.U.C. Names	Common Names
$CH_3 \cdot SH$	methanethiol	methyl mercaptan, methyl hydrosulfide, methyl thioalcohol
$CH_3 \cdot CH_2 \cdot SH$	ethanethiol	ethyl mercaptan, ethyl hydrosulfide, ethyl thioalcohol
$CH_3 \cdot CH_2 \cdot CH_2 \cdot SH$	1-propanethiol	n-propyl mercaptan
$CH_3 \cdot CH(SH) \cdot CH_3$	2-propanethiol	isopropyl mercaptan
$CH_3 \cdot CH_2 \cdot CH_2 \cdot CH_2 \cdot SH$	1-butanethiol	n-butyl mercaptan

C. Preparations:

THE ALKANETHIOLS MAY BE PREPARED:

1. By the action of sodium or potassium hydrogen sulfide on (a) alkyl halides, or (b) the salts of alkyl sulfates:

a. $R \cdot X + Na \cdot SH$, alc. $\rightarrow R \cdot SH + NaX$, and

$CH_3 \cdot CH_2 \cdot I + Na \cdot SH$, alc. $\rightarrow CH_3 \cdot CH_2 \cdot SH$ (ethanethiol, **ethyl mercaptan**) $+ NaI$

b. $R \cdot O \cdot SO_2 \cdot ONa + Na \cdot SH$, distil $\rightarrow R \cdot SH + Na_2SO_4$

2. *By the reduction of sulfonyl chlorides:*

$R \cdot SO_2 \cdot Cl + 6 Na/(Hg) + 3 H_2O \rightarrow R \cdot SH + NaCl + 5 NaOH$, and
$CH_3 \cdot CH_2 \cdot SO_2 \cdot Cl + 6 Na/(Hg) + 3 H_2O \rightarrow CH_3.CH_2 \cdot SH + NaCl + 5 NaOH$

3. *By the action of phosphorus pentasulfide on alcohols:*

$5 R \cdot OH + P_2S_5 \rightarrow 5 R \cdot SH$ (very poor yields) $+ P_2O_5$

4. *By the action of hydrogen sulfide on ethanol in the vapor phase to give ethanethiol (commercial):*

$CH_3 \cdot CH_2 \cdot OH + H \cdot SH/(ThO_2, 360°C) \rightarrow CH_3 \cdot CH_2 \cdot SH + H_2O$

D. Physical Properties:

The alkanethiols (**thioalcohols**) are generally colorless liquids which are slightly soluble in water, readily soluble in strong basic solutions, and possess extremely disagreeable odors. The boiling points are much lower than those of the corresponding alcohols due to the lack, perhaps, of highly associated molecules.

E. Chemical Properties:

THE ALKANETHIOLS REACT UNDER PROPER CONDITIONS:

1. *By replacement of* (*a*) *the hydrogen atom of the hydrosulfide group, and* (*b*) *the hydrosulfide group, as indicated below when treated with:*

a. Active metals $R \cdot S$ | $H +$ | | Na (mercaptides)

Mercuric oxide, alc. $2 R \cdot S$ | $H +$ | O | Hg (mercaptides)

Mercuric chloride $R \cdot S$ | $H +$ | Cl | $Hg \cdot Cl$

Acyl halides $R \cdot S$ | $H +$ | Cl | $OC \cdot R$ (thioesters)

Organic acids $R \cdot S$ | $H +$ | HO | $OC \cdot R$ (thioesters)

Aldehydes $2 R \cdot S$ | $H +$ | O | $C(H)R$ (bisalkylthioalkanes, **mercaptals**)

Ketones $2 R \cdot S$ | $H +$ | O | CR_2 (bisalkylthioalkanes, mercaptones, mercaptols)

b. Reducing agents R | $S \cdot H +$ | H | H

    In the Frasch process for the refining of petroleum, the **mercaptans** are converted to less objectionable compounds by the following reactions:

$2 R \cdot S \cdot H + CuO \rightarrow (R \cdot S)_2Cu + H_2O$, then $(R \cdot S)_2Cu + heat \rightarrow R_2S + CuS$, or $(R \cdot S)_2Cu + CuO \rightarrow R \cdot S \cdot S \cdot R + Cu_2O$. In the Doctor process, the following reactions are involved: $2 R \cdot SH + Pb(ONa)_2 \rightarrow Pb(SR)_2 + 2 NaOH$, then $Pb(SR)_2 + S$ (as an oxidizing agent) $\rightarrow R \cdot S \cdot S \cdot R + PbS$.

    Condensation of ethanethiol with **acetone**, followed by permanganate oxidation, yields sulfonal ($[CH_3]_2C[SO_2 \cdot CH_2 \cdot CH_3]_2$) which represents an important class of hypnotics.

2. *By oxidation with (a) nitric acid to give sulfonic acids, or (b) concentrated sulfuric acid, sulfuryl chloride, or iodine to give disulfides:*

   a. $R \cdot SH + 2 HNO_3 \rightarrow R \cdot SO_2OH + 2 NO + H_2O$, and

   $CH_3 \cdot CH_2 \cdot SH + 2 HNO_3 \rightarrow CH_3 \cdot CH_2 \cdot SO_2 \cdot OH$ (ethanesulfonic acid) + 2 NO + H_2O

   b. $2 R \cdot SNa + I_2 \rightarrow R \cdot S \cdot S \cdot R$ (ethyl disulfide) + 2 NaI

3. *By addition to ene linkages to give sulfides:*

   $R \cdot HC:CH_2 + R' \cdot S \cdot H \rightarrow R \cdot HC(S \cdot R') \cdot CH_3$ (an alkyl sulfide)

II. Alkylthioalkanes, Thioethers or Alkyl Sulfides

A. Introduction:

The alkylthioalkanes (**thioethers**) may be regarded as dialkyl derivatives of hydrogen sulfide, or as derivatives of the ethers in which the oxygen atom has been replaced by a sulfur atom. The general type formula is $R \cdot S \cdot R$.

Allyl sulfide ($[H_2C:CH \cdot CH_2]_2S$) is an ingredient of the oil of garlic, but most of the alkylthioalkanes are synthetic products.

The thioethers have a limited use in organic synthesis.

The alkylthioalkanes exhibit chain and position isomerism among themselves and are functional isomers of the alkanethiols containing the same carbon content.

B. Nomenclature:

Structural Formulas	I.U.C. Names	Common Names
$CH_3 \cdot S \cdot CH_3$	methylthiomethane	methyl sulfide
$CH_3 \cdot CH_2 \cdot S \cdot CH_2 \cdot CH_3$	ethylthioethane	ethyl sulfide
$CH_3 \cdot S \cdot CH_2 \cdot CH_2 \cdot CH_3$	1-(methylthio)propane	methyl propyl sulfide
$CH_3 \cdot CH_2 \cdot S \cdot CH_2 \cdot CH_2 \cdot CH_3$	1-(ethylthio)propane	ethyl propyl sulfide

C. Preparations:

THE ALKYLTHIOALKANES MAY BE PREPARED:

1. *By the action of (a) sodium or potassium sulfides on alkyl halides or the salts of alkyl sulfates, or (b) cadmium sulfide at 320°C. on the alkanethiols:*

   a. 1. $R \cdot X + Na \cdot S \cdot Na + X \cdot R \rightarrow R \cdot S \cdot R + 2 NaX$, and

   $2 CH_3 \cdot CH_2 \cdot Cl + Na_2S \rightarrow (CH_3 \cdot CH_2)_2S$ (ethylthioethane) + 2 NaCl.

   2. $R \cdot O \cdot SO_2 \cdot ONa + Na \cdot S \cdot Na + NaO \cdot SO_2 \cdot O \cdot R$, distil $\rightarrow R \cdot S \cdot R + 2 Na_2SO_4$

   b. $2 R \cdot SH + (CdS, 320°C.) \rightarrow R \cdot S \cdot R + H_2S$, and

   $2 CH_3 \cdot CH_2 \cdot SH + (CdS, 320°C.) \rightarrow (CH_3 \cdot CH_2)_2S$ (**ethyl sulfide**) + H_2S

   These two methods should be correlated, respectively, with the silver oxide and the sulfuric acid methods for the preparation of ethers.

2. *By the action of sodium or potassium mercaptides on alkyl halides:*

   $R \cdot S \cdot Na + X \cdot R \rightarrow R \cdot S \cdot R + NaX$, and

   $CH_3 \cdot S \cdot Na + I \cdot CH_2 \cdot CH_3 \rightarrow CH_3 \cdot S \cdot CH_2 \cdot CH_3$ (methylthioethane) + NaI

   This reaction is analogous to the Williamson synthesis, and is especially useful for the preparation of mixed sulfides.

3. By the action of phosphorus pentasulfide on ethers:

$5 R \cdot O \cdot R + P_2S_5 \rightarrow 5 R \cdot S \cdot R + P_2O_5$, and

$5 (CH_3 \cdot CH_2)_2O + P_2S_5 \rightarrow 5 (CH_3 \cdot CH_2)_2S$ (ethylthioethane) $+ P_2O_5$

4. By heating the lead mercaptides:

$(R \cdot S)_2Pb + heat \rightarrow R \cdot S \cdot R + PbS$

5. By the passage of ethene into (a) sulfur monochloride, or (b) hypochlorous acid, followed by sodium sulfide and acid treatment to give bis-2-chloroethyl sulfide (mustard gas):

a. $H_2C{:}CH_2 + S_2Cl_2 + H_2C{:}CH_2 \rightarrow Cl \cdot CH_2 \cdot CH_2 \cdot S \cdot CH_2 \cdot CH_2Cl$ (**mustard gas**) $+ S$

b. $H_2C{:}CH_2 + HO \cdot Cl \rightarrow HO \cdot CH_2 \cdot CH_2 \cdot Cl$ (**ethylene chlorohydrin**), then

$2 HO \cdot CH_2 \cdot CH_2 \cdot Cl + Na_2S \rightarrow (HO \cdot CH_2 \cdot CH_2)_2S + 2 NaCl$, and

$(HO \cdot CH_2 \cdot CH_2)_2S + 2 HCl \rightarrow (Cl \cdot CH_2 \cdot CH_2)_2S + 2 H_2O$

Mustard gas was used extensively as a liquid vesicant and poison gas during the World War.

Mustard gas is now being produced commercially by: (1) the treatment of ethylene oxide with hydrogen sulfide, and (2) the treatment of the thiodiglycol thus obtained with hydrogen chloride near the locality where the mustard gas is to be used, as:

1. $2 H_2C \overset{\displaystyle O}{\overbrace{ }} CH_2 + H_2S \rightarrow HO \cdot CH_2 \cdot CH_2 \cdot S \cdot CH_2 \cdot CH_2 \cdot OH$, bis-2-hydroxyethyl sulfide, thiodiglycol, and

2. $(HO \cdot CH_2 \cdot CH_2)_2S + 2 HCl \rightarrow (Cl \cdot CH_2 \cdot CH_2)_2S + 2 H_2O$.

This process has a decided advantage in that the thiodiglycol is safe to transport, and is readily converted to the mustard gas as needed.

D. Physical Properties:

The alkylthioalkanes (**thioethers**) are colorless liquids, and are soluble in alcohol and ether, but insoluble in water. When pure, they are characterized by an ethereal odor.

E. Chemical Properties:

THE ALKYLTHIOALKANES REACT UNDER PROPER CONDITIONS:

1. To give addition products when treated with:

   a. Salts $R_2S + HgCl_2 \rightarrow R_2S \cdot HgCl_2$

   b. Alkyl iodides $R_2S + R \cdot I \rightarrow R_2S(R)I$ (**sulfonium iodides, sulfine iodides**)

   c. Halogen $R_2S + Br_2 \rightarrow R_2S \cdot Br_2$

The sulfonium iodides react with alkalies to give **sulfonium hydroxides** which are comparatively strong bases: $R_3S \cdot I + NaOH \rightarrow R_3S \cdot OH + NaI$.

2. To give, upon oxidation, (a) alkylsulfinylalkanes (sulfoxides) with hydrogen peroxide, or (b) alkylsulfonylalkanes (sulfones) with nitric acid:

   a. $R \cdot S \cdot R + H_2O_2 \rightarrow R \cdot SO \cdot R$ (**sulfoxides**) $+ H_2O$

   b. $3 R \cdot S \cdot R + 4 HNO_3 \rightarrow 3 R \cdot SO_2 \cdot R$ (**sulfones**) $+ 4 NO + 2 H_2O$

3. To give trialkylsulfonium halides when heated with halogen acids;

$2 R_2S + HX$, heat $\rightarrow R_3S \cdot X + R \cdot SH$

III. Sulfonic Acids

A. Introduction:

The alkanesulfonic acids may be considered as derivatives of the hydrocarbons in which a hydrogen atom has been replaced by the sulfonic acid group ($\cdot SO_2 \cdot OH$).   The general type formula is $R \cdot SO_2 \cdot OH$.

2-Aminoethanesulfonic acid (taurine) is an ingredient of bile.

Industrial applications of the alkanesulfonic acids are rather limited, but the corresponding aryl derivatives are decidedly important.

The alkanesulfonic acids are functional isomers of the alkyl sulfites, and the alkyl radical gives rise to chain isomerism.

B. Nomenclature:

Type Formulas	I.U.C. Names	Common Names
$CH_3 \cdot SO_2 \cdot OH$	methanesulfonic acid	methylsulfonic acid
$CH_3 \cdot CH_2 \cdot SO_2 \cdot OH$	ethanesulfonic acid	ethylsulfonic acid
$CH_3 \cdot CH_2 \cdot CH_2 \cdot SO_2 \cdot OH$	1-propanesulfonic acid	n-propylsulfonic acid
$CH_3 \cdot CH(SO_2 \cdot OH) \cdot CH_3$	2-propanesulfonic acid	isopropylsulfonic acid

C. Preparations:

THE ALKANESULFONIC ACIDS MAY BE PREPARED:

1. By the action of metallic sulfites on alkyl halides:

$$R \cdot X + Na_2SO_3 \rightarrow R \cdot SO_2 \cdot ONa + NaX, \text{ then}$$
$$R \cdot SO_2 \cdot ONa + H^+ \rightarrow R \cdot SO_2 \cdot OH + Na^+$$

2. By the oxidation of alkanethiols (thioalcohols):

$$R \cdot SH + 2 HNO_3 \rightarrow R \cdot SO_2 \cdot OH + 2 NO + H_2O$$

3. By the action of sulfuric acid or SO_2/Cl_2 on hydrocarbons:

$$R_3C \cdot H + HO \cdot SO_2 \cdot OH \rightarrow R_3C \cdot SO_2 \cdot OH + H_2O$$

This reaction, although of great importance in the chemistry of the aromatic compounds, is still relatively unimportant in aliphatic chemistry.

$$R \cdot CH_2 \overset{\cdot}{R} + SO_2/Cl_2 + H_2O \rightarrow R \cdot CHSO_3H \cdot R + HCl$$

D. Physical Properties:

The alkanesulfonic acids are viscous liquids or crystalline solids which are quite soluble in water.

E. Chemical Properties:

THE ALKANESULFONIC ACIDS REACT UNDER PROPER
CONDITIONS:

1. To give derivatives in which (a) the hydrogen atom of the sulfonic group is replaced, and (b) the OH group is replaced, when treated with:

a. Alkalies $R \cdot SO_2 \cdot OH + NaOH \rightarrow R \cdot SO_2 \cdot ONa + H_2O$

Alkalies, then alkyl halides $R \cdot SO_2 \cdot ONa + R \cdot I \rightarrow R \cdot SO_2 \cdot O \cdot R + NaI$

b. Phosphorus pentachloride $R \cdot SO_2 \cdot OH + PCl_5 \rightarrow R \cdot SO_2Cl + POCl_3 + HCl$

2. *To give, upon reduction, (a) sulfinic acids, or (b) thioalcohols, when treated with:*

   a. PCl₅, then Zn/(H⁺) $R \cdot SO_2 \cdot Cl + Zn/2H^+$, aq. $\rightarrow R \cdot SO \cdot OH +$
                                       $Zn^{++} + H_2O + HCl$

   b. PCl₅, then Na·Hg/H₂O $R \cdot SO_2 \cdot Cl + 6 \ Na/(Hg) + 3 \ H_2O \rightarrow$
                                       $R \cdot SH + 5 \ NaOH + NaCl$

3. *To give alcohols when fused with sodium or potassium hydroxide:*

$R \cdot SO_2 \cdot OH + NaOH \rightarrow R \cdot SO_2 \cdot ONa$, then $+ NaOH$, fuse $\rightarrow R \cdot OH + Na_2SO_3 + H_2O$

These reactions have indicated that the alkanesulfonic acids have rather pronounced acidic properties.

IV. Alkyl and Alkyl Metallic Sulfates

Ethyl sulfate $(CH_3 \cdot CH_2 \cdot O \cdot SO_2 \cdot O \cdot CH_2 \cdot CH_3)$ is obtained as a by-product in the commercial preparation of ether by the sulfuric acid process, and finds use as an ethylating agent. A number of the alkyl sulfates are known, but are not commercially available. They are highly corrosive to the skin and must be handled with extreme care.

The alkyl metallic sulfates, containing an alkyl group of about ten or more carbon atoms, are important detergent and wetting agents. Sodium lauryl sulfate, which is the essential ingredient in dreft, is produced commercially from lauryl alcohol, sulfuric acid, and alkali. The lauryl alcohol is obtained by the catalytic reduction of coconut oil.

The alkyl metallic sulfates are discussed more fully in the chapter on Detergents and Wetting Agents, p. 280.

V. Other Aliphatic Sulfur Compounds

Some other sulfur compounds in the aliphatic series, and their relation to the corresponding oxygen series, are indicated below:

$R \cdot CHO$	$R \cdot CHS$	$(CH_3 \cdot CHS)_3$
alkanal	alkanethial	trimeric ethanethial
aldehyde	**thioaldehyde**	**trithioacetaldehyde**
$R \cdot CO \cdot R$	$R \cdot CS \cdot R$	$CH_3 \cdot CS \cdot CH_3$
alkanone	alkanethione	propanethione
ketone	**thioketone**	**thioacetone**
$R \cdot CO \cdot OH$	$R \cdot CO \cdot SH$	$CH_3 \cdot CO \cdot SH$
alkanoic acid	alkanethiolic acid	ethanethiolic acid
fatty acid	**thio acid**	**thioacetic acid**
$R \cdot CO \cdot OH$	$R \cdot CS \cdot OH$	$CH_3 \cdot CS \cdot OH$
alkanoic acid	alkanethionic acid	ethanethionic acid
$R \cdot CO \cdot OH$	$R \cdot CS \cdot SH$	$CH_3 \cdot CS \cdot SH$
alkanoic acid	alkanethionothiolic acid	ethanethionothiolic acid
$R \cdot CO \cdot NH_2$	$R \cdot CS \cdot NH_2$	$CH_3 \cdot CS \cdot NH_2$
alkanamide	alkanethiamide	ethanethiamide
amide	**thiamide**	**thioacetamide**
$R \cdot O \cdot C \colon N$	$R \cdot S \cdot C \colon N$	$CH_3 \cdot S \cdot C \colon N$
alkyl cyanate	alkyl thiocyanate	methyl thiocyanate
(unknown)		

R·N:C:O **alkyl isocyanate**	R·N:C:S **alkyl isothiocyanate**	CH₃·N:C:S **methyl isothiocyanate**
H₂N·CO·NH₂ **urea**	H₂N·CS·NH₂ **thiourea**	

O:C:S **carbon oxysulfide**	R·SO·R **alkylsulfinylalkane sulfoxide**	CH₃·CH₂·SO·CH₂·CH₃ **ethylsulfinylethane ethyl sulfoxide**
HO·CS·SH **dithiocarbonic acid**	R·SO₂·R **alkylsulfonylalkane sulfone**	CH₃·SO₂·CH₂·CH₃ **methylsulfonylethane methyl ethyl sulfone**
H₂N·CS·SH **(unknown)**	R₂C(SO₂·R)₂ **x,x-bis(alkylsulfonyl)- alkane disulfone**	(CH₃)₂C(SO₂·CH₂·CH₃)₂ **2,2-bis(ethylsulfonyl)- propane sulfonal**
H·S·C:N **thiocyanic acid**	R·O·CS·S·M **xanthate**	CH₃·CH₂·O·CS·S·Na **sodium ethylxanthate**

Four important rubber accelerators should be mentioned. Two of these are "Tuads" ([CH₃]₂N·CS·S·S·SC·N [CH₃]₂) and "Thionex" ([CH₃]₂N·CS·S·SC·N[CH₃]₂). The reactions for their preparation are:

$$(CH_3)_2N·H + CS_2/NaOH \rightarrow (CH_3)_2N·CS·S·Na + H_2O, \text{ then}$$

$$2 (CH_3)_2N·CS·S·Na + H_2O_2 \rightarrow (CH_3)_2N·CS·S·S·SC·N·(CH_3)_2 + 2 NaOH,$$
and $(CH_3)_2N·CS·S·S·SC·N(CH_3)_2 + NaCN, aq. \rightarrow (CH_3)_2N·CS·S·SC·N·(CH_3)_2 + NaNCS$

2-Mercaptothiazoline (2 MT) is obtained by heating ethylamine with carbon disulfide and sulfur, and 2-mercaptobenzothiazole (MBT) is prepared by heating aniline with carbon disulfide and sulfur at about 300°C. at 800 pounds pressure.

2 MT   MBT

The thiocyanates are used somewhat as insecticides.

CHAPTER XIII

ALIPHATIC NITROGEN DERIVATIVES

I. Nitroalkanes or Nitroparaffins

A. Introduction:

The nitroalkanes are saturated hydrocarbon derivatives in which one hydrogen atom has been replaced by the nitro group ($\cdot NO_2$). The general type formulas are $C_nH_{2n+1}NO_2$ or $R\cdot NO_2$. They may be classified, according to structure, as primary ($R\cdot CH_2\cdot NO_2$), secondary ($R_2CH\cdot NO_2$), and tertiary ($R_3C\cdot NO_2$).

Nitromethane is an explosive, but the higher members of the series appear to be comparatively stable. Development of commercial methods for the production of the nitroalkanes will make feasible their use as solvents, and as intermediates in organic synthesis.

The reduction of the nitroalkanes to the corresponding amines indicates that the nitrogen atom is linked directly to a carbon atom. The methods of preparation and the other chemical properties of the nitroalkanes lead to the same conclusion, and they are usually represented by the general type formula, $R\cdot NO_2$. There is some evidence, however, to the effect that the two oxygen atoms are not exactly equivalent and the nitro compounds might be represented more accurately by $R\cdot N(\rightarrow O){:}O$.

Position isomerism is illustrated by 1-nitropropane ($CH_3\cdot CH_2\cdot CH_2\cdot NO_2$) and 2-nitropropane ($CH_3\cdot CH[NO_2]\cdot CH_3$), and functional isomerism with the alkyl nitrites by nitroethane ($CH_3\cdot CH_2\cdot NO_2$) and ethyl nitrite ($CH_3\cdot CH_2\cdot O\cdot NO$).

B. Nomenclature:

The nitroalkanes are named as substitution products of the corresponding hydrocarbon: $CH_3\cdot NO_2$, nitromethane; $CH_3\cdot CH_2\cdot NO_2$, nitroethane; $CH_3\cdot CH_2\cdot CH_2\cdot NO_2$, 1-nitropropane; and $CH_3\cdot CH(NO_2)\cdot CH_3$, 2-nitropropane.

C. Preparations:

THE NITROALKANES MAY BE PREPARED:

1. By the action of silver nitrite on alkyl iodides:

$$R\cdot I + AgNO_2 \rightarrow R\cdot NO_2 \ (+ \text{ some } R\cdot O\cdot NO) + AgI, \text{ and}$$

$$CH_3\cdot CH_2\cdot I + AgNO_2 \rightarrow CH_3\cdot CH_2\cdot NO_2 \text{ (nitroethane)} + AgI$$

This reaction may be interpreted as follows:

$$x\ R\cdot I + x\ Ag\cdot O\cdot NO \rightarrow \begin{cases} y\ R\cdot O\cdot NO \text{ (alkyl nitrite)} + y\ AgI \\ z\ R\cdot N\ (:O)\cdot O\cdot Ag \rightarrow z\ R\cdot NO_2 + z\ AgI \\ \quad \overset{\pm}{I} \end{cases}$$

132

2. By the action of nitric acid on certain hydrocarbons:

$$(CH_3)_3C \cdot H + HO \cdot NO_2 \ (34\%, \ 150°C., \ 20 \ minutes) \rightarrow$$
$$(CH_3)_3C \cdot NO_2 + H_2O$$

In this reaction, in addition to obtaining the possible isomers, the hydrocarbon chain undergoes some cleavage at higher temperatures which results in the formation of simpler nitro derivatives. The ease of substitution decreases according to the type of hydrogen atom involved in the order $R_3C \cdot H$, $R_2CH \cdot H$, $R \cdot CH_2 \cdot H$. Direct vapor phase nitration of the alkanes is being used commercially.

3. By the action of sodium nitrite on the salt of chloroacetic acid, followed by hydrolysis, to give nitromethane:

$$Cl \cdot CH_2 \cdot CO \cdot ONa + Na \cdot NO_2 \rightarrow O_2N \cdot CH_2 \cdot CO \cdot ONa + NaCl, \ then$$

$$O_2N \cdot CH_2 \cdot CO \cdot ONa + H_2O, \ heat \rightarrow O_2N \cdot CH_3 + NaHCO_3$$

Dinitroalkanes and nitroalkenes have been prepared, but they are relatively less important than the nitroalkanes.

D. Physical Properties:

The nitroalkanes are colorless, pleasant-smelling, dense liquids which are sparingly soluble in water. Their densities, melting points, and boiling points are higher, in general, than those of the corresponding alkyl nitrites. These values for nitromethane are D. $1.130^{2\frac{0}{4}}$, m.p. -29.2, b.p. 101.9, whereas for methyl nitrite they are D.0.991^{15}, m.p. -17, b.p. -12.

E. Chemical Properties

THE NITROALKANES REACT UNDER PROPER CONDITIONS:

1. To give salts, of the primary and secondary compounds, when treated with alkalies:

$$R \cdot CH_2 \cdot NO_2 \rightleftharpoons R \cdot CH{:}N(\rightarrow O) \cdot OH + NaOH \rightarrow R \cdot CH{:}N(\rightarrow O) \cdot ONa$$
$$+ H_2O, \ or$$

$$R_2CH \cdot NO_2 \rightleftharpoons R_2C{:}N(\rightarrow O) \cdot OH + NaOH \rightarrow R_2C{:}N(\rightarrow O) \cdot ONa$$
$$+ H_2O$$

The organic products are the salts of the nitronic acids.

2. To give substitution products, of the primary and secondary compounds, when treated with (a) halogen, or (b) nitrous acid:

a. $R \cdot CH_2 \cdot NO_2 + Br_2 \rightarrow R \cdot CHBr \cdot NO_2 + HBr$, and

$R \cdot CHBr \cdot NO_2 + Br_2 \rightarrow R \cdot CBr_2 \cdot NO_2 + HBr$, or

$R_2CH \cdot NO_2 + Br_2 \rightarrow R_2CBr \cdot NO_2 + HBr$

b. $R \cdot CH_2 \cdot NO_2 + HO \cdot NO \rightarrow R \cdot C({:}N \cdot OH) \cdot NO_2 + H_2O$, or

$R_2CH \cdot NO_2 + HO \cdot NO \rightarrow R_2C(\cdot NO) \cdot NO_2 + H_2O$

3. To give, by condensation with aldehydes, (a) nitroalcohols, or (b) nitroölefins, depending on the conditions:

a. $R \cdot CH_2 \cdot NO_2 + R' \cdot CHO/(KHCO_3) \rightarrow R' \cdot CH(OH)$, then +

$$\qquad\qquad\qquad\qquad\qquad\qquad\qquad\qquad R \cdot \overset{.}{C} \cdot NO_2$$

$$\dfrac{R' \cdot CHO}{(KHCO_3)} \rightarrow \begin{array}{l} R' \cdot CH(OH) \\ R \cdot \overset{.}{C} \cdot NO_2 \\ R' \cdot \overset{.}{C}H(OH) \end{array} \qquad \overset{.}{H}$$

Only one mole of aldehyde condenses with a mole of a secondary nitro compound, but three moles condense with a mole of nitromethane. Treatment of product one with alkali and then acid yields sugars.

b. $R \cdot CH_2 \cdot NO_2 + R' \cdot CHO / (ZnCl_2) \rightarrow R \cdot C(:CHR') \cdot NO_2 + H_2O$

The tertiary nitro compounds do not enter into either of these condensations.

4. To give, by cleavage, (a) acids and hydroxylamine salts from the primary nitro compounds, and (b) aldehydes and ketones respectively from the primary and secondary compounds:

a. $R \cdot CH_2 \cdot NO_2 + H_2SO_4 + H_2O$, reflux $\rightarrow R \cdot CO \cdot OH + NH_2OH \cdot H_2SO_4$

b. $R \cdot CH_2 \cdot NO_2 + NaOH$ (about 10%) $\rightarrow R \cdot CH:N(\rightarrow O) \cdot O \cdot Na + H_2O$

Then add solution dropwise to 25% H_2SO_4:

$2 \ R \cdot CH:N(\rightarrow O) \cdot O \cdot Na + 2 \ H_2SO_4$, aq. $\rightarrow 2 \ R \cdot CHO + N_2O + 2 \ NaHSO_4 + H_2O$

5. To give, by acid reduction, the corresponding amines:
$R \cdot NO_2 + 9 \ Fe / (FeCl_2, H^+) + 4 \ H_2O \rightarrow 4 \ R \cdot NH_2 + 3 \ Fe_3O_4$

This reaction is catalyzed by ferrous chloride, and it is likely that the mechanism is rather complex.

II. The Ammonia System of Compounds

I. INTRODUCTION

Water is so common that its significance as a chemical compound is often overlooked. Although chemically pure water is rarely obtained, its solutions are abundant and widely distributed. Because of its abundance, because of its extraordinary solvent properties, and because of its tendency to react with elements and compounds, water has become the basis of the common system of chemistry.

Its dielectric properties make water a highly efficient ionizing agent. Its physical properties have been determined accurately, and they are indispensable in the study of the physical sciences.

The inorganic hydroxides and oxides may be considered as derivatives of water in which one or both of the hydrogen atoms have been replaced by other elements. Organic compounds (alcohols, ethers, esters, aldehydes, ketones, acids, and peroxides), likewise, may be thought of as derivatives of water. It might be noted that hydroxides were once thought to be "hydrated oxides" and that the carbohydrates were once thought of as "hydrated carbon."

Although water has become the basis of our common chemical system, other compounds have been and are being studied from the same viewpoint. It has been noted previously that many of the sulfur compounds might be considered as derivatives of hydrogen sulfide. Still another system of chemistry, due to the work of Franklin, Cady, Kraus, and others, is being developed in which ammonia is the basic unit. It is the purpose of this division to trace the relationship between the "water system" and the "ammonia system" of compounds.

Water and liquid ammonia are comparable in many respects. The physical constants for ammonia are, in general, lower than the corresponding constants for water. Ammonia is less convenient to use, therefore, as a reactant or as a medium in which reactions may occur.

TABLE OF PHYSICAL CONSTANTS

Constant	Liquid Ammonia	Water
Boiling point.............	$-33.35°C$........................	$100°C$
Critical temperature......	$132.°C$	$360°C$
Density at triple point.....	0.735	1.00
Dielectric constant........	18.	78.
Freezing point............	$-77.7°C$	$0.°C$
Heat of vaporization......	327 (calories per gram)................	539.
Specific heat..............	1.07 (calories/gram $-$degree).......	1.

II. TYPES OF COMPOUNDS

The following type formulas will serve to illustrate the analogy between the water and ammonia systems of compounds:

1. Interrelated derivatives:

   a. Hydrogen hydroxide, $H·OH$ (water)

      Hydrogen amine, $H·NH_2$ (ammonia)

   b. Hydroxyl hydroxide, $HO·OH$ (hydrogen peroxide)

      Aminoamine, $H_2N·NH_2$ (hydrazine)

   c. Hydroxylamine, $HO·NH_2$

2. Metallic derivatives (M is a metal):

   a. Hydroxides, $M·OH$ (as $Na·OH$) or $M(OH)_2$ (as $HO·Hg·OH$)

      Amides, $M·NH_2$ (as $Na·NH_2$) or $M(NH_2)_2$ (as $H_2N·Hg·NH_2$)

      Hydroxymetal amides, $HO·M·NH_2$ (as $HO·Hg·NH_2$)

   b. Oxides, $M·O·M'$ (as $Li·O·Li$)

      Imides, $M·NH·M'$ (as $Li·NH·Li$). Also, M_3N (as Li_3N); no oxygen analog.

3. Hydrocarbyl derivatives:

Hydrocarbyl is the radical obtained by the loss of a hydrogen atom from any hydrocarbon:

   a. Alcohols, $R·OH$ (as $CH_3·CH_2·OH$) or phenols, $Ar·OH$ (as $C_6H_5·OH$)

      Primary amines, $R·NH_2$ (as $CH_3·CH_2·NH_2$, ethylamine)

   b. Ethers, $R·O·R$ (as $CH_3·O·CH_3$, methyl ether)

      Secondary amines, $R·NH·R'$ (as $CH_3·CH_2·NH·CH_3$, ethylmethyl amine). Also, tertiary amines, R_3N, for which there is no oxygen analog.

c. Oxonium compounds, $R_2O \cdot HX$ (as $CH_3 \cdot CH_2 \cdot O \cdot CH_2 \cdot CH_3$ [HCl])

Ammonium compounds, $R \cdot NH_2 \cdot HX$, $R_2NH \cdot HX$, $R_3N \cdot HX$, $R_4N \cdot X$, and $R_4N \cdot OH$

4. *Acyl derivatives:*

    a. Hydrates, $R \cdot C(OH)_2 \cdot H$ (as $CCl_3 \cdot C(OH)_2 \cdot H$, **chloral hydrate**)

        Ammoniates, $R \cdot C(NH_2)_2 \cdot H$

    b. Aldehydes, $R \cdot CO \cdot H$ (as $CH_3 \cdot CO \cdot H$, **acetaldehyde**)

        Ammono aldehydes, $R \cdot C(:NH) \cdot H$ (as $CH_3 \cdot C(:NH) \cdot H$

    c. Ketones, $R \cdot CO \cdot R'$ (as $CH_3 \cdot CO \cdot CH_2 \cdot CH_3$, **ethyl methyl ketone**)

        Ammono ketones, $R \cdot C(:NH) \cdot R$ (as $CH_3 \cdot C(:NH) \cdot CH_2 \cdot CH_3$)

    d. Carboxylic acids, $R \cdot CO \cdot OH$ (as $CH_3 \cdot CO \cdot OH$, **acetic acid**)

        Carboxazylic acids (amides), $R \cdot CO \cdot NH_2$ (as $CH_3 \cdot CO \cdot NH_2$, **acetamide**)

    e. Esters, $R \cdot CO \cdot O \cdot R$ (as $CH_3 \cdot CO \cdot O \cdot CH_2 \cdot CH_3$, **ethyl acetate**)

        N-Alkyl amides, $R \cdot CO \cdot NH \cdot R$ (as $CH_3 \cdot CO \cdot NH \cdot CH_2 \cdot CH_3$, *N*-ethylacetamide). Also, $R \cdot CO \cdot NR_2$, for which there is no oxygen analog. Likewise, $R \cdot C(:NH) \cdot NHR$ and $R \cdot C(:NH) \cdot NR_2$ have no oxygen analogs.

5. *Substituted acids:*

Hydroxy acids, $R \cdot CHOH \cdot CO \cdot OH$ (as $CH_3 \cdot CHOH \cdot CO \cdot OH$, **lactic acid**)

Amino acids, $R \cdot CHNH_2 \cdot CO \cdot OH$ (as $CH_3 \cdot CHNH_2 \cdot CO \cdot OH$, **alanine**)

6. *Carbonic acid derivatives:*

Carbonic acid, $HO \cdot CO \cdot OH$

Carbamic acid, $HO \cdot CO \cdot NH_2$

Urea, $H_2N \cdot CO \cdot NH_2$. Also, $H_2N \cdot C(:NH) \cdot NH_2$ (guanidine), $HO \cdot CN$ (cyanic acid), and $H_2N \cdot CN$ (cyanamide), which have no oxygen analogs.

These examples have illustrated some of the more important relationships between the "water system" and the "ammonia system" of compounds. The rest of this chapter will be devoted to aliphatic nitrogen compounds, and these compounds should be correlated, as far as possible, with the corresponding oxygen compounds.

III. Amides

A. Introduction:

The amides may be regarded as derivatives of the hydrocarbons in which a hydrogen atom has been replaced by the amide group ($\cdot CO \cdot NH_2$), as derivatives of ammonia in which one or more hydrogen atoms in ammonia (NH_3) have been replaced by the acyl group ($R \cdot CO \cdot$), or as derivatives of the acids ($R \cdot CO \cdot OH$) in which the OH group has been replaced by the amino (NH_2) or substituted amino groups.

The primary amides, which are most important, may be represented by the type formula $R \cdot CO \cdot NH_2$.

The amides, with few exceptions, are produced only by organic synthesis. Urea occurs as one of the products of the catabolism of proteins.

The amides are important synthetic intermediates, the Hofmann reaction serving as a means of degrading the carbon chain. They are also very good organic solvents, though their solvent use seems to have been somewhat neglected.

The preparations and reactions of the amides will indicate that they may be regarded as acid derivatives in which the OH of the carboxyl group has been replaced by an NH_2 group, and that their structure, in general, may be represented by the type formula $R \cdot CO \cdot NH_2$. Certain of their reactions, however, suggest the possibility of two tautomeric forms: $R \cdot CO \cdot NH_2 \rightleftarrows R \cdot C$ $(:NH) \cdot OH$. The monoalkyl amides $(R \cdot CO \cdot NHR)$ are derivatives of the first form, but the imido esters $(R \cdot C(:NH) \cdot OR')$ are derivatives of the second form. The metallic derivatives (cf. E. 2. b. and c.) are doubtless the salts of the second form $(R \cdot C(:NH) \cdot O \cdot Na)$.

B. Nomenclature:

Structural Formulas	I.U.C. Names	Common Names
$H \cdot CO \cdot NH_2$	methanamide	formamide
$CH_3 \cdot CO \cdot NH_2$	ethanamide	acetamide
$CH_3 \cdot CH_2 \cdot CO \cdot NH_2$	propanamide	propionamide
$CH_3 \cdot CH_2 \cdot CH_2 \cdot CO \cdot NH_2$	butanamide	butyramide, butyric amide
$H_2N \cdot CO \cdot NH_2$	urea	urea, carbamide
$H_2N \cdot OC \cdot CO \cdot NH_2$	ethanediamide	oxamide, oxalamide
$H_2N \cdot OC \cdot CH_2 \cdot CO \cdot NH_2$	propanediamide	malonamide
$H_2N \cdot OC \cdot CH_2 \cdot CH_2 \cdot CO \cdot NH_2$	butanediamide	succinamide

C. Preparations:

THE AMIDES MAY BE PREPARED:

1. By the dry distillation of the corresponding ammonium salts:

$R \cdot CO \cdot O \cdot NH_4$, distil $\rightarrow R \cdot CO \cdot NH_2 + H_2O$, and

$CH_3 \cdot CO \cdot O \cdot NH_4$, distil $\rightarrow CH_3 \cdot CO \cdot NH_2$ (ethanamide, **acetamide**) $+ H_2O$

The higher amides are readily prepared by bubbling ammonia through the hot alkanoic acids.

2. By the action of ammonia on (a) esters, (b) anhydrides, or (c) acid halides:

a. $R \cdot CO \cdot OR' + NH_3 \rightarrow R \cdot CO \cdot NH_2 + R' \cdot OH$, and

$CH_3 \cdot CO \cdot O \cdot CH_2 \cdot CH_3 + NH_3 \rightarrow CH_3 \cdot CO \cdot NH_2 + CH_3 \cdot CH_2 \cdot OH$

b. $R \cdot CO \cdot O \cdot OC \cdot R + 2 NH_3 \rightarrow R \cdot CO \cdot NH_2 + R \cdot CO \cdot O \cdot NH_4$

c. $R \cdot CO \cdot X + 2 NH_3 \rightarrow R \cdot CO \cdot NH_2 + NH_4X$

3. By the hydration of acid nitriles:

$R \cdot C : N + H_2O / (\text{conc. } H_2SO_4 \text{ as a catalyst}) \rightarrow R \cdot CO \cdot NH_2$, and

$CH_3 \cdot CH_2 \cdot C : N + H_2O / (\text{conc. } H_2SO_4 \text{ as a catalyst}) \rightarrow CH_3 \cdot CH_2 \cdot CO \cdot NH_2$

These three reactions are applicable, in general, to the preparation of the diamides of the corresponding diacids. Urea and its derivatives will be considered in the next division of this chapter.

D. Physical Properties:

Methanamide (formamide, b.p. 210.7) is a colorless liquid, but ethanamide (acetamide, m.p. 81, b.p. 222) is a crystalline solid. The amides, in general, are crystalline solids and are soluble in alcohol and ether. The lower members of the series are water-soluble and may be distilled without decomposition.

E. Chemical Properties:

THE AMIDES REACT UNDER PROPER CONDITIONS:

1. *By the formation of unstable salts when treated with:*
   Some Acids: $\qquad$ $2\ CH_3 \cdot CO \cdot NH_2 + HCl \rightarrow (CH_3 \cdot CO \cdot NH_2)_2 \cdot HCl$

2. *By substitution of one or more of the hydrogen atoms of the NH_2 group when heated with:*

   a. Halogen $\qquad$ $R \cdot CO \cdot NH$ | H + Br | Br
   b. Active metals $\qquad$ $R \cdot CO \cdot NH$ | H + | Na
   c. Some metallic oxides $\qquad$ $2\ R \cdot CO \cdot NH$ | H + O | Hg

3. *By the replacement of the NH_2 group when treated with:*

   a. Dilute acids (hydrolysis) $\qquad$ $R \cdot CO$ | NH₂ + HCl/H | OH
   b. Dilute bases (hydrolysis) $\qquad$ $R \cdot CO$ | NH₂ + H | ONa
   c. Nitrous acid $\qquad$ $R \cdot CO$ | N | H₂ + O | N | OH

4. *By the loss of water when treated with:*

   | O H₂ + P₂O₅, P₂S₅, or PCl₅ |

   Strong dehydrating agents, heat $\quad R \cdot C \cdot N$

5. *By the loss of CO as carbonate when treated with (Hofmann reaction):*

   Bromine, then alkali $R \cdot CO \cdot NH_2 \rightarrow R \cdot CO \cdot NHBr \rightarrow R \cdot N$ | C:O | + NaO | H

   | + NaO | H

IV. Urea

A. Introduction:

Historically, biologically, academically, industrially, and politically, urea (carbamide) is a very significant compound. The announcement by Wöhler in 1828 of the relationship between urea and ammonium cyanate was a milestone in the development of organic chemistry. Biologically, urea is one of the products of catabolism. It may be considered as the diamide ($H_2N \cdot CO \cdot NH_2$) of carbonic acid. Unlike other amides, however, it has weak basic properties.

Urea, a product of animal catabolism, is a normal ingredient of the urine. An adult excretes about 25 g. of urea per day.

Urea is used extensively as a fertilizer, the ammonia being liberated by hydrolysis as indicated under E.l.*a*. It is used also as a precipitant for aluminum and ferric compounds in analytical chemistry.

The reaction of carbonyl chloride (C.3) with ammonia, and the interrelationship between ammonium cyanate and urea (C.2), together with the other reactions given below, indicate that the structure of urea is $H_2N \cdot CO \cdot NH_2$, and that it is a tautomeric form of ammonium cyanate. Guanidine ($H_2N \cdot C\ [:NH] \cdot NH_2$) is closely related to urea structurally.

B. Nomenclature:

The common name, also adopted by the I.U.C., is urea. The acyl derivatives of urea are known as ureides. $H_2N \cdot CO \cdot NH \cdot OC \cdot CH_3$, for example, is ethanoylurea (**acetylurea**). N, N-Propanedioylurea (**malonylurea**) is commonly known as **barbituric acid**.

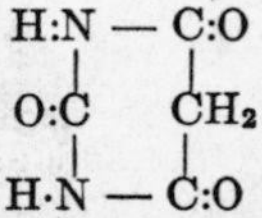

$$H:N - C:O$$
$$O:C \qquad CH_2$$
$$H \cdot N - C:O$$

barbituric acid

C. Preparations:

UREA MAY BE PREPARED:

1. *By isolation from urine in the form of its nitrate which is decomposed with barium carbonate, and the urea extracted with alcohol:*

   $H_2N \cdot CO \cdot NH_2$ (in concentrated urine) $+ HNO_3 \rightarrow H_2N \cdot CO \cdot NH_2 \cdot HNO_3$ (urea nitrate), then

   $2 H_2N \cdot CO \cdot NH_2 \cdot HNO_3 + BaCO_3$, aq. $\rightarrow 2 H_2N \cdot CO \cdot NH_2 + Ba(NO_3)_2 + CO_2 + H_2O$

2. *By heating ammonium cyanate (Wöhler synthesis):*

   $NH_4 \cdot O \cdot CN$, aq., slow evaporation $\rightarrow H_2N \cdot CO \cdot NH_2$ (urea)

3. *By the ammonolysis of (a) phosgene, or (b) ethyl carbonate:*

   a. $Cl \cdot CO \cdot Cl + 4 NH_3 \rightarrow H_2N \cdot CO \cdot NH_2 + 2 NH_4Cl$

   b. $CH_3 \cdot CH_2 \cdot O \cdot CO \cdot O \cdot CH_2 \cdot CH_3 + 2 NH_3 \rightarrow H_2N \cdot CO \cdot NH_2 + 2 CH_3 \cdot CH_2 \cdot OH$

4. *By the hydration of cyanamide:*

   $H_2N \cdot CN + H_2O/(H^+) \rightarrow H_2N \cdot CO \cdot NH_2$

5. *By the partial dehydration of ammonium carbamate:*

   $CO_2 + 2 NH_3$, 140°C., pressure $\rightleftarrows H_2N \cdot CO \cdot O \cdot NH_4$ (ammonium carbamate, commercial), then, simultaneously,

   $H_2N \cdot CO \cdot O \cdot NH_4$, heat, pressure $\rightleftarrows H_2N \cdot CO \cdot NH_2 + H_2O$

   By proper control, good yields of urea are obtained. Ammonium carbonate is a member of this equilibrium system also.

D. Physical Properties:

Urea is a white solid (m.p. 132) and is very soluble in water. It decomposes when heated above its melting point to give biuret and cyanuric acid.

E. Chemical Properties:

UREA REACTS UNDER PROPER CONDITIONS:

1. *To give typical derivatives of the amide group* (cf. Ch. XIII, Div. III, Sec. E):

   a. $H_2N \cdot CO \cdot NH_2 + 2 H_2O/2 H^+ \rightarrow CO_2 + H_2O + 2 NH_4^+$, or

   $H_2N \cdot CO \cdot NH_2 + 2 H_2O/2 OH^- \rightarrow CO_3^- + 2 NH_3 + 2 H_2O$

   Hydrolysis is catalyzed also by the enzyme *urease*.

   b. $H_2N \cdot CO \cdot NH_2 + 2 HO \cdot NO/(NaNO_2/HCl) \rightarrow CO_2 + 2 N_2 + 3 H_2O$, or

   (Van Slyke method for the determination of aliphatic *amino nitrogen*)

   $H_2N \cdot CO \cdot NH_2 + 3 Br_2/8 NaOH \rightarrow Na_2CO_3 + 6 NaBr + N_2 + 6 H_2O$

   This equation represents a modified Hofmann reaction.

   c. $3 H_2N \cdot CO \cdot NH_2 + (ZnCl_2)$, 220° C. $\rightarrow 3 HO \cdot CN \rightarrow (HOCN)_3$ (cyanuric acid) $+ 3 NH_3$

d. $H_2N \cdot CO \cdot NH_2 + R \cdot CO \cdot Cl \rightarrow H_2N \cdot CO \cdot NH \cdot OC \cdot R$ (alkanoyl urea) $+ HCl$

2. *To give, by the loss of ammonia when heated strongly, biuret and some cyanuric acid:*

$$H_2N \cdot CO \cdot NH \boxed{H + H_2N} CO \cdot NH_2 \rightarrow H_2N \cdot CO \cdot NH \cdot CO \cdot NH_2$$

(biuret) $+$

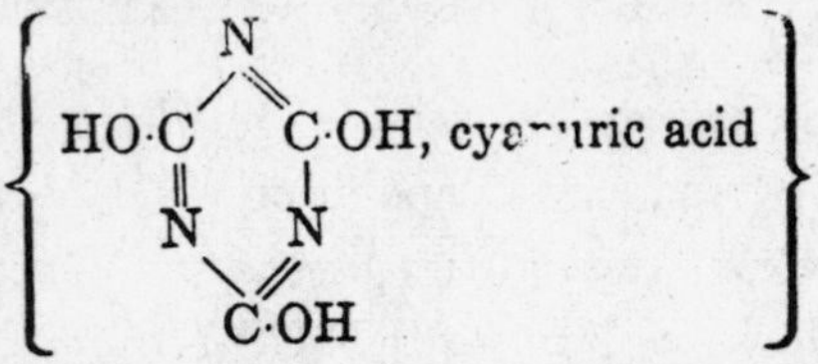

$$\left\{ \begin{array}{c} \text{N} \\ HO \cdot C \quad\quad C \cdot OH, \text{ cyan\underline{\ }\ ric acid} \\ N \quad\quad N \\ C \cdot OH \end{array} \right\} + NH_3$$

Biuret is detected by a violet-red coloration in the presence of OH^- ion and a trace of Cu^{++} ion.

3. *To give, by replacement of a hydrogen atom, nitrourea:*

$H_2N \cdot CO \cdot NH_2 \cdot HNO_3$, added to $c.H_2SO_4$, $- 3°C. \rightarrow$
$H_2N \cdot CO \cdot NH \cdot NO_2 + H_2SO_4 \cdot H_2O$

Nitrourea, upon electrolytic reduction, gives **semicarbazide** ($H_2N \cdot CO \cdot NH \cdot NH_2$) which, in the form of the sulfate or hydrochloride, is sold as a reagent for aldehydes and ketones.

V. Some Derivatives of Urea

I. UREIDES

Urea forms derivatives with the anhydrides, chlorides, or esters of various acids—acetic, oxalic, malonic, mesoxalic, succinic, etc.:

1. *Mono- and diacetylurea (ureides):*

$H_2N \cdot CO \cdot NH_2 + (CH_3 \cdot CO)_2O \rightarrow H_2N \cdot CO \cdot NH \cdot OC \cdot CH_3$ (**acetureide**) $+$ $CH_3 \cdot CO \cdot OH$, or

$H_2N \cdot CO \cdot NH_2 + 2 (CH_3 \cdot CO)_2O \rightarrow (CH_3 \cdot CO \cdot NH)_2C:O$ (**sym-diacetylurea**) $+ 2 CH_3 \cdot CO \cdot OH$

2. *Ureido acids and oxalyl urea (a cyclic ureide):*

$H_2N \cdot CO \cdot NH_2 + HO \cdot OC \cdot CO \cdot OEt/(Et \cdot ONa) \rightarrow H_2N \cdot CO \cdot NH \cdot OC \cdot CO \cdot OH$ (**oxaluric acid**) $+ Et \cdot OH$

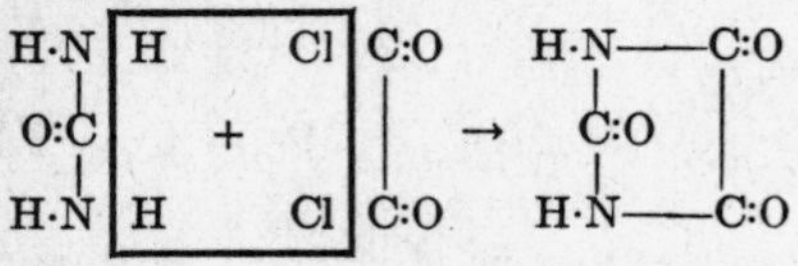

$$\begin{array}{ll} H \cdot N \boxed{H \quad Cl} C:O & H \cdot N\text{——}C:O \\ O:C \quad + \quad | & \rightarrow \quad C:O \quad | \\ H \cdot N \boxed{H \quad Cl} C:O & H \cdot N\text{——}C:O \end{array}$$ (**parabanic acid**, oxalylurea) $+ 2 HCl$

3. *Malonylurea (barbituric acid) and the Fischer-Dilthey condensation:*

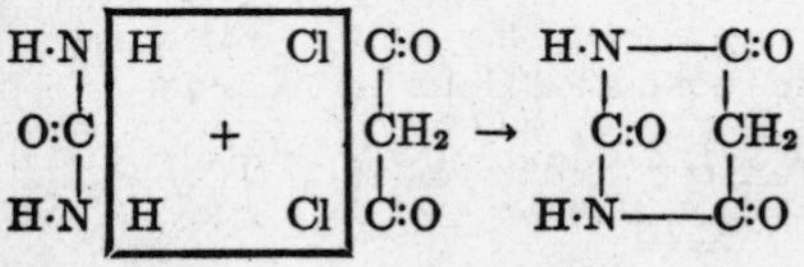

$$\begin{array}{ll} H \cdot N \boxed{H \quad Cl} C:O & H \cdot N\text{——}C:O \\ O:C \quad + \quad CH_2 & \rightarrow \quad C:O \quad CH_2 \\ H \cdot N \boxed{H \quad Cl} C:O & H \cdot N\text{——}C:O \end{array}$$ (**barbituric acid**, malonylurea) $+ 2 HCl$

By starting with alkyl derivatives of malonic ester, an analogous reaction takes place (Fischer-Dilthey condensation):

$$
\begin{array}{ll}
\text{H·N} & \text{H} \quad\quad \text{Et·O} \quad \text{C:O} \\
\text{O:C} & \quad + \\
\text{H·N} & \text{H} \quad\quad \text{Et·O} \quad \text{C:O}
\end{array}
\quad
\begin{array}{l}
\text{C:O} \\
\text{CR}_2 \rightarrow \\
\text{C:O}
\end{array}
\quad
\begin{array}{l}
\text{H·N——C:O} \\
\text{C:O} \quad \text{CR}_2 \\
\text{H·N——C:O}
\end{array}
\quad
\begin{array}{l}
\text{(dialkylbarbituric acid)} \\
+\,2\,\text{Et·OH}
\end{array}
$$

If both alkyl groups are ethyl, **barbital** (veronal) is obtained; if one is ethyl and the other phenyl, the product is **luminal**; and if one is ethyl and the other isoamyl, the resulting compound is **amytal**. These compounds are used extensively as soporifics.

II. PURINES

When barbituric acid is treated with nitrous acid, reduced, condensed with cyanic acid, and dehydrated, uric acid is obtained:

$$
\begin{array}{l}
\text{H·N——C:O} \\
\text{C:O} \quad \text{C——N·H} \\
\text{H·N——C——N·H}
\end{array}
> \text{C:O}
\qquad
\begin{array}{l}
{}_1\text{N}\!\!=\!\!{}_6\text{C·H} \\
{}_2\text{C·H} \quad {}_5\text{C——}{}_7\text{N·H} \\
{}_3\text{N——}{}_4\text{C——}{}_9\text{N}
\end{array}
> {}_8\text{C·H}
\qquad
\begin{array}{l}
\text{CH}_3\text{·N——C:O} \quad\quad \text{CH}_3 \\
\text{C:O} \quad \text{C——N} \\
\text{CH}_3\text{·N——C——N}
\end{array}
\!\!\!> \!\!\text{C·H}
$$

uric acid $\qquad\qquad$ purine $\qquad\qquad\qquad$ **caffeine**

The purine nucleus is found in a number of compounds which are known as purine derivatives. It contains residues from two molecules of **urea** and one molecule of **malonic acid**.

Purine may be prepared by treating **uric acid** with phosphorus pentachloride and subsequent reduction.

Caffeine and **theobromine** are two common alkaloids belonging to the purine group. The former, as indicated by the formula, is 1, 3, 7-trimethyl-2,6-dioxypurine, and the latter is 3, 7-dimethyl-2, 6-dioxypurine.

VI. Amines

A. Introduction:

The amines may be regarded as derivatives of the hydrocarbons in which a hydrogen atom has been replaced by an amine group, or as derivatives of ammonia in which one or more hydrogen atoms have been replaced by alkyl groups. They may be classified, according to structure, as primary ($R·CH_2·NH_2$), secondary ($R_2N·H$), and tertiary (R_3N) amines.

In a similar way ammonium hydroxide may have one or more hydrogen atoms of the NH_4 group replaced by alkyl groups to give alkylammonium ($RNH_3·OH$), dialkylammonium ($R_2NH_2·OH$), trialkylammonium ($R_3NH·OH$), and tetraalkylammonium ($R_4N·OH$) hydroxides.

Methylamine, along with other members of the series, is a degradation product of nitrogenous plant and animal substances. Decaying fish have a distinctive amine odor. The primary, secondary, and tertiary methylamines are constituents of herring brine. They are obtained, also, by the destructive distillation of the residues from the sugar beet industry.

The amines are used as condensing agents and catalysts in organic synthesis, and in the production of synthetic medicinals. Dimethylamine finds use in the compounding of rubber accelerators. The ethanolamines, alone or combined with a fatty acid, are excellent emulsifying agents and find use in dry cleaning. Benzyltrimethylammonium hydroxide is a solvent for cellulose.

Metamerism is illustrated by ethylamine ($CH_3 \cdot CH_2 \cdot NH_2$) and dimethylamine ($CH_3 \cdot NH \cdot CH_3$), or by butylamine ($C_4H_9 \cdot NH_2$) and methylpropylamine ($CH_3 \cdot NH \cdot C_3H_7$).

B. Nomenclature:

The amines are named, as indicated above, by suffixing the word "amine" to the name of the alkyl radicals involved: $CH_3 \cdot NH_2$, methylamine; $(CH_3)_2$ NH, dimethylamine; $CH_3 \cdot CH_2(CH_3)NH$, ethylmethylamine; $(CH_3)_3N$, trimethylamine; and $CH_3 \cdot CH_2(CH_3)N \cdot CH_2 \cdot CH_2 \cdot CH_3$, ethylmethylpropylamine.

The ammonium derivatives are named by suffixing the words "ammonium hydroxide" to the name of the alkyl radicals involved: $C_6H_5 \cdot CH_2(CH_3)_3$ N^+OH, benzyltrimethylammonium hydroxide.

C. Preparations:

I. THE ALKYLAMINES MAY BE PREPARED:

1. *By replacing the halogen atom in an alkyl halide by the amine group:*

   a. $R \cdot X + 2 NH_3$, heat in sealed tube $\rightarrow R \cdot NH_2$ (primary amine) $+ NH_4X$, then

   b. $R \cdot X + R \cdot NH_2$, heat in sealed tube $\rightarrow R_2NH \cdot HX$, then

   $R_2NH \cdot HX + NaOH$, aq., distil $\rightarrow R_2NH$ (secondary amine) $+ NaX + H_2O$, or

   c. $R \cdot X + R_2NH$, heat in sealed tube $\rightarrow R_3N \cdot HX$, then

   $R_3N \cdot HX + NaOH$, aq., distil $\rightarrow R_3N$ (tertiary amine) $+ NaX + H_2O$, or

   $R \cdot X + R_3N$, heat in sealed tube $\rightarrow R_4N \cdot X$, then

   $2 R_4N \cdot X + Ag_2O/H_2O$ (or $NaOH/ROH$)$\rightarrow 2 R_4N^+OH$ (tetraalkylammonium hydroxide) $+ 2 AgX$

   As indicated by these equations, this reaction gives mixtures of primary, secondary, tertiary, and quaternary compounds. Although methods have been devised for the separation of these different types from the mixture, the procedure is somewhat unsatisfactory.

2. *By the reduction of the appropriate compounds by catalytic hydrogenation or the use of certain reducing agents:*

   a. 1. $R \cdot CN + 4 Na/4 CH_3 \cdot CH_2 \cdot OH \rightarrow R \cdot CH_2 \cdot NH_2 + 4 CH_3 \cdot CH_2 \cdot O \cdot Na$

   2. $R \cdot C(:N \cdot OH) \cdot H + 4 Na/4 CH_3 \cdot CH_2 \cdot OH \rightarrow R \cdot CH_2 \cdot NH_2 + 4 CH_3 \cdot CH_2 \cdot O \cdot Na + H_2O$

   3. $R \cdot C(:N \cdot NH_2) \cdot H + 4 Na/4 CH_3 \cdot CH_2 \cdot OH \rightarrow R \cdot CH_2 \cdot NH_2 + 4 CH_3 \cdot CH_2 \cdot O \cdot Na + NH_3$

   4. $4 R \cdot CH_2 \cdot NO_2 + 9 Fe/(FeCl_2, H^+) + 4 H_2O \rightarrow 4 R \cdot CH_2 \cdot NH_2$ (primary amine) $+ 3 Fe_3O_4$

  b. $R \cdot NC + 4 Na / 4 CH_3 \cdot CH_2 \cdot OH \rightarrow R \cdot NH \cdot CH_3 +$
  $4 CH_3 \cdot CH_2 \cdot O \cdot Na$

3. By the action of (a) bromine and alkali on amides (Hofmann reaction), or (b) alkyl halides and alkali on potassium phthalimide (Gabriel synthesis), to give in each case primary amines:

  a. $R \cdot CO \cdot NH_2 + Br_2 / NaOH, aq. \rightarrow R \cdot CO \cdot NHBr + NaBr + H_2O$, then

  $R \cdot CO \cdot NHBr + NaOH, aq. \rightarrow R \cdot N:C:O$ (alkyl isocyanate) $+ NaBr + H_2O$, and

  $R \cdot N:C:O + 2 NaOH, aq. \rightarrow R \cdot NH_2$ (primary amine) $+ Na_2CO_3$

  b.

4. By the action of methanol on ammonia, in the presence of a catalyst, to give a mixture of the methyl amines (commercial):

  a. $CH_3 \cdot OH + NH_3 / (catalyst) \rightarrow CH_3 \cdot NH_2 + H_2O$, then

  b. $CH_3 \cdot OH + CH_3 \cdot NH_2 / (catalyst) \rightarrow (CH_3)_2NH + H_2O$, and

  c. $CH_3 \cdot OH + (CH_3)_2NH / (catalyst) \rightarrow (CH_3)_3N + H_2O$

II. THE ETHANOLAMINES MAY BE PREPARED:

By the action of ammonia on ethylene oxide:

  a. $H_2C \overset{O}{\cdot} CH_2 + NH_3 \rightarrow HO \cdot CH_2 \cdot CH_2 \cdot NH_2$ (ethanolamine), and

  b. $2 H_2C \overset{O}{\cdot} CH_2 + NH_3 \rightarrow (HO \cdot CH_2 \cdot CH_2)_2N \cdot H$ (diethanolamine), or

  c. $3 H_2C \overset{O}{\cdot} CH_2 + NH_3 \rightarrow (HO \cdot CH_2 \cdot CH_2)_3N$ (triethanolamine)

D. Physical Properties:

At room temperature, the lower members of the series are gases, propylamine to dodecylamine are liquids, and the higher members are solids. Those below hexylamine are readily soluble in water and more basic than ammonia. The amines possess a fishy odor, and their vapors are flammable. They are all colorless in the pure state. The quaternary ammonium compounds possess poisonous properties.

E. Chemical Properties:

THE ALKYL AMINES REACT UNDER PROPER CONDITIONS:

1. By (a) the formation of addition products, (b) the step-wise replacement of one or more of the hydrogen atoms of the amine group, (c) the simultaneous replacement of both hydrogen atoms of the NH_2 group, and (d) the replacement of the NH_2 group, as indicated below, when treated with:

Reactants:		$R \cdot NH_2+$	R_2NH+	R_3N+
a. 1. HX	→	$R \cdot NH_2 \cdot HX$	$R_2NH \cdot HX$	$R_3N \cdot HX$
2. $R \cdot X$	→	$R_2NH \cdot HX$	$R_3N \cdot HX$	$R_4N \cdot X$
b. 1. $R' \cdot CO \cdot Cl$	→	$R \cdot NH \cdot OC \cdot R'$	$R_2N \cdot OC \cdot R'$	....
2 $R \cdot CO \cdot Cl$	→	$R \cdot N(OC \cdot R')_2$	....	....
2. $R' \cdot CO \cdot O \cdot OC \cdot R'$	→	$R \cdot NH \cdot OC \cdot R'$	$R_2N \cdot OC \cdot R'$	....
2 $R' \cdot CO \cdot O \cdot OC \cdot R'$	→	$R \cdot N(OC \cdot R')_2$	....	....
c. $CHCl_3$	→	$R \cdot NC$	....	....
d. HONO (poor yields)	→	$R \cdot OH$	$R_2N \cdot NO$	$R_3N \cdot HNO_2$

The equations for the reaction of ethylamine with each of these reagents, take the form:

a. $CH_3 \cdot CH_2 \cdot NH_2 + HCl$ (or HNO_2, etc.) → $CH_3 \cdot CH_2 \cdot NH_2 \cdot HCl$ (ethylamine hydrochloride)

b. $2 CH_3 \cdot CH_2 \cdot NH_2 + CH_3 \cdot CO \cdot Cl$ → $CH_3 \cdot CO \cdot NH \cdot CH_2 \cdot CH_3$ (N-ethylethanamide) + $CH_3 \cdot CH_2 \cdot NH_2 \cdot HCl$

c. $CH_3 \cdot CH_2 \cdot NH_2 + CHCl_3/3NaOH$ → $CH_3 \cdot CH_2 \cdot NC$ (ethylcarbylamine, ethyl isocyanide) + $3 NaCl + 3 H_2O$

d. $CH_3 \cdot CH_2 \cdot NH_2 + HONO/(NaNO_2/HCl)$ → $CH_3 \cdot CH_2 \cdot OH$ (ethanol) + $N_2 + H_2O$

2. By the formation of ammonium bases with water:

$$R \cdot NH_2 + H_2O \rightleftarrows (R \cdot NH_3 \cdot OH) \rightleftarrows R \cdot NH_3^+ + OH^-, \text{ and}$$
$$R_2NH + H_2O \rightleftarrows (R_2NH_2 \cdot OH) \rightleftarrows R_2NH_2^+ + OH^-, \text{ or}$$
$$R_3N + H_2O \rightleftarrows (R_3NH \cdot OH) \rightleftarrows R_3NH^+ + OH^-.$$

The tetraalkylammonium hydroxides ($R_4N \cdot OH$) are strong bases. When subjected to pyrolysis, they undergo dissociation to give (a) tertiary amines and alcohols, or (b) tertiary amines, olefins, and water:

a. $(CH_3)_4N \cdot OH$, heat → $(CH_3)_3N + CH_3 \cdot OH$, but

b. $(CH_3 \cdot CH_2)_4N \cdot OH$, heat → $(CH_3 \cdot CH_2)_3N + H_2C{:}CH_2 + H_2O$, and

$CH_3 \cdot CH_2 (CH_3)_3N \cdot OH$, heat → $(CH_3)_3N + H_2C{:}CH_2 + H_2O$

VII. Amino Acids

A. Introduction:

The amino acids may be considered as amino and carboxyl derivatives of the alkanes, or as amino-substituted alkanoic acids. The general type formula is $R \cdot CH(NH_2)(CH_2)_x \cdot CO \cdot OH$, where x is usually zero.

The amino acids serve as the building units of the proteins. Over twenty different amino acids have been isolated from the hydrolytic products of protein material. These are, without exception, of the *alpha* type ($R \cdot CH[NH_2] \cdot CO \cdot OH$).

The α-amino acids are the building and repair units for the protein material of animal and vegetable tissues. They find a limited use in organic synthesis.

The amino acids, as indicated by their properties, are derivatives of the saturated hydrocarbons which contain both an amino and a carboxyl group. They exhibit chain, optical, position, and dynamic isomerism.

B. Nomenclature:

Structural Formulas	I.U.C. Names	Common Names
$H_2N \cdot CH_2 \cdot CO \cdot OH$	aminoethanoic acid	glycine, glycocoll
$CH_3 \cdot CH(NH_2) \cdot CO \cdot OH$	2-aminopropanoic acid	alanine
$HO \cdot CH_2 \cdot CH(NH_2) \cdot CO \cdot OH$	2-amino-3-hydroxy-propanoic acid	serine
$C_6H_5 \cdot CH_2 \cdot CH(NH_2) \cdot CO \cdot OH$	2-amino-3-phenyl-propanoic acid	phenylalanine
$p\text{-}HO \cdot C_6H_4 \cdot CH_2 \cdot CH(NH_2) \cdot CO \cdot OH$	2-amino-3-(*p*-hydrox-yphenyl) propanoic acid	tyrosine

C. Preparations:

THE AMINO ACIDS MAY BE PREPARED:

1. *By the ammonolysis of halogen-substituted acids, followed by conversion to a slightly soluble salt and subsequent treatment with an appropriate acid:*

$$R \cdot CHX \cdot CO \cdot OH + 3 \ NH_3 / \text{excess} \rightarrow R \cdot CHNH_2 \cdot CO \cdot O \cdot NH_4 + NH_4X, \text{ then}$$

$$2 \ R \cdot CHNH_2 \cdot CO \cdot O \cdot NH_4 + CuSO_4, \text{ aq.} \rightarrow (R \cdot CHNH_2 \cdot CO \cdot O)_2Cu + (NH_4)_2SO_4, \text{ and}$$

$$(R \cdot CHNH_2 \cdot CO \cdot O)_2Cu + H_2S, \text{ aq.} \rightarrow 2 \ R \cdot CHNH_2 \cdot CO \cdot OH + CuS \ (\text{insoluble})$$

2. *By the reaction of potassium phthalimide with esters of halogenated acids, followed by hydrolysis* (cf. Gabriel synthesis, Ch. XIII, Div. VI, Sec. C):

$$C_6H_4(CO)_2NNa + X(CH_2)_nCO \cdot OR \rightarrow C_6H_4(CO)_2N(CH_2)_nCO \cdot OR + NaX, \text{ then}$$

$$C_6H_4(CO)_2N(CH_2)_nCO \cdot OR + 3 \ H_2O/(H^+), \ 200°C. \rightarrow H_2N(CH_2)_n CO \cdot OH + C_6H_4(CO \cdot OH)_2 \ (\text{phthalic acid}) + R \cdot OH \ (\text{an alcohol})$$

Since the hydrolysis is carried out in acid solution, the excess acid will react with the amino group to give the salt.

3. *By the addition of hydrogen cyanide, followed by ammonolysis and subsequent hydrolysis, to (a) aldehydes, or (b) ketones to give alpha-amino acids:*

   a. $R \cdot CHO + H \cdot CN \rightarrow R \cdot CHOH \cdot CN$, then

      $R \cdot CHOH \cdot CN + NH_3 \rightarrow R \cdot CHNH_2 \cdot CN + H_2O$, then

         $R \cdot CHNH_2 \cdot CN + 2 H_2O/HX \rightarrow R \cdot CHNH_2 \cdot CO \cdot OH + NH_4X$

      The hydrogen cyanide and the ammonia for this reaction are produced by $Na \cdot CN + H_2O \rightleftarrows H \cdot CN + NaOH$, then $NaOH + NH_4Cl \rightleftarrows NaCl + NH_3 + H_2O$

   b. $R_2C{:}O + H \cdot CN$ ($Na \cdot CN/NH_4 \cdot Cl$, aq.) $\rightarrow R_2C(OH) \cdot CN$, then

      $R_2C(OH) \cdot CN + NH_3 \rightarrow R_2C(NH_2) \cdot CN + H_2O$, then

         $R_2C(NH_2) \cdot CN + 2 H_2O/HX \rightarrow R_2C(NH_2) \cdot CO \cdot OH + NH_4X$

D. Physical Properties:

Aminoethanoic acid (**glycine**) is a crystalline solid which melts with decomposition at 233°C., and is soluble in water but insoluble in alcohol and ether. The other members are, in general, crystalline, are usually sweet, and are readily soluble in water but insoluble in alcohol and ether.

E. Chemical Properties:

THE AMINO ACIDS REACT UNDER PROPER CONDITIONS:

1. *By neutralization to give (a) inner salts, (b) basic salts, or (c) acidic salts:*

   a. $R \cdot CH(NH_2) \cdot CO \cdot OH \rightleftarrows R \cdot HC(NH_3{}^+) \cdot CO \cdot O^-$ (betaine structure, inner salt)

   b. $R \cdot CH(NH_2) \cdot CO \cdot OH + NaOH$, aq. $\rightarrow R \cdot CH(NH_2) \cdot CO \cdot O \cdot Na$ (basic salt) $+ H_2O$

   c. $R \cdot CH(NH_2) \cdot CO \cdot OH + HCl$, aq. $\rightarrow R \cdot CH(NH_2 \cdot HCl) \cdot CO \cdot OH$ (acidic salt)

2. *By replacement of (a) one or more hydrogen atoms of the NH_2 group, or (b) the NH_2 group:*

   a. 1. $R \cdot CH(NH_2) \cdot CO \cdot OH + R \cdot CO \cdot X \rightarrow R \cdot CH(NH \cdot OC \cdot R) \cdot CO \cdot OH + HX$, and

      $R \cdot CH(NH \cdot OC \cdot R) \cdot CO \cdot OH + R \cdot CO \cdot X \rightarrow R \cdot CH(N[OC \cdot R]_2) \cdot CO \cdot OH + HX$

     2. $R \cdot CH(NH_2) \cdot CO \cdot OH + H \cdot CHO \rightarrow R \cdot CH(N{:}CH_2) \cdot CO \cdot OH$ (Schiff bases) $+ H_2O$

       The Schiff bases (or azomethine compounds) are stable to alkali, and the acidic group may be titrated (Sörensen formol titration).

   b. $R \cdot CH(NH_2) \cdot CO \cdot OH + HONO$ ($NaNO_2/HCl$), aq. $\rightarrow R \cdot CH(OH) \cdot CO \cdot OH + N_2 + H_2O$

3. *By replacement of the OH of the carboxyl group:*

$R \cdot CH(NH_2) \cdot CO \cdot OH + R \cdot OH/dry\ HCl \rightarrow R \cdot CH(NH_2 \cdot HCl) \cdot CO \cdot OR + H_2O$

4. *By the loss, when heated, of (a) water by the 2-amino acids, (b) ammonia by the 3-amino acids, or (c) water by the 4- or 5-amino acids:*

$$\overset{O}{\overset{\cdots}{C}}\boxed{OH} + \boxed{H}\overset{H}{\underset{\cdots}{N}}\cdot\overset{R}{\underset{\cdots}{C}}\cdot H,\ \text{heat} \rightarrow \overset{O}{\overset{\cdots}{C}}\text{———}\overset{H}{\underset{\cdots}{N}}\text{———}\overset{R}{\underset{\cdots}{C}}\cdot H + 2\ H_2O,\ \text{and}$$

a. $H\cdot\overset{R}{\underset{H}{\overset{\cdots}{C}}}\cdot N\boxed{H}\quad\boxed{HO}\overset{O}{\underset{\cdots}{C}}$ $H\cdot\overset{R}{\underset{H}{C}}\text{———}\overset{H}{\underset{\cdots}{N}}\text{———}\overset{O}{\underset{\cdots}{C}}$

$$H\cdot\overset{CH_3}{\underset{\overset{\cdots}{C}}{\overset{\cdots}{C}}}\text{———}\overset{H}{N}\boxed{H + HO}\overset{O}{\overset{\cdots}{C}}$$

$$\overset{O}{\underset{O}{\overset{\cdots}{C}}}\boxed{OH}\quad\boxed{H}\overset{H}{N}\text{———}\overset{CH_3}{\underset{\cdots}{C}}\cdot H,\ \text{heat} \rightarrow H\cdot\overset{CH_3}{\underset{\overset{\cdots}{C}}{\overset{\cdots}{C}}}\text{———}\overset{H}{N}\text{———}\overset{O}{\overset{\cdots}{C}}$$

$$\overset{\cdots}{C}\text{———}\overset{\cdots}{N}\text{———}\overset{\cdots}{C}\cdot H$$
$$\overset{\,}{O}\quad\ \overset{\,}{H}\quad\ \overset{\,}{CH_3}$$

(2-aminopropionic bimolecular cyclic anhydride, or 2,5-diketo-3,6-dimethylpiperazine)

b. $R\cdot CHNH_2\cdot CH_2\cdot CO\cdot OH$, heat $\rightarrow R\cdot CH:CHCO\cdot OH + NH_3$, and

$CH_3CHNH_2\cdot CH_2\cdot CO\cdot OH$, heat$\rightarrow CH_3CH:CH\cdot CO\cdot OH$ (2-butenoic acid) $+ NH_3$

The 2, 3- unsaturated acids are always obtained.

c. $R\cdot CH\cdot CH_2\cdot CH_2\cdot CO\cdot OH$, heat $\rightarrow R\cdot CH\cdot CH_2CH_2CO + H_2O$, and
$H\cdot N\cdot H$ $H\cdot N$————⌐

$CH_3\cdot CH\cdot CH_2\cdot CH_2\cdot CO\cdot OH$, heat$\rightarrow CH_3\cdot CH\cdot CH_2\cdot CH_2\cdot CO$ (4-aminopentanoic
$H\cdot N\cdot H$ $H\cdot N$————⌐
lactone) $+ H_2O$

VIII. Proteins[1]

A. Introduction:

Proteins are built up from the alpha-amino acids, which in some cases contain complex substituents. A great many different proteins exist and they are all very complex, the molecular weight varying from about 30,000 (egg albumin is about 34,000) to perhaps several million. All contain carbon, hydrogen, oxygen, and nitrogen. Most of them contain sulfur and many contain phosphorus. A few contain other elements.

The proteins are divided into three large divisions; simple, conjugated, and derived. The simple proteins, as egg albumin, on hydrolysis yield primarily α-amino acids; the conjugated proteins yield besides α-amino acids one or more other substances. Casein, from milk, and hemoglobin from blood are examples of conjugated proteins. Derived proteins as proteoses and peptones are those which have been formed by the partial hydrolysis of naturally occurring proteins. These three large categories are classified still further into many other divisions, according to properties, chiefly physical.

Nothing definite can be given concerning type formulas and isomerism because the structures of these very complex molecules are not known. However, one can use the building units of proteins, the α-amino acids, to synthesize polypeptides which, while they are much simpler, do illustrate, it is believed, the nature of the linkage in the protein molecule:

$2\ H_2N\cdot CH_2\cdot CO\cdot OH \rightarrow H_2N\cdot CH_2CO\cdot HN\cdot CH_2CO\cdot OH$ (glycylglycine) $+ H_2O$, or

$4\ \ H_2N\cdot CH_2\cdot CO\cdot OH \rightarrow H_2N\cdot CH_2CO\cdot HN\cdot CH_2CO\cdot HN\cdot CH_2CO\cdot HN\cdot CH_2\cdot CO\cdot OH$ (triglycylglycine) $+ 3\ H_2O$

[1] Cf. Protein Structure, Max Bergmann, Seventh National Organic Chemistry Symposium. p. 68 (1937).

The protein linkage doubtless involves the loss of a hydrogen atom from the amino group and an "OH" from the carboxyl group to form water. On hydrolysis, these polypeptides form amino acids just as the proteins do.

The proteins are constituents of all living animal and plant cells. The number of distinct proteins which exist is almost unbelievably large.

The one essential use of proteins is as food. Some have various industrial uses. Casein, for example, is used as a food, for sizing paper, for the preparation of synthetic resins, and for other purposes. Zein, the protein of corn, is one of the promising new raw materials for plastics.

As previously noted, the complexity of the proteins prevents, as yet, definite formulation of their structure. The different possible linkages between the different amino-acid building units, makes possible the existence of an infinitely large number of proteins.

B. Nomenclature:

The names are individual with no well formulated system.

C. Preparation:

Proteins have not been synthesized but they are readily isolated from natural material by dialysis, precipitation, or similar physical methods.

D. Physical Properties:

The proteins are colloidal substances. They are most sensitive to coagulation at their isoelectric point. They are usually coagulated by dehydrating agents as alcohol, acetone, or concentrated solutions of salts (sodium chloride, magnesium sulfate, ammonium sulfate, etc.), or by heat.

E. Chemical Properties:

1. The chemical properties of the proteins are dependent upon their ultimate building units, the amino acids. They are amphoteric (both acidic and basic in character).

2. The principal tests used in characterizing and identifying protein material are:

   a. Biuret. Protein material + c.$NaOH$/H_2O + trace $CuSO_4$ → violet-pink color.

   b. Xanthoproteic. Protein + c.HNO_3, heat → yellow coloration, then add NH_4OH → orange coloration (if $NaOH$ is used instead of NH_4OH, a red coloration results).

   c. Millon. Protein + $Hg(NO_3)_2$, heat → brick red color or precipitate.

   d. Hopkins-Cole. Protein + **glyoxylic acid,** then add H_2SO_4 → violet ring.

   e. Molisch. Protein containing the carbohydrate glucosamine + α-naphthol, then add H_2SO_4 → violet ring. This is essentially a carbohydrate test.

3. Precipitation may be effected by the use of:

a. Cations of the heavy metals if the pH is greater than that of the isoelectric point.

b. Complex anions as from phosphotungstic acid, phosphomolybdic acid, tannic acid, picric acid, and similar reagents if the pH is less than that of the isoelectric point.

IX. Nitriles or Alkyl Cyanides

A. Introduction:

The nitriles may be regarded as alkane derivatives in which the three hydrogen atoms linked to a primary carbon atom have been replaced by a nitrogen atom, as acid derivatives in which the hydrogen and oxygen atoms of the carboxyl group have been replaced by a nitrogen atom, or as derivatives of ammonia in which the three hydrogen atoms have been replaced by a carbon atom. They are represented by the general formulas $R \cdot C : N$ or $C_nH_{2n+1}C:N$, and $(CH_2)_x(CN)_2$.

The nitriles occur to a small extent in bone oil and coal tar. They find their principal use in organic synthesis.

The nitriles are functional and dynamic (especially at elevated temperatures) isomers of the isonitriles ($R \cdot CN \rightleftarrows R \cdot NC$).

B. Nomenclature (cf. Ch. XXIV, Div II):

Structural Formulas	I.U.C. Names	Common Names
$H \cdot CN$	methanenitrile	hydrogen cyanide, prussic acid
$CH_3 \cdot CN$	ethanenitrile	acetonitrile, methyl cyanide
$CH_3 \cdot CH_2 \cdot CN$	propanenitrile	propionitrile, ethyl cyanide
$CH_3 \cdot CH_2 \cdot CH_2 \cdot CN$	butanenitrile	butyronitrile, propyl cyanide
$(CH_3)_2CH \cdot CN$	2-methylpropanenitrile	isobutyronitrile, isopropyl cyanide

C. Preparations:

THE ALKANE NITRILES MAY BE PREPARED:

1. By the dehydration of (a) the ammonium salts of alkanoic acids, (b) the amides, or (c) the aldoximes:

a. $R \cdot CO \cdot ONH_4 + 2 P_2O_5$, distil $\rightarrow R \cdot CN + 4 HPO_3$, and

$CH_3 \cdot CO \cdot ONH_4 + 2 P_2O_5$, distil $\rightarrow CH_3 \cdot CN$ (ethanenitrile) $+ 4 HPO_3$

b. $R \cdot CO \cdot NH_2 + P_2O_5$, distil $\rightarrow R \cdot CN + 2 HPO_3$

c. $R \cdot C(:N \cdot OH) \cdot H + CH_3 \cdot CO \cdot O \cdot OC \cdot CH_3$, heat $\rightarrow R \cdot CN + 2 CH_3 \cdot CO \cdot OH$

Closely related to this reaction is the following:

$R \cdot C(:N \cdot Cl) \cdot H$, loss of HCl $\rightarrow R \cdot CN + HCl$

2. By the replacement of a halogen atom in an alkyl halide, sulfate, or polyhalogenated hydrocarbon by the CN group:

$R \cdot X + Na \cdot CN$ (or $K \cdot CN$) $\rightarrow R \cdot CN$ (also small amounts of $R \cdot NC$) $+ NaX$

All of these methods are applicable to the preparation of the dinitriles. Butanedinitrile, for example, may be prepared by $Br \cdot CH_2 \cdot CH_2 \cdot Br + 2\ NaCN \rightarrow NC \cdot CH_2 \cdot CH_2 \cdot CN + 2\ NaBr$

Methanenitrile (**hydrogen cyanide**) may be prepared by the action of a mineral acid on a metallic cyanide: $NaCN + HX \rightarrow H \cdot CN + NaX$

D. Physical Properties:

The nitriles, up to $C_{14}H_{29} \cdot CN$ (m.p. 23) are liquids and are comparatively insoluble in water. They distil without decomposition. **Hydrogen cyanide** (perhaps in its isomeric form, HNC) is very poisonous and, in general, the nitriles, or their isomeric forms which are usually present, possess toxic properties.

E. Chemical Properties:

THE NITRILES REACT UNDER PROPER CONDITIONS:

1. By addition, (a) hydrolytic, (b) reductive, or (c) simple, when treated with the following:

a.	Excess H_2O/H^+	$R \cdot CN + 2\ H_2O/H^+ \rightarrow R \cdot CO \cdot OH$ (acids) $+ NH_4^+$
b.	$Na/CH_3 \cdot CH_2 \cdot OH$	$R \cdot CN + 4\ Na/4\ Et \cdot OH \rightarrow R \cdot CH_2 \cdot NH_2$ (amines) $+ 4\ EtONa$
c. 1.	Grignard reagent	$R \cdot CN + R \cdot MgX \rightarrow R_2C:N \cdot MgX$, then $R_2C:N \cdot MgX + H_2O/2\ HX \rightarrow R_2C:O + MgX_2 + NH_4X$
2.	H_2O/conc. H_2SO_4	$R \cdot CN + H_2O/(c.H_2SO_4) \rightarrow R \cdot CO \cdot NH_2$
3.	Halogen acids	$R \cdot CN + HX \rightarrow R \cdot C(X):N \cdot H$ (imide halides) $R \cdot C(X):N \cdot H + HX \rightarrow R \cdot CX_2 \cdot NH_2$ (amide halides)
4.	Hydrogen sulfide	$R \cdot CN + H_2S \rightarrow R \cdot CS \cdot NH_2$ (thioamides, poor yields)
5.	Alcohols/dry HCl	$R \cdot CN + CH_3 \cdot CH_2 \cdot OH/(dry\ HCl) \rightarrow R \cdot C(:NH) \cdot OCH_2 \cdot CH_3$
6.	Fatty acids Anhydrides	$R \cdot CN + R \cdot CO \cdot OH \rightarrow (R \cdot CO)_2N \cdot H$ (diamides) $R \cdot CN + (R \cdot CO)_2O \rightarrow (R \cdot CO)_3N$ (triamides)
7.	Ammonia Amines	$R \cdot CN + NH_3 \rightarrow R \cdot C(:NH) \cdot NH_2$ (amidines) $R \cdot CN + R \cdot NH_2 \rightarrow R \cdot C(:NH) \cdot NHR$ (alkyl amidines)
8.	Hydroxylamine	$R \cdot CN + H_2NOH \rightarrow R \cdot C(:N \cdot OH) \cdot NH_2$ (amidoximes)

2. By polymerization, when treated with metallic sodium in ether:

$2\ CH_3 \cdot CN + (Na/dry\ ether) \rightarrow CH_3 \cdot C(:NH) \cdot CH_2 \cdot CN$ (3-iminobutanenitrile)

If the polymerization is produced by sodium in the absence of ether, cyanethine (a pyrimidine derivative) results.

X. Carbylamines, Isocyanides or Isonitriles

A. Introduction:

The carbylamines may be regarded as derivatives of the hydrocarbons in which one hydrogen atom has been replaced by the NC group. They may be represented by the general formula $R \cdot NC$.

The carbylamine test is used to identify the presence of primary amines, but, due to its sensitivity, the test has no quantitative value as secondary and tertiary amines usually contain sufficient traces of primary amines to give a positive test.

The additive properties of the carbylamines (R·NC) indicate that bivalent carbon may be involved in the linkage, but the most recent work has indicated that they contain a triple covalent bond and one electrovalent bond. Methylcarbylamine, for example, may be represented *electronically* by $H_3C:N:::C:$, and ethanenitrile by $H_3C:C:::N:$. They are functional and dynamic (especially at elevated temperatures) isomers of the nitriles.

B. Nomenclature:

Structural Formulas	I.U.C. Names	Common Names
$CH_3 \cdot NC$	methylcarbylamine	methyl isocyanide
$CH_3 \cdot CH_2 \cdot NC$	ethylcarbylamine	ethyl isocyanide
$CH_3 \cdot CH_2 \cdot CH_2 \cdot NC$	propylcarbylamine	propyl isocyanide
$(CH_3)_2 CH \cdot NC$	isopropylcarbylamine	isopropyl isocyanide

C. Preparations:

THE CARBYLAMINES MAY BE PREPARED:

1. *By the replacement of the halogen atom in an alkyl halide by the carbylamine group* $(\cdot NC)$:

   a. $R \cdot X + Ag \cdot CN \rightarrow (R \cdot N:C\ [Ag]\ X) \rightarrow R \cdot NC$ (also, some $R \cdot CN$) $+ AgX$

   b. $R \cdot X + NaCN$ (or KCN) $\rightarrow R \cdot NC$ (but mostly $R \cdot CN$) $+ NaX$

2. *By the action of chloroform, in the presence of alcoholic sodium or potassium hydroxide, on primary amines (Hofmann carbylamine reaction):*

$$R \cdot N \begin{array}{|cc|} \hline H & Cl \\ & + \\ H & Cl \\ \hline \end{array} C \begin{array}{|c|} \hline H \\ \\ Cl \\ \hline \end{array} + 3\,NaOH, \text{ alc.} \rightarrow R \cdot NC + 3\,NaCl + 3\,H_2O$$

D. Physical Properties:

The alkylcarbylamines are colorless liquids which distil without decomposition. They are slightly soluble in water, but readily soluble in alcohol and ether. They possess a very disagreeable odor and are quite poisonous.

E. Chemical Properties:

THE ALKYLCARBYLAMINES REACT UNDER PROPER CONDITIONS:

1. *By addition, (a) hydrolytic, (b) reductive, (c) simple, or (d) oxidative, when treated with the following:*

   a. Excess $H_2O/180°C$. $R \cdot NC + 2\,H_2O \rightarrow R \cdot NH_2$ (primary amine) $+ H \cdot CO \cdot OH$

   b. $Na/CH_3 \cdot CH_2 \cdot OH$ $R \cdot NC + 4\,Na/4\,Et \cdot OH \rightarrow R \cdot NH \cdot CH_3$ (sec. amine) $+ 4\,Et \cdot ONa$

   c. 1. Grignard reagent $R \cdot NC + R \cdot MgX \rightarrow R \cdot N:C\,(R) \cdot MgX$, then $R \cdot N:C(R) \cdot MgX + H_2O/HX \rightarrow R \cdot NH_2 + R \cdot CHO + MgX_2$

2. Halogens $R \cdot NC + X_2 \rightarrow R \cdot N:CX_2$

3. Halogen acids $R \cdot NC + HX \rightarrow R \cdot N:C(H)X$

4. Oxygen $2 R \cdot NC + O_2 \rightarrow 2 R \cdot N:C:O$ (an isocyanate)

5. Sulfur $R \cdot NC + S \rightarrow R \cdot N:C:S$ (an isothiocyanate)

d. Mercuric oxide $R \cdot NC + HgO \rightarrow R \cdot N:C:O + Hg$

2. By tautomerization, when heated to 250°C.

$R \cdot NC$, heat to about 250°C. $\rightleftarrows R \cdot CN$

XI. Hydrogen Cyanide and Derivatives

A. Introduction:

Methanenitrile (**formonitrile**), better known as **hydrogen cyanide or hydrocyanic acid,** is of importance because of a number of important compounds that may be regarded as its derivatives. **Hydrogen cyanide** is represented by the formula $H \cdot C \vdots N$, but it doubtless exists in an equilibrium system such as $H \cdot C \vdots N \rightleftarrows H \cdot \overset{+}{N} \vdots \overset{-}{C}$ (cf. Ch. XIII, Div. IX, Sec. A). The poisonous properties of hydrogen cyanide are attributed to the latter form.

Hydrogen cyanide is obtained from the hydrolysis of amygdalin which occurs in the leaves of the cherry and laurel, in peach kernels, and in bitter almonds. Spectroscopic analysis indicates the presence of cyanogen ($NC \cdot CN$) in the tails of comets.

Hydrogen cyanide is used as a fungicide, as an insecticide, and for fumigating ships, orange groves, and flour mills. It has found a limited use in medicine in the treatment of some heart and respiratory diseases, but such use is decreasing. Cyanogen chloride found a limited use as a poisonous gas during the war, and seems to be replacing **hydrogen cyanide** for fumigating ships. Cyanamide serves as a source of organic nitrogen, and finds use in synthesis. Calcium cyanamide is used as a fertilizer. The metallic cyanides, principally sodium cyanide, are used in extracting silver and gold from low grade ores. They are used in the preparation of **hydrogen cyanide.** Solutions of $NaAg(CN)_2$, $NaAu(CN)_2$, and $NaNi(CN)_3$ are used extensively in electroplating. Prussian blue and potassium ferricyanide are used in calico printing, in dyeing, and as reagents. Prussian blue finds some use as a pigment. Some blue prints are made by precipitating Prussian blue in the paper. Mercury fulminate is used in percussion caps as a detonator, although its use has been largely abandoned in recent years due to the commercial production of lead azide (N_3PbN_3) and diazodinitrophenol. Silver fulminate is used in toy torpedoes. Metallic cyanates, cyanic acid, cyanuric acid, and **cyanogen** have a limited use in synthesis.

The methods of preparation and the chemical properties given below will indicate that the structures assigned to these derivatives are correct.

B. Nomenclature:

The names of some of the more important derivatives of hydrogen cyanide will be considered in connection with their preparations and chemical properties.

C. Preparations:

1. OF HYDROGEN CYANIDE:

a. $NaCN + H_2SO_4$, aq. $\rightarrow H \cdot CN$ (hydrogen cyanide) $+ NaHSO_4$

b. $2 K_4Fe(CN)_6 + 3 H_2SO_4$, aq. $\rightarrow 6 H \cdot CN + FeK_2Fe(CN)_6 + 3 K_2SO_4$

c. $K \cdot CN$ (or $Ca(CN)_2$) + CO_2/H_2O, from the air $\rightarrow$ $H \cdot CN$ + $KHCO_3$

d. N_2 + $H_2/2$ C (carbon terminal, electric arc) $\rightarrow$ 2 $H \cdot CN$

e. $(CH_3)_3N$, heated strongly $\rightarrow$ $H \cdot CN$ + 2 CH_4

f. $H \cdot CO \cdot ONH_4$, heat $\rightarrow$ $H \cdot CO \cdot NH_2$ + H_2O, then
  $H \cdot CO \cdot NH_2$ + P_2O_5, heat $\rightarrow$ $H \cdot CN$ + 2 HPO_3

2. OF CYANOGEN CHLORIDE:

a. $H \cdot CN$ + Cl_2 $\rightarrow$ $Cl \cdot CN$ (cyanogen chloride) + HCl

b. $NaZn(CN)_3$ + Cl_2 $\rightarrow$ $Cl \cdot CN$ + $Zn(CN)_2$ + $NaCl$

3. OF CYANAMIDE ($H_2 \cdot N \cdot CN \rightleftarrows H \cdot N:C:N \cdot H$):

a. $Cl \cdot CN$ + 2 NH_3 $\rightarrow$ $H_2N \cdot CN$ (cyanamide) + NH_4Cl

b. $H_2N \cdot C(:N \cdot H) \cdot NH_2$ (guanidine) + (KOH, fuse) $\rightarrow$ $H_2N \cdot CN$ + NH_3

c. $H_2N \cdot CO \cdot NH_2$ (urea) + $SOCl_2$ $\rightarrow$ $H_2N \cdot CN$ + SO_2 + 2 HCl

Calcium cyanamide ($Ca:N \cdot CN$) is prepared by nitrogen fixation as follows: CaC_2 + $N_2/(Fe)$, $1,000°C.$ $\rightarrow$ $Ca:N \cdot CN$ + C
The crude product is known as "lime nitrogen" or Nitrolime.

4. OF METALLIC CYANIDES:

a. 2 $H \cdot CN$ (liquid, trace of H_2O) + CaC_2 $\rightarrow$ $Ca(CN)_2$ (calcium cyanide) + $H \cdot C:C \cdot H$

b. BaC_2 + N_2, heat $\rightarrow Ba(CN)_2$ (barium cyanide)

c. 2 $K \cdot CN$ + $CuSO_4$ $\rightarrow$ $Cu(CN)_2$ (copper cyanide, unstable, see 5.c.) + K_2SO_4

d. $K_4Fe(CN)_6$, heated strongly $\rightarrow$ 4 $K \cdot CN$ (potassium cyanide) + FeC_2 + N_2, or
  $K_4Fe(CN)_6$ + 2 Na, heat $\rightarrow 4$ $K \cdot CN$ + 2 $Na \cdot CN$ + Fe, or
  $NC \cdot CN$ + 2 KOH, aq. $\rightarrow$ $K \cdot CN$ + $K \cdot O \cdot CN$ + H_2O

e. 2 $Na \cdot NH_2$ (sodamide) + 2 C, $400°C.$ $\rightarrow$ 2 $Na \cdot CN$ (sodium cyanide) + 2 H_2, or
  Na_2CO_3 + 4 C + $N_2/(Fe)$, $1,000°C.$ $\rightarrow$ 2 $Na \cdot CN$ + 3 CO (Bucher Process), or $Ca:N \cdot CN$ + 2 NaCl + C, heat strongly $\rightarrow$ 2 $Na \cdot CN$ + $CaCl_2$

Sodium cyanide may also be prepared from sugar beet residues.

5. OF COMPLEX CYANIDES:

a. $Ag \cdot CN$ + $K \cdot CN$, aq. $\rightarrow KAg(CN)_2$ (potassium silver cyanide, ionic)

b. $Au \cdot CN$ + $K \cdot CN$, aq. $\rightarrow KAu(CN)_2$ (potassium aurocyanide, ionic)

c. 2 Cu^{++} + 4 $K \cdot CN$, aq. $\rightarrow 2Cu(CN)_2$ (cupric cyanide, unstable) + 4 K^+, then
  2 $Cu(CN)_2$ $\rightarrow$ 2 CuCN (cuprous cyanide, insoluble) + $NC \cdot CN$, and
    2 CuCN + 4 $K \cdot CN$, aq. $\rightarrow$ 2 $K_2Cu(CN)_3$ (potassium cuprous cyanide)

d. Cd^{++} + 2 $K \cdot CN$ $\rightarrow$ $Cd(CN)_2$ (cadmium cyanide) + 2 K^+, then
  $Cd(CN)_2$ + 2 $K \cdot CN$ $\rightarrow$ $K_2Cd(CN)_4$ (potassium cadmium cyanide)

e. 4 M^+ + Fe^{++} + 6 CN^- $\rightarrow$ $M_4Fe(CN)_6$ (a metallic ferrocyanide), or in the old industrial process nitrogenous waste material was mixed with potassium carbonate and iron and heated to give potassium ferrocyanide ($K_4Fe(CN)_6$).

f. $3 M^+ + Fe^{+++} + 6 CN^- \rightarrow M_3Fe(CN)_6$ (a metallic ferricyanide), and

$2 K_4Fe(CN)_6 + Cl_2$, aq. $\rightarrow 2 K_3Fe(CN)_6$ (potassium ferricyanide) + $2 KCl$

g. $3 K_4Fe(CN)_6 + 4 Fe^{+++}$, aq. $\rightarrow Fe_4(Fe[CN]_6)_3$ (ferric ferrocyanide) + $12 K^+$

The product is Prussian blue, sometimes known as Berlin blue.

h. $2 K_3Fe(CN)_6 + 3 Fe^{++}$, aq. $\rightarrow Fe_3(Fe [CN]_6)_2$ (ferrous ferricyanide) + $6 K^+$

The product is known as Turnbull blue, but recent work has indicated that Prussian blue and Turnbull blue are identical. They are both thought to be equilibrium mixtures of the ferric ferrocyanide and the ferrous ferricyanide.

6. OF FULMING ACID (HO·NC), AND ITS DERIVATIVES:

a. $Ag·O·NC + HCl$, conc. $\rightarrow HO·NC$ (fulminic acid) $+ AgCl$, and

$Hg (O·NC)_2 + HCl$, conc. $\rightarrow 2 HO·NC + HgCl_2$, or

$H_2N·C (:N·OH)·H$, aq., by loss of $NH_3 \rightarrow HO·NC + NH_3$

b. $Ag + HNO_3$, conc., then add $CH_3·CH_2·OH \rightarrow Ag·O·NC$ (silver fulminate) $+ ?$

c. $Hg + HNO_3$, conc., then add $CH_3·CH_2·OH \rightarrow Hg (O·NC)_2$ (mercuric fulminate) $+ ?$

$(H_2C:NO·O)_2Hg$. aq., heat $\rightarrow Hg (O·NC)_2$ (mercuric fulminate) $+ H_2O$

7. OF METALLIC CYANATES:

$Cl·CN + 2 KOH \rightarrow KO·CN$ (potassium cyanate) $+ KCl + H_2O$

$2 K·CN + O_2/(air) \rightarrow 2 KO·CN$ (potassium cyanate)

$K·CN + PbO \rightarrow KO·CN$ (potassium cyanate) $+ Pb$

8. OF CYANURIC ACID AND METALLIC CYANATES:

a. $(HO·CN)_3$ (cyanuric acid), heat $\rightarrow 3 HO·CN$ (cyanic acid)

$H_2N·CO·NH_2$ (urea), heat $\rightarrow HO·CN$ (small yields) $+ NH_3$ (and biuret)

b. $3 HO·CN$, above $0°C. \rightarrow$

$$\begin{array}{c} N \\ \diagup\!\!\diagup \diagdown \\ HO·C \quad C·OH \text{ (cyanuric acid), or} \\ | \quad\quad || \\ N \quad\quad N \\ \diagdown\!\!\diagdown \diagup \\ C·OH \end{array}$$

$(Cl·CN)_3 + 3 H_2O \rightarrow \quad\quad C·OH + 3 HCl$

c. $3 Cl·CN$, at room temperature $\longrightarrow$

$$\begin{array}{c} N \\ \diagup\!\!\diagup \diagdown \\ Cl·C \quad C·Cl \text{ (cyanuric chloride)} \\ | \quad\quad || \\ N \quad\quad N \\ \diagdown\!\!\diagdown \diagup \\ C·Cl \end{array}$$

d. $Ag·N:C:O + R·X \rightarrow R·N:C:O$ (alkyl isocyanate) $+ AgX$

9. OF CYANOGEN:

$H_4N·O·OC·CO·O·NH_4$ (ammonium oxalate) $+ 4 P_2O_5$, heat $\rightarrow NC·CN$ (cyanogen) $+ 8 HPO_3$

$2 C + N_2$, electric arc $\rightarrow NC·CN$ (cyanogen)

$Hg(CN)_2$ (mercuric cyanide), heat → NC·CN (**cyanogen**) + Hg

2 $Cu(CN)_2$ (cupric cyanide), aq. → NC·CN (**cyanogen**) + 2 CuCN (cuprous cyanide)

D. Physical Properties:

1. Hydrogen cyanide is a colorless liquid, solidifying at −15 and boiling at 26. It is miscible with water in all proportions and it is a weak acid. It, or its isomer, is extremely poisonous.

2. Cyanogen chloride is a very poisonous liquid, b.p. 14.5.

3. Cyanamide is a deliquescent, colorless solid, m.p. 41-42. It is soluble in water, alcohol, and ether. Its amphoteric properties may be due to the tautomeric system, $H_2N·CN \rightleftarrows H·N:C:N·H$.

4. The metallic cyanides of the alkali metals are soluble in water, slightly basic, and their solutions are extremely poisonous. The simple cyanides of the heavy metals are somewhat insoluble.

5. Some of the complex cyanides are quite soluble in water; they are usually slightly basic, and they are poisonous.

6. Fulminic acid is unstable, quite volatile, and extremely poisonous. When dry, its salts of mercury and silver are powerful explosives.

7. Ammonium cyanate can be obtained as a white crystalline powder. It is soluble in water, but upon evaporation of an aqueous solution of ammonium cyanate, it is changed into urea. Potassium cyanate is a colorless, crystalline powder. It is soluble in water and dilute alcohol, but the aqueous solution is unstable: $KO·CN + 2 H_2O \rightarrow KHCO_3 + NH_3$.

8. Cyanic acid can be obtained as an unstable liquid below $0°$, but with rise of temperature it polymerizes to cyamelide $(H·NCO)_3$.

9. Cyanogen is a colorless, poisonous gas, with a faint odor of peaches. It can be condensed to a colorless liquid, b.p. −21. At −34 it freezes to a white solid. It dissolves readily in water, and burns with a violet-colored flame, forming carbon dioxide and nitrogen.

E. Chemical Properties:

1. OF HYDROGEN CYANIDE:

a. $H·CN + 4 Na/4 Et·OH \rightarrow CH_3·NH_2$ (methylamine) + 4 Et·ON

b. $H·CN + 2 H_2O/HCl \rightarrow H·CO·OH$ (formic acid) + NH_4Cl

c. $H·CN + Cl_2 \rightarrow Cl·CN$ (cyanogen chloride) + HCl

d. 2 H·CN (liquid) + CaC_2/trace $H_2O \rightarrow Ca(CN)_2$ + H·C:C·H

2. OF CYANOGEN CHLORIDE:

a. $Cl·CN + 2 NH_3 \rightarrow H_2N·CN$ (cyanamide) + NH_4Cl

b. 3 Cl·CN, at room temperature ⟶

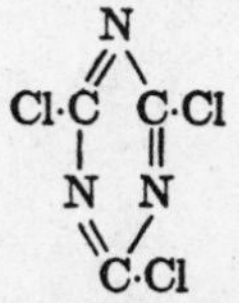

$$\begin{array}{c} N \\ \diagup\!\!\diagup \quad \diagdown \\ Cl·C \qquad C·Cl \\ | \qquad\quad || \\ N \qquad\quad N \\ \diagdown\!\!\diagdown \quad \diagup \\ C·Cl \end{array}$$

(cyanuric chloride)

c. $Cl·CN + 2 KOH$, aq. → KO·CN (potassium cyanate) + KCl + H_2O

3. OF CYANAMIDE:

a. $H_2N \cdot CN + H_2O/H^+ \rightarrow HO \cdot CN$ (cyanic acid) $+ NH_4^+$, and

$HO \cdot CN + H_2O \rightarrow CO_2 + NH_3$

b. $3\ H_2N \cdot CN$, heat at 150 °C. $\rightarrow$ $H_2N \cdot C\ \ C \cdot NH_2$ (cyanuramide, **melamine**)

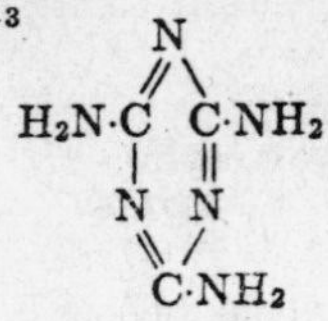

The following reactions of

the calcium salt of cyanamide (calcium cyanamide) are important:

$Ca{:}N \cdot CN + 3\ H_2O$, slow hydrolysis $\rightarrow CaCO_3 + 2\ NH_3$ (cf. C.*3.c*)

$2\ Ca{:}N \cdot CN + 4\ H_2O$, boil $\rightarrow (H_2N \cdot CN)_2$ (dicyanamide) $+ 2\ Ca(OH)_2$

$Ca{:}N \cdot CN + C/2\ NaCl$, heat $\rightarrow 2\ NaCN$ (sodium cyanide) $+ CaCl_2$

4. OF METALLIC CYANIDES:

a. $NH_4 \cdot CN + (NH_4)_2S_x \rightarrow NH_4 \cdot S \cdot CN$ (ammonium thiocyanate) $+ (NH_4)_2\ S_{x-1}$

b. 1. $K \cdot CN + R \cdot I \rightarrow R \cdot CN$ (alkanenitrile, alkyl cyanide) $+ KI$, or

$K \cdot CN + R \cdot O \cdot SO_2 \cdot OH \rightarrow R \cdot CN + KHSO_4$

2. $2\ K \cdot CN + O_2$ (air) $\rightarrow 2\ KO \cdot CN$ (potassium cyanate), or

$K \cdot CN + PbO$, aq., heat $\rightarrow KO \cdot CN + Pb$

3. $K \cdot CN + S$, heat $\rightarrow KS \cdot CN$ (potassium thiocyanate)

4. $2\ K \cdot CN + CuSO_4 \rightarrow Cu(CN)_2$ (cupric cyanide, unstable) $+ K_2SO_4$, and

$2\ Cu(CN)_2$, aq., heat $\rightarrow NC \cdot CN$ (cyanogen) $+ 2\ CuCN$ (insoluble)

5. $2\ K \cdot CN + Fe^{++} \rightarrow Fe(CN)_2$ (ferrous cyanide) $+ 2\ K^+$, and

$Fe(CN)_2 + 4\ K \cdot CN \rightarrow K_4Fe(CN)_6$(potassium ferrocyanide), or

$K \cdot CN + Ag^+$, aq. $\rightarrow AgCN$ (insoluble) $+ K^+$, and

$Ag \cdot CN + K \cdot CN$, aq. $\rightarrow KAg(CN)_2$ (potassium silver cyanide, soluble), or $4\ K \cdot CN + 2\ Au/2\ H_2O/O_2$ (air) $\rightarrow 2\ KAu(CN)_2 + 2\ KOH, + H_2O_2$, then

$H_2O_2 + 4\ K \cdot CN + 2\ Au \rightarrow KAu(CN)_2$ (potassium aurocyanide) $+ 2\ KOH$

6. $K \cdot CN + 2\ H_2O$, slow hydrolysis $\rightarrow H \cdot CO \cdot OK + NH_3$

c. $Hg\ (CN)_2$, heat $\rightarrow NC \cdot CN$ (**cyanogen**) $+ Hg$

d. $Ag \cdot CN + R \cdot I \rightarrow R \cdot NC$ (alkylcarbylamine, **alkyl isocyanide**) $+ AgI$

e. $Na \cdot CN + H_2SO_4$, aq. $\rightarrow H \cdot CN$ (hydrogen cyanide) $+ NaHSO_4$

5. OF COMPLEX CYANIDES:

a. $KAg(CN)_2 + HCl$, aq. $\rightarrow Ag \cdot CN$ (silver cyanide) $+ H \cdot CN$ (**hydrogen cyanide**) $+ KCl$

b. $2\ K_4Fe(CN)_6 + 6\ H^+$, aq. $\rightarrow 6\ H \cdot CN + K_2Fe_2(CN)_6 + 6\ K^+$

c. $K_4Fe(CN)_6 + 4\ H^+$, aq. $\rightarrow H_4Fe(CN)_6$ (ferrocyanic acid, white ppt.) $+ 4\ K^+$

d. $3\ K_4Fe(CN)_6 + 4\ Fe^{+++}$, aq. $\rightarrow Fe_4(Fe[CN]_6)_3$ (ferric ferrocyanide, **Prussian blue**) $+ 12\ K^+$, or

$3\ Fe_3(Fe[CN]_6)_2 + Cl_2$, aq. $\rightarrow 2\ Fe_4(Fe[CN]_6)_3 + FeCl_2$

e. $2 K_3Fe(CN)_6 + 3 Fe^{++}$, aq. $\rightarrow$ $Fe_3(Fe[CN]_6)_2$ (ferrous ferricyanide, Turnbull blue) $+ 6 K^+$

f. $K_4Fe(CN)_6$, fusion $\rightarrow 4 K \cdot CN + FeC_2$ (ferrous carbide) $+ N_2$

6. OF FULMINIC ACID (HO·NC) AND ITS DERIVATIVES:

a. HO·NC (fulminic acid), is very unstable.

b. $Hg(O \cdot NC)_2$ (mercuric fulminate), explodes on slight detonation, and Ag·O·NC (silver fulminate), explodes even more readily.

c. $Ag \cdot O \cdot NC + HI \rightarrow Ag \cdot O \cdot N:C(H)I$

7. OF METALLIC CYANATES (alkyl cyanates are not known):

a. $K \cdot O \cdot CN + 2 H_2O/H^+ \rightarrow KHCO_3 + NH_4^+$

b. $NH_4 \cdot O \cdot CN$, heat $\rightarrow H_2N \cdot CO \cdot NH_2$ (urea)

c. $Ag \cdot O \cdot CN + R \cdot X \rightarrow R \cdot N:C:O$ (alkyl isocyanate) $+ AgX$

8. OF CYANIC ACID AND ITS DERIVATIVES:

a. 1. 3 HO·CN, polymerizes violently, 5°C $\rightarrow$

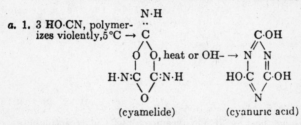

(cyamelide)   (cyanuric acid)

   2. $HO \cdot CN + H_2O/H^+ \rightarrow CO_2 + NH_4^+$

   $HO \cdot CN + 2 H_2O/H^+ \rightarrow H \cdot CO \cdot OH + HONH_2^+$

   3. $HO \cdot CN + Cl_2 \rightarrow Cl \cdot CN$ (cyanogen chloride) $+ HOCl$

b. 1. $H \cdot N:C:O + CH_3 \cdot CH_2 \cdot OH \rightarrow CH_3 \cdot CH_2 \cdot O \cdot OC \cdot NH_2$ (ethyl carbamate, urethane)

   2. $H \cdot N:C:O + NH_3 \rightarrow H_2N \cdot CO \cdot NH_2$ (urea)

   3. $H \cdot N:C:O + R \cdot NH_2 \rightarrow R \cdot HN \cdot CO \cdot NH_2$ (an alkylurea)

c. 1. $(HO \cdot CN)_3$ (cyamelide) $+ 3 H_2O$ (from air, rapid hyd.) $\rightarrow 3 CO_2 + 3 NH_3$

   2. $(HO \cdot CN)_3$ (cyanuric acid) $+ 3 H_2O$ (from air, slow hyd.) $\rightarrow 3 CO_2 + 3 NH_3$

d. $(Cl \cdot CN)_3$ (cyanuric chloride) $+ 3 H_2O \rightarrow (HO \cdot CN)_3$ (cyanuric acid) $+ 3 HCl$

9. OF CYANOGEN:

a. $NC \cdot CN + 4 H_2O/2 H^+ \rightarrow HO \cdot OC \cdot CO \cdot OH$ (oxalic acid) $+ 2 NH_4^+$

b. $NC \cdot CN + 2 KOH$, aq. $\rightarrow K \cdot CN + KO \cdot CN + H_2O$

c. $NC \cdot CN + $ oxidation $\rightarrow 2 CO_2 + N_2$

XII. Fixation of Nitrogen

One of the most notable accomplishments of modern chemistry is the conversion of atmospheric nitrogen into useful compounds, particularly nitric acid and ammonia. With some variations and modifications, five primary processes have been developed.

I. The Arc Process:

This process was studied by Bradley and Lovejoy in the United States, Schoenher in Germany, Pauling in Austria, and was developed most extensively by Birkeland and Eyde in Norway during the early part of the present century.

Nitrogen and oxygen in appropriate proportions are passed between the poles of a huge electric arc. Combination takes place and the products are absorbed in water to form nitric acid. The reactions involved may be represented by the following equations: $N_2 + O_2 \rightarrow 2\,NO$, then $2\,NO + O_2 \rightarrow 2\,NO_2$, and, finally, $3\,NO_2 + H_2O \rightarrow 2\,HNO_3 + NO$

II. The Direct Ammonia Process:

Hydrogen and nitrogen, in the presence of various catalysts (usually metallic iron) and under varying conditions of temperature and pressure, combine to form ammonia. The reaction may be represented by the following equation:

$3\,H_2 + N_2 /$ (appropriate catalysts, heat and pressure) $\rightarrow 2\,NH_3$

The pressure varies from 200 to 1,000 atmospheres and the temperature range is about 400-550 °C., depending on the catalyst and the pressure employed.

This is the most important of the nitrogen fixation processes, and has been utilized in the United States, France, Italy, and in Germany where Haber invented and developed the process.

III. The Cyanamide Process:

Calcium oxide and coke are heated together to produce calcium carbide which, in turn, is caused to react with nitrogen to produce calcium cyanamide. Calcium cyanamide, upon hydrolysis, yields ammonia. The reactions are:

$CaO + 3\,C$, heat $\rightarrow CaC_2 + CO$, then $CaC_2 + N_2$, heat $\rightarrow Ca{:}N{\cdot}CN + C$

The mixture of calcium cyanamide and carbon is marketed as *Nitrolime*, which hydrolyzes in the soil as follows:

$Ca{:}N{\cdot}CN + 3\,H_2O \rightarrow CaCO_3 + 2\,NH_3$

IV. The Cyanide Process:

Sodium carbonate, carbon, and nitrogen are heated together to give sodium cyanide, which, upon subsequent hydrolysis, yields sodium formate and ammonia. The equations for these reactions are:

$Na_2CO_3 + 4\,C + N_2$, heat $\rightarrow 2\,Na{\cdot}CN + 3\,CO$ (Bucher process), then

$Na{\cdot}CN + 2\,H_2O \rightarrow H{\cdot}CO{\cdot}O{\cdot}Na + NH_3$

V. The Nitride Process:

Certain metals combine readily with nitrogen to form nitrides which, in turn, may be hydrolyzed to produce ammonia. To illustrate:

$3\,Mg + N_2$, heat $\rightarrow Mg_3N_2$, then $Mg_3N_2 + 6\,H_2O \rightarrow 3\,Mg\,(OH)_2 + 2\,NH_3$

In some cases the oxides are employed:

$Al_2O_3 + 3\,C + N_2$, heat $\rightarrow 2\,AlN + 3\,CO$, then

$AlN + 3\,H_2O \rightarrow Al\,(O)\,OH{\cdot}H_2O + NH_3$

This latter reaction has been used extensively by Serpek in France as a step in the purification of alumina for production of aluminum.

CHAPTER XIV

FOODS AND METABOLISM

I. Introduction

Food consists of carbohydrates, fats, proteins, vitamins, mineral salts, water, and oxygen. This discussion will be confined to the metabolism of the first four.

Metabolism, in biochemistry, is a term used to include all of the reactions to which the materials of the body are subjected. Anabolism applies specifically to the biochemical synthetic reactions, whereas catabolism refers to the processes of degradation.

Carbohydrates and fats are oxidized stepwise in the body to carbon dioxide and water. Proteins yield the same products, but in addition they give incompletely oxidized nitrogen compounds, such as urea, uric acid and creatinine. These nitrogenous waste products are eliminated, for the most part, in the urine.

II. Carbohydrates

A. Digestion and anabolism:

The saliva, the pancreatic juice, and the intestinal juices contain specific enzymes for certain carbohydrate materials. The principal carbohydrate enzymes and their functions are indicated below:

Sucrose + H_2O /sucrase → D-glucose + D^--fructose,

Maltose + H_2O /maltase → D-glucose + D-glucose,

Lactose + H_2O /lactase → D-glucose + D-galactose, and

Alpha dextrans + H_2O /amylases → Maltose

These hydrolytic products are absorbed through the walls of the small intestine into the blood, are oxidized immediately, or are stored as glycogen in the liver or muscle, or are stored as fat. Glycogen, an *alpha* dextran, is sometimes spoken of as animal starch. Glycogen also occurs in small amounts in the blood, brain, kidneys, pancreas, and spleen.

B. Catabolism:

When body fuel is needed, some of the glycogen may be hydrolyzed. Regardless of the fact that D-glucose, D^--fructose (levulose) and D-galactose are changed to glycogen in the body, the only hydrolytic product of glycogen is D-glucose. The oxidation of sugar in the body affords a readily available source of energy. Possible steps in this oxidation may be summarized as follows:

$CH_2OH \cdot HCOH \cdot HCOH \cdot HOCH \cdot HCOH \cdot CHO$ (pentahydroxyhexanal.++⁻+,
D-glucose), etc.   ⇅

$CH_2OH \cdot HCOH \cdot CHO$ ⇌ $CH_3 \cdot CO \cdot CHO$ ⇌ $CH_3 \cdot CHOH \cdot CO \cdot OH$
2, 3-dihydroxypropanal   2-ketopropanal   2-hydroxypropanoic acid
(glyceraldehyde)   (pyruvic aldehyde)   (lactic acid)
                                    ↓
                    $CH_3 \cdot CO \cdot CO \cdot OH$ (2-ketopropanoic acid, **pyruvic
                    ↓   acid**)
                    $CH_3 \cdot CHO$ (ethanal, **acetaldehyde**) + CO_2
                    ↓
                    $CH_3 \cdot CO \cdot OH$ (ethanoic acid, **acetic acid**)
                    ↓
                    $CO_2 + H_2O$ (carbon dioxide and water)

The excess carbon dioxide is eliminated through the lungs and the water is disposed of through the lungs, skin, and kidneys.

The role of lactic acid in the animal organism, according to Carlson,[1] has to do with both an anaerobic reaction (glycogen ⇌ lactic acid + energy, which is liberated rapidly) and an aerobic reaction (lactic acid + O_2 → CO_2 + H_2O + energy). The energy from the anaerobic reaction is utilized in muscular contraction, whereas the energy derived from the aerobic reaction is utilized in converting about four-fifths of the lactic acid formed in the body into glycogen, that is, in displacing the anaerobic reaction to the left.

III. Fats

A. Digestion and anabolism:

The fats are hydrolyzed by water, catalyzed by the lipases in the alimentary tract, into glycerol and salts of the fatty acids. Of these acids, oleic, stearic, and palmitic are the most abundant. Non-emulsified fats are hydrolyzed very slowly; emulsified fats much more readily. The alimentary tract contains bile salts whose function is the emulsification of the fats.

These hydrolytic products are absorbed through the walls of the small intestine, resynthesized to fats during absorption, and carried by the lymph by way of the thoracic duct into the systemic blood stream. Fat that is not needed for immediate energy requirements is stored in adipose tissue.

B. Catabolism:

When sufficient food is not being absorbed, stored fats are mobilized. The glycerol is oxidized to 2, 3-dihydroxypropanal (**glyceraldehyde**), the subsequent metabolism of which has been indicated under the discussion on glucose.

It is believed that the oxidation of the acids to carbon dioxide and water, as suggested by Knoop in his *beta* oxidation theory,

---

[1] Carlson, *The Machinery of the Human Body*, p. 65, University of Chicago Press (1937).

is a stepwise degradation. This process may be illustrated by the following diagrammatic representation of the oxidation of hexanoic acid. It should be noted that, according to this theory, the oxidation occurs on the *beta* carbon atom.

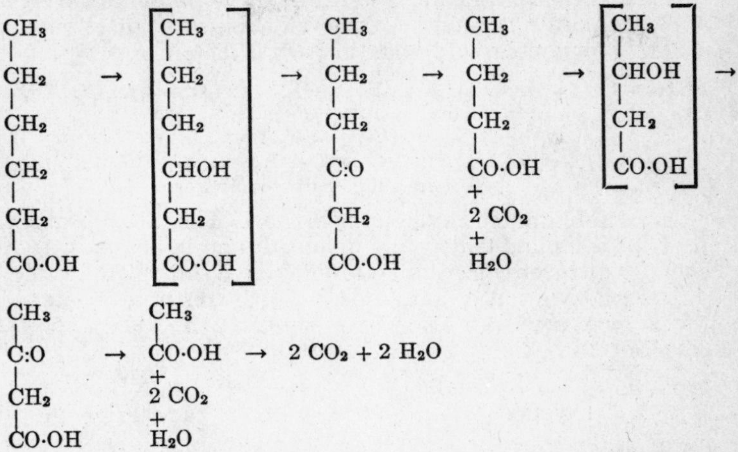

$$
\begin{array}{ccccccccc}
CH_3 & & \begin{bmatrix}CH_3 \\ CH_2 \\ CH_2 \\ CHOH \\ CH_2 \\ CO\cdot OH\end{bmatrix} & & CH_3 & & CH_3 & & \begin{bmatrix}CH_3 \\ CHOH \\ CH_2 \\ CO\cdot OH\end{bmatrix} \\
CH_2 & & & & CH_2 & & CH_2 & & \\
CH_2 & \rightarrow & & \rightarrow & CH_2 & \rightarrow & CH_2 & \rightarrow & \rightarrow \\
CH_2 & & & & C{:}O & & CO\cdot OH & & \\
CH_2 & & & & CH_2 & & +\ 2\ CO_2 & & \\
CO\cdot OH & & & & CO\cdot OH & & +\ H_2O & &
\end{array}
$$

$$
\begin{array}{ccccc}
CH_3 & & CH_3 & & \\
C{:}O & \rightarrow & CO\cdot OH & \rightarrow & 2\ CO_2 + 2\ H_2O \\
CH_2 & & +\ 2\ CO_2 & & \\
CO\cdot OH & & +\ H_2O & &
\end{array}
$$

There is some question as to whether or not the formulas shown in brackets represent actual intermediates. If, however, carbohydrates are not being metabolized at the same time, the oxidation of the butyric acid does not go to completion as indicated above but yields instead the products which are commonly known as the "acetone bodies." These include β-hydroxybutyric acid, acetoacetic acid, and acetone. The acetone is the decarboxylation product of acetoacetic acid. The "acetone bodies" are excreted as such, but their acidic and toxic character account for much of the danger in diabetes mellitus.

IV. Proteins

A. Digestion and anabolism:

The proteins are hydrolyzed by the proteases in the digestive tract to alpha amino acids. These are absorbed through the walls of the small intestine, carried by the blood to the different parts of the body, and resynthesized into specific body proteins or subjected to immediate deamination.

The deamination, which occurs largely in the liver, may be oxidative (alanine → pyruvic acid), hydrolytic (alanine → lactic acid), direct (alanine → acrylic acid or pyruvic aldehyde), and, in certain cases, reductive (alanine → propionic acid).
$CH_3 \cdot CH(NH_2) \cdot CO \cdot OH$ + direct deamination → $CH_3 \cdot CO \cdot CO \cdot H$ (pyruvic aldehyde) + NH_3

The stepwise oxidation of pyruvic aldehyde has been indicated under the discussion of glucose.

The nitrogen of the amino acids is excreted largely as urea. While the exact mechanism of the formation of urea is still unknown, the theory of Krebs is widely accepted.[1]

The ammonia formed by deamination may combine with water and carbon dioxide to form either ammonium carbonate or ammonium carbamate. Either of these substances may lose water to form urea. In equation form, the older view, is:

$$2\,NH_3 + CO_2 + H_2O \rightarrow H_4N \cdot O \cdot CO \cdot O \cdot NH_4 - 2\,H_2O \rightarrow H_2N \cdot CO \cdot NH_2$$
$$\text{(ammonium carbonate)}$$
$$+ H_2O \updownarrow - H_2O$$
$$\text{or } 2\,NH_3 + CO_2 \rightarrow H_2N \cdot CO \cdot ONH_4 - H_2O \rightarrow H_2N \cdot CO \cdot NH_2 \; (urea).$$
$$\text{(ammonium carbamate)}$$

Urea is eliminated through the kidneys. The catabolism of the proteins is undoubtedly more involved than indicated here, for complex nitrogen compounds are eliminated through the kidneys.

Decarboxylation of amino acids by putrefactive bacteria in the alimentary tract leads to the formation of an important class of toxins, the ptomaines, *thus:*

$$H_2N \cdot (CH_2)_4 \cdot CH\,(NH_2) \cdot COOH \rightarrow H_2N \cdot (CH_2)_5 \cdot NH_2 + CO_2$$
$$\text{lysine} \qquad\qquad\qquad\qquad \text{cadaverine}$$

V. Metabolic Interrelationships of Fats, Carbohydrates and Proteins

It is known that excess carbohydrate is converted to fat and stored as such. Various theories have been advanced to explain this phenomenon, a plausible representation being as follows:

$$CH_3 \cdot CHO + CH_3 \cdot CO \cdot CO \cdot OH \rightarrow CH_3 \cdot CH(OH) \cdot (H)CH \cdot CO \cdot CO \cdot OH$$

$$- H_2O \rightarrow CH_3 \cdot HC{:}CH \cdot CO \cdot CO \cdot OH,$$

then $CH_3 \cdot HC{:}CH \cdot CO \cdot CO \cdot OH - CO_2 \rightarrow CH_3 \cdot HC{:}CH \cdot CHO$, and

$$CH_3 \cdot HC{:}CH \cdot CHO + \text{biochemical hydrogenation} \rightarrow$$
$$CH_3 \cdot CH_2 \cdot CH_2 \cdot CHO$$

The butanal, in turn, may condense with pyruvic acid. In this manner two carbon atoms may be added to the chain at a time. This theory, therefore, can account for the fact that all the natural occurring fatty acids, without significant exceptions, contain an even number of carbon atoms.

The biochemical interrelationship of fats, carbohydrates, and proteins, may be shown by the following:

$$CH_2OH \cdot HCOH \cdot HCOH \cdot HOCH \cdot HCOH \cdot CHO$$
$$(D\text{-glucose})$$
$$\updownarrow$$

$$\begin{array}{l} H_2C \cdot OH \\ | \\ H \cdot C \cdot OH \\ | \\ H_2C \cdot OH \\ \text{(glycerol)} \end{array} \rightleftarrows \begin{array}{l} CH_2OH \cdot HCOH \cdot CHO \\ (d\text{-glyceraldehyde}) \\ \\ \updownarrow \end{array}$$

[1] Krebs and Henseleit. *Z. physiol. Chem.*, *210*, 33(1932). See Bodansky, 4th. ed., p. 411 (1938).

$$\text{CH}_3\cdot\text{CO}\cdot\text{CHO} \quad \rightleftarrows (\text{CH}_3\cdot\text{HOCH}\cdot\text{CO}\cdot\text{OH}) \rightarrow \text{CH}_3\cdot\text{CHNH}_2\cdot\text{CO}\cdot\text{OH, etc}$$
$$\text{(pyruvic aldehyde)} \qquad \text{(lactic acid)} \qquad \text{(alanine)}$$

$$\text{HO}\cdot\text{OC}\cdot\text{R} \;\leftarrow\quad \text{CH}_3\cdot\text{CO}\cdot\text{CO}\cdot\text{OH} \qquad\qquad\qquad \updownarrow$$
$$\text{(acids)} \qquad\qquad \text{(pyruvic acid)} \qquad\qquad\qquad \text{(proteins)}$$

$$\text{H}_2\text{C}\cdot\text{O}\cdot\text{OC}\cdot\text{R}$$
$$\text{H}\cdot\text{C}\cdot\text{O}\cdot\text{OC}\cdot\text{R} \qquad \text{CH}_3\cdot\text{CHO}$$
$$\text{(acetaldehyde)}$$
$$\text{H}_2\text{C}\cdot\text{O}\cdot\text{OC}\cdot\text{R} \qquad \text{CH}_3\cdot\text{CO}\cdot\text{OH} \qquad \rightarrow \qquad \text{CO}_2 \qquad + \qquad \text{H}_2\text{O}$$
$$\text{(fats)} \qquad\qquad \text{(acetic acid)} \qquad\qquad \text{(carbon dioxide)} \qquad \text{(water)}$$

VI. Enzymes

Enzymes are biochemical catalysts that enable living tissues to carry out readily processes that the chemist can bring about only by the use of rigorous conditions as high temperatures, strong reagents, or protracted reaction. When separated from the tissue forming them, they may continue to be active. They are rather easily destroyed by heat or by acids, alkalies, or in some cases by heavy metal salts. While their chemical nature is not definitely established, some have been obtained as or closely associated with crystalline proteins.

VII. Vitamins[1]

Vitamin A, which maintains the integrity of the epithelial tissue, is distributed in close association with colored substances as in green leaves and colored vegetables. It is replaceable in the animal body by three carotene isomers, to which it seems to be closely related structurally.

Beta carotene, $C_{40}H_{56}$, forms vitamin A

Vitamin A, provisional formula

Vitamin B appears to be a mixture of a number of substances with important biochemical functions. Among these are nicotinic acid (P. P. factor), riboflavin (B_2 or G), thiamine hydrochloride (B_1), 5-hydroxy-6-methyl-3, 4-pyridinedicarbinol (H), and the important B_{12} group which protect against pernicious anemia.

Vitamin C, which protects against scurvy, is found in fresh fruits and vegetables and actively growing plant tissue. Its synthesis has become commercial.

[1] Cf. The Weston-Levine Vitamin Chart, by Roe E. Remington, 280 Calhoun Street, Charleston, S. C. Also, News Ed., Ind. Eng. Chem., *18*, 670 (1940).

$$CH_2OH \cdot CH \cdot CH \quad C{:}O \rightleftarrows CH_2OH \cdot CH \cdot CH \quad C{:}O$$

Vitamin C (*l*-ascorbic acid.   Trade name, cevitamic acid).

Vitamin D, which protects against rickets, occurs in considerable amounts in the liver oils of many species of fish.   It is also found in butter, eggs, and the liver tissue of some mammals.   It, or a substance similar to it in biochemical potency, may be prepared by subjecting ergosterol to irradiation with ultraviolet light. While the chemical nature is not definitely known, it presumably has a configuration quite similar to that of ergosterol.

Ergosterol, $C_{28}H_{44}O$, a provisional formula

Vitamin E, which is necessary for the reproduction of the white rat, is found in the oil of the germs of the cereal grains.   Its chemical nature has been shown to be:

Vitamin K, the blood clotting factor, is:

β-Methylanthraquinone has vitamin K activity.

VIII. Other Factors

ACTH, Cortisone, and other factors aid materially in maintaining normal health.

CHAPTER XV

NON-METALLIC AND METALLIC ALKYL DERIVATIVES

I. Non-Metallic Alkyl Derivatives

A. Introduction:

Ammonia (NH_3), phosphine (PH_3), and arsine (AsH_3) form corresponding derivatives in which one or more of the hydrogen atoms are replaced by alkyl groups. The derivatives are known respectively as primary, secondary, or tertiary (depending on the number of hydrogen atoms replaced) amines, phosphines, or arsines.

The mono- and dialkyl arsines are not very stable, and only tertiary stibines and bismuthines are known.

Nitrogen, phosphorus, arsenic, and antimony, all form alkyl onium compounds of the types $R \cdot NH_3I$, R_2NH_2I, R_3NHI, and R_4NI.

The methods of preparation and the chemical properties of the alkyl derivatives of phosphorus and arsenic will indicate their similarity to the corresponding derivatives of ammonia. They possess the same types of isomerism.

Some of the arylchloroarsines were used as poisonous gases during the World War. Iron cacodylate is used in the treatment of anemia.

B. Nomenclature:

$CH_3 \cdot PH_2$	methylphosphine	$CH_3 \cdot AsH_2$	methylarsine
$(CH_3)_2PH$	dimethylphosphine	$(CH_3)_2AsH$	dimethylarsine
$(CH_3)_3P$	trimethylphosphine	$(CH_3)_3As$	trimethylarsine
$(CH_3)_4P \cdot I$	tetramethylphosphonium iodide	$(CH_3)_4As \cdot I$	tetramethylarsonium iodide
$(CH_3)_4P \cdot OH$	tetramethylphosphonium hydroxide	$(CH_3)_4As \cdot OH$	tetramethylarsonium hydroxide

C. Preparations:

I. THE ALKYL DERIVATIVES OF PHOSPHORUS MAY BE PREPARED:

1. *By heating alkyl iodides with phosphonium iodide in the presence of zinc oxide, followed by hydrolysis, to give primary phosphines:*

$$2 \, R \cdot I + 2 \, PH_4I \,/ZnO, \, 150° \rightarrow 2 \, R \cdot PH_3I + ZnI_2 + H_2O, \text{ then}$$

$$R \cdot PH_3I + H_2O \rightarrow R \cdot PH_2 \text{ (alkylphosphine)} + HI + H_2O$$

2. *By heating two moles of alkyl iodide with one mole of phosphonium iodide in the presence of zinc oxide, followed by hydrolysis with hot alkali, to give a secondary phosphine:*

$$2 \, R \cdot I + PH_4I \,/ZnO, \, 150° \rightarrow R_2PH_2I + ZnI_2 + H_2O, \text{ then}$$

$$R_2PH_2I + NaOH, \text{ aq., boil} \rightarrow R_2PH \text{ (dialkylphosphine)} + NaI + H_2O$$

165

3. *By the reaction of alkyl iodides on* (a) *phosphonium iodide in the presence of zinc oxide, followed by hydrolysis, or* (b) *calcium phosphide, to give in both cases tertiary phosphines:*

a. $6 \text{ R·I} + 2 \text{ PH}_4\text{I}/3 \text{ ZnO, } 180° \rightarrow 2 \text{ R}_3\text{PHI} + 3 \text{ ZnI}_2 + 3 \text{ H}_2\text{O,}$

   $\text{R}_3\text{PHI} + \text{NaOH, aq., heat} \rightarrow \text{R}_3\text{P (trialkylphosphine)} + \text{NaI} + \text{H}_2\text{O}$

b. $6 \text{ R·I} + \text{Ca}_3\text{P}_2 \rightarrow 2 \text{ R}_3\text{P} + 3 \text{ CaI}_2$

   The tertiary phosphines may also be made by the reaction of phosphorus trichloride on dialkylzinc:

   $3 \text{ R}_2\text{Zn} + 2 \text{ PCl}_3 \rightarrow 2 \text{ R}_3\text{P} + 3 \text{ ZnCl}_2$

4. *By the action of alkyl iodides on phosphonium iodide in the presence of zinc oxide to give* (a) *tetraalkylphosphonium iodide or* (b) *tetraalkylphosphonium hydroxide when subjected to hydrolysis:*

a. $4 \text{ R·I} + \text{PH}_4\text{I}/2 \text{ ZnO, } 180° \rightarrow \text{R}_4\text{P·I} + 2 \text{ ZnI}_2 + 2 \text{ H}_2\text{O}$, then

b. $\text{R}_4\text{P·I} + \text{NaOH, aq.} \rightarrow \text{R}_4\text{P·OH} + \text{NaI}$

II. THE ALKYL DERIVATIVES OF ARSENIC MAY BE PREPARED:

1. *By the action of* (a) *alkyl halides on sodium arsenide, or* (b) *dialkyl zinc on arsenic trichloride, to give in each case the tertiary arsines:*

a. $3 \text{ R·I} + \text{Na}_3\text{As} \rightarrow \text{R}_3\text{As (trialkylarsine)} + 3 \text{ NaI}$

b. $3 \text{ R}_2\text{Zn} + 2 \text{ AsCl}_3 \rightarrow 2 \text{ R}_3\text{As} + 3 \text{ ZnCl}_2$

   The primary and secondary arsines are quite unstable.

2. *By the distillation of arsenic trioxide with potassium ethanoate to give bisdimethylarsenic oxide (cacodyl oxide):*

   $\text{As}_2\text{O}_3 + 4 \text{ CH}_3\text{·CO·OK, distil} \rightarrow ((\text{CH}_3)_2\text{As})_2\text{O} + 2 \text{ K}_2\text{CO}_3 + 2 \text{ CO}_2$

3. *By the following reactions to give chloromethylarsines:*

a. $(\text{CH}_3)_3\text{As} + \text{Cl}_2 \rightarrow (\text{CH}_3)_3\text{AsCl}_2$ (trimethylarsonium dichloride)

b. $(\text{CH}_3)_3\text{AsCl}_2$, heat $\rightarrow (\text{CH}_3)_2\text{AsCl}$ (chlorodimethylarsine) $+ \text{CH}_3\text{·Cl}$

c. $(\text{CH}_3)_2\text{AsCl} + \text{Cl}_2 \rightarrow (\text{CH}_3)_2\text{AsCl}_3$ (dimethylarsonium trichloride)

d. $(\text{CH}_3)_2\text{AsCl}_3$, heat $\rightarrow \text{CH}_3\text{·AsCl}_2$ (dichloromethylarsine) $+ \text{CH}_3\text{·Cl}$

4. *By the action of arsenic trichloride on acetylene to give dichloro-2-chlorovinylarsine (Lewisite):*

$$\text{AsCl}_3 + \text{H·C:C·H} \rightarrow \overset{\text{H H}}{\text{Cl·C:C·AsCl}_2}$$

D. Physical Properties:

The alkylphosphines are colorless, strongly refractive, volatile liquids with a disagreeable odor.   Methylarsine (b.p. 2) and ethylarsine (b.p. 36) are both colorless, poisonous compounds with a disagreeable odor.   These derivatives are all quite insoluble in water but are soluble in alcohol and ether.

The alkylchloroarsines are colorless liquids with a disagreeable, stupefying odor, and they tend to irritate the mucous membrane. Bisdimethylarsenic oxide (cacodyl oxide, b.p. 120) is a liquid with a stupefying odor which produces nausea and severe irritation of the nasal mucous membrane. It is quite soluble in water.

E. Chemical Properties:

I. THE ALKYLPHOSPHINES REACT UNDER PROPER CONDITIONS:

1. To give addition products when treated with:

a. Acids $R_2PH + HX \rightarrow R_2PH \cdot HX$ (dialkylphosphonium halide)

b. Halogens $R_3P + X_2 \rightarrow R_3PX_2$ (trialkylphosphine dihalide)

c. Sulfur $R_3P + S \rightarrow R_3PS$ (trialkylphosphine sulfide)

d. Carbon disulfide $R_3P + CS_2 \rightarrow R_3P\overset{S}{\cdots}C{:}S$ (trialkylphosphine carbon disulfide)

2. To give (a) rather stable oxidized derivatives if carefully oxidized, or (b) spontaneous combustion products in contact with air:

   a. 1. $CH_3 \cdot PH_2 + 2\ HNO_3$, aq. $\rightarrow CH_3 \cdot PO\ (OH)_2 + 2\ NO + H_2O$

       2. $3\ (CH_3)_2PH + 2\ HNO_3$, aq. $\rightarrow 3\ (CH_3)_2P \cdot OH + 2\ NO + H_2O$

       3. $3\ (CH_3)_3P + 2\ HNO_3 \rightarrow 3\ (CH_3)_3P(\rightarrow O) + 2\ NO + H_2O$

   b. $2\ (CH_3)_3\ P + 13\ O_2$, burning $\rightarrow 6\ CO_2 + 2\ H_3PO_4 + 6\ H_2O$

3. To give very slightly basic solutions with water:

$$R_3P + H_2O \rightleftarrows R_3PH^+ + OH^-$$

Since phosphine is less basic than ammonia, the alkyl derivatives of phosphine should be less basic than the corresponding derivatives of ammonia, but, like the amines, the phosphines become more basic with the introduction of additional alkyl groups.

The following reactions should be contrasted:

$(CH_3)_4P \cdot OH$, heat $\rightarrow (CH_3)_3P(\rightarrow O) + CH_4$, and

$(CH_3)_4N \cdot OH$, heat $\rightarrow (CH_3)_3N + CH_3 \cdot OH$

II. THE ALKYL ARSINES REACT UNDER PROPER CONDITIONS:

1. To give addition reactions when treated with:

a. Halogens $R_3As + X_2 \rightarrow R_3AsX_2$ (trialkylarsine dihalide)

b. Sulfur $R_3As + S \rightarrow R_3As(\rightarrow S)$ (trialkylarsine sulfide)

c. Oxygen $2\ R_3As + O_2 \rightarrow 2\ R_3As(\rightarrow O)$ (trialkylarsine oxide)

2. To give practically neutral solutions with water:

Even the trialkylarsines have such weak basic properties that they do not form salts with acids.

3. To give the following reactions of the dichloro derivatives:

   a. $CH_3AsCl_2 + H_2S \rightarrow CH_3 \cdot As{:}S$ (methylarsenic sulfide) $+ 2\ HCl$

   b. $CH_3 \cdot AsCl_2 + Na_2CO_3$, aq. $\rightarrow CH_3 \cdot As{:}O$ (methylarsenic oxide) $+ 2\ NaCl + CO_2$

II. Alkyl Metallic Derivatives

A. Introduction:

The alkyl metallic derivatives are compounds in which a metallic atom is linked directly to one or more alkyl groups. They may be represented by the general type formulas $R \cdot M \cdot X$ or $R_2 M$, where R is an alkyl group, M is a metallic atom, and X is a halogen atom.

Frankland, in his search for free radicals in 1849, treated alkyl iodides with zinc and obtained the first organometallic compound, ethylzinc iodide $(CH_3 \cdot CH_2 \cdot Zn \cdot I)$. This was the beginning of his important syntheses with zinc alkyls.

In 1901 Victor Grignard, at the suggestion of Barbier, studied the reaction between an ethereal solution of a ketone, methyl iodide, and magnesium. As a result, he discovered the reagent $R \cdot Mg \cdot X$ which bears his name. Because the technic involved in the use of the Grignard reagent is comparatively simple, the use of alkylmagnesium compounds have almost replaced the use of alkylzinc compounds in organic synthesis.

Aluminum, beryllium, cadmium, germanium, lead, mercury, potassium, silicon, sodium, and tin alkyls have been prepared. The derivatives increase in stability with a decrease in metallic properties. Silicon, for example, is a border line element and its alkyl derivatives are almost as stable as the corresponding derivatives of carbon. The organomercury compounds, however, that are of biomedicinal interest are of other types.

These compounds are all synthetic intermediates, although the closely related metallic carbides occur naturally to some extent. One former theory postulated the formation of petroleum from natural carbides.

The methods of preparation and the chemical properties will indicate that a carbon atom is linked directly to a metallic atom. A quantitative analysis of the concentration of a Grignard reagent may be made by the use of the reaction: $R \cdot Mg \cdot X + H \cdot OH \rightarrow R \cdot H + Mg (OH) X$. If the hydrocarbon formed in the reaction is a gas it may be measured volumetrically, otherwise the basic magnesium halide is titrated with standard acid.

The metallic alkyls find their principal use in organic syntheses, but tetraethyllead is used, along with 1, 2-dibromoethane and 1, 2-dichloroethane, in ethyl gasoline as an anti-knock.

B. Nomenclature:

These compounds are named by prefixing the name of the alkyl radical to the name of the metal. $CH_3 \cdot Mg \cdot Br$, for example, is methylmagnesium bromide; $CH_3 \cdot CH_2 \cdot CH_2 \cdot Zn \cdot I$, propylzinc iodide; and $(CH_3 \cdot CH_2)_2 Zn$, diethylzinc.

C. Preparations:

I. THE ORGANOMETALLIC HALIDES MAY BE PREPARED:

1. *By the action of an alkyl halide on (a) zinc dust, or (b) magnesium in dry ether or benzene:*

   a. $R \cdot X + Zn$, finely divided, reflux $\rightarrow R \cdot Zn \cdot X$ (Frankland Reagent)

   b. $R \cdot X + Mg$, dry ether $\rightarrow R \cdot Mg \cdot X$ (Grignard Reagent)

   The Grignard reagent is always prepared in an anhydrous ether solution free from alcohol. If benzene or xylene is used as the solvent, a catalyst is required. The reagent is usually used in the solvent in which it is prepared, but by evaporation, when ether is used as the solvent, the reagent may be obtained as a colorless, crystalline solid which contains one or two moles of ether of crystallization.

In laboratory practice, the alkyl bromides are usually employed since the reaction goes less readily with the chlorides and more side reactions are encountered by the use of the iodides. Decreasing yields are obtained by the use of secondary and tertiary alkyl halides, and by increasing the relative size of the alkyl group.

2. By the action of alkylmagnesium or alkylzinc halides on the metallic halides of the less active metals:

$$R \cdot Mg \cdot X / dry\ ether + HgX_2 \rightarrow R \cdot Hg \cdot X + MgX_2$$

II. THE ORGANOMETALLIC COMPOUNDS MAY BE PREPARED:

1. By the action of metallic sodium on (a) dialkylzinc or dialkylmercury in dry benzene to give alkylsodium or (b) diphenylmercury in pure thiophene to give phenylsodium:

   a. $(CH_3 \cdot CH_2)_2 Zn + 2\ Na$, dry benzene $\rightarrow 2\ CH_3 \cdot CH_2 \cdot Na + Zn$

   b. $(C_6H_5)_2 Hg + 2\ Na$, dry thiophene $\rightarrow 2\ C_6H_5 \cdot Na + Hg$

2. By the reaction of the metallic halides of the less reactive metals with (a) alkylmagnesium halides, or (b) dialkylzinc:

   a. 1. $4\ R \cdot Mg \cdot X$, dry ether $+ 2\ PbCl_2 \rightarrow R_4Pb + Pb + 2MgX_2 + 2MgCl_2$

      2. $2\ R \cdot Mg \cdot X$, dry ether $+ HgCl_2 \rightarrow R_2Hg + MgX_2 + MgCl_2$

   b. R_2Zn, inert solvent $+ HgCl_2 \rightarrow R_2Hg + ZnCl_2$

3. By the action of intermetallic compounds with alkyl halides:

   a. $2\ R \cdot I + Na_2Hg \rightarrow R_2Hg + 2\ NaI$

   b. $4\ R \cdot X + 4\ NaPb \rightarrow R_4Pb + 4\ NaX + 3\ Pb$, and

      $4\ CH_3 \cdot CH_2 \cdot Cl + 4\ NaPb \rightarrow (CH_3 \cdot CH_2)_4 Pb + 4\ NaCl + 3\ Pb$

This reaction is employed commercially in the production of tetraethyllead, about 20 million pounds being used annually.

   c. $2\ C_6H_5 \cdot Br + 2\ Na \cdot Hg$, dry toluene $\rightarrow (C_6H_5)_2 Hg + Hg + 2\ NaBr$

A trace of ethyl acetate is employed in this reaction as a catalyst.

D. Physical Properties:

Methylsodium and methylpotassium are colorless amorphous solids. Dimethylzinc (b.p. 46) is a colorless, mobile, strongly refractive liquid. Dimethylmercury, trimethylaluminum, tetramethyltin, and tetramethyllead, are volatile liquids that distil without decomposition. Tetraethyllead (b.p. 198-202) is also a liquid.

E. Chemical Properties:

I. THE GRIGNARD REAGENT REACTS UNDER PROPER CONDITIONS:

1. To give hydrocarbons when treated with reagents containing labile hydrogen atoms as:

a. Acids	R	MgX	+	R·CO·O	H
b. Water	R	MgX	+	HO	H
c. Alcohols	R	MgX	+	RO	H
d. Thioalcohols	R	MgX	+	RS	H
e. Amides	R	MgX	+	R·CO·NH	H
f. Amines	R	MgX	+	R·NH	H
g. Ammonia	R	MgX	+	H_2N	H

That is, $R \cdot MgX + R \cdot CO \cdot OH \rightarrow R \cdot H + R \cdot CO \cdot O \cdot MgX$

2. *To give condensation products when treated with reagents containing labile halogen atoms as:*

a. Phosgene	2 R	MgX	+	Cl_2	C:O
b. Acid halides	R	MgX	+	X	OC·R
c. Alkyl halides	R	MgX	+	X	R
d. Aryl halides	R	MgX	+	X	Ar
e. Halo derivatives	R	MgX	+	X	$CH_2 \cdot O \cdot R$
	R	MgX	+	X	$CH_2 \cdot HC{:}CH_2$

For example, $2 \ R \cdot MgX + Cl_2C{:}O \rightarrow R_2C{:}O + MgX_2 + MgCl_2$

3. *To give addition products with reagents containing (a) the carbonyl group, or (b) other active unsaturated linkages as:*

a. R⁻ · · · ·   ·C:O   ·CH_2 ·CHR   ·CR_2 ·C(OR) R   ·C(NR_2)R

X·Mg – .   ·O   ·O   ·O   ·O   ·O   ·O

The alkyl group always adds directly to the carbon atom and the MgX residue becomes linked to the oxygen atom.

b. R⁻ · · · ·   ·S:O   ·C:S   ·C·R   ·C:N·R   ·$CH_2 \cdot CH_2$ ·C:N·Ar

X·Mg – .   ·O   ·S   ·N   · | · · · · ·O   ·O

The alkyl group links to carbon or its equivalent, whereas the Mg·X residue adds to the more negative of the two groups.

4. *To give double decomposition products with:*

a. Organic anhydrides	R	MgX	+·	R·CO·O	OC·R
b. Sulfonic acid esters	R	MgX	+	$ArSO_2 \cdot O$	R
c. Inorganic halides as $AsCl_3$, $BiCl_3$,	R	MgX	+	Cl	$AsCl_2$
$GeCl_4$, $HgCl_2$, PCl_3, $SbCl_3$, and $SiCl_4$	3R	MgX	+	Cl_3	As

II. THE ORGANOMETALLIC HALIDES AND ORGANOMETALLIC COMPOUNDS REACT:

In general, to give the same reactions that have been indicated above for alkylmagnesium halides.

CHAPTER XVI

ALICYCLIC COMPOUNDS: THE CYCLOALKANES AND CYCLOALKENES

A. Introduction:

The cycloalkanes constitute a saturated, cyclic, homologous series which is isomeric with the alkenes; the unsaturated, cyclic, homologous series known as the cycloalkenes is isomeric with the alkynes. Representative members are:

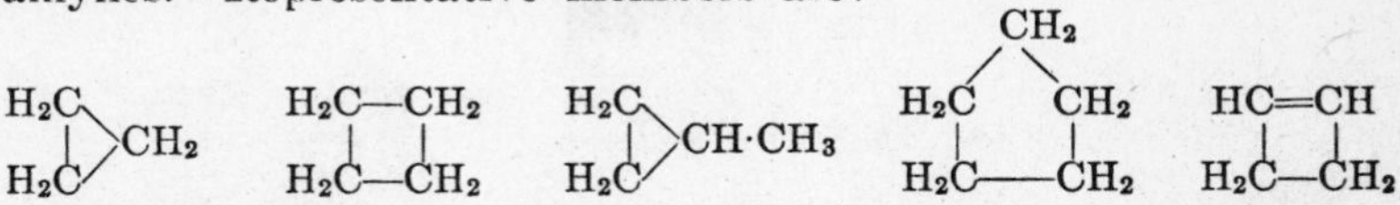

$$
\begin{array}{ccccc}
H_2C{\diagdown}CH_2 & H_2C{-}CH_2 & H_2C{\diagdown}CH{\cdot}CH_3 & H_2C{\diagup}^{CH_2}{\diagdown}CH_2 & HC{=}CH \\
H_2C{\diagup} & H_2C{-}CH_2 & H_2C{\diagup} & H_2C{-}CH_2 & H_2C{-}CH_2 \\
\text{cyclo-} & \text{cyclo-} & \text{methylcyclo-} & \text{cyclo-} & \text{cyclo-} \\
\text{propane} & \text{butane} & \text{propane} & \text{pentane} & \text{butene}
\end{array}
$$

Cyclohexanes and cyclopentanes occur in appreciable amounts in the petroleum obtained from Russia, California, and Texas, and in traces in most petroleum. These cyclic compounds are known as naphthenes.

The cycloalkanes and cycloalkenes are ingredients of gasoline, and cyclo-alkanes are doubtless the principal ingredients in lubricating oils. Cyclo-propane is used as an anesthetic.

One can predict, qualitatively at least, the relative stability of a given cyclic structure by means of the Baeyer Strain Theory. This theory assumes that the four valence forces of a carbon atom are symmetrically directed from the center of the atom to the four corners of an imaginary tetrahedron. According to this concept the angle between any two valence forces is 109° 28′. The minimum strain should exist in a cyclohydrocarbon, therefore, when the angle between adjacent valence forces is of this order.

In the two-dimensional representation of the formulas shown below, it is seen that cyclopentane approaches these optimum conditions. Cyclic structures with more or less than five carbon atoms have their valence forces displaced from the normal to a greater degree and are, consequently, subjected to greater strains.

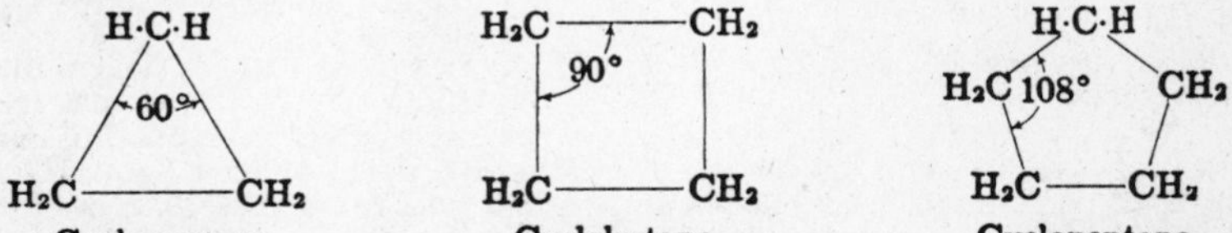

Cyclopropane

Cyclobutane

Cyclopentane

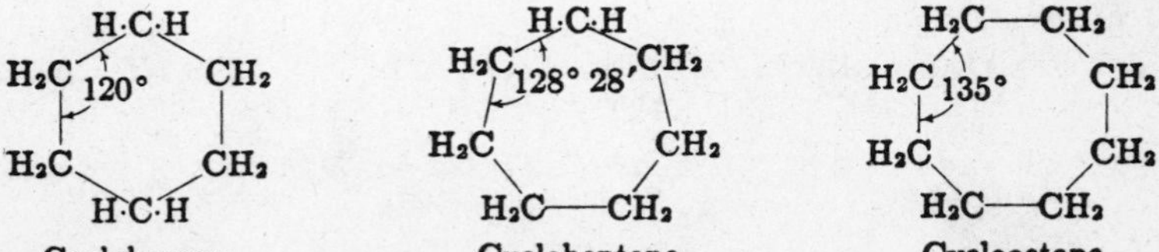

Cyclohexane

Cycloheptane

Cyclooctane

This explanation is only qualitatively true as indicated by the fact that cyclohexane is of the same order of stability as cyclopentane. It should be noted, however, that the angular measurements indicated for these examples have been calculated on the false assumption that all of the carbon atoms comprising the ring lie in the same plane. In compounds containing more than four carbon atoms in the ring, there is a tendency for distortion from a single plane. These angular measurements are, consequently, only an approximation. Cycloparaffins containing as many as 32 methylene groups have been prepared, however, and found to be comparatively stable.

B. Nomenclature:

These compounds are named as cycloalkanes, cycloalkenes, or cycloalkadienes. Specific examples are given above.

C. Preparations:

THE CYCLOALKANES AND THE CYCLOALKENES MAY BE PREPARED:

1. *By the action of active metals (sodium, magnesium, or zinc) on certain dihalogenated hydrocarbons:*

   a. $X \cdot (CH_2)_x \cdot X + 2 M(\text{or } M) \rightarrow (CH_2)_x$ (a cycloalkane) $+ 2$ MBr (or MBr$_2$), and

   $$Br \cdot CH_2 \cdot CH_2 \cdot CH_2 \cdot Br + Zn \rightarrow \begin{array}{c} H_2C \\ | \\ H_2C \end{array}\!\!\!\!> CH_2 \text{ (cyclopropane) } + ZnBr_2$$

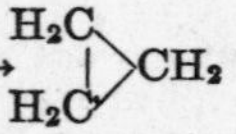

   b. $$Br \cdot CH_2 \cdot HC:CH \cdot CH_2 \cdot Br + Zn \rightarrow \begin{array}{c} HC=CH \\ | \quad | \\ H_2C-CH_2 \end{array} \text{ (cyclobutene) } + ZnBr_2\,[1]$$

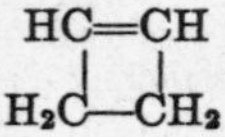

2. *By the partial or complete reduction of benzene in the presence of heated nickel to give cyclohexadienes, cyclohexene, or cyclohexane:*

   $$C_6H_6 + 3 H_2 / \text{(finely divided nickel, heat below } 300°) \rightarrow C_6H_8 \rightarrow C_6H_{10} \rightarrow C_6H_{12}$$

   The cyclohexadienes and cyclohexene are not readily obtained from the above reaction, but may be obtained by the fractionation of the naphthenes which occur in certain petroleum distillates since they are formed by cracking reactions.

D. Physical Properties:

Cyclopropane and cyclobutane are both colorless gases, but cyclopentane is a colorless liquid, b.p. 50.5 °C. The boiling points of the cycloalkanes are usually about 10 to 20° higher than those of the corresponding alkanes. Their density is less than one, but increases with increase in molecular weight. The physical constants for the cycloalkenes are, as a rule, rather close to those for the corresponding cycloalkanes. Both of these series of compounds are insoluble in water, but soluble in alcohol and ether.

E. Chemical Properties:

THE CYCLOALKANES REACT, DEPENDING ON THE NUMBER OF CARBON ATOMS IN THE RING:

1. *To give addition products, as illustrated by the action of bromine on cyclopropane to give 1, 3-dibromopropane:*

---

[1] Ostromuisslenski, *J. Chem. Soc., 110*, I, 242. The reaction gives butadiene also.

$$\begin{matrix} H_2C \\ \quad | \\ H_2C \end{matrix} \! \! > \! \! CH_2 + Br_2 \rightarrow Br \cdot CH_2 \cdot CH_2 \cdot CH_2 \cdot Br \quad (1,\, 3\text{-dibromopropane})$$

2. To give substitution products, as illustrated by the action of bromine on cyclopentane to give bromocyclopentane:

$$\begin{matrix} CH_2 \cdot CH_2 \\ \quad | \\ CH_2 \cdot CH_2 \end{matrix} \! \! > \! \! CH_2 + Br_2 \rightarrow \begin{matrix} CH_2 \cdot CH_2 \\ \quad | \\ CH_2 \cdot CH_2 \end{matrix} \! \! > \! \! CH \cdot Br \quad \begin{matrix} \text{(bromocyclopentane)} \\ + \; HBr \end{matrix}$$

3. To give oxidation products, when treated for some time with alkaline permanganate solution.

In passing from a given member of one series to the corresponding member of another series, the chemical activity increases in general in the order alkanes, cycloalkanes, cycloalkenes, alkenes.

THE CYCLOALKENES REACT:

1. To give, as a rule, addition products as their first derivatives:

C_6H_8 (cyclohexadiene) + 2 H_2/(Ni, heat) → C_6H_{10} (cyclohexene) → C_6H_{12} (cyclohexane).

2. To give, in certain cases, substitution products.

The synthesis of cycloöctatetraene from pseudopelletierine has assumed the status of a questionable classic in organic chemistry because some of its properties resembled those of styrene, its monomer, more than those of benzene. Recent studies, however, have confirmed the previous syntheses of both Willstätter and Reppe, and cycloöctatetraene is now accepted as a known member of the aromatic series, containing a conjugated system of double bonds, perfectly analogous to the structure of benzene.

CHAPTER XVII

AROMATIC HYDROCARBONS: BENZENE, NAPHTHALENE, AND ANTHRACENE

A. Introduction:

Most of the compounds considered previously belong to the open-chain series and might be considered as derivatives of methane (CH_4). They constitute the aliphatic series. The cycloalkanes and cycloalkenes, although they are ring compounds, closely resemble the aliphatic compounds and are usually classified as such. They may be considered, however, as transition compounds between the aliphatic and aromatic series. Most of the aromatic compounds may be considered as derivatives of benzene (C_6H_6).

The relative ease of halogenation, nitration, and sulfonation, and their solubility in methyl sulfate, serve to distinguish the aromatic hydrocarbons from the aliphatic hydrocarbons.

The limited amount of hydrogen in the benzene molecule would suggest that the compound is highly unsaturated. The tendency of benzene to undergo substitution much more readily than addition, has caused much speculation as to its structure. As a result, various formulas have been proposed in the attempt to account for the valences of the carbon atoms without resorting to the use of unsaturated linkages. Some of the proposed formulas are:

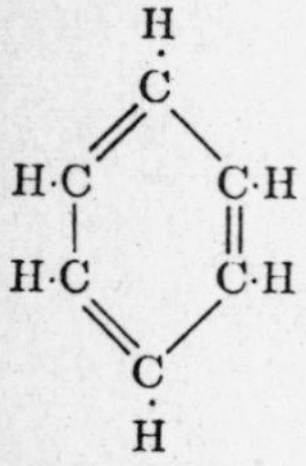

Kekulé 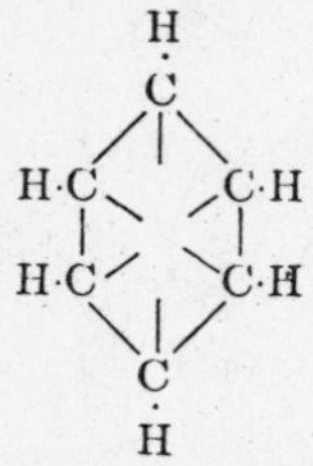   Armstrong-Baeyer 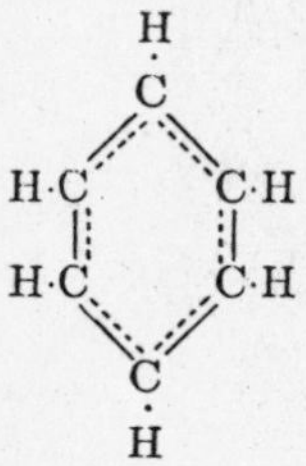   Thiele 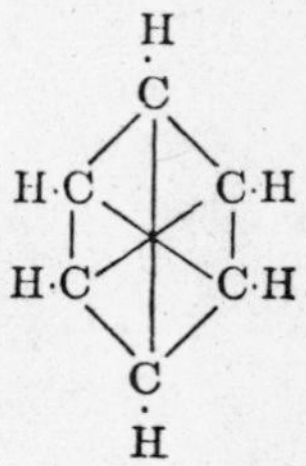   Claus

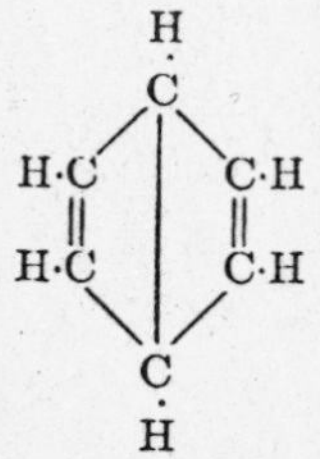

Dewar 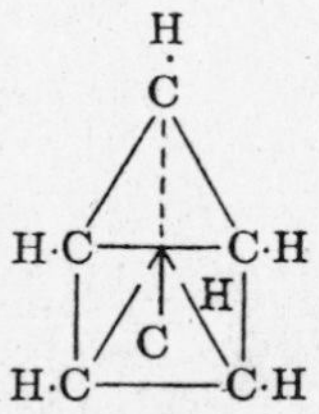   Ladenburg

Kekulé utilized the fourth valence of each of the carbon atoms by alternate single and double bonds. Later he suggested the existence of a dynamic system, that is, the unsaturated linkages oscillate from one position to the other. The Ladenburg formula must be rejected because it cannot be used to explain the structures of the oxidized derivatives of the three phthalic acids. While each of the other proposed formulas lends itself to an explanation of certain properties of benzene, the Kekulé formula is usually used. Throughout this outline, the skeletonized formula for benzene will be used. In this formula, a "C·H" group is understood to be located at each corner of the hexagon if no other group is indicated and alternate linkages in the ring are unsaturated.

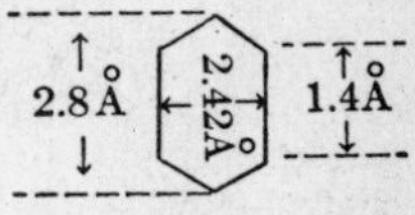

skeleton formula

The formulas for naphthalene, anthracene, and phenanthrene are:

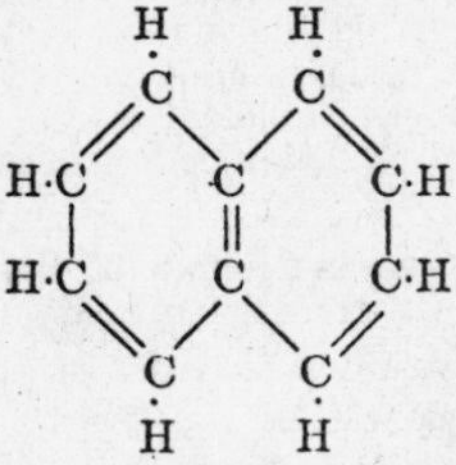

naphthalene 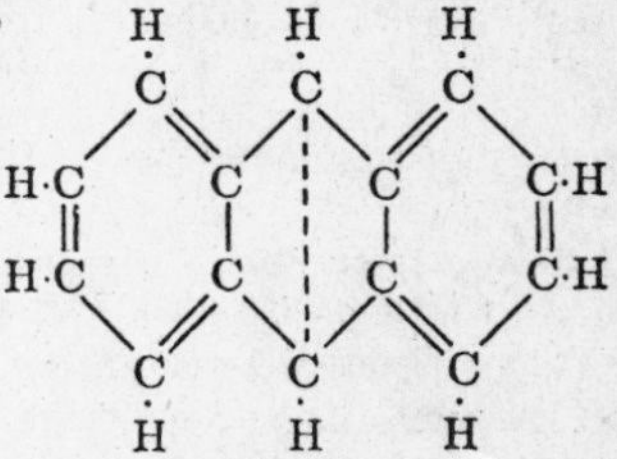   anthracene

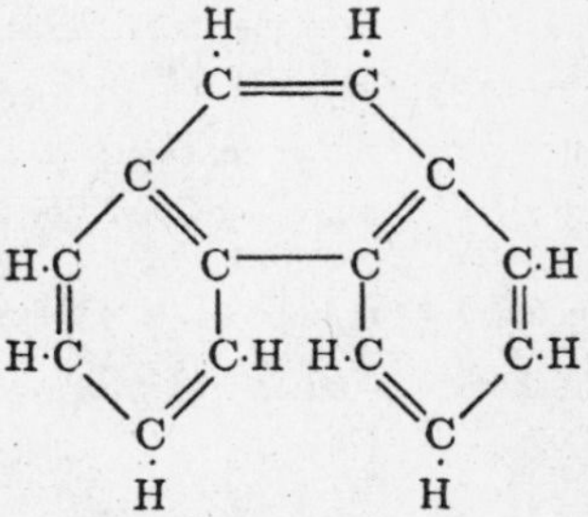

phenanthrene

The corresponding skeleton formulas are commonly used.

A ton of bituminous coal, through destructive distillation, yields approximately the following: coal gas, 11,000 cubic feet; ammonia gas, 2 to 6 cubic feet; light oil, 3 to 4 gallons; coal tar, 70 to 120 pounds; and coke, 1200 to 1500 pounds. The coal gas contains appreciable quantities of the more volatile aromatic hydrocarbons. From the light oils is obtained benzene, **toluene, xylene,** naphthalene, and other raw materials. Destructive distillation of the coal tar yields such important products as phenol, cresols, naphthalene, anthracene, carbazole, and phenanthrene.

Petroleum, as previously noted, is the main source of the aliphatic hydrocarbons. By use of the Bergius hydrogenation process, 100 gallons of petroleum yield as much as 105 gallons of gasoline.

Benzene serves as an excellent solvent and extracting medium in both industry and the laboratory.   Although benzene is present in most gasolines, it is used in blending some motor fuels.   Large quantities of benzene are used in a great variety of organic syntheses.   Toluene and xylene are used in the synthesis of a large number of useful products, such as dyes, explosives, medicinals, and photographic developers.   Toluene is used in the lacquer industry as a diluent.   Naphthalene, in the form of moth balls, is used to protect furs and woolens from moths.   It is used for the fumigation of greenhouses and in the manufacture of various dyes and dye intermediates.   It is the basic raw material for the preparation of phthalic anhydride.   Tetrahydronaphthalene (Tetralin) is a good solvent for fats and waxes and is claimed to be valuable in motor fuels as a carbon remover.   Anthracene is used in the manufacture of dyes.   Formerly, it was the basic material for the manufacture of anthraquinone, now made from phthalic anhydride.   The phenanthrene nucleus may have an important biological function as it is found in cholesterol, in the sex hormones, in certain carcinogenic compounds, in the digitalis glucosides, and in the venom of the toad.

The establishment of the structure of an aromatic compound is accomplished by a line of proof analogous to that given below for benzene and naphthalene:

a. The structure of benzene is indicated by the following:

   1. Analysis and synthesis, together with molecular weight determinations, lead to the conclusion that the molecular formula for benzene is correctly represented by C_6H_6.   This indicates a high degree of unsaturation.

   2. Each hydrogen atom in the benzene nucleus is equivalent, that is, the benzene structure is symmetrical as shown by:

      a. The absence of isomeric monosubstituted derivatives.

      b. The synthesis and structure of mesitylene.

      c. The synthesis and structure of the phthalic acids.

      d. The reduction of the phthalic acids by Baeyer.

      e. The application of the Körner orientation theory.

      f. The X-ray analysis of graphite by Bragg.

   3. The structure of benzene is dynamic, as shown by:

      a. The absence of isomeric o-disubstituted derivatives.

      b. The application of the Körner orientation theory.

b. The structure of naphthalene is indicated by its synthesis and the following set of reactions:

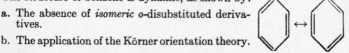

Oxidation of 1-nitronaphthalene indicates that the substituted ring is a true benzene ring, whereas the reduction and subsequent oxidation of the same nucleus indicates that the unsubstituted ring is a true benzene ring. Since other reactions show that one of the rings is alicyclic and the other aromatic, naphthalene, like benzene, must possess a dynamic structure.

The resonating systems of naphthalene and anthracene may be represented respectively by three and four formulas.

B. Nomenclature:

Aromatic compounds are named either as derivatives of the

corresponding hydrocarbon nucleus or by common names of historical origin. Examples are:

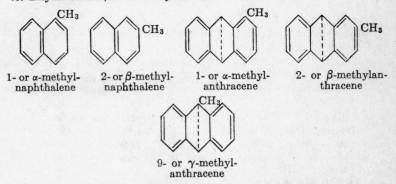

1,2-dimethyl-
benzene, *o-*
xylene

1, 3-dimethyl-
benzene, *m-*
xylene

1, 4-dimethyl-
benzene, *p-*
xylene

1, 2, 3- or *v-*
trimethyl-
benzene

1 2, 4- or *u-*
trimethyl-
benzene

1, 3, 5- or *s-*tri-
methylbenzene, **mesitylene**

The contraction *o-* is for ortho-, *m-* for meta-, *p-* for para-, *v-* for vicinal-, *u-* for unsymmetrical-, and *s-* for symmetrical-.

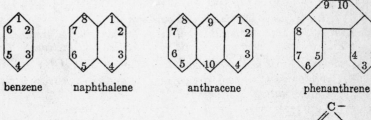

1- or α-methyl-
naphthalene

2- or β-methyl-
naphthalene

1- or α-methyl-
anthracene

2- or β-methylan-
thracene

9- or γ-methyl-
anthracene

In the I.U.C. system of nomenclature, the position of a substituent group is always indicated by arabic numerals, and the order of numbering the respective carbon atoms should be as indicated below:

benzene   naphthalene   anthracene   phenanthrene

The phenyl radical, which like the methyl radical is known only as an unstable reaction intermediate, is shown at the right.

C. Preparations:

I. Benzene and its alkyl derivatives

a. BENZENE MAY BE PREPARED:

1. *By the passage of ethyne (acetylene), methane, ethane, propane, butane, or hexane through hot tubes:*
   - a. 3 H·C∶C·H, pass through tube, 580°C → C_6H_6, 40% yield
   - b. $CH_3·CH_2·CH_2·CH_2·CH_2·CH_3/(Cr_2O_3,$ hot tower) → C_6H_6
     This reaction is employed commercially.

2. *By the fusion of sodium benzoate with soda lime:*

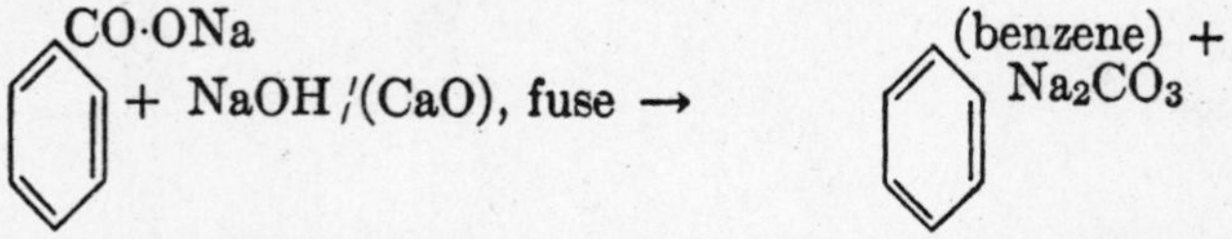

   + NaOH/(CaO), fuse → (benzene) + Na_2CO_3

   This reaction is analogous to the one used in the preparation of methane from sodium acetate.

3. *By heating phenol with zinc dust:*

   $C_6H_5·OH$ + Zn powder, distil → C_6H_6 (benzene) + ZnO

4. *By the hydrolysis of benzenesulfonic acid with superheated steam:*

   $C_6H_5·SO_2·OH$ + H_2O, superheated steam → C_6H_6 + H_2SO_4

   This hydrolysis proceeds more readily (a) in the presence of a catalyst such as phosphoric acid, or (b) if other substituent groups are present in the ring.

5. *By (a) scrubbing the coal gas with mineral seal oil from which benzene, toluene, and other volatile aromatic hydrocarbons are recovered by fractionation, or (b) the fractional distillation of the light-oil fraction obtained in the dry distillation of coal (commercial).*

b. TOLUENE MAY BE PREPARED:

1. *By the action of sodium on a mixture of a halobenzene and a halomethane (Wurtz-Fittig Reaction, cf. Wurtz Reaction):*

   $C_6H_5·Br$ + 2 Na + Br·CH_3, ether → $C_6H_5·CH_3$ + 2 NaBr
   Biphenyl and ethane are also obtained.

2. *By the action of an alkyl halide on benzene in the presence of anhydrous aluminum chloride (Friedel-Crafts Reaction):*

   $C_6H_5·H$ + Br·$CH_3/(AlCl_3,$ anhydrous) → $C_6H_5·CH_3$ + HBr

3. *By recovery from coal gas as in a.5, or from cyclization and dehydrogenation of a heptane.*

c. THE XYLENES MAY BE PREPARED:

1. *By the action of a halomethane on toluene in the presence of anhydrous aluminum chloride to give a mixture of the ortho, meta, and para isomers (Friedel-Crafts Reaction), or from dehydrogenation and cyclization of octane:*

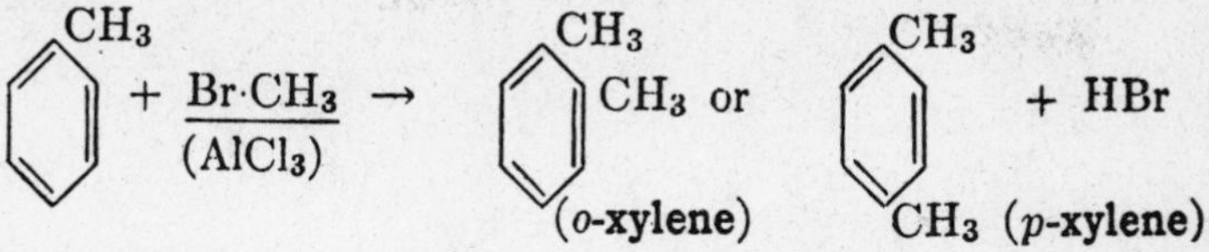

$$\text{(CH}_3\text{-benzene)} + \frac{\text{Br·CH}_3}{\text{(AlCl}_3)} \rightarrow \text{(o-xylene)} \text{ or } \text{(p-xylene)} + \text{HBr}$$

2. *By the following series of reactions to give the meta isomer:*

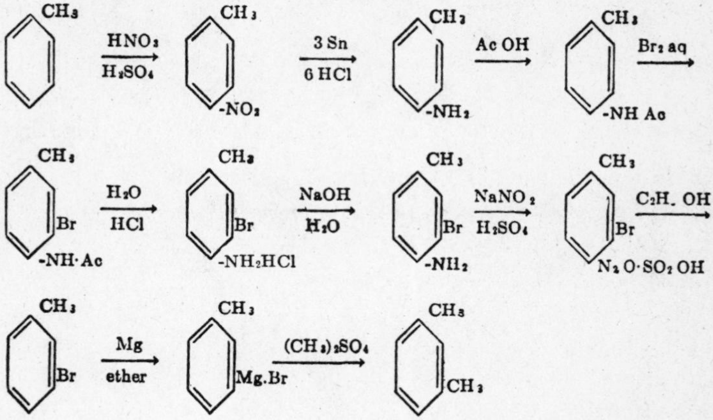

$$\xrightarrow[\text{H}_2\text{SO}_4]{\text{HNO}_3} \xrightarrow[\text{6 HCl}]{\text{3 Sn}} \xrightarrow{\text{Ac OH}} \xrightarrow{\text{Br}_2 \text{ aq}}$$

$$\xrightarrow[\text{HCl}]{\text{H}_2\text{O}} \xrightarrow[\text{H}_2\text{O}]{\text{NaOH}} \xrightarrow[\text{H}_2\text{SO}_4]{\text{NaNO}_2} \xrightarrow{\text{C}_2\text{H}_5\text{. OH}}$$

$$\xrightarrow[\text{ether}]{\text{Mg}} \xrightarrow{\text{(CH}_3\text{)}_2\text{SO}_4}$$

II. Naphthalene, Anthracene, and Phenanthrene

a. NAPHTHALENE MAY BE PREPARED:

1. *By passing benzene and ethyne (acetylene) through a hot tube:*

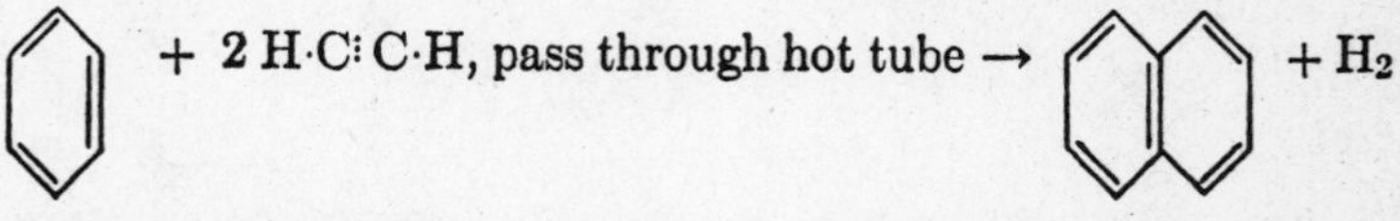

$$+ 2 \text{ H·C}\vdots\text{C·H, pass through hot tube} \rightarrow + \text{H}_2$$

2. *By the action of zinc (or sodium) on a mixture of 1, 2-bis-bromomethylbenzene (α, α'-dibromo-o-xylene) and 1, 1, 2, 2-tetrabromoethane (acetylene tetrabromide):*

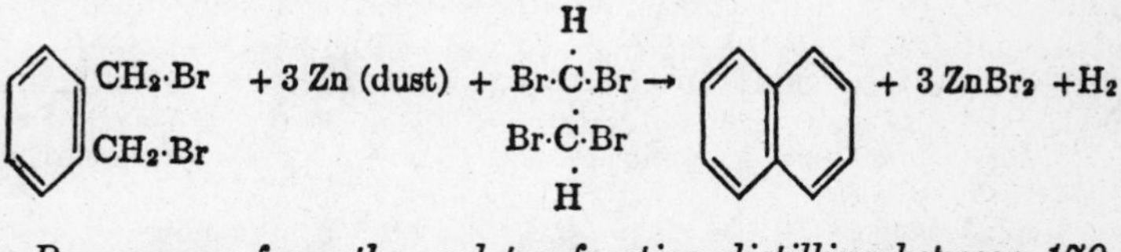

$$\begin{array}{c} \text{H} \\ \cdot \\ \text{CH}_2\text{·Br} \\ \text{CH}_2\text{·Br} \end{array} + 3 \text{ Zn (dust)} + \begin{array}{c} \text{Br·C·Br} \\ \text{Br·C·Br} \\ \cdot \\ \text{H} \end{array} \rightarrow + 3 \text{ ZnBr}_2 + \text{H}_2$$

3. *By recovery from the coal tar fraction distilling between 170–230°C. (commercial).*

b. ANTHRACENE MAY BE PREPARED:

1. *By the reaction of benzene with 1, 1, 2, 2-tetrabromoethane in the presence of anhydrous aluminum chloride (Friedel–Crafts):*

$$\text{◯} + \overset{\text{H}}{\underset{\text{H}}{\text{Br·C·Br}}}_{\text{Br·C·Br}} + \text{◯} \quad / \ (\text{AlCl}_3) \rightarrow \text{◯◯◯} + 4\,\text{HBr}$$

2. *By the action of zinc dust on phenyl o-tolyl ketone:*

$$\text{◯}\overset{\overset{\text{O}}{\text{‖}}}{-\text{C}-}\text{◯} + \text{Zn (dust, heat)} \rightarrow \text{◯◯◯} + \text{ZnO} + \text{H}_2$$

3. *By recovery from the coal tar fraction above 270°C. (commercial).*

c. PHENANTHRENE MAY BE PREPARED:

1. *By the action of sodium on 1-bromo-2-(bromomethyl)benzene:*

$$\text{◯}\overset{\text{CH}_2\text{·Br}}{\underset{\text{Br}}{}} + 4\,\text{Na} + \text{Br·H}_2\text{C}\,\text{◯} \rightarrow \text{◯◯◯} \quad \text{or}$$

$$\text{◯◯◯}$$

$$+ 4\,\text{NaBr} + \text{H}$$

Anthracene is obtained also (Fittig Reaction):

2. *By recovery from the coal tar fraction boiling above 270°C. (commercial). Carbazole is also obtained from this fraction.*

carbazole

D. Physical Properties:

Benzene (D. 0.8941$\frac{2°}{4}$, b.p. 80.08, m.p. 5.4) and toluene (D. 0.866$\frac{2°}{4}$, b.p. 110.8, m.p. −95.1) are both colorless liquids which are insoluble in water. Benzene, on chilling, yields hexagonal crystals. *o*-Xylene (D.0.8745$\frac{2°}{4}$, b.p. 144, m.p. −29) and *m*-xylene (D· 0.8684^{15}, b.p. 138.8, m.p. −53.6) are both colorless liquids and are insoluble in water. *p*-Xylene (D. 0.8612$\frac{2°}{4}$, b.p. 138, m.p. 15-6) forms colorless monoclinic crystals. These hydrocarbons are all soluble in alcohol and ether.

Naphthalene (D. 1.145, m.p. 80.22, b.p. 217.9) crystallizes in glistening white plates. It sublimes quite readily and has a characteristic odor. Anthracene (D. 1.25$\frac{2°}{4}$, m.p. 217, b.p. 354-5) crystallizes in colorless leaflets. Phenanthrene (D. 1.025, m.p. 100, b.p. 340.2), an isomer of anthracene, crystallizes in colorless leaflets. These three hydrocarbons are insoluble in water, but somewhat soluble in alcohol and ether.

Pyrene (benzo[def]phenanthrene, m.p. 150) and retene (7-isopropyl-1-methylphenanthrene, m.p. 98.5) are derivatives of phenanthrene.

Benzene and some of its derivatives are toxic and should be handled with care.

E. Chemical Properties:

I. Benzene and its alkyl derivatives

a. BENZENE REACTS UNDER PROPER CONDITIONS:

1. By substitution when treated with:

a. Bromine/ (Fe, P, S, I, FeBr$_3$, or AlCl$_3$) C_6H_5 | H + Br | Br

Chlorine (Fe, P, S, I, FeBr$_3$, or AlCl$_3$) C_6H_5 | H + Cl | Cl

b. HNO$_3$, conc./H$_2$SO$_4$, conc., 60°C. C_6H_5 | H + HO | NO$_2$

c. H$_2$SO$_4$/7% SO$_3$, 50°C. C_6H_5 | H + HO | SO$_2$·OH

d. R·X/(AlCl$_3$, anhydrous) C_6H_5 | H + X | R

e. R·CO·Cl/(AlCl$_3$, anhydrous) C_6H_5 | H + Cl | OC·R

f. (AcO)$_2$Hg/AcOH, heat C_6H_5 | H + AcO | Hg·OAc

2. By addition when treated with:

a. H$_2$/ (Pt, room temp. or Ni at 180°) $C_6H_6 + 3 H_2 \rightarrow C_6H_{12}$

b. Bromine/ (sunlight, no catalyst) $C_6H_6 + 3 Br_2 \rightarrow C_6H_6Br_6$

c. Ozone $C_6H_6 + 3 O_3 \rightarrow C_6H_6 (O_3)_3$

The graphic formulas for these addition products are:

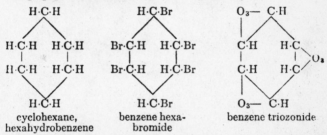

cyclohexane, hexahydrobenzene   benzene hexa-bromide   benzene triozonide

The addition of the first two hydrogen atoms to benzene absorbs 5 kilogram calories per mole, whereas the second and third additions are accompanied by the liberation of 26 and 28 kilogram calories per mole, respectively.

3. By cleavage when passed through a hot tube to give ethyne (acetylene) and other products: C_6H_6, through hot tube $\rightarrow$ x C_2H_2

4. By oxidation when treated with:

a. Air/ (V$_2$O$_5$, 450°C.) $C_6H_6 + $ air/ (V$_2$O$_5$, 450°C.) $\rightarrow$
O:C·HC:CH·C:O

b. Chloric acid $C_6H_6 + HClO_3 \rightarrow$
Cl$_3$C·CO·HC:CH·CO·OH

c. Vigorous reagents $C_6H_6 +$ vig. oxi. $\rightarrow$
CO$_2$, H$_2$O, H·CO·OH, etc.

d. Burning $2 C_6H_6 + (15\text{-}x)O_2 \rightarrow (12\text{-}x)CO_2 + 6 H_2O + x C$

The above reactions apply in general to the alkyl derivatives of benzene (toluene, xylene, etc.) with the following modifications:

1. Substitution by halogen at an elevated temperature in the presence or absence of sunlight occurs in the side chain, but at room temperature in diffused light or no light, and the presence of a catalyst a nuclear hydrogen atom is replaced: $C_6H_5 \cdot CH_3 +$ Cl_2, heat, sunlight $\rightarrow C_6H_5 \cdot CH_2 \cdot Cl + HCl$.

2. The position taken up by a second group entering the ring is dependent, to an appreciable extent, upon the electronic nature of the original substituent atom. The directing tendencies of the substituents permit of the following classification:

   a. Ortho-para directing groups are NR_2, NHR, NH_2, $NH \cdot CO \cdot R \cdot$ OH, OR, $O \cdot CO \cdot R$, SH, SR, R, CH_2Cl, C_6H_5, F, Cl, Br, and I,

   b. Meta directing groups are NO_2, CN, $SO_2 \cdot OH$, $SO_2 \cdot Cl$, $SO_2 \cdot R$, $SO_2 \cdot NH_2$, CHO, $CO \cdot R$, $CO \cdot OH$, $CO \cdot O \cdot R$, CCl_3, $NH_2 \cdot HX$, $NHR \cdot HX$, $NR_2 \cdot HX$, and NR_3X.

   These observations have been formulated into a number of rules, one of which states that *if the original substituent atom is negative, the next element or group will tend to substitute in the o- or p-position; otherwise, in the m-position.* Crum Brown-Gibson expressed the concept as follows: *If the hydrogen compound HX, of the substituent X in the ring, is not directly oxidizable to HOX, then o- and p-derivatives predominate on further substitution; if so, m-derivatives predominate.* According to Vorländer, saturated substituent atoms direct o- and p- whereas unsaturated groups direct to the m-position. More recently, Hammick and Illingworth have assigned the orienting effect to Y and YW groups. *If YW is the substituent group, the next group entering the ring tends to go to the m-position if W is an element occurring in a later group in the periodic table, or if it is an element with a lower atomic weight in the same periodic group; otherwise, it tends to direct o- and p-. When only Y is present, the orientation tends to the o- and p-positions.*

3. The alkyl derivatives of benzene, when subjected to oxidation, are first attacked on the side chains. The alkyl group is oxidized back to a carboxyl group attached to the nucleus before the nucleus itself is attacked. For example: $C_6H_5 \cdot CH_2 \cdot CH_3 +$ vigorous oxidation $\rightarrow C_6H_5 \cdot CO \cdot CH_3 \rightarrow C_6H_5 \cdot CO \cdot OH \rightarrow$ nuclear cleavage.

II. Naphthalene, Anthracene, and Phenanthrene

a. NAPHTHALENE REACTS UNDER PROPER CONDITIONS:

1. *By substitution when treated with:*

   a. HNO_3, conc./H_2SO_4, conc., 50°C. $C_{10}H_7$ | H (1) + HO | NO_2
   b. H_2SO_4, conc., 80°C. $C_{10}H_7$ | H (1) + HO | $SO_2 \cdot OH$
       H_2SO_4, conc., 160°C. $C_{10}H_7$ | H (2) + HO | $SO_2 \cdot OH$
   c. Bromine, boiling temperature $C_{10}H_7$ | H (1) + Br | Br
       Chlorine, boiling temperature $C_{10}H_7$ | H (1) + Cl | Cl

The numerals indicate the position of the replaceable hydrogen atom. Note that the 1- or α-derivatives predominate.

2. *By addition when treated with:*

a. 2 Na/2 $C_2H_5 \cdot OH$, boil $\qquad$ $C_{10}H_8 + 2$ Na/2 $C_2H_5 \cdot OH \rightarrow C_{10}H_{10}$
(1, 4)

$\quad$ 4 Na/4 $C_5H_9 \cdot OH$, boil $\qquad$ $C_{10}H_8 + 4$ Na/4 $C_5H_9 \cdot OH \rightarrow C_{10}H_{12}$
(1, 2, 3, 4)

$\quad$ 2 H_2/Ni powder, 180-200°C. $\quad$ $C_{10}H_8 + 2$ H_2/Ni, 180°C. $\rightarrow C_{10}H_{12}$
(1, 2, 3, 4)

$\quad$ 5 H_2/ catalyst, heat. $\qquad$ $C_{10}H_8 + 5$ H_2/ cat., heat $\rightarrow C_{10}H_{18}$

b. Cl_2/ ($KClO_3$/HCl), 20°C. $\qquad$ $C_{10}H_8 + Cl_2 \rightarrow C_{10}H_8Cl_2$ (1, 4)

$\quad$ 2 Cl_2/$KClO_3$/HCl), 20°C. $\qquad$ $C_{10}H_8 + 2$ $Cl_2 \rightarrow C_{10}H_8Cl_4$ (1, 2, 3, 4)

$\quad$ 2 Br_2, 20°C. $\qquad$ $C_{10}H_8 + 2$ $Br_2 \rightarrow C_{10}H_8Br_4$ (1, 2, 3,4)

3. *By oxidation when treated with air/(V_2O_5, 400°C.) or other appropriate catalysts:*

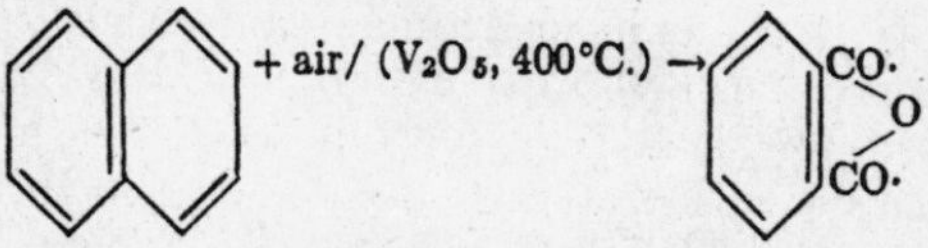

$+$ air/ (V_2O_5, 400°C.) $\rightarrow$ (phthalic anhydride) + other products

Because of the aliphatic character of the one ring in the naphthalene nucleus, it enters into substitution and addition reactions more readily than does benzene until the aliphatic ring has become saturated. The remaining ring, now typically aromatic, reacts much as a benzene ring. Naphthalene burns with a very smoky flame when ignited.

b. ANTHRACENE REACTS UNDER PROPER CONDITIONS:

1. *By substitution when treated with sulfuric acid to give 1-, 2-, or 9-anthracenesulfonic acid* ($C_{14}H_9 \cdot SO_2 \cdot OH$).

2. *By addition with (a) two hydrogen atoms when treated with hydrogen iodide in the presence of red phosphorus to give 9, 10-dihydroanthracene, or (b) two bromine or chlorine atoms at about 100°C. to give 9, 10-anthracene dibromide or 9, 10-anthracene dichloride.*

The 9,10-anthracene dihalides are unstable and lose HX when treated with sodium hydroxide to give the 9-haloanthracene.

3. *By oxidation when treated with air (V_2O_5, heat) or other oxidizing agents such as nitric acid:*

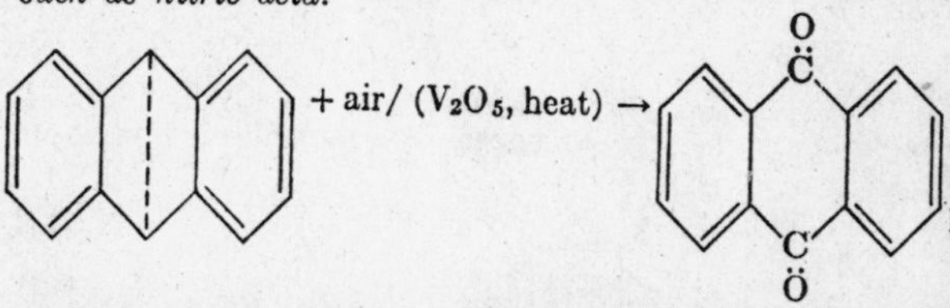

$+$ air/ (V_2O_5, heat) $\rightarrow$ (anthraquinone, old commercial process)

It should be noted that whereas nitric acid effects substitution in benzene and naphthalene, it effects the oxidation of anthracene. This may be attributed to the highly alicyclic character of the one ring in anthracene.

4. *By polymerization in the presence of sunlight to give para-anthracene*
$C_{14}H_{10} + $ sunlight $\rightarrow (C_{14}H_{10})_2$

c. PHENANTHRENE, ISOMERIC WITH ANTHRACENE, RESEMBLES ANTHRACENE IN ITS CHEMICAL PROPERTIES.

CHAPTER XVIII

AROMATIC HALOGEN COMPOUNDS

A. Introduction:

Typical aromatic halogen derivatives such as chlorobenzene ($C_6H_5 \cdot Cl$) are obtained by the substitution reaction of the halogens on the nucleus of the aromatic hydrocarbons. If substitution occurs on a side chain the halo derivatives, of which benzyl chloride ($C_6H_5 \cdot CH_2 \cdot Cl$) is an example, are aliphatic in character in so far as the type of reactivity of the halogen atoms is concerned. Since the phenyl group is more electronegative than the alkyl group, $C_6H_5 \cdot CH_2 \cdot Cl$ contains a more reactive chlorine than does $R \cdot Cl$.

The principal use of the aromatic halides is in organic synthesis. Chlorobenzene is used in the commercial production of aniline, phenol, sulfur black, and picric acid. p-Dichlorobenzene is sold under the trade name Dichlorocide as a moth killer. Compounds such as Gamexane (cis-1, 3-benzene hexachloride), DDT, and 2, 4-D are important insecticides.

Position isomerism exists whenever two or more substituent groups are involved. The establishment of the positions of two like groups substituted on the benzene nucleus is readily accomplished by the use of Körner's Orientation Theory. p-Dichlorobenzene, for example, will give only one trichloro derivative, but o-dichlorobenzene will give two trichloro derivatives, and m-dichlorobenzene will give three trichloro derivatives.

B. Nomenclature:

The aromatic halides are named as substitution products of the corresponding hydrocarbon nucleus. Examples are:

fluoro-benzene; 1, 2- (or o-) dichlorobenzene; 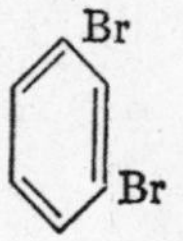 1, 3- (or m-) dibromobenzene;  1, 4- (or p-) diiodobenzene; 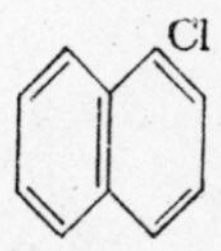 1- (or α-) chloronaphthalene

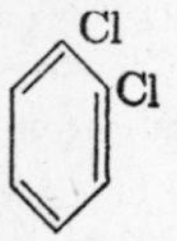

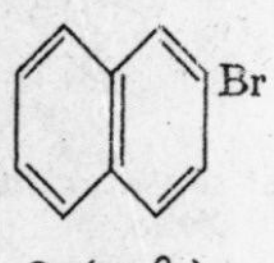

2- (or β-) bromonaphthalene; 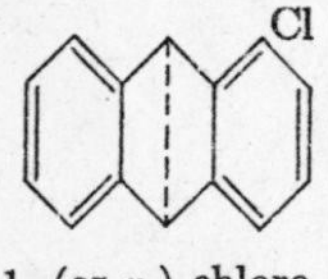 1- (or α-) chloroanthracene; 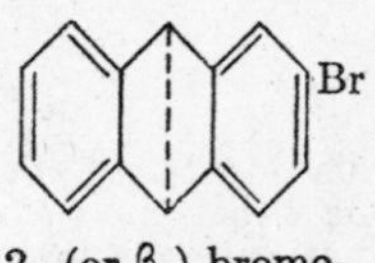 2- (or β-) bromoanthracene; 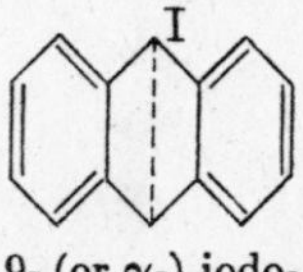 9- (or γ-) iodoanthracene

Where the substituent halogen atom is in a side chain, the compound may be named as a derivative of a radical or as a substituted hydrocarbon. Examples are afforded by the following:

$C_6H_5 \cdot CH_2 \cdot Cl$	benzyl chloride	α-chlorotoluene
$C_6H_5 \cdot CHCl_2$	benzal chloride	α-dichlorotoluene
$C_6H_5 \cdot CCl_3$	benzotrichloride	α-trichlorotoluene
$C_6H_5 \cdot CH_2 \cdot CH_2 \cdot Cl$	β-chloroethylbenzene	1-chloro-2-phenylethane
$C_6H_5 \cdot CHCl \cdot CH_3$	α-chloroethylbenzene	1-chloro-1-phenylethane

C. Preparations:

I. Halo Derivatives of Benzene and Toluene

a. THE HALO DERIVATIVES OF BENZENE MAY BE PREPARED:

1. *By direct halogenation with bromine or chlorine (a) at room temperature with diffused or no light and the presence of a catalyst (FeCl₃, FeBr₃, AlCl₃, Fe, P, S, or I) to effect substitution, or (b) at an elevated temperature in bright light without a catalyst to effect addition:*

   a. Substitution:

   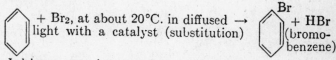

   Iodobenzene may be prepared by the following reaction:

   $$5\ C_6H_6\ +\ 2\ I_2/HIO_3\ \rightarrow\ 5\ C_6H_5 \cdot I\ +\ 3\ H_2O$$

   b. Addition:

   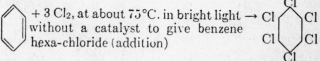

2. *By the reaction of benzenediazonium chloride with (a) HF, (b) HCl/Cu₂Cl₂, or HBr/Cu₂Br₂ (Sandmeyer reaction), or (c) HI or KI, all at a temperature of 0° to about 5°C.:*

   a. $C_6H_5 \cdot N(Cl):N\ +\ HF/\ (0°\text{-}5°C.)\ \rightarrow\ C_6H_5 \cdot F\ +\ N_2\ +\ HCl$

   b. $C_6H_5 \cdot N(Cl):N\ +\ HCl/\ (Cu_2Cl_2,\ 0°\text{-}5°C.)\ \rightarrow\ C_6H_5 \cdot Cl\ +\ N_2\ +\ HCl$
   $C_6H_5 \cdot N(Cl):N\ +\ HBr/\ (Cu_2Br_2,\ 0°\text{-}5°C.)\ \rightarrow\ C_6H_5 \cdot Br\ +\ N_2\ +HCl$

   c. $C_6H_5 \cdot N(Cl):N\ +\ KI/\ (0°\text{-}5°C.)\ \rightarrow\ C_6H_5 \cdot I\ +\ N_2\ +\ KCl$

The use of phosphorus halides, which are employed advantageously in the preparation of the alkyl halides from the alcohols, has no importance as a reagent in analogous reactions in the aromatic series when the hydroxyl group is linked to the nucleus.

b. THE HALO DERIVATIVES OF TOLUENE MAY BE PREPARED:

1. *By direct halogenation with bromine or chlorine (a) at room temperature with diffused or no light and the presence of a catalyst (FeCl₃, FeBr₃, AlCl₃, Fe, P, S, or I) to effect substitution on the nucleus, or (b) at an elevated temperature in bright light without a catalyst to effect substitution on the side chain:*

a.

at about 20°C. in diffused light, catalyst

b. 1. $C_6H_5 \cdot CH_3 + Cl_2$ light, boil $\rightarrow C_6H_5 \cdot CH_2 \cdot Cl + HCl$

2. $C_6H_5 \cdot CH_3 + 2 Cl_2$ light, boil $\rightarrow C_6H_5 \cdot CHCl_2 + 2 HCl$

3. $C_6H_5 \cdot CH_3 + 3 Cl_2$ light, boil $\rightarrow C_6H_5 \cdot CCl_3 + 3 HCl$

The light supplies the quantum (hv) of energy required for the dissociation of X_2 into X^+ and X^-.

2. By the action of phosphorus pentachloride or bromide on (a) benzyl alcohol to give benzyl chloride, or (b) benzaldehyde to give benzal chloride:

a. $C_6H_5 \cdot CH_2 \cdot OH + PCl_5 \rightarrow C_6H_5 \cdot CH_2 \cdot Cl + POCl_3 + HCl$

b. $C_6H_5 \cdot CHO + PCl_5 \rightarrow C_6H_5 \cdot CHCl_2 + POCl_3$

By an analogous reaction, only small yields of chlorotoluene are obtained; for an OH group linked to the nucleus is not readily replaced. Likewise, only small yields of benzotrichloride are obtained by the action of phosphorus pentachloride on benzoic acid ($C_6H_5 \cdot CO \cdot OH$).

II. Halo Derivatives of Naphthalene and Anthracene

a. THE HALO DERIVATIVES OF NAPHTHALENE MAY BE PREPARED:

1. By the action of bromine or chlorine (a) on boiling naphthalene by substitution to give 1-halonaphthalene, or (b) on naphthalene at room temperature by addition to give naphthalene dihalide or naphthalene tetrahalide:

a.

b.

2. By the use of the Sandmeyer Reaction on the appropriate naphthalenediazonium chloride to give 1- or 2-halonaphthalene:

2-naphthalenediazonium chloride　　2-chloronaphthalene

b. THE HALO DERIVATIVES OF ANTHRACENE MAY BE PREPARED:

1. By the substitution reaction of chlorine on anthracene at about 100°C. to give 9,10-dichloroanthracene:

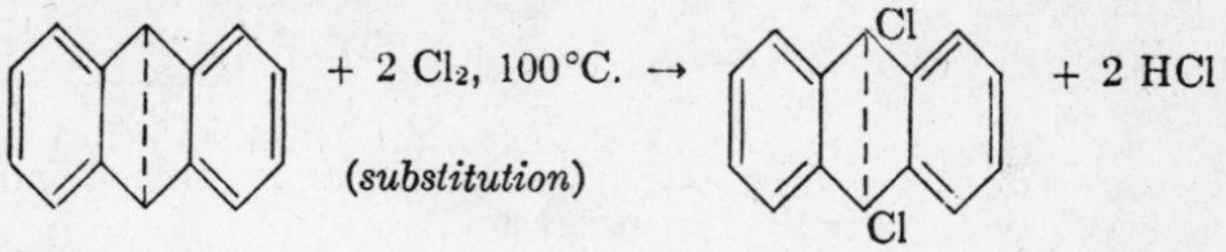

$$+ 2 Cl_2, 100°C. \rightarrow \qquad + 2 HCl$$

(substitution)

2. By the addition reaction of bromine or chlorine on anthracene at about 0°C. to give 9, 10-anthrace dihalide:

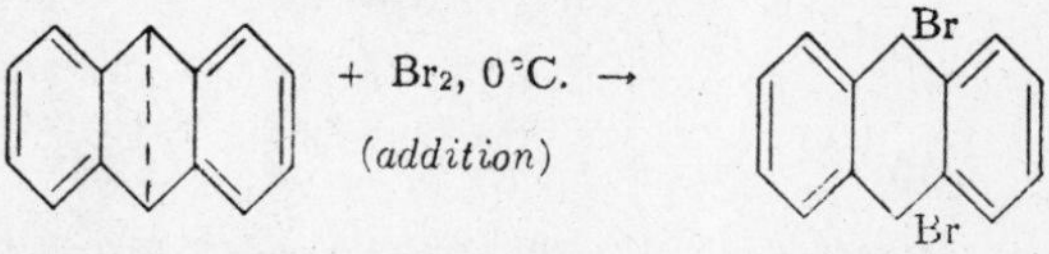

$$+ Br_2, 0°C. \rightarrow$$

(addition)

3. By the loss of HX from 9, 10-anthracene dihalide on treatment with alkali:

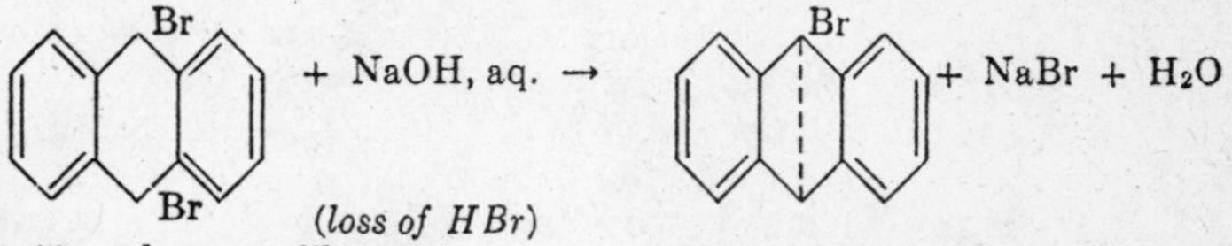

$$+ NaOH, aq. \rightarrow \qquad + NaBr + H_2O$$

(loss of HBr)

9, 10-anthracene dibromide   9-bromoanthracene

D. Physical Properties:

Fluorobenzene (b.p. 86, m.p. —41.2, D. 1.024 $\frac{?}{4}$), chlorobenzene (b.p. 132, m.p. —45, D. 1.1066$\frac{2.0}{4}$), bromobenzene (b.p. 155-6, m.p. –30.6, D. 1.499$\frac{1}{15}$), and iodobenzene (b.p. 188.6, m.p. –31.4, D. 1.832$\frac{2.0}{4}$), are all colorless liquids which are practically insoluble in water but very soluble in alcohol and ether. A similar relationship is found among the halo derivatives of toluene and xylene.

1-Chloronaphthalene (b.p. 263, D. 1.1938$\frac{?}{4}$) is a colorless liquid, but 2-chloronaphthalene (b.p. 264-6, m.p. 55-6) is a colorless solid.

E. Chemical Properties:

I. Halo Derivatives of Benzene, Toluene, and Xylene

a. THE HALO DERIVATIVES OF BENZENE REACT UNDER PROPER CONDITIONS:

1. By replacement of the halogen atom when treated with:

a. Excess NH$_3$, aq., heat, pressure C_6H_5 | X + H | NH$_2$
b. NaOH, aq., heat, pressure C_6H_5 | X + Na | OH

c. R·X/Na or Ar.X/Na in dry ether or benzene C_6H_5 | X + 2Na + X | R

d. R·Mg·X or Ar·Mg·X in dry ether ‹ benzene   H$_5$ | X +   XMg | R

e. Mg in dry ether, as C_6H_5·Br + Mg/(ether) $\rightarrow C_6H_5$·Mg·Br

2. *By replacement of the hydrogen atoms when treated with:*

a. Halogen/ (diffused light, catalyst, 20°C.) $\quad$ X·C$_6$H$_4$ | H + X | X

b. HNO$_3$, conc./H$_2$SO$_4$, conc., 50-60°C. $\quad$ X·C$_6$H$_4$ | H + HO | NO$_2$

c. H$_2$SO$_4$/SO$_3$ (sp. gr. 1.88) $\quad$ X·C$_6$H$_4$ | H + HO | SO$_2$·OH

In each of these reactions, the substitution occurs in the *o-* and *p-*positions to give a mixture of *o-* and *p-*isomers. A second and a third group may be introduced in each case.

b. THE HALO DERIVATIVES OF TOLUENE AND XYLENE REACT:

1. *To give additional substitution in the nucleus when treated under appropriate conditions with halogen, nitric acid, sulfuric acid, or the Friedel-Crafts reagents.*

2. *To give additional substitution in the side chain up to the maximum when treated under proper conditions with halogen.*

3. *To give replacement of the halogen atom in the side chain when treated with the standard reagents employed for effecting reactions with the alkyl halides (cf. Ch. IV, Div. 1, Sec. E).*

II. Halo Derivatives of Naphthalene and Anthracene

a. The halo derivatives of naphthalene enter into the same general type of reactions as those listed above for the halo derivatives of benzene.

b. The halo derivatives of anthracene enter into the same general reactions that have been listed for the halo derivatives of benzene, but, as already noted, 9, 10-anthracene dibromide readily loses HBr when treated with alkali to give 9-bromoanthracene.

CHAPTER XIX

AROMATIC OXYGEN DERIVATIVES

I. Phenols and Aromatic Alcohols

A. Introduction:

The phenols are compounds in which one or more hydrogen atoms in an aromatic nucleus have been replaced by a hydroxyl group (OH) as illustrated by phenol ($C_6H_5 \cdot OH$), while the aromatic alcohols are compounds in which a hydroxyl group has replaced a hydrogen atom in a side chain as in benzyl alcohol ($C_6H_5 \cdot CH_2 \cdot OH$).

Phenol is a constituent of the tars of both coal (0.7%) and wood. Pyrocatechol is found in raw beet sugar, in some of the natural resins, and in the sap and leaves of certain plants. The middle fraction from coal tar furnishes o-, m-, and p-cresols along with other phenols. Traces of p-cresol are found in human urine. Pyrogallol is prepared from gallic acid which is obtained from tannic acid. The hydroxy derivatives of naphthalene and anthracene are synthetic products. Benzyl alcohol occurs in the form of its benzoic and cinnamic esters in the Peru and Tolu balsams and in storax. The volatile oil of cherry laurel contains small amounts of benzyl alcohol.

Several million pounds of phenol are used annually in medicinals, dyes, resins, and other commercial products. As a disinfectant, phenol is being replaced by other products. 4-Hexyl-1,3-benzenediol (hexylresorcinol, S.T. 37) has superior antiseptic properties and appears, unlike phenol, to be noninjurious to body tissues. Pyrogallol, hydroquinone, and phloroglucinol, are used quite extensively as photographic developers. An alkaline solution of pyrogallol is used to determine oxygen in gas analysis, and 2-naphthol is used in the preparation of a solution of its copper salt for use in the quantitative determination of carbon monoxide. Lysol is a mixture of the three sodium cresolates emulsified in a soap solution. 2, 4-Dinitrophenol has been used to some extent as a fat reducer as it tends to speed up metabolism, but its use has not been approved by the American Medical Association.[1] Picric acid is used as a drug, as a dye, and, especially in the form of its ammonium salt, as an explosive. Benzyl alcohol, 2-phenylethanol (phenethyl alcohol), cinnamyl alcohol, and some of their derivatives are used in the synthetic perfume industry.

The reactions will indicate that the hydroxyl groups in the typically phenolic compounds are linked to nuclear carbon atoms, whereas in the aromatic alcohols the hydroxyl groups are linked to carbon atoms in a side chain. Position isomerism exists whenever two or more substituent groups are involved.

B. Nomenclature:

Although the phenols should be named as substitution products of the parent compound, a number of them are better known by names of historical origin.

---

[1] 2, 4-Dinitrophenol is definitely toxic.

Structural Formulas	Named as Derivatives	Common Names
$C_6H_5 \cdot OH$	hydroxybenzene	phenol, carbolic acid
o-$C_6H_4(OH)_2$	1, 2-benzenediol	pyrocatechol, catechol
m-$C_6H_4(OH)_2$	1, 3- benzenediol	resorcinol, resorcin
p-$C_6H_4(OH)_2$	1, 4-benzenediol	hydroquinone, quinol
1, 2, 3-$C_6H_3(OH)_3$	1, 2, 3-benzenetriol	pyrogallol, pyrogallic acid
1, 3, 5-$C_6H_3(OH)_3$	1, 3, 5-benzenetriol	phloroglucinol
o-$CH_3 \cdot C_6H_4 \cdot OH$	2-hydroxytoluene	o-cresol
m-$HO \cdot C_6H_4 \cdot Br$	3-bromophenol	m-bromophenol
p-$HO \cdot C_6H_4 \cdot NO_2$	4-nitrophenol	p-nitrophenol
1, 2, 4-$HO \cdot C_6H_3(NO_2)_2$	2, 4-dinitrophenol	dinitrophenol
1,2,4, 6-$HO \cdot C_6H_2(NO_2)_3$	2, 4, 6-trinitrophenol	picric acid
o-$HO \cdot C_6H_4 \cdot NH_2$	2-aminophenol	o-aminophenol
1-$C_{10}H_7 \cdot OH$	1-naphthol	α-naphthol
$C_6H_5 \cdot CH_2 \cdot OH$	phenylcarbinol	benzyl alcohol
$C_6H_5 \cdot HC{:}CH \cdot CH_2 \cdot OH$	3-phenyl-2-propenol	cinnamyl alcohol
$(C_6H_5)_2CH \cdot OH$	diphenylcarbinol	benzohydrol
$(C_6H_5)_3C \cdot OH$	triphenylcarbinol	triphenylcarbinol

C. Preparations:

I. The Phenols

a. PHENOL MAY BE PREPARED:

1. By the (a) hydrolysis of chlorobenzene, (b) alkali fusion of benzenesulfonic acid, or (c) hydrolysis of benzenediazonium chloride:

a. 1. $C_6H_5 \cdot Cl + 2\,NaOH$, aq., 300°C. $\rightarrow C_6H_5 \cdot ONa + NaCl + H_2O$, then

$2\,C_6H_5 \cdot ONa + CO_2/H_2O \rightarrow 2\,C_6H_5 \cdot OH + Na_2CO_3$ (Dow Process)

The caustic solution is about 15% sodium hydroxide.

The process is continuous, and yields about 10% diphenyl oxide as a by-product through interaction between $C_6H_5 \cdot ONa$ and $C_6H_5 \cdot Cl$.

2. $C_6H_5 \cdot Cl + H_2O/$ (Cu, silica gel) $\rightarrow C_6H_5 \cdot OH + HCl$

b. $C_6H_5 \cdot SO_2 \cdot OH + 3\,NaOH$, fuse $\rightarrow C_6H_5 \cdot ONa + Na_2SO_3 + 2\,H_2O$, then

$2\,C_6H_5 \cdot ONa + CO_2/H_2O \rightarrow 2\,C_6H_5 \cdot OH + Na_2CO_3$

This is the older commercial process, but the by-product sodium sulfite is commercially important in the removal of oxygen from boiler feed water.

c. $C_6H_5 \cdot N\,(Cl){:}N + H_2O$, warm $\rightarrow C_6H_5 \cdot OH + N_2 + HCl$ (poor yields)

2. By recovery from coal tar distillates (commercial).

b. CATECHOL MAY BE PREPARED:

By the (a) hydrolysis of o-chlorophenol, or (b) alkali fusion of o-hydroxybenzenesulfonic acid:

a. o-HO·C$_6$H$_4$·Cl + 3 NaOH, aq., heat → o-NaO·C$_6$H$_4$·ONa + NaCl + 2 H$_2$O, then

    o-NaO·C$_6$H$_4$·ONa + CO$_2$/H$_2$O → o-HO·C$_6$H$_4$·OH + Na$_2$CO$_3$

b. o-HO·C$_6$H$_4$·SO$_2$·OH + 4 NaOH, 350°C. → o-NaO·C$_6$H$_4$·ONa + Na$_2$SO$_3$ + 3 H$_2$O, then

    o-NaO·C$_6$H$_4$·ONa + CO$_2$/H$_2$O → o-HO·C$_6$H$_4$·OH + Na$_2$CO$_3$

c. *RESORCINOL MAY BE PREPARED:*

By the fusion of m-benzenedisulfonic acid with alkali:

    m-C$_6$H$_4$(SO$_2$·OH)$_2$ + 6 NaOH, fuse → m-C$_6$H$_4$(ONa)$_2$ + 2 Na$_2$SO$_3$ + 4 H$_2$O, then

    m-C$_6$H$_4$ (ONa)$_2$ + CO$_2$/H$_2$O → m-HO·C$_6$H$_4$·OH + Na$_2$CO$_3$

d. *HYDROQUINONE MAY BE PREPARED:*

By the reduction of p-benzoquinone with sulfur dioxide:

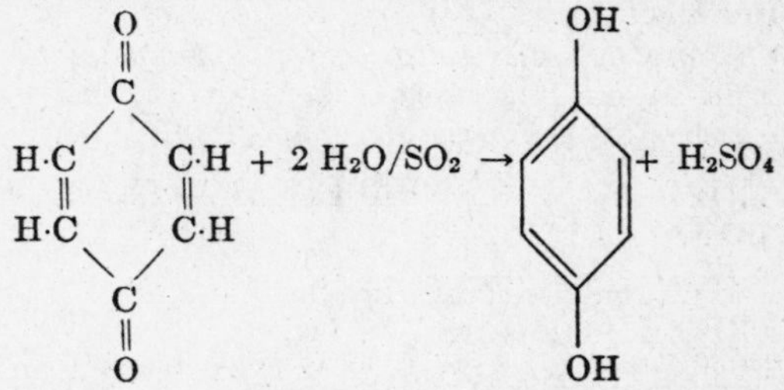

e. *PYROGALLOL MAY BE PREPARED:*

By the dry distillation of gallic (3, 4, 5-trihydroxybenzoic) acid:

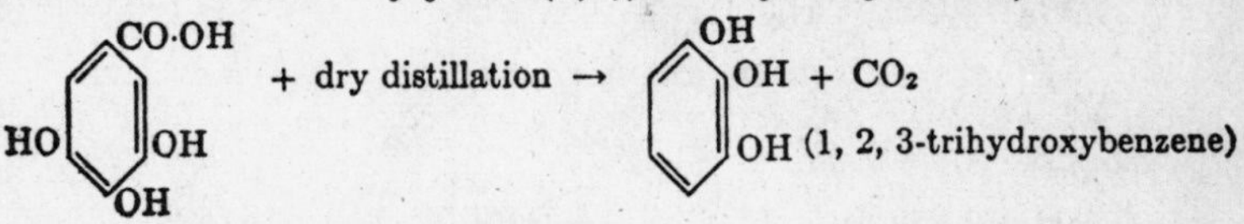

f. *THE 1- AND 2-NAPHTHOLS MAY BE PREPARED:*

1. *By heating the corresponding naphthalenesulfonic acids with sodium hydroxide at about 300°C.:*

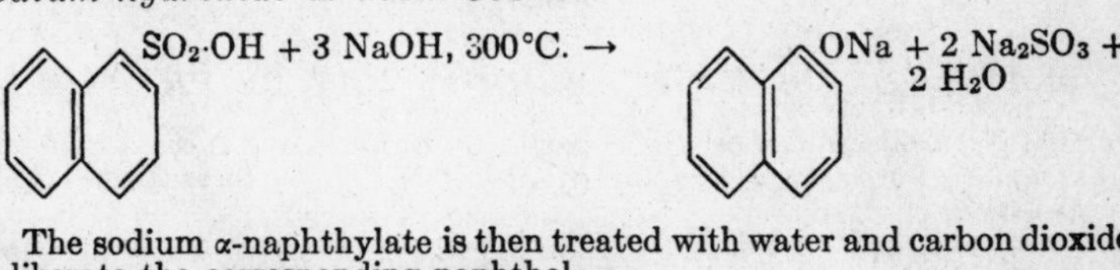

    The sodium α-naphthylate is then treated with water and carbon dioxide to liberate the corresponding naphthol.

2. *By the hydrolysis of the corresponding (a) diazonium chloride, or (b) amine:*
   a. C$_{10}$H$_7$·N (Cl):N + H$_2$O, warm → C$_{10}$H$_7$·OH + N$_2$ + HCl
   b. C$_{10}$H$_7$·NH$_2$ + H$_2$O/ (H$_2$SO$_4$), heat → C$_{10}$H$_7$·OH + NH$_3$

II. The Aromatic Alcohols

a. BENZYL ALCOHOL MAY BE PREPARED:

1. *By the reduction of benzaldehyde with sodium amalgam / H_2O:*

$$C_6H_5 \cdot CHO + 2\,Na/(Hg)\ /2\,H_2O \rightarrow C_6H_5 \cdot CH_2 \cdot OH + 2\,NaOH$$

2. *By the hydrolysis of benzyl chloride in the presence of Ca $(OH)_2$:*

$$2\,C_6H_5 \cdot CH_2 \cdot Cl + Ca\,(OH)_2,\ aq.,\ 100\text{-}110°C. \rightarrow 2\,C_6H_5 \cdot CH_2OH + CaCl_2$$

3. *By the autoxidation of benzaldehyde in an alkaline solution (the Cannizzaro reaction):*

$$2\,C_6H_5 \cdot CHO + NaOH,\ aq. \rightarrow C_6H_5 \cdot CH_2 \cdot OH + C_6H_5 \cdot CO \cdot ONa$$

b. THE PHENYLETHANOLS MAY BE PREPARED:

1. *By (a) the reduction of ethyl phenylethanoate, (b) the hydrolysis of (2-chloroethyl)benzene, (c) the reaction of phenylmagnesium halides with epoxyethane, followed by hydrolysis, or (d) the extraction from neroli oil, roses, jasmine, or similar sources, to give in each case 2-phenylethanol (phenethyl alcohol):*

   a. $C_6H_5 \cdot CH_2 \cdot CO \cdot OEt + 4\,Na/3\,Et \cdot OH \rightarrow C_6H_5 \cdot CH_2 \cdot CH_2 \cdot OH + 4\,Et \cdot ONa$

   b. $C_6H_5 \cdot CH_2 \cdot CH_2 \cdot Cl + NaOH/(soap),\ aq. \rightarrow$
   $C_6H_5 \cdot CH_2 \cdot CH_2 \cdot OH + NaCl$

   c. $C_6H_5 \cdot Mg \cdot Br + H_2C \overset{O}{\underset{}{\cdot}} CH_2 \rightarrow C_6H_5 \cdot CH_2 \cdot CH_2 \cdot O \cdot Mg \cdot Br,$
   then

   $C_6H_5 \cdot CH_2 \cdot CH_2 \cdot O \cdot Mg \cdot Br + HX,\ aq. \rightarrow$
   $C_6H_5 \cdot CH_2 \cdot CH_2 \cdot OH + X \cdot Mg \cdot Br$

2. *By (a) the reduction of acetophenone, or (b) the hydrolysis of (1-chloroethyl)benzene, to give in each case 1-phenylethanol (α-methylbenzyl alcohol):*

   a. $C_6H_5 \cdot CO \cdot CH_3 + 2\,KOH/Zn/(Et \cdot OH) \rightarrow C_6H_5 \cdot CHOH \cdot CH_3 + K_2ZnO_2$

   b. $C_6H_5 \cdot CHCl \cdot CH_3 + NaOH/(soap),\ aq. \rightarrow$
   $C_6H_5 \cdot CHOH \cdot CH_3 + NaCl$

c. THE POLYPHENYLCARBINOLS MAY BE PREPARED:

1. *By (a) the reduction of benzophenone, or (b) treating benzaldehyde with phenylmagnesium bromide and subsequent hydrolysis, to give in each case diphenylcarbinol:*

   a. $C_6H_5 \cdot CO \cdot C_6H_5 + 2\,KOH/Zn/(Et \cdot OH) \rightarrow (C_6H_5)_2CH \cdot OH + K_2ZnO_2$

   b. $C_6H_5 \cdot CHO + C_6H_5 \cdot MgBr/(dry\ ether) \rightarrow (C_6H_5)_2CH \cdot OMgBr,$ then

   $(C_6H_5)_2CH \cdot OMgBr + HCl,\ aq. \rightarrow (C_6H_5)_2CH \cdot OH + Cl \cdot MgBr$

   This is an illustration of the Grignard reaction.

2. By (a) treating an alkyl benzoate or benzophenone with phenylmagnesium bromide and subsequent hydrolysis, or (b) the careful oxidation of triphenylmethane, to give in each case triphenylcarbinol:

   a. $(C_6H_5)_2C{:}O + C_6H_5{\cdot}MgBr/$ (dry ether) $\rightarrow (C_6H_5)_3\,C{\cdot}OMgBr$, then

       $(C_6H_5)_3C{\cdot}OMgBr + HCl$, aq. $\rightarrow (C_6H_5)_3C{\cdot}OH + Cl{\cdot}MgBr$

   b. $3\ (C_6H_5)_3C{\cdot}H + Cr_2O_7 ={/}$ (Ac·OH) $+ 4\ H_2O \rightarrow 3\ (C_6H_5)_3C{\cdot}OH + 2\ Cr^{+++} + 8\ OH^-$

D. Physical Properties:

Phenol (m.p. 41, b.p. 182) is a colorless, crystalline solid, which is slightly soluble in water. It is very corrosive and poisonous. Pyrocatechol (m.p. 105, b.p. 240) is more soluble in water than phenol. Resorcinol (m.p. 110, b.p. 276.5) and pyrogallol (m.p. 133-4, b.p. 309) are very soluble in water. Hydroquinone (m.p. 170.5, b.p. 286.2), it might be noted, is much less soluble in water than pyrocatechol. There is some correlation between the melting point of a compound and its solubility: the higher the melting point, the lower the solubility in water. 1-Naphthol (m.p. 96, b.p. 288) is a yellow solid but 2-naphthol (m.p. 122, b.p. 294.8) is a colorless solid, and both are slightly soluble in hot water. All of these compounds are quite soluble in alcohol and ether.

Benzyl alcohol (b.p. 205.2, m.p. —15.3), in contrast with methanol, is almost insoluble in water. 1-Phenylethanol (b.p. 205) and 2-phenylethanol (b.p. 219-21) are practically insoluble in water. Diphenylcarbinol (m.p. 68-9, b.p. 298.5) and triphenylcarbinol (m.p. 162.5) are insoluble in water. All of these are quite soluble in alcohol and ether.

E. Chemical Properties:

I. The Phenols

a. PHENOL REACTS UNDER PROPER CONDITIONS:

1. By replacement of the hydrogen atom of the hydroxyl group when treated with:

a. Acid anhydrides	$C_6H_5{\cdot}O$	H +	R·CO·O	OC·R
b. Acid halides	$C_6H_5{\cdot}O$	H +	Cl	OC·R
c. Alcohols, over hot ThO_2	$C_6H_5{\cdot}O$	H +	HO	R
d. Alkalies, aqueous	$C_6H_5{\cdot}O$	H +	HO	Na
e. Phosphorus oxyhalides	$3\ C_6H_5{\cdot}O$	H +	Cl_3	P:O
f. Phosphorus pentahalides	$C_6H_5{\cdot}O$	H +	Cl	PCl_4

Somewhat analogous to the above reactions are:

$$C_6H_5{\cdot}ONa + (CH_3)_2SO_4 \rightarrow C_6H_5{\cdot}OCH_3 + CH_3O{\cdot}SO_2{\cdot}ONa,\ \text{and}$$
$$C_6H_5{\cdot}ONa + X{\cdot}R \rightarrow C_6H_5{\cdot}OR + NaX$$

2. By replacement of the hydroxyl group when treated with:

   a. Zinc dust: $C_6H_5{\cdot}OH + Zn$, distil $\rightarrow C_6H_6 + ZnO$

   b. PCl_5: $C_6H_5{\cdot}OH + PCl_5$, heat $\rightarrow C_6H_5{\cdot}Cl + POCl_3 + HCl$

This last reaction, in contrast to its use in the aliphatic series, gives *very* poor yields.

3. By typical substitution on the nucleus when treated with:

a. Bromine/H_2O	$HO{\cdot}C_6H_4$	H (2 or 4) +	Br	Br
b. Chlorine/H_2O	$HO{\cdot}C_6H_4$	H (2 or 4) +	Cl	Cl
c. Nitric acid, dilute	$HO{\cdot}C_6H_4$	H (2 or 4) +	HO	NO_2
d. Sulfuric acid, cold	$HO{\cdot}C_6H_4$	H (2 or 4) +	HO	$SO_2{\cdot}OH$
e. Nitrous acid	$HO{\cdot}C_6H_4$	H (4) +	HO	N:O

The presence of the hydroxyl group activates the hydrogen atoms in the *o*- and *p*-positions as indicated by the concentrations of the reagents used above in contrast to those used in effecting the same substitutions in benzene. *The tribromophenol forms readily with excess bromine water,* but the introduction of the second and third nitro and sulfonic acid groups is accomplished less readily.

The numbers in parenthesis indicate the positions of the replaceable hydrogen atoms.

4. By condensation when treated with:

a.	Aldehydes	$2 HO \cdot C_6H_4$	H (4)	+	O	C (H)·R
b.	Diazo compounds	$HO \cdot C_6H_4$	H (4)	+	HO	N:N·Ar
c.	Chloroform/alkali	$HO \cdot C_6H_4$	H (2 or 4)	+	Cl	CHCl$_2$
d.	Carbon tetrachloride/alkali	$HO \cdot C_6H_4$	H (2 or 4)	+	Cl	CCl$_3$
e.	Phthalic anhydride, warm	$2 HO \cdot C_6H_4$	H (4)	+	O	:C·O·C:O

The numbers in parenthesis indicate the positions of the replaceable hydrogen atoms. In c and d, only the first step of the reaction is indicated. Excess alkali removes the other chlorine atoms to give respectively an aldehyde group and an acid group (Reimer-Tiemann reaction).

5. By addition when treated with hydrogen in the presence of finely divided nickel at about 200°C.:

$$C_6H_5 \cdot OH + 3 H_2 / (Ni, \text{ finely divided, } 200°C.) \rightarrow C_6H_{11} \cdot OH$$

6. By oxidation (a) to give 1,2- or 1,4-dihydroxybenzene when treated with persulfuric acid, or (b) to give meso-tartaric acid when treated with alkaline permanganate solution.

b. THE AROMATIC HYDROXY COMPOUNDS REACT, IN GENERAL, AS INDICATED BY THE EXAMPLES GIVEN ABOVE.

The hydroxyl group in the naphthols is even more reactive than it is in phenol: $C_{10}H_7 \cdot OH + HO \cdot CH_3/H_2SO_4 \rightarrow C_{10}H_7 \cdot O \cdot CH_3 + H_2SO_4 \cdot H_2O$, and $2 C_{10}H_7 \cdot OH + (NH_3)_2 \cdot CaCl_2 \rightarrow 2 C_{10}H_7 \cdot NH_2 + CaCl_2 \cdot 2H_2O$.

II. The Aromatic Alcohols

a. THE AROMATIC ALCOHOLS REACT UNDER PROPER CONDITIONS:

1. To give typical derivatives of the aliphatic hydroxyl group (cf. Ch. V, Div. I, Sec. E).

2. To give typical nuclear substitution reactions provided the hydroxyl group is properly "protected" by the formation of ethers, esters, or other comparatively stable derivatives.

II. Aromatic Ethers

A. Introduction:

The aromatic ethers may contain an aryl (Ar) and an alkyl (R) radical, or two aryl radicals.

The diaryl ethers are important products of organic synthesis. Anisaldehyde (p-$CH_3 \cdot O \cdot C_6H_4 \cdot CHO$), guaiacol ($o$-$CH_3 \cdot O \cdot C_6H_4 \cdot OH$), and many of the alkaloids contain an ether group.

The aromatic ethers are used in the synthetic perfume industry. "Dowtherm A", a heat transfer medium, is 75% diphenyl ether and 25% biphenyl.

The methods of preparation and the chemical properties of the aryl-alkyl and diaryl ethers, will indicate their structure. Position isomers exist where two or more substituent groups are involved.

B. Nomenclature:

Structural Formulas	I.U.C. Names	Common Names
$C_6H_5 \cdot O \cdot CH_3$	methoxybenzene	anisole, methyl phenyl ether
$C_6H_5 \cdot O \cdot CH_2 \cdot CH_3$	ethoxybenzene	phenetole, ethyl phenyl ether
o-$Br \cdot C_6H_4 \cdot O \cdot CH_3$	1-bromo-2-methoxybenzene	o-bromophenyl methyl ether
$C_6H_5O \cdot C_6H_5$	phenoxybenzene	phenyl ether
α-$C_{10}H_7 \cdot O \cdot CH_3$	1-methoxynaphthalene	α-methyl naphthyl ether

C. Preparations:

a. THE ARYL-ALKYL ETHERS MAY BE PREPARED:

1. By the action of the potassium or sodium salts of phenolic compounds on (a) alkyl halides (cf. Williamson reaction), or (b) alkyl sulfates:

   a. $Ar \cdot O \cdot Na + X \cdot R$, heat $\rightarrow Ar \cdot O \cdot R + NaI$, and

   $C_6H_5 \cdot ONa + I \cdot CH_3$, heat $\rightarrow C_6H_5 \cdot O \cdot CH_3$ (anisole) $+ NaI$

   b. $Ar \cdot O \cdot Na + R_2SO_4$, heat $\rightarrow Ar \cdot O \cdot R + RO \cdot SO_2 \cdot ONa$

2. By the action of absolute methanol on (a) benzenediazonium chloride to give methoxybenzene (anisole), or (b) 1- or 2-naphthol in the presence of sulfuric acid to give 1- or 2-methoxynaphthalene:

   a. $C_6H_5 \cdot N(Cl) \vdots N + HO \cdot CH_3 \rightarrow C_6H_5 \cdot O \cdot CH_3 + N_2 + HCl$

   b. $C_{10}H_7 \cdot OH + HO \cdot CH_3 / H_2SO_4 \rightarrow C_{10}H_7 \cdot O \cdot CH_3 + H_2SO_4 \cdot H_2O$

b. PHENYL ETHER, A DIARYL ETHER, MAY BE PREPARED:

By heating (a) phenol with zinc chloride or anhydrous aluminum chloride, or (b) chlorobenzene with sodium phenolate (commercial):

   a. $C_6H_5 \cdot OH + ZnCl_2 + HO \cdot C_6H_5$, heat $\rightarrow C_6H_5 \cdot O \cdot C_6H_5 + ZnCl_2 \cdot H_2O$

   b. $C_6H_5 \cdot Cl + Na \cdot O \cdot C_6H_5 \rightarrow C_6H_5 \cdot O \cdot C_6H_5 + NaCl$

D. Physical Properties:

Anisole (b.p. 155, m.p. —37.5) and phenetole (b.p. 172, m.p. —30.2) are insoluble in water but soluble in alcohol and ether. Phenyl ether (m.p. 28, b.p. 259) has a geranium-like odor. It is insoluble in water, but soluble in alcohol and ether.

E. Chemical Properties:

THE ARYL-ALKYL AND DIARYL ETHERS ARE CHARACTERIZED BY THEIR STABILITY, BUT:

1. *They undergo normal nuclear substitution, even with the Friedel-Crafts reaction.*

2. *The aryl-alkyl ethers are split by HX, especially by HI:*
   $Ar \cdot O \cdot R + HI$, conc., heat $\rightarrow Ar \cdot OH + R \cdot I$

3. *Rearrangement of the alkyl group to the open o- or p-position can, under certain conditions, be effected.*

III. Aromatic Aldehydes and Ketones, and Quinones

A. Introduction:

The aromatic aldehydes may be divided into the aryl type $(Ar \cdot CHO)$ and the aryl-alkanal type $(Ar(CH_2)_x CHO)$. The aromatic ketones, likewise, may be divided into the aryl type $(Ar \cdot CO \cdot Ar)$ and the aryl-alkyl type $(Ar \cdot CO \cdot R)$. The quinones are compounds in which two doubly bonded oxygen atoms are linked to nuclear carbon atoms in the o- or p-positions.

Benzaldehyde occurs in "oil of bitter almonds," and salicylaldehyde is found in the oils of certain flowers. Vanillin is obtained from the vanilla bean. Piperonal is the principal ingredient in delicate perfume of the heliotrope and is manufactured commercially from oil of sassafras. Cinnamaldehyde has a cinnamon-like odor and occurs in the oils of cassia and cinnamon.

Aromatic ketones and quinones are synthetic products, although the quinoid structure occurs in some natural pigments.

Benzaldehyde finds extensive use as a flavoring extract, as well as in perfumes, in the manufacture of dyes, and in organic synthesis. Salicylaldehyde is used in the preparation of coumarin, and in perfumery. Vanillin is an ingredient of vanilla extract. Piperonal, phenylacetaldehyde, and cinnamaldehyde are used in perfumery. Acetophenone is used in organic synthesis and as a sleep producer (hypnone). α-Chloroacetophenone has lachrymatory properties and it was used as a tear gas during the World War. It is useful in quelling rioters and dispersing mobs. Benzophenone is used in synthesis, and Michler ketone is an important dye intermediate. Hydroquinone is used as a mild reducing agent, and naphthoquinone and anthraquinone are used in the dye industry.

B. Nomenclature:

Structural Formulas	Named as Derivatives	Common Names
$C_6H_5 \cdot CHO$	benzenecarbonal	benzaldehyde
$1,2\text{-}C_6H_4(CHO)OH$	2-hydroxybenzaldehyde	salicylaldehyde
$1,3,4\text{-}C_6H_3(CHO)(OCH_3)OH$	4-hydroxy-3-methoxy-benzaldehyde	vanillin
$3\text{-}4,1\text{-}CH_2 \overset{O}{\underset{O}{\diagup \diagdown}} C_6H_3 \cdot CHO$	3,4-methylenedioxy-benzaldehyde	piperonal
$C_6H_5 \cdot HC{:}CH \cdot CHO$	3-phenylpropenal	cinnamaldehyde
$C_6H_5 \cdot CO \cdot C_6H_5$	phenyl ketone	benzophenone
$p,p'\text{-}(Me_2N \cdot C_6H_4)_2C{:}O$	bis-(4-dimethylamino-phenyl) ketone	Michler ketone
$C_6H_5 \cdot CO \cdot CH_3$	methyl phenyl ketone	acetophenone
$C_6H_5 \cdot CO \cdot CH_2 \cdot CH_3$	1-phenyl-1-propanone	propiophenone

The more important quinones are:

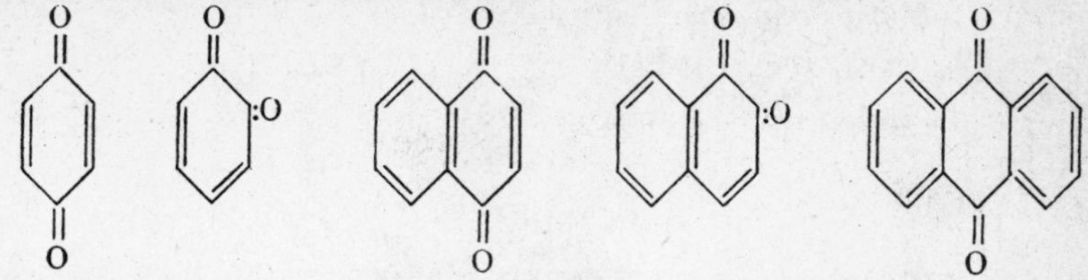

p-quinone   o-quinone   α(or 1,4)-naph-   β(or 1,2)-naph-   anthraquinone
                           thoquinone   thoquinone

C. Preparations:

I. The Aldehydes

a. *BENZALDEHYDE MAY BE PREPARED:*

1. *By the partial oxidation of (a) toluene with manganese dioxide or with chromyl chloride (Etard reaction), or (b) benzyl alcohol with air in the presence of vanadium pentoxide or with calcium hypochlorite:*

     a. 1. $C_6H_5 \cdot CH_3 + 2\ MnO_2 / 2\ H_2SO_4$ (65%), 180°C. →
            $C_6H_5 \cdot CHO + 2\ MnSO_4 + 3\ H_2O$

         2. $3\ C_6H_5 \cdot CH_3 + 6\ CrO_2Cl_2 \rightarrow 3\ C_6H_5 \cdot CH_3 \cdot 2CrO_2Cl_2$, then $+ H_2O$,
           excess $\rightarrow 3\ C_6H_5 \cdot CHO + 2\ H_2CrO_4 + 4\ CrCl_3 + 2\ H_2O$

     b. 1. $2\ C_6H_5 \cdot CH_2OH + O_2 / (V_2O_5, 400°C.) \rightarrow 2\ C_6H_5 \cdot CHO +$
            $2\ H_2O$

         2. $C_6H_5 \cdot CH_2OH + Ca(Cl)OCl, 40°C., aq. \rightarrow C_6H_5 \cdot CHO + H_2O + CaCl_2$

2. *By a modified Friedel-Crafts reaction on benzene:*

     $C_6H_5 \cdot H + CO + HCl, dry / (AlCl_3 / Cu_2Cl_2) \rightarrow C_6H_5 \cdot CHO +$
     HCl

3. *By the hydrolysis of benzal chloride under pressure:*

     $C_6H_5 \cdot CHCl_2 + Ca(OH)_2, aq., heat, pressure \rightarrow C_6H_5 \cdot CHO +$
     $CaCl_2 + H_2O$

b. *THE o- AND p-HYDROXYBENZALDEHYDES MAY BE PREPARED:*

1. *By heating phenol with chloroform in the presence of alkali to give salicylaldehyde and some of the p-isomer (Reimer-Tiemann reaction):*

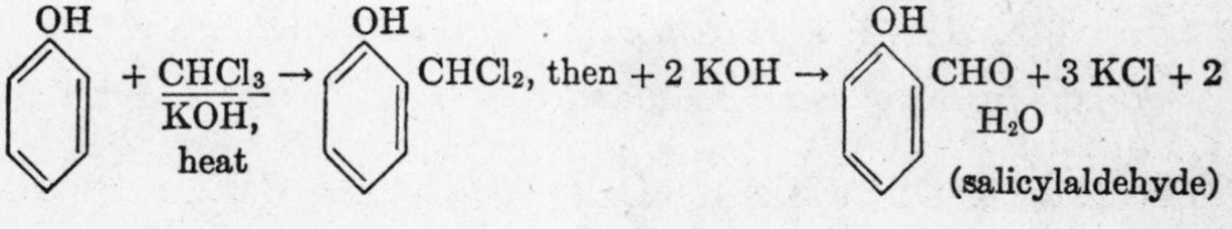

2. *By a modified Friedel-Crafts reaction on phenol to give p-hydroxy-benzaldehyde and some salicylaldehyde:*

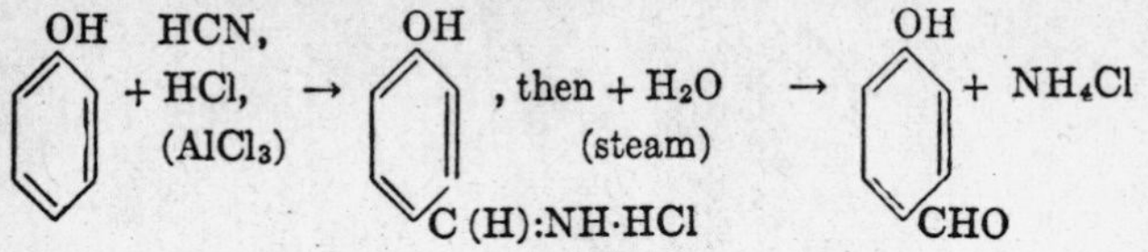

c. PIPERONAL AND VANILLIN MAY BE PREPARED AS FOLLOWS:

1.

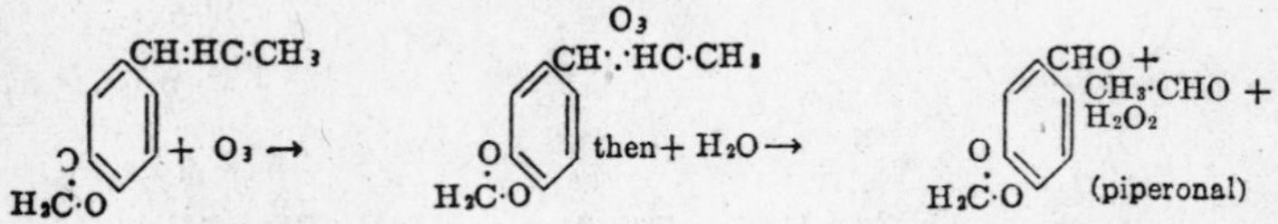

2.

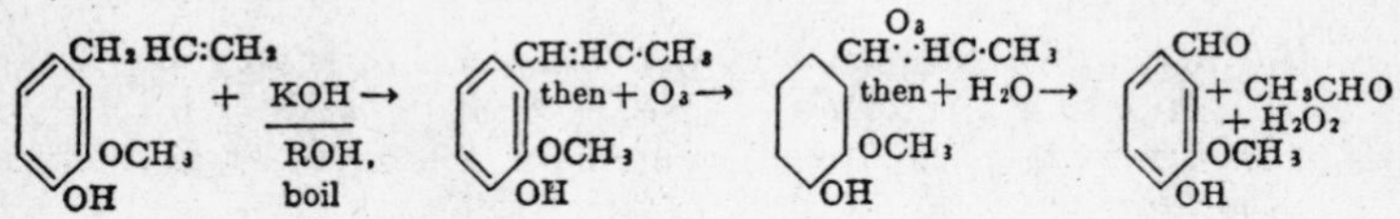

d. CINNAMALDEHYDE MAY BE PREPARED:

By condensing benzaldehyde with ethanal in alkali (Claisen reaction, cf. Aldol Condensation):

$$C_6H_5 \cdot CHO + CH_3 \cdot CHO / (\text{dilute alkali}) \rightarrow C_6H_5 \cdot HC:CH \cdot CHO + H_2O$$

II. The Ketones

a. ACETOPHENONE MAY BE PREPARED:

1. *By the action of ethanoyl chloride or ethanoic anhydride on benzene in the presence of anhydrous aluminum chloride (Friedel-Crafts):*

$$C_6H_5 \cdot H + Cl \cdot OC \cdot CH_3 / (AlCl_3) \rightarrow C_6H_5 \cdot CO \cdot CH_3 \text{ (aceto-phenone)} + HCl$$

2. *By the oxidation of 1-phenylethanol:*

$$3\ C_6H_5 \cdot CHOH \cdot CH_3 + Cr_2O_7 = / H_2O \rightarrow 3\ C_6H_5 \cdot CO \cdot CH_3 + 2\ Cr^{+++} + 8\ OH^-$$

3. *By passing the vapors of benzoic and ethanoic acids over manganous oxide at about 600°C.:*

$$C_6H_5 \cdot CO \cdot OH + CH_3 \cdot CO \cdot OH / (MnO, 600°C.) \rightarrow C_6H_5 \cdot CO \cdot CH_3 + CO_2 + H_2O$$

b. BENZOPHENONE MAY BE PREPARED:

1. *By the fusion of calcium benzoate (cf. Ch. VI, Div. I, Sec. C. II, 2):*

$$(C_6H_5 \cdot CO \cdot O)_2Ca, \text{ fuse} \rightarrow C_6H_5 \cdot CO \cdot C_6H_5 + CaCO_3$$

2. *By treating benzene, in the presence of anhydrous aluminum chloride, with (a) carbonyl chloride, or (b) benzoyl chloride (both Friedel-Crafts reactions):*

   a. $2 C_6H_5 \cdot H + Cl_2C \colon O / (AlCl_3) \rightarrow (C_6H_5)_2C \colon O$ (benzophenone) $+ 2 HCl$

   b. $C_6H_5 \cdot H + Cl \cdot OC \cdot C_6H_5 / (AlCl_3) \rightarrow C_6H_5 \cdot CO \cdot C_6H_5 + HCl$

c. MICHLER KETONE MAY BE PREPARED:

By treating dimethylaniline with $COCl_2/AlCl_3$ (Friedel-Crafts):

$$2 (CH_3)_2N \cdot C_6H_4 \cdot H + Cl_2C \colon O / (AlCl_3) \rightarrow$$
$$((CH_3)_2N \cdot C_6H_4)_2C \colon O + 2 HCl$$

III. The Quinones

a. THE BENZOQUINONES MAY BE PREPARED:

1. *By the careful oxidation of (a) aniline, (b) p-hydroxyaniline, or (c) 1, 4-benzenediol, to give p-benzoquinone:*

   a. $3 C_6H_5 \cdot NH_2 + 2 Cr_2O_7 = /8 H_2O \rightarrow 3 p\text{-}O \colon C_6H_4 \colon O + 4 Cr^{+++} + 16 OH^- + 3 NH_3$

   b. $3 p\text{-}HO \cdot C_6H_4 \cdot NH_2 + Cr_2O_7 = /4 H_2O \rightarrow 3 p\text{-}O \colon C_6H_4 \colon O + 2 Cr^{+++} + 8 OH^- + 3 NH_3$

   c. $3 p\text{-}HO \cdot C_6H_4 \cdot OH + Cr_2O_7 = /4 H_2O \rightarrow 3 p\text{-}O \colon C_8H_4 \colon O + 2 Cr^{+++} + 8 OH^- + 3 H_2O$

2. *By the careful oxidation of (a) o-hydroxyaniline, or (b) catechol, to give o-benzoquinone.*

b. THE NAPHTHOQUINONES MAY BE PREPARED:

1. *By the careful oxidation of (a) 1-naphthylamine, (b) 1-amino-4-naphthol, or (c) 1, 4-naphthalenediol, to give α-naphthoquinone:*

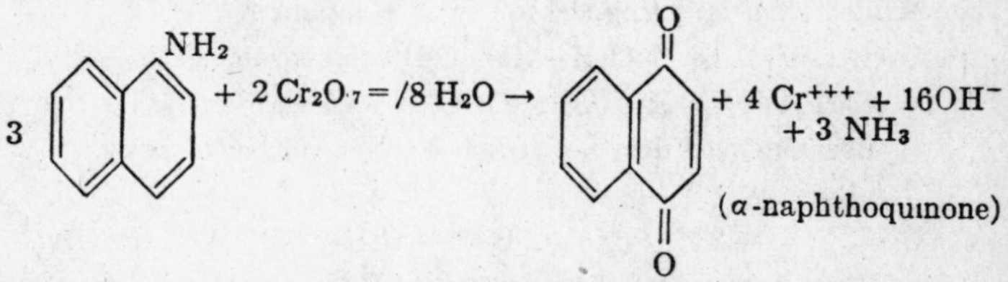

(α-naphthoquinone)

2. *By the careful oxidation of (a) 1-amino-2-naphthol, or (b) 1, 2- naphthalenediol, to give β-naphthoquinone.*

c. ANTHRAQUINONE MAY BE PREPARED:

1. *By treating benzene with phthalic anhydride in the presence of anhydrous aluminum chloride to give o-benzoylbenzoic acid which loses water on treatment with sulfuric acid:*

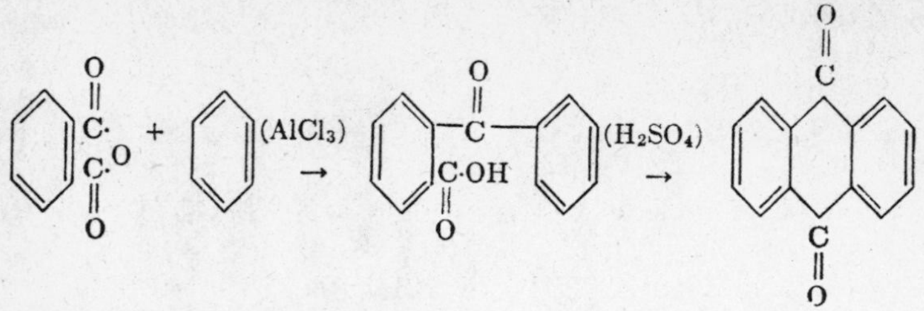

2. By the mild oxidation of anthracene.

D. Physical Properties:

Benzaldehyde (b.p. 179.5, m.p. −26) gives to almonds their characteristic odor. Salicylaldehyde (b.p. 196.5, m.p. −7) reacts with the skin to give a yellow coloration. Vanillin (m.p. 81-2, b.p. 285), piperonal (m.p. 37, b.p. 263), and cinnamaldehyde (b.p. 251, m.p. −7.5) all possess faint odors. These are all slightly soluble in water, but very soluble in alcohol and ether.

Acetophenone (b.p. 202.3, m.p. 19.7), benzophenone (b.p. 306), hydroquinone (m.p. 170.5, b.p. 286.2), 1, 4-naphthoquinone (yellow, m.p. 125, b.p. sublimes at 100), 1, 2-naphthoquinone (yellow to red, m.p. decomposes at 115-20), and anthraquinone (yellow, m.p. 286, sublimes, b.p. 379-81), are all quite insoluble in water but are soluble in alcohol and ether.

E. Chemical Properties:

I. The Aldehydes

a. *BENZALDEHYDE REACTS UNDER PROPER CONDITIONS:*

1. *To give derivatives characteristic of the carbonyl group (cf. Ch. VI, Div. I, Sec. F) when treated to effect:*

   a. Oxidation:

$$3\ C_6H_5 \cdot CHO + Cr_2O_7{}^- / 4\ H_2O \rightarrow 3\ C_6H_5 \cdot CO \cdot OH + 2\ Cr^{+++} + 8\ OH^-$$

The ease with which benzaldehyde is oxidized by air may be explained by postulating the following reactions:

$$C_6H_5 \cdot CHO + O_2 \rightarrow C_6H_5 \cdot CO \cdot O \cdot O \cdot H \text{ (perbenzoic acid), then}$$

$$C_6H_5 \cdot CO \cdot O \cdot O \cdot H + C_6H_5 \cdot CHO \rightarrow 2\ C_6H_5 \cdot CO \cdot OH \text{ (benzoic acid)}$$

Benzaldehyde does not reduce Fehling solution readily.

   b. Reduction:

$$C_6H_5 \cdot CHO + 2\ Na/\ (Hg)\ /2\ H_2O \rightarrow C_6H_5 \cdot CH_2OH + 2\ NaOH$$

Closely related to these two types, is the Cannizzaro reaction which involves autooxidation and reduction.

$$2\ C_6H_5 \cdot CHO/\ (alkali, aq.) \rightarrow C_6H_5 \cdot CO \cdot OH + C_6H_5 \cdot CH_2 \cdot OH$$

   c. Addition:

     1. $C_6H_5 \cdot CHO + H \cdot CN$, aq. $\rightarrow C_6H_5 \cdot CHOH \cdot CN$ (*mandelonitrile*)

     2. $2\ C_6H_5 \cdot CHO/(KCN)$, aq. $\rightarrow C_6H_5 \cdot CHOH \cdot CO \cdot C_6H_5$ (benzoin)

This equation illustrates the benzoin condensation.

d. Substitution:

1. $C_6H_5 \cdot CHO + H_2N \cdot OH \rightarrow C_6H_5 \cdot C(H){:}N \cdot OH$ (benzaldoxime) $+ H_2O$

   Both cis- (m.p. 130) and trans- (m.p. 35) forms are known.

2. $C_6H_5 \cdot CHO + CH_3 \cdot CHO/(NaOH, aq.) \rightarrow C_6H_5 \cdot HC{:}CH \cdot CHO + H_2O$, or

   $C_6H_5 \cdot CHO + CH_3 \cdot CO \cdot C_6H_5/$ (NaOH, aq.) $\rightarrow$
   $C_6H_5 \cdot HC{:}CH \cdot CO \cdot C_6H_5 + H_2O$

   These equations are examples of the Claisen reaction.

   $C_6H_5 \cdot CHO + CH_3 \cdot CO \cdot ONa/Ac_2O \rightarrow C_6H_5 \cdot HC{:}CH \cdot CO \cdot ONa$
   (sodium cinnamate) $+ 2$ AcOH

   This is an example of the Perkin reaction.   In each of these reactions, the carbonyl oxygen atom has been replaced.   Addition may actually precede the loss of water, in which event they might be regarded as addition reactions.

3. $C_6H_5 \cdot CO \cdot H + Cl_2$, boil $\rightarrow C_6H_5 \cdot CO \cdot Cl$ (benzoyl chloride) $+$ HCl

   This reaction is employed commercially, but analogous reactions are of little value in the aliphatic series.

e. Condensation:

$$C_6H_5 \cdot CH \boxed{O + H_2} N \boxed{\begin{matrix} H \\ \\ H \end{matrix}} + O \begin{matrix} H \\ C \cdot C_6H_5 \end{matrix} \rightarrow$$

$$C_6H_5 \cdot CH \boxed{O + H_2} N \boxed{H}$$

$$\begin{matrix} C_6H_5 \cdot CH{:}N \\ \\ C_6H_5 \cdot CH{:}N \\ \text{zamide)} \end{matrix} \begin{matrix} \\ CH \cdot C_6H_5 + \\ 3\ H_2O \\ \text{(hydroben-} \end{matrix}$$

2. *To give characteristic nuclear substitutions with* (*a*) *halogen,* (*b*) *nitric acid, or* (*c*) *sulfuric acid, the substituent entering principally at the m-position:*

$$\underset{\text{CHO}}{\bigcirc} + Br_2/ \text{ (Fe, about 20 °C.)} \rightarrow \underset{\text{CHO} \quad \text{Br}}{\bigcirc} + \text{HBr, and isomers}$$

b. THE AROMATIC ALDEHYDES REACT, IN GENERAL:

1. To give characteristic carbonyl derivatives.

2. To give characteristic nuclear substitutions.

II. The Ketones

a. ACETOPHENONE REACTS UNDER PROPER CONDITIONS:

1. To give characteristic carbonyl derivatives when treated to effect (*a*) *oxidation,* (*b*) *reduction,* (*c*) *addition,* (*d*) *substitution, or* (*e*) *condensation:*

a. $C_6H_5 \cdot CO \cdot CH_3 + 8 HNO_3$, aq. $\rightarrow C_6H_5 \cdot CO \cdot OH + CO_2 + 8 NO_2 + 5 H_2O$

   The side chain is oxidized to yield benzoic acid.

b. $C_6H_5 \cdot CO \cdot CH_3 + H_2/$ (Pt Black) $\rightarrow C_6H_5 \cdot CHOH \cdot CH_3$ (1-phenylethanol)

c. $C_6H_5 \cdot CO \cdot CH_3 + R \cdot Mg \cdot X /ether \rightarrow$
$C_6H_5 \cdot C(R)(CH_3) \cdot O \cdot MgX$, then

$C_6H_5 \cdot C(R)(CH_3) \cdot O \cdot MgX + HX$, aq. $\rightarrow$
$C_6H_5 \cdot C(OH)(R)CH_3 + MgX_2$

d. $C_6H_5 \cdot CO \cdot CH_3 + PCl_5$, heat $\rightarrow C_6H_5 \cdot CCl_2 \cdot CH_3 + POCl_3$

e. $C_6H_5 \cdot CHO + CH_3 \cdot CO \cdot C_6H_5 /$ (alkali, aq.) $\rightarrow$
$C_6H_5 \cdot CH \colon HC \cdot CO \cdot C_6H_5 + H_2O$

2. *To give normal substitution on the nucleus with (a) halogen, (b) nitric acid, or (c) sulfuric acid, the substituent entering principally at the m-position.*

b. THE ARYL-ALKYL AND THE DIARYL KETONES REACT IN GENERAL:

To give regular carbonyl and nuclear derivatives.

III. The Quinones

a. p-BENZOQUINONE REACTS UNDER PROPER CONDITIONS:

1. *By reduction with sulfur dioxide to give 1, 4-benzenediol:*
$p\text{-}O \colon C_6H_4 \colon O + SO_2 /2 H_2O \rightarrow p\text{-}HO \cdot C_6H_4 \cdot OH + H_2SO_4$

2. *By oxidation with permanganate to give meso-tartaric acid.*

3. *By the addition of bromine in chloroform to give p-benzoquinone dibromide and p-benzoquinone tetrabromide.*

4. *By substitution for (a) the oxygen atoms with hydroxylamine to give p-benzoquinone monoxime and dioxime, or (b) the hydrogen atoms with chlorine to give tetrachloroquinone:*

b. $p\text{-}O \colon C_6H_4 \colon O + 4 Cl_2 \rightarrow p\text{-}O \colon C_6Cl_4 \colon O + 4 HCl$ (a substitution)

b. NAPHTHOQUINONE AND ANTHRAQUINONE REACT, IN GENERAL:

To give the same types of derivatives as those indicated above.

IV. Aromatic Acids

A. Introduction:

Aromatic acids are compounds in which one or more hydrogen atoms of an aromatic nucleus have been replaced by the carboxyl group (CO·OH). Closely related to these in most of their properties are the substituted aliphatic acids which contain an aromatic nucleus as a substituent group.

Benzoic acid occurs in coal tar and in cranberries. As its benzyl ester, it occurs in gum benzoin and in Peru and Tolu balsams. Hippuric acid (C_6H_5·CO·NH·CH_2·CO·OH) is found as a waste product in horse urine. Cinnamic acid occurs in some of the tree balsams. Mellitic acid is an ingredient of the mineral honeystone ($C_{12}O_{12}Al_2$·18 H_2O). Benzoic acid and its derivatives are far more abundant in the natural excretions and waste products of plants and animals than are the other aromatic acids and their derivatives.

Benzoic acid is employed in seasoning tobacco, and sodium benzoate is used as a food preservative. Ethyl benzoate is an ingredient of a number of artificial perfumes and flavors. Coumarin, which is prepared by the action of sodium ethanoate on salicylaldehyde in the presence of acetic anhydride, is used as a substitute for vanillin in perfumes. Phthalic anhydride is a raw material in the synthetic resin industry. It has been used to distinguish between primary and secondary alcohols by the esterification reaction. The rate of esterification of alcohols with organic acids decreases in the order R·CH_2·OH, R_2CH·OH, R_3C·OH.

The reactions employed for the preparation of these acids, together with their chemical properties, indicate their respective structures. Isomerism exists whenever two or more substituent groups are involved.

B. Nomenclature:

Structural Formulas	Named as Derivatives	Common Names
C_6H_5·CO·OH	benzenecarboxylic acid	benzoic acid
C_6H_5·CH_2·CO·OH	phenylethanoic acid	α-toluic acid, phenylacetic acid
C_6H_5·CH_2·CH_2·CO·OH	3-phenylpropanoic acid	hydrocinnamic acid
C_6H_5·HC:CH·CO·OH	*trans*-3-phenylpropenoic acid	cinnamic acid
C_6(CO·OH)$_6$	benzenehexacarboxylic acid	mellitic acid
o-C_6H_4(CH_3)CO·OH	2-methylbenzenecarboxylic acid	o-toluic acid
m-C_6H_4(CH_3)CO·OH	3-methylbenzenecarboxylic acid	m-toluic acid
p-C_6H_4(CH_3)CO·OH	4-methylbenzenecarboxylic acid	p-toluic acid
o-C_6H_4(CO·OH)$_2$	1, 2-benzenedicarboxylic acid	phthalic acid
m-C_6H_4(CO·OH)$_2$	1, 3-benzenedicarboxylic acid	isophthalic acid
p-C_6H_4(CO·OH)$_2$	1, 4- benzenedicarboxylic acid	terephthalic acid
α-$C_{10}H_7$·CO·OH	1-naphthalenecarboxylic acid	α-naphthoic acid
β-$C_{10}H_7$·CO·OH	2-naphthalenecarboxylic acid	β-naphthoic acid
α-$C_{14}H_9$·CO·OH	1-anthracenecarboxylic acid	α-anthroic acid
β-$C_{14}H_9$·CO·OH	2-anthracenecarboxylic acid	β-anthroic acid

C. Preparations:

I. The Aromatic Acids

a. BENZOIC ACID MAY BE PREPARED:

1. *By the oxidation of (a) toluene, or (b) benzyl alcohol:*
   a. $C_6H_5 \cdot CH_3 + Cr_2O_7^- /3 H_2O \rightarrow C_6H_5 \cdot CO \cdot OH + 2 Cr^{+++} + 8 OH^-$
   b. $3 C_6H_5 \cdot CH_2OH + 2 Cr_2O_7^- /5 H_2O \rightarrow 3 C_6H_5 \cdot CO \cdot OH + 4 Cr^{+++} + 16 OH^-$

2. *By the hydrolysis of (a) benzonitrile, or (b) benzotrichloride:*
   a. $C_6H_5 \cdot CN + 2 H_2O /H_2SO_4$ (85%), 110°C. $\rightarrow C_6H_5 \cdot CO \cdot OH + NH_4 \cdot HSO_4$
   b. $C_6H_5 \cdot CCl_3 + 4 NaOH/$ soap, aq. $\rightarrow C_6H_5 \cdot CO \cdot ONa + 3 NaCl + 2 H_2O$, then
      $C_6H_5 \cdot CO \cdot ONa + H_2SO_4$, aq. $\rightarrow C_6H_5 \cdot CO \cdot OH + NaHSO_4$

3. *By the reaction of benzene with (a) carbonyl chloride/ (AlCl₃) and subsequent hydrolysis, or (b) carbon dioxide/(AlCl₃) under pressure:*
   a. $C_6H_5 \cdot H + Cl \cdot CO \cdot Cl / (AlCl_3) \rightarrow C_6H_5 \cdot CO \cdot Cl + HCl$, then
      $C_6H_5 \cdot CO \cdot Cl + H_2O \rightarrow C_6H_5 \cdot CO \cdot OH + HCl$
   b. $C_6H_5 \cdot H + CO_2 / (AlCl_3$, pressure$) \rightarrow C_6H_5 \cdot CO \cdot OH$

4. *By the reaction of phenylmagnesium bromide with carbon dioxide and subsequent hydrolysis:*
   $C_6H_5 \cdot Mg \cdot Br + CO_2 \rightarrow C_6H_5 \cdot CO \cdot O \cdot MgBr$, then
      $2 C_6H_5 \cdot CO \cdot O \cdot MgBr + H_2SO_4$, aq. $\rightarrow 2 C_6H_5 \cdot CO \cdot OH + MgBr_2 + MgSO_4$

5. *By the loss of CO₂ from o-phthalic acid:*
   $o\text{-}C_6H_4(CO \cdot OH)_2 + (CuSO_4$, catalyst$)$, heat $\rightarrow C_6H_5 \cdot CO \cdot OH + CO_2$

b. THE TOLUIC ACIDS (o-, m-, and p-) MAY BE PREPARED:

By the partial oxidation of the corresponding xylene:
$p\text{-}C_6H_4 (CH_3)_2 + 6 HNO_3$, aq. $\rightarrow p\text{-}CH_3 \cdot C_6H_4 \cdot CO \cdot OH + 6 NO_2 + 4 H_2O$

c. THE PHTHALIC ACIDS MAY BE PREPARED:

By the oxidation of (a) naphthalene with air in the presence of vanadium pentoxide to give phthalic anhydride, or (b) by the oxidation of the corresponding xylenes:

a:

$+ x O_2/$ (air/V_2O_5, 400-450°C.) $\rightarrow$ + other products

The reaction is carried out in a Downs reactor, which consists of square iron tubes about one inch in diameter surrounded by mercury as the heat exchange medium. By refluxing the mercury under carefully controlled pressure, the temperature is accurately maintained.

b. $p\text{-}C_6H_4(CH_3)_2 + 2\ Cr_2O_7^- /6\ H_2O \rightarrow p\text{-}C_6H_4(CO\cdot OH)_2 + 4\ Cr^{+++} + 16\ OH^-$

d. THE NAPHTHOIC AND ANTHROIC ACIDS MAY BE PREPARED:

By the hydrolysis of the corresponding nitriles.

II. The Aryl-substituted Aliphatic Acids

a. PHENYLETHANOIC ACID MAY BE PREPARED:

By treating benzyl chloride with sodium cyanide and hydrolysis:
$C_6H_5\cdot CH_2\cdot Cl + Na\cdot CN \rightarrow C_6H_5\cdot CH_2\cdot CN + NaCl$, then

$C_6H_5\cdot CH_2\cdot CN + 2\ H_2O/H_2SO_4$, aa. $\rightarrow C_6H_5\cdot CH_2\cdot CO\cdot OH + NH_4HSO_4$

b. CINNAMIC ACID MAY BE PREPARED:

By treating benzaldehyde with sodium ethanoate in ethanoic anhydride (Perkin reaction):
$C_6H_5\cdot CHO + CH_3\cdot CO\cdot ONa/Ac_2O \rightarrow C_6H_5\cdot HC{:}CH\cdot CO\cdot ONa + 2\ Ac\cdot OH$, then

$C_6H_5\cdot HC{:}CH\cdot CO\cdot ONa + H_2SO_4$, aq. $\rightarrow C_6H_5\cdot HC{:}CH\cdot CO\cdot OH + NaHSO_4$

c. HYDROCINNAMIC ACID MAY BE PREPARED:

By reducing cinnamic acid with the appropriate reagents:
$C_6H_5\cdot HC{:}CH\cdot CO\cdot OH + 2\ Na/2\ Et\cdot OH \rightarrow$
$C_6H_5\cdot CH_2\cdot CH_2\cdot CO\cdot OH + 2\ Et\cdot ONa$

D. Physical Properties:

Benzoic acid (m. p. 122, b.p. 249), the toluic acids (m.p. o-103.7, m- 108.75 p-179.6), the phthalic acids (m.p. o- 206-8, m- 330, p- sublimes), the naphthoic acids (m.p. 1- 160, 2-185), and the anthroic acids (m.p. 1- 245, 2- 281, 9- 217 d.) are all insoluble in water but quite soluble in alcohol and ether.

Phenylethanoic acid (m.p. 76.7), cinnamic acid (m.p. 133), and hydrocinnamic acid (m.p. 48.6) are slightly soluble in water and quite soluble in alcohol and ether.

E. Chemical Properties:

I. THE AROMATIC ACIDS REACT UNDER PROPER CONDITIONS:

1. To give typical derivatives of the carboxyl group (cf. Ch. VII, Div. I, Sec. E).

2. To give typical nuclear substitutions (cf. Ch. XVII, Div. E).

3. To give typical side chain reactions (cf. Ch. VII, Div. I, Sec. E).

In the oxidation of a side chain, the chain is quite readily oxidized to give a carboxyl group linked to the nucleus.

CHAPTER XX

AROMATIC SULFUR DERIVATIVES

The Sulfonic Acids

A. Introduction:

The aromatic sulfonic acids are derivatives of the aromatic hydrocarbons in which one or more hydrogen atoms of an aromatic hydrocarbon have been replaced by the sulfonic acid group ($\cdot SO_2 \cdot OH$). If both of the hydroxyl groups in sulfuric acid are replaced by aryl radicals, sulfones result ($Ar \cdot SO_2 \cdot Ar$). In most of the sulfonation reactions, small amounts of sulfones are formed due to presence of the SO_3, along with the formation of the sulfonic acids.

Aromatic sulfonic acids occur in protein degradation products, urine, and similar waste products from metabolism.

The benzene-, toluene-, and xylenesulfonic acids are important synthetic intermediates. Until the development of the Dow process for producing phenol from chlorobenzene, most of the synthetic phenol was produced by the alkali fusion of benzenesulfonic acid. α- and β-naphthol are prepared from the corresponding sulfonic acids, and α- and β-anthraquinonesulfonic acids are important dye intermediates.

B. Nomenclature:

The aromatic sulfonic acids are named by adding "sulfonic acid" to the name of the parent compound, or they are known by common or trade names.

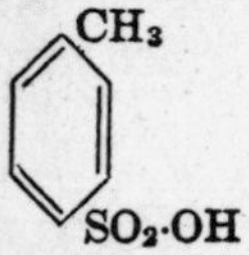

p-toluenesulfonic acid 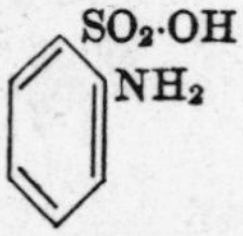   o-aminobenzene-sulfonic acid 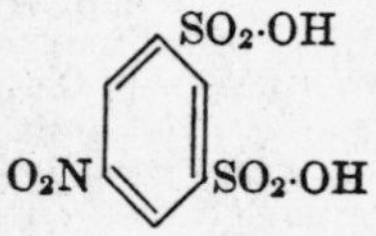   5-nitro-1, 3-benzene-disulfonic acid 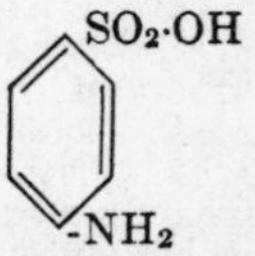   **sulfanilic acid**

C. Preparations:

1. BENZENE- AND TOLUENESULFONIC ACIDS MAY BE PREPARED:

1. By the reaction of one, two, or three moles of sulfuric acid (sp.gr. 1.88, 7% SO_3) with benzene to give the mono-, di- or tri- sulfonic acid derivatives:

    a. $C_6H_5 \cdot H + HO \cdot SO_2 \cdot OH / (7\% SO_3)$, heat $\rightarrow C_6H_5 \cdot SO_2 \cdot OH + H_2O$

206

b.

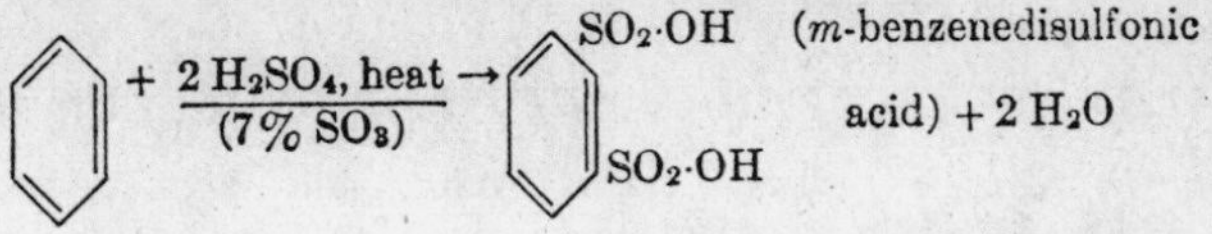

$$\text{(benzene)} + 2\,H_2SO_4, \text{heat} \xrightarrow{(7\%\ SO_3)} \text{(ring with } SO_2\cdot OH \text{ groups)} \quad (\textit{m-}\text{benzenedisulfonic acid)} + 2\,H_2O$$

c.

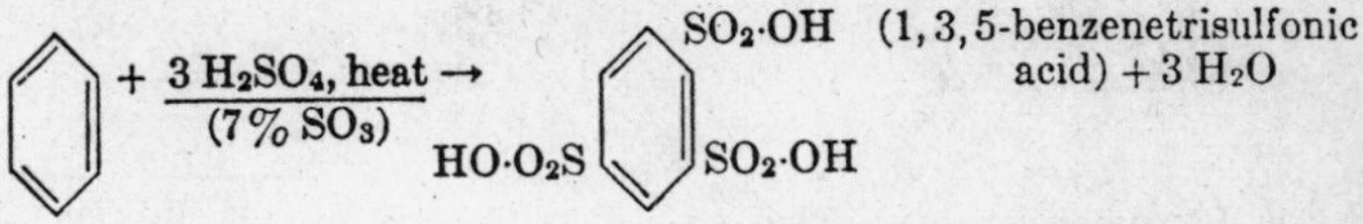

$$\text{(benzene)} + 3\,H_2SO_4, \text{heat} \xrightarrow{(7\%\ SO_3)} \text{(ring)} \quad (1,3,5\text{-benzenetrisulfonic acid)} + 3\,H_2O$$

2. *By the reaction of one, two, or three moles of sulfuric acid (sp.gr. 1.88, 7% SO₃) with toluene to give the mono-, di-, or trisulfonic acid derivatives:*

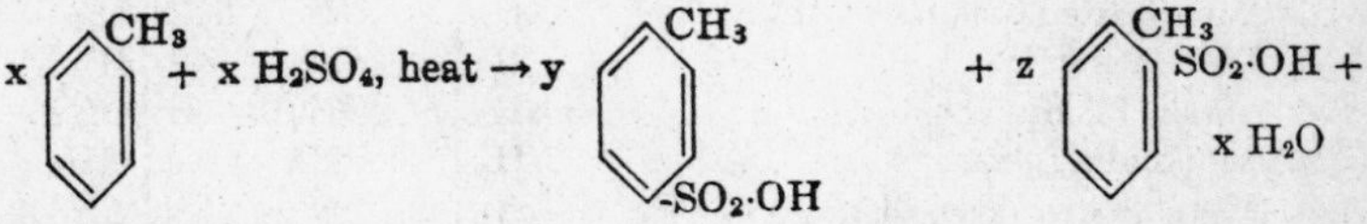

$$x\ \text{(toluene)} + x\,H_2SO_4, \text{heat} \rightarrow y\ \text{(ring)} + z\ \text{(ring)} + x\,H_2O$$

The yield of the *p*-isomer is increased from about 53% at 0°C. to about 79% at 100°C., but the yield of the *o*-isomer (42%) is best at 0°C. $Cl\cdot SO_2\cdot OH$ is a useful sulfonation agent. The introduction of the second and third sulfonic acid groups into an aromatic nucleus takes place somewhat less readily than does the initial sulfonation. Substituents on the nucleus govern, to some extent, the ease of sulfonation.

2. NAPHTHALENE- AND ANTHRACENESULFONIC ACIDS MAY BE PREPARED:

1. *By the action of sulfuric acid on naphthalene (a) at 80-100°C. to give mostly α-naphthalenesulfonic acid, or (b) at 160°C. to give β-naphthalenesulfonic acid as the principal product:*

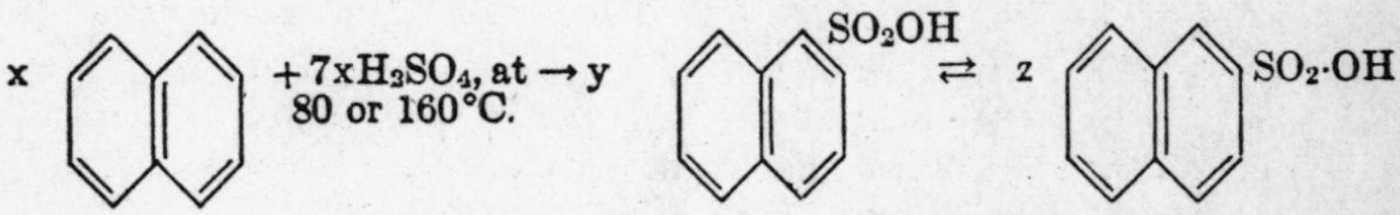

$$x\ \text{(naphthalene)} + 7x\,H_2SO_4, \text{at} \atop 80 \text{ or } 160\,°C. \rightarrow y\ \text{(ring } SO_2OH) \rightleftarrows z\ \text{(ring } SO_2\cdot OH)$$

2. *By the action of (a) dilute sulfuric acid on anthracene to give 2-anthracenesulfonic acid, or (b) concentrated sulfuric acid on anthracene to give 1, 5- or 1, 8-anthracenedisulfonic acid.*

D. Physical Properties:

Benzenesulfonic acid crystallizes with water of crystallization to give colorless leaflets ($(C_6H_5\ SO_2\cdot OH)_2\cdot3\ H_2O$, m.p. 46, anyhydrous, m.p. 50) which are quite soluble in water and alcohol. The calcium, barium, and lead salts are appreciably soluble in water. The introduction of a sulfonic acid group into an aromatic compound, markedly increases its solubility in water and its crystalline properties. Toluenesulfonic acid, the xylenesulfonic acids and α- and β-naphthalenesulfonic acids are all solids and are fairly soluble in water.

E. Chemical Properties:

THE AROMATIC SULFONIC ACIDS REACT, IN GENERAL:

1. *To give typical derivatives of the hydroxyl group when treated with:*

    a. Alkalies, aqueous $C_6H_5 \cdot SO_2 \cdot O$ | H + HO | Na

    b. Alcohols, reflux $C_6H_5 \cdot SO_2 \cdot O$ | H + HO | R

    c. Phosphorus pentahalides $C_6H_5 \cdot SO_2$ | O| H + Cl |PCl$_3$ | Cl

2. *To give replacement of the sulfonic acid group when treated with:*

    a. Superheated steam, 150°C. C_6H_5 | SO_2OH + HO | H

    b. Distillation C_6H_5 | $SO_2 \cdot O$ - - - - | H

    c. Alkali, fusion C_6H_5 | SO_2ONa + H | ONa

    d. Sodamide, fusion C_6H_5 | SO_2ONa + Na | NH_2

    e. Metallic cyanides, distil C_6H_5 | SO_2ONa + Na | CN

    f. Sodium formate, fusion C_6H_5 | $SO_2 \cdot ONa$ + H | $CO \cdot ONa$

    g. Phosphorus pentachloride, 210°C.,

$C_6H_5 \cdot SO_2 \cdot Cl + PCl_5, 210°C. \rightarrow C_6H_5 \cdot Cl + SOCl_2 + POCl_3$

    Benzenesulfonyl chloride also reacts with (1) ammonia, (2) alcoholates, and (3) activated hydrogen:

    1. $C_6H_5 \cdot SO_2 \cdot Cl + 2 NH_3 \rightarrow C_6H_5 \cdot SO_2 \cdot NH_2 + NH_4Cl$

    2. $C_6H_5 \cdot SO_2 \cdot Cl + Na \cdot OR \rightarrow C_6H_5 \cdot SO_2 \cdot OR + NaCl$

    3. $C_6H_5 \cdot SO_2 \cdot Cl + 6 Na/5 ROH \rightarrow C_6H_5 \cdot SH + NaCl + 5 RONa + 2 H_2O$

3. *To give additional substitution in the m-position when treated with (a) chlorine or bromine in diffused or no light at room temperature in the presence of a catalyst, (b) concentrated nitric acid when heated in the presence of concentrated sulfuric acid, or (c) fuming sulfuric acid when heated to about 250°C., or when treated with* $Cl \cdot SO_2 \cdot OH$ *or* $Cl \cdot SO_2 \cdot Cl$:

    a.

    b. $C_6H_5 \cdot SO_2 \cdot OH + HNO_3 / (H_2SO_4) \rightarrow m\text{-}HO \cdot SO_2 \cdot C_6H_4 \cdot NO_2 + H_2O$

    c. $C_6H_5 \cdot SO_2 \cdot OH + H_2SO_4 / (SO_3, 250°C.) \rightarrow m\text{-}HO \cdot O_2S \cdot C_6H_4 \cdot SO_2 \cdot OH + H_2O$

    In the commercial production of picric acid, phenol is sulfonated to give *p*-hydroxybenzenesulfonic acid and the disulfonic acid derivative, which are then nitrated to give 2,4,6-trinitrophenol (picric acid).

CHAPTER XXI

AROMATIC NITROGEN DERIVATIVES

1. Aromatic Nitro Compounds

A. Introduction:

The aromatic nitro compounds are substances in which one or more hydrogen atoms on the aromatic nucleus have been replaced by nitro groups ($\cdot NO_2$) as in nitrobenzene ($C_6H_5 \cdot NO_2$) or o-, m-, or p-**dinitrobenzene** ($C_6H_4 (NO_2)_2$). Similar derivatives of toluene, xylene, naphthalene, anthracene, and phenanthrene are readily formed.

Nitrobenzene is used extensively in the commercial production of aniline and other derivatives and to a limited extent in shoe and metal polishes. The nitrotoluenes and the nitroxylenes are important synthetic intermediates. 2, 4, 6-Trinitrotoluene (T.N.T.) is a valuable explosive as it requires detonation by either shock or heat. The explosive "amatol" is a mixture of T.N.T. and ammonium nitrate. 1-Nitronaphthalene is used in the preparation of 1-naphthylamine and other derivatives.

The general type formula may be represented by $Ar(NO_2)_{x+1} - H_x$ Position isomerism is illustrated by o-, m-, and p- **dinitrobenzene** and by 1- and 2-nitronaphthalene. The nitro compounds are functional isomers of the nitrites.

B. Nomenclature:

The nitro compounds are named as substitution products by prefixing "nitro" to the name of the parent compound. When isomers are possible, suitable numerals or letters are used to indicate the relative positions of the substituted groups. Some specific examples are:

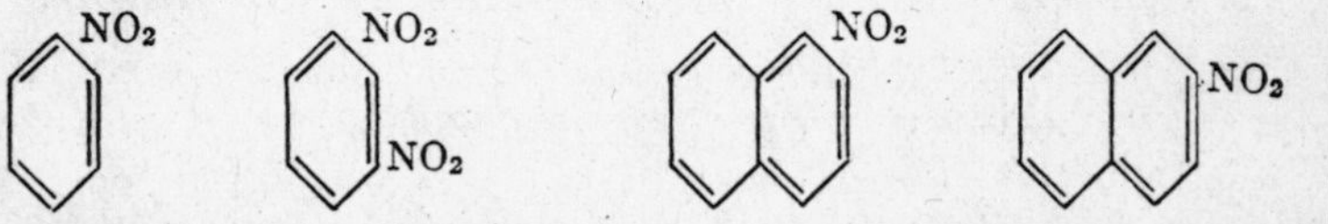

nitrobenzene   1, 3- (m-) dinitrobenzene   1- (α-) nitronaphthalene   2- (β) nitronaphthalene

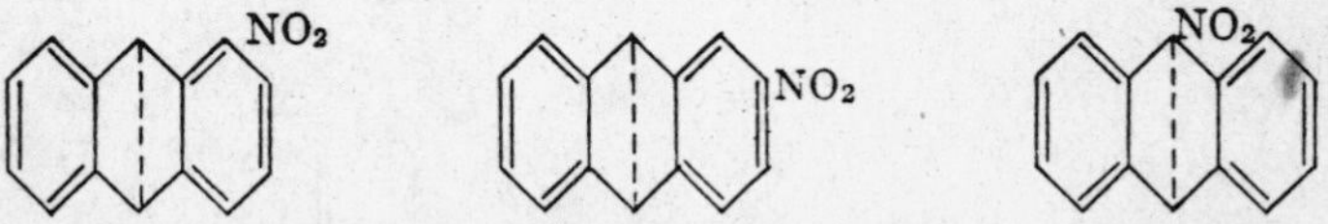

1- (α-) nitroanthracene   2- (β-) nitroanthracene   9- (γ-) nitroanthracene

C. Preparations:

1. NITROBENZENES, NITROTOLUENES, AND NITROXYLENES MAY BE PREPARED:

By the action of concentrated nitric acid, in the presence of concentrated sulfuric acid, on the parent compound at the appropriate temperature:

a. $C_6H_5 \cdot H + HO \cdot NO_2/H_2SO_4$, 50-60°C. $\rightarrow C_6H_5 \cdot NO_2 + H_2SO_4 \cdot H_2O$, and

$C_6H_5 \cdot NO_2 + HNO_3/H_2SO_4$, 90°C. $\rightarrow m\text{-}C_6H_4(NO_2)_2 + H_2SO_4 \cdot H_2O$, and

$m\text{-}C_6H_4(NO_2)_2 + HNO_3/H_2SO_4$, 120°C. $\rightarrow s\text{-}C_6H_3(NO_2)_3 + H_2SO_4 \cdot H_2O$

The second and third groups are substituted less readily with decreasing yields. The principal function of the sulfuric acid is the reproduction of nitrogen ions, as indicated by: $H_2SO_4 + HNO_3 \rightarrow NO_2^+ + HSO_4^- + H_2O$.

b.

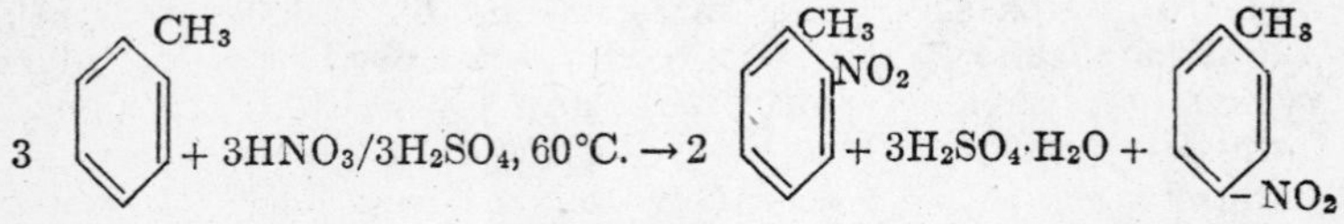

The products are *o*-nitrotoluene (60%) and *p*-nitrotoluene (30%). If these products are renitrated with a similar acid mixture at 70-90°C., the *o*-isomer yields 2, 4- and 2, 6-dinitrotoluene and the *p*-isomer yields 2, 4-dinitrotoluene, and both isomers on continued nitration yield 2, 4, 6-trinitrotoluene (T.N.T.).

c.

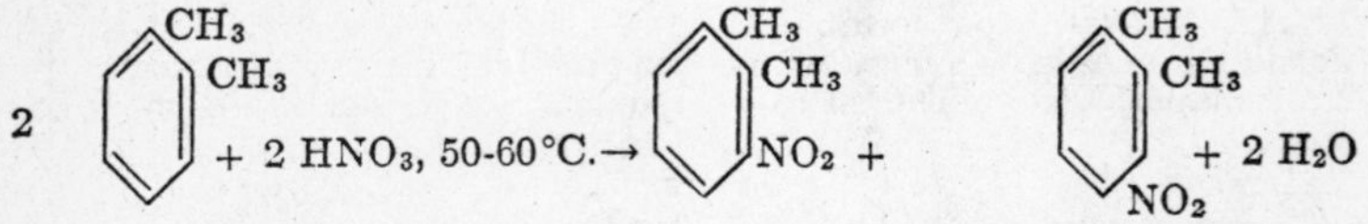

The products are 1, 2-dimethyl-3-nitrobenzene and 1, 2-dimethyl-4-nitrobenzene. The *m*- and *p*-xylenes may be nitrated by a similar procedure, and the continued nitration will result in the substitution of additional nitro groups. Toluene, the xylenes, and most of the substituted benzenes nitrate more readily than benzene (cf. nitration of phenol).

2. THE NITRONAPHTHALENES MAY BE PREPARED:

By (a) the direct nitration of naphthalene to give 1-nitronaphthalene, or (b) indirect methods to give 2-nitronaphthalene:

a.

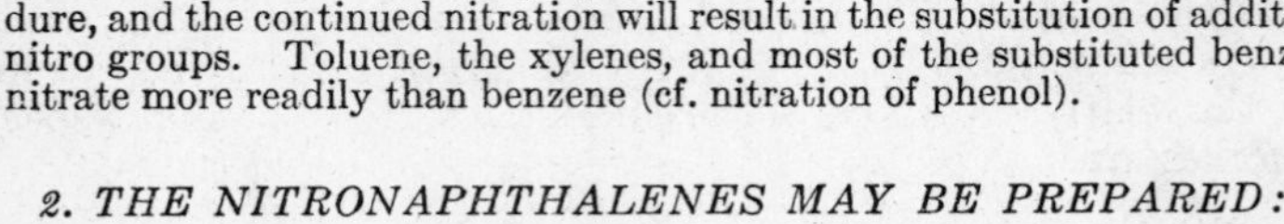

b.

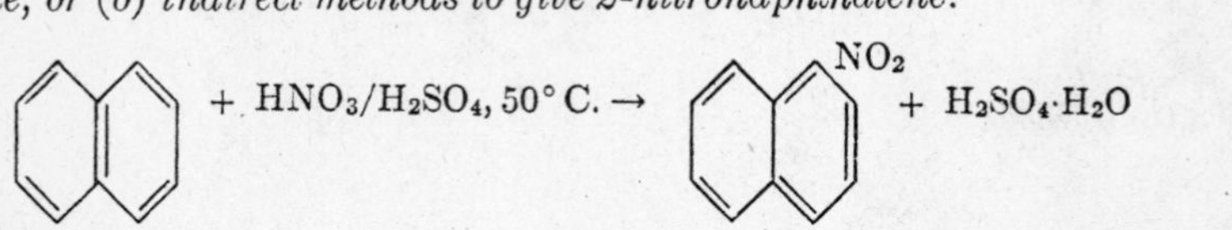

Then, $C_{10}H_7 \cdot SO_2 \cdot OH + 2\ NaOH,\ 300°C. \rightarrow C_{10}H_7 \cdot ONa + NaHSO_3 + H_2O$, and

$C_{10}H_7 \cdot ONa + CO_2/H_2O \rightarrow C_{10}H_7 \cdot OH$ (β-naphthol) $+ NaHCO_3$, then

$C_{10}H_7 \cdot OH + H \cdot NH_2/\ ((NH_4)_2SO_3,\ aq.) \rightarrow C_{10}H_7 \cdot NH_2 + H_2O$, and

$C_{10}H_7 \cdot NH_2 + NaNO_2/2\ HCl,\ aq. \rightarrow C_{10}H_7 \cdot N(\vdots N)Cl + NaCl + H_2O$, then

$C_{10}H_7 \cdot N(\vdots N)Cl + NaNO_2/\ (CuO,\ aq.),\ cold \rightarrow C_{10}H_7 \cdot NO_2 + N_2 + NaCl$

The final product is the β-isomer. The hydroxyl group in β-naphthol may be replaced by an NH_2 group equally well by treatment with $CaCl_2 \cdot 2\ NH_3$ or $ZnCl_2 \cdot 2NH_3$.

The nitro derivatives of anthracene and phenanthrene are relatively unimportant.

D. Physical Properties:

Nitrobenzene (b.p. 210.9, m.p. 5.9, sp.gr. 1.207) is a light yellow liquid which is slightly soluble in water but soluble in ether. It is an excellent solvent. The dinitrobenzenes are all solids. The *o*- and *m*-nitrotoluene (b.p. 222.3, 238) are yellow liquids, but the *p*-isomer is a solid (m.p. 51.3). Trinitrotoluene (T.N.T., m.p. 80.7) is a colorless solid which boils at 240°C. with explosive decomposition. Both the α- and β-nitronaphthalenes are solids which are insoluble in water but are soluble in alcohol and ether.

E. Chemical Properties:

1. NITROBENZENE REACTS UNDER PROPER CONDITIONS:

1. *By reduction when treated with (a) tin or iron and hydrochloric acid to give aniline, (b) sodium methoxide to give azoxybenzene, (c) zinc in the presence of ammonium chloride solution at an elevated temperature to give phenylhydroxylamine (d) zinc in alcoholic sodium hydroxide to give azobenzene, and (e) zinc and concentrated aqueous sodium hydroxide at 60°C. to give 1, 2-diphenylhydrazine (hydrazobenzene):*

a. $4\ C_6H_5 \cdot NO_2 + 9\ Fe\ /\ (FeCl_2,\ H^+) + 4\ H_2O \rightarrow 4\ C_6H_5 \cdot NH_2 + 3\ Fe_3O_4$

This reduction is more complicated than the equation indicates, and only small amounts of hydrochloric acid are required. By the use of ammonium sulfide as the reducing agent, *m*-dinitrobenzene may be reduced to *m*-nitroaniline:

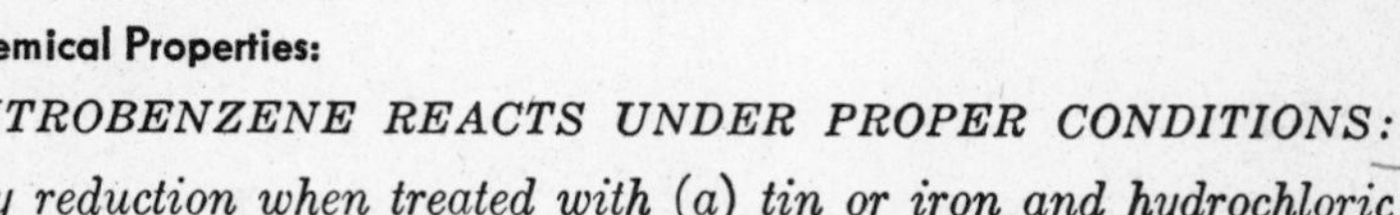

b. $4\ C_6H_5 \cdot NO_2 + 3\ NaOCH_3/\ (CH_3 \cdot OH),\ boil \rightarrow 2\ C_6H_5 \cdot N(\rightarrow O){:}N \cdot C_6H_5$ (azoxybenzene) $+ 3\ H \cdot CO \cdot ONa + 3\ H_2O$ (alkaline reduction)

c. $C_6H_5 \cdot NO_2 + 2\ Zn/3\ H_2O/(NH_4Cl),\ heat \rightarrow C_6H_5 \cdot NH \cdot OH + 2\ Zn(OH)_2$

d. $2\ C_6H_5 \cdot NO_2 + 4\ Zn/8\ NaOH/\ (C_2H_5OH) \rightarrow C_6H_5 \cdot N{:}N \cdot C_6H_5 + 4\ Na_2ZnO_2 + 4\ H_2O$ (alkaline reduction)

e. $2\ C_6H_5 \cdot NO_2 + 5\ Zn/10\ NaOH,\ aq.,\ 60°C. \rightarrow C_6H_5 \cdot NH \cdot HN \cdot C_6H_5$ (**hydrazobenzene**) $+ 5\ Na_2ZnO_2 + 4\ H_2O$ (alkaline reduction)

2. *By additional substitution on the nucleus when halogenated, nitrated, or sulfonated, yielding meta derivatives (cf. Ch. XVII, E.I.a. 1).*

The amino, hydroxyl, and halogen groups in substituted nitro compounds are activated when they are in the ortho or para position.

II. THE NITRONAPHTHALENES REACT UNDER PROPER CONDITIONS:

1. *By reduction with mild reducing agents to give the corresponding naphthylamine:*

$$4\ C_{10}H_7 \cdot NO_2 + 9\ Fe/(FeCl_2, H^+) + 4\ H_2O \rightarrow 4\ C_{10}H_7 \cdot NH_2 + 3\ Fe_3O_4$$

2. *By additional substitution in the nucleus:*

Concentrated sulfuric acid, for example, yields a mixture of 5-nitro-1-naphthalenesulfonic acid and 5-nitro-2-naphthalenesulfonic acid:

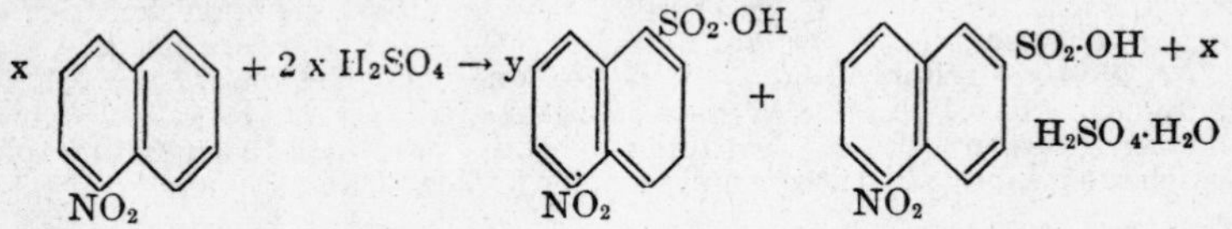

$$x \quad \text{[structure]} + 2\ x\ H_2SO_4 \rightarrow y \quad \text{[structure]}\ SO_2 \cdot OH$$
$$+ \quad \text{[structure]}\ SO_2 \cdot OH + x$$
$$NO_2 \qquad NO_2 \qquad NO_2 \qquad H_2SO_4 \cdot H_2O$$

Additional substitution on the nucles may be effected.   Anthracene enters into similar reactions.

II. Aromatic Amines

A. Introduction:

The aromatic primary amines are compounds in which one or more of the hydrogen atoms on an aromatic nucleus have been replaced by the amino group ($\cdot NH_2$).

Aniline and a large number of the aromatic amines are used extensively in the preparation of dye intermediates, dyes, indicators, medicinals, rubber accelerators, photographic developers, and similar products. Aniline is used extensively as a solvent, and phenylenediamines have been used by the imprudent in cosmetics.

The primary, secondary, and tertiary aromatic amines may be represented by the type formulas, $Ar \cdot NH_2$, Ar_2NH, and Ar_3N. There are also numerous mixed amines of the types $Ar \cdot NH \cdot R$, $Ar \cdot NR_2$, and $Ar_2N \cdot R$. The amines exhibit position isomerism.

B. Nomenclature:

The aromatic amines may be named as derivatives of the parent compound, but most of them are known by trade names. Examples are:

| aminobenzene, phenylamine, **aniline** | 2-aminotoluene, *o*-tolylamine, ***o*-toluidine** | 1-aminonaphthalene, 1-naphthylamine, α-naphthylamine | 1,3-benzenediamine, 1,3-diaminobenzene, *m*-phenylenediamine |

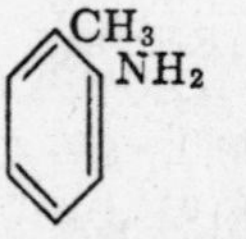

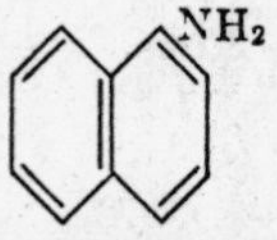

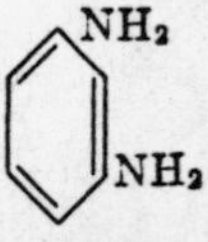

C. Preparations:

I. OF THE AMINO DERIVATIVES OF BENZENE, TOLUENE, AND XYLENE:

1. By (a) *the reduction of nitrobenzene with tin or iron and acid, or* (b) *the action of aqueous ammonia on chlorobenzene at a temperature of about 200°C. in the presence of a catalyst, to give aniline:*

    a. $4\ C_6H_5\cdot NO_2 + 9\ Fe\ /\ (FeCl_2,\ H^+) + 4\ H_2O \rightarrow 4\ C_6H_5\cdot NH_2 + 3\ Fe_3O_4$

    b. $C_6H_5\cdot Cl + 2\ NH_3$, aq. /(catalyst, pressure, 200°C.) $\rightarrow$ $C_6H_5\cdot NH_2 + NH_4Cl$ (Dow Process, commercial)

Finely divided copper, oxides of metals, or copper sulfate may be used as the catalyst in this reaction.

2. By the reduction of (a) *nitrotoluenes to give toluidines, or* (b) *nitroxylenes to give xylidines:*

    a. $CH_3\cdot C_6H_4\cdot NO_2 + 3\ Sn\ /6\ HCl$, aq. $\rightarrow CH_3\cdot C_6H_4\cdot NH_2 + 3\ SnCl_2 + 2\ H_2O$

    b. $(CH_3)_2C_6H_3\cdot NO_2 + 3\ Sn\ /6\ HCl$, aq. $\rightarrow (CH_3)_2C_6H_3\cdot NH_2 + 3\ SnCl_2 + 2\ H_2O$

Tin is commonly used in laboratory reductions of this type, but iron is used commercially.

3. By heating aniline in an autoclave (a) *with aniline hydrochloride at about 240°C. to give diphenylamine, or* (b) *with methyl chloride or methyl alcohol and hydrogen chloride to give either mono- or dimethylaniline hydrochloride:*

    a. $C_6H_5\cdot NH_2 + C_6H_5\cdot NH_2\cdot HCl$, 240°C. $\rightarrow (C_6H_5)_2N\cdot H + NH_4Cl$

    b. 1. $C_6H_5\cdot NH_2 + CH_3\cdot Cl$, heat, pressure $\rightarrow$ $C_6H_5\cdot NH(HCl)\cdot CH_3$

       2. $C_6H_5\cdot NH_2 + 2\ CH_3\cdot Cl$, heat, pressure $\rightarrow$ $C_6H_5\cdot N(HCl)(CH_3)_2 + HCl$

When methyl alcohol is used with hydrogen chloride it is likely that methyl chloride is first formed.

II. OF THE AMINO DERIVATIVES OF NAPHTHALENE AND ANTHRACENE:

1. By (a) *the reduction of 1-nitronaphthalene to give 1-naphthylamine, or* (b) *the action of aqueous ammonia on 2-naphthol in the presence of a catalyst to give 2-naphthylamine:*

a.

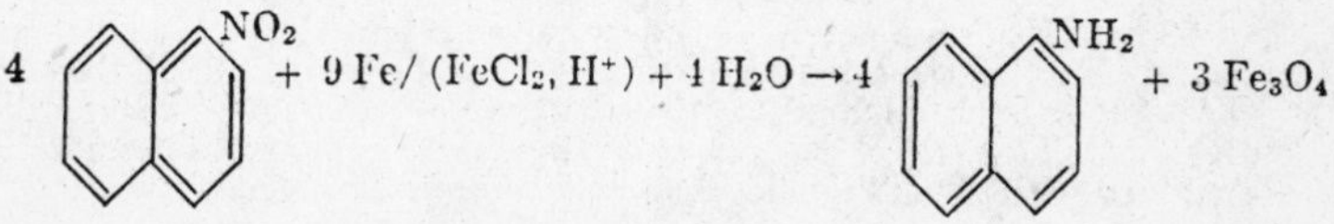

$4 \quad \text{[naphthalene-NO}_2\text{]} + 9\ Fe/\ (FeCl_2,\ H^+) + 4\ H_2O \rightarrow 4 \quad \text{[naphthalene-NH}_2\text{]} + 3\ Fe_3O_4$

b.

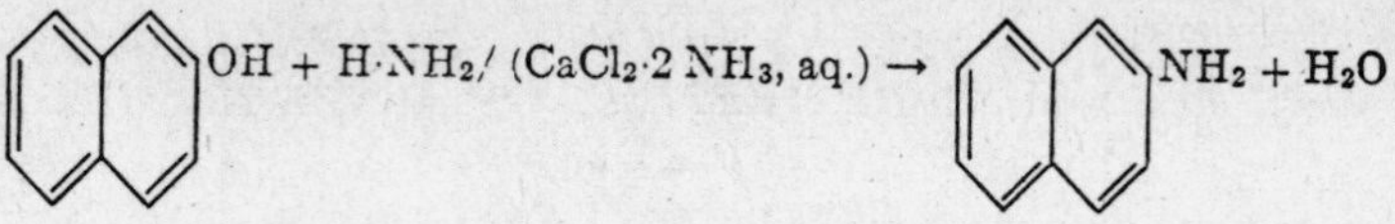

$$\text{(naphthol)}\ OH + H \cdot NH_2 / (CaCl_2 \cdot 2\,NH_3,\ aq.) \rightarrow \text{(naphthylamine)}\ NH_2 + H_2O$$

Other catalysts that may be used are $(NH_4)_2SO_3$ and $ZnCl_2 \cdot 2NH_3$. It should be noted that the hydroxyl group in naphthol is replaced more readily than is the corresponding group in phenol.

2. *By the reduction of nitroanthracenes to give aminoanthracenes:*

$$4\ \alpha\text{-}C_{14}H_9 \cdot NO_2 + 9\ Fe / (FeCl_2,\ H^+) + 4\ H_2O \rightarrow 4\ \alpha\text{-}C_{14}H_9 . NH_2 + 3\ Fe_3O_4$$

D. Physical Properties:

Aniline (b.p. 184.4, m.p. —6.2, D. $1.022\frac{20}{4}$) is a colorless oil, with a faint characteristic odor, which is sparingly soluble in water but readily soluble in alcohol and ether. It gradually darkens when exposed to air and light. The benzenediamines are all solids. The o- and m-toluidines (b.p. 197, 199) are both oils, but the p-isomer is a crystalline solid (m.p. 45, b.p. 198). N-Methyl- and N, N-dimethylaniline (b.p. 195.7, 193.5) are both yellow liquids, but the naphthylamines (α-, m.p. 50, b.p. 301; β- m.p. 110.2, b.p. 306.1) and 9-aminoanthracene (m.p. 145-50) are all yellow solids.

E. Chemical Properties:

THE AROMATIC AMINES REACT UNDER PROPER CONDITIONS:

1. *By substitution of one or both hydrogen atoms of the NH_2 group when treated with:*

a. 1. **Acetic acid**	Ar·NH	H +		HO	OC·CH₃
2. **Acetic anhydride**	Ar·NH	H +	CH₃·CO·O		OC·CH₃
3. **Acyl halides**	Ar·NH	H +		Cl	OC·CH₃
4. **Carbon disulfide**	2 Ar·NH	H +		S	C:S
b. 1. **Chloroform/alkali**	Ar·N	H₂ +		Cl₂	C (H)Cl
2. **Aldehydes/alkali**	Ar·N	H₂ +		O	C (H)·C₆H₅

2. *By substitution of one or more hydrogen atoms on the nucleus when treated with:*

a. *Bromine water to give 2, 4, 6-tribromoaniline:*

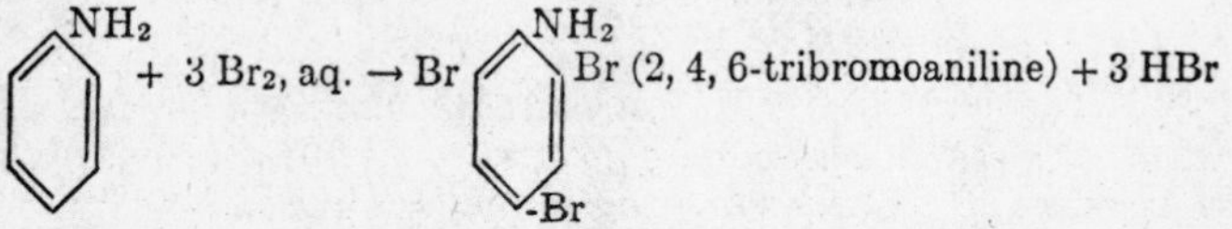

$$\text{(aniline)}\ NH_2 + 3\ Br_2,\ aq. \rightarrow Br\ \text{(ring)}\ Br\ (2, 4, 6\text{-tribromoaniline}) + 3\ HBr$$

The NH_2 group activates the hydrogen atoms in the o- and p- positions and the tribromoaniline forms almost instantly. If, however, acetanilide is brominated, a mixture of the o- and p- bromoacetanilides result. These may be separated by fractional crystallization and warmed with alkali to yield the o- and p- bromoanilines:

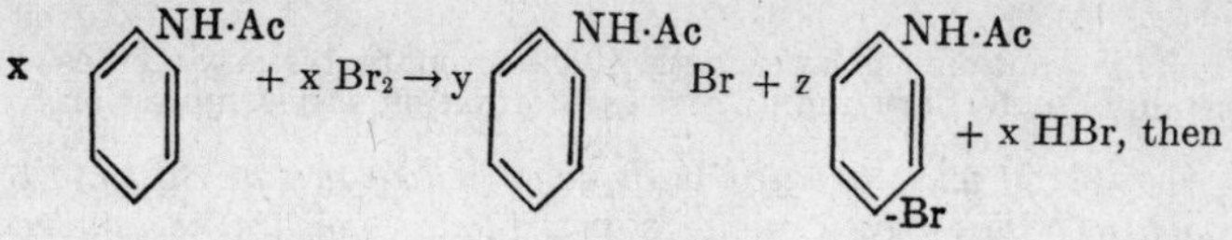

x [structure: NH·Ac benzene] $+ x \, Br_2 \rightarrow y$ [structure: NH·Ac benzene] $Br + z$ [structure: NH·Ac benzene -Br] $+ x \, HBr$, then

$p\text{-}Br\cdot C_6H_4\cdot NH\cdot Ac + NaOH$, aq., heat $\rightarrow p\text{-}Br\cdot C_6H_4\cdot NH_2 + Ac\cdot ONa$

b. Concentrated sulfuric acid to give aniline acid sulfate which, upon heating at about 180-190°C., gives sulfanilic acid:

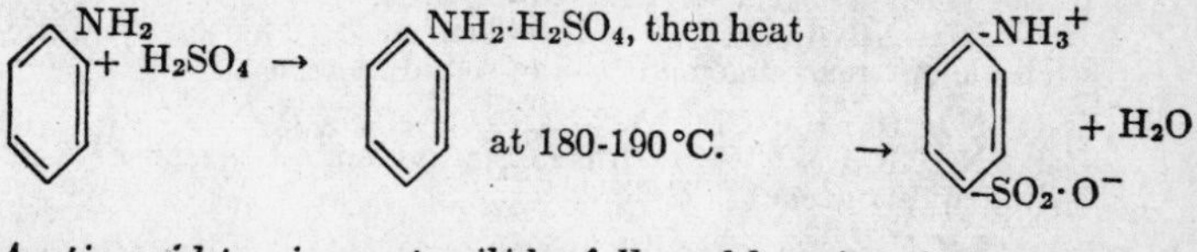

[structure: NH_2 benzene] $+ H_2SO_4 \rightarrow$ [structure: NH_2·H_2SO_4 benzene], then heat at 180-190°C. $\rightarrow$ [structure: NH_3^+ benzene -SO_2·O^-] $+ H_2O$

c. Acetic acid to give acetanilide, followed by nitration and subsequent hydrolysis, to yield o- and p-nitroaniline:

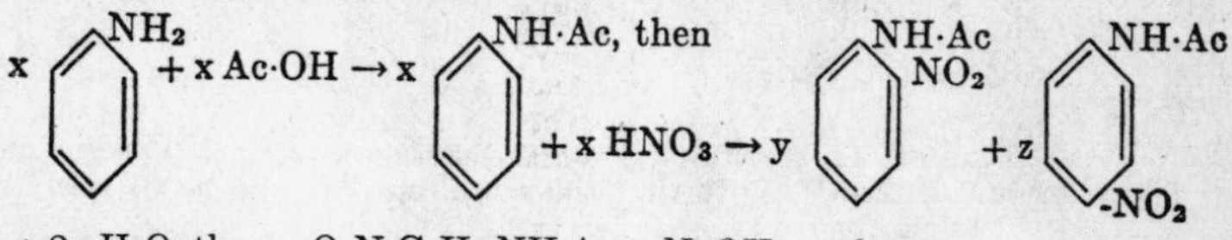

x [structure: NH_2 benzene] $+ x \, Ac\cdot OH \rightarrow x$ [structure: NH·Ac benzene], then $+ x \, HNO_3 \rightarrow y$ [structure: NH·Ac NO_2 benzene] $+ z$ [structure: NH·Ac benzene -NO_2]

$+ 2x \, H_2O$, then $p\text{-}O_2N\cdot C_6H_4\cdot NH\cdot Ac + NaOH$, aq. heat $\rightarrow$
$p\text{-}O_2N\cdot C_6H_4\cdot NH_2$ (*p*-nitroaniline) $+ Ac\cdot ONa$

The NH$_2$ group is acetylated to protect it from oxidation by the nitric acid. This is an example of a "blocked side chain reaction." The *o*- and *p*-nitroacetanilides are separated by fractional crystallization. *m*-Nitroaniline may be prepared by the nitration of aniline acid sulfate, followed by alkaline hydrolysis.

$C_6H_5\cdot NH_2\cdot H_2SO_4 + HNO_3 \rightarrow m\text{-}O_2N\cdot C_6H_4\cdot NH_2\cdot H_2SO_4 + H_2O$

Likewise, *m*-chloroaniline (*m*-Cl·C$_6$H$_4$·NH$_2$) may be prepared by the chlorination of aniline acid sulfate, followed by alkaline hydrolysis.

3. By diazotization or substitution, depending on the type of the amine, when treated with nitrous acid:

a. [structure: NH_2 benzene] $+ HONO/HCl$, aq. (0-5°C.) $\rightarrow$ [structure: N(:N)Cl benzene] $\rightleftarrows$ [structure: N:N·Cl benzene] $+ 2 \, H_2O$

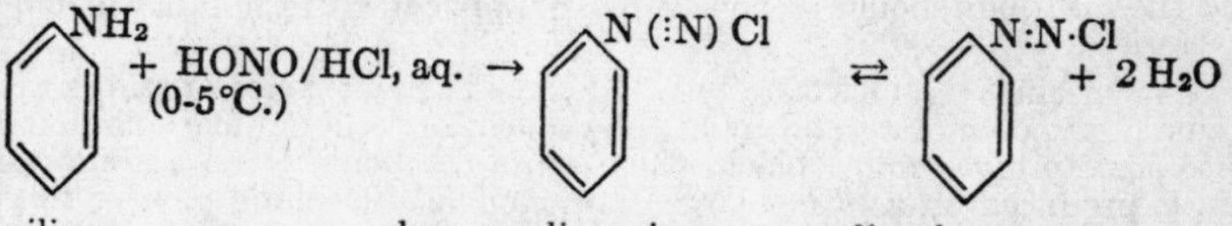

aniline   benzenediazonium chloride   diazobenzene chloride

The nitrous acid is formed in the solution by the action of hydrochloric acid on sodium nitrite. Primary aromatic amines, in general, are quite readily diazotized.

b. $C_6H_5\cdot N(CH_3)\cdot H + HONO \rightarrow C_6H_5\cdot N(CH_3)\cdot NO$ (N-nitroso-N-methylaniline) $+ H_2O$. This reaction is characteristic of secondary amines.

c. [structure: N(CH_3)_2 benzene] $+ HONO \rightarrow$ [structure: N(CH_3)_2 benzene -NO] (*p*-nitroso-N-dimethylaniline) $+ H_2O$

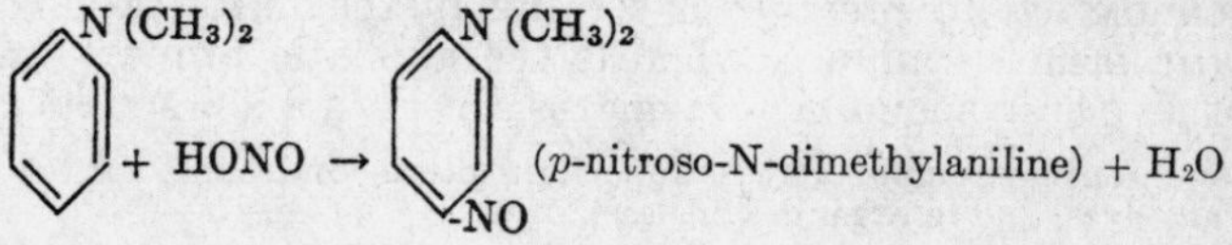

This reaction is characteristic of the aromatic tertiary amines whereas the aliphatic tertiary amines give salt formation or decomposition.

4. *By coupling when treated with benzenediazonium chloride to give derivatives that are characteristic of the type of amine involved:*

 a. $C_6H_5 \cdot NH \cdot H$ + $Cl(N:)N \cdot C_6H_5/(AcONa, 0°C)$ →
 $C_6H_5 \cdot NH \cdot N:N \cdot C_6H_5$ (diazoaminobenzene or 1, 3 - diphenyltriazene) + HCl, then

 $C_6H_5 \cdot NH \cdot N:N \cdot C_6H_5$ + $(C_6H_5 \cdot NH_2 \cdot HCl, 40°C.)$ →
 $H_2N \cdot C_6H_4 \cdot N:N \cdot C_6H_5$ (*p*-aminoazobenzene). This last equation is an example of a rearrangement of the semidine type.

 b. $C_6H_5 \cdot N(CH_3) \cdot H$ + $Cl(N:)N \cdot C_6H_5/(AcONa, 0°C.)$ →
 $C_6H_5 \cdot N(CH_3) \cdot N:N \cdot C_6H_5$ (diazo(methylamino)benzene or 3-methyl-1, 3-diphenyl-triazene) + HCl.

 c.

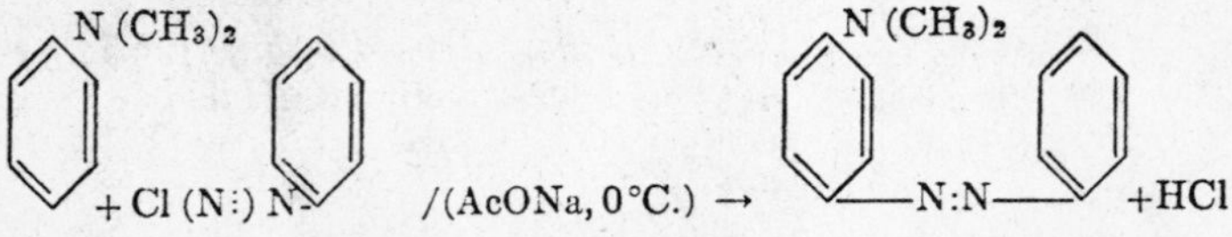

The product is *p*-(dimethylamino)azobenzene or N, N-dimethyl-*p*-phenylazoaniline. Note that the coupling is in the *p*-position.

5. *By addition when treated with halogen acids, sulfuric acid, or alkyl halides:*

 a. $C_6H_5 \cdot NH_2$ + HCl → $C_6H_5 \cdot NH_2 \cdot HCl$ (phenylammonium chloride or aniline hydrochloride).

 b. $C_6H_5 \cdot NH \cdot CH_3$ + HCl → $C_6H_5 \cdot NH(CH_3) \cdot HCl$ (methylphenylammonium chloride or methylaniline hydrochloride).

 c. $C_6H_5 \cdot N(CH_3)_2$ + HCl → $C_6H_5 \cdot N(CH_3)_2 \cdot HCl$ (dimethylphenylammonium chloride or dimethylaniline hydrochloride).

 Diphenylamine is such a weak base that its salts are decomposed by water. Triphenylamine is still less basic, and forms no salts.

Aniline and substituted anilines have been used, for the most part, in the above reactions to illustrate the chemical properties of the aromatic amines and the reactions should be considered as applicable to the other members of the series.

Two specific reactions of aniline that should be mentioned are those with (1) ammonium perchlorate and hydrochloric acid or sodium dichromate and sulfuric acid to give aniline black which contains about eleven aniline residues and is produced on a rather large scale, and (2) bleaching powder to give a characteristic violet coloration.

III. Aromatic Diazo and Diazonium Compounds

A. Introduction:

Aromatic diazonium compounds are very unstable, reactive compounds that are prepared in solution by the diazotization of the aromatic primary amines, whereas the aliphatic primary amines do not give corresponding derivatives except in a few special cases.

The diazonium compounds are important intermediates in the production of the azo dyes, and in organic synthesis.

They are represented by the type formula $Ar \cdot N_2 \cdot X$, in which "Ar" is an aromatic nucleus and "X" is an inorganic group such as halogen, nitrate, cyanide, or hydroxide. They were discovered by Griess (1858, in an investigation suggested by Kolbe) and the reaction for their preparation (diazotization) bears his name (Griess diazo reaction). Griess, Kekulé, and Blomstrand each, in turn, assigned formulas to these compounds:

| Griess | Kekulé | Kekulé | Blomstrand |

Some believe that the structures assigned by Kekulé and Blomstrand exist in tautomeric equilibrium as indicated above. Electrical conductivity supports the Blomstrand formula, but the existence of two potassium diazoates tends to support the Kekulé formula as a member of the equilibrium. Acid shifts the equilibrium to the diazonium structure and alkali favors the diazo form, thus indicating the amphoteric character of the free diazonium bases. The diazonium compounds are unstable and colorless, whereas the azo compounds ($Ar \cdot N:N \cdot Ar$) are stable and colored.

Substituted diazonium compounds show position isomerism.

B. Nomenclature:

$C_6H_5 \cdot N \,(:N) \, Cl$	benzenediazonium chloride
$C_6H_5 \cdot N:N \cdot O \cdot Na$	sodium benzenediazoate
$1\text{-}C_{10}H_7 \cdot N \,(:N) \, Cl$	1-naphthalenediazonium chloride
$1\text{-}C_{10}H_7 \cdot N:N \cdot O \cdot Na$	Sodium 1-naphthalenediazoate, sodium α-naphthalenediazoate.

C. Preparation:

AROMATIC DIAZONIUM COMPOUNDS MAY BE PREPARED:

By the diazotization of the salts of primary amines at 0-5°C. with (a) sodium nitrite and hydrochloric acid, or (b) with an alcoholic solution of butyl or amyl nitrite and hydrogen ion:

a. $Ar \cdot NH_2 \cdot HX + HONO / (NaNO_2/HCl)$, aq., 0-5°C. →
$Ar \cdot N(:N) \, X + 2 \, H_2O$

b. $Ar \cdot NH_2 \cdot HX + HONO / (RONO/H^+)$, alc., 0-5°C. →
$Ar \cdot N(:N) \, X + 2 \, H_2O$

The last reaction is employed when the isolation of the diazonium salt is desirable.

Often an azo dye can be decomposed to regenerate the diazonium salt $O_2N \cdot C_6H_4 \cdot N:N \cdot C_6H_4 \cdot O \cdot CH_3 + 3 \, HNO_3 \rightarrow O_2N \cdot C_6H_4 \cdot N \, (:N) \cdot NO_3 + (O_2N)_2$ $C_6H_3 \cdot OCH_3 + 2 \, H_2O$. Such reactions are of interest in studies on structure although the reduction of a dye is usually a better way of establishing structure.

When both groups in a diamine are diazotized simultaneously, the reaction is known as tetrazotization. Tetrazotization occurs much more readily if both of the amino groups are not on the same ring.

At 0-5°C. these derivatives are fairly stable in aqueous solution, but at higher temperatures they react with water to form hydroxy derivatives, nitrogen, and the mineral acid.

D. Physical Properties:

Benzenediazonium chloride, nitrate and most of the diazonium salts of the mineral acids, are colorless solids which are very soluble in water, quite soluble in alcohol, but insoluble in ether.   Many of these salts, especially the nitrates, are explosive when dry.   Even moist benzenediazonium perchlorate explodes violently when rubbed.

E. Chemical Properties:

THE DIAZONIUM SALTS REACT UNDER PROPER CONDITIONS:

1. By reactions involving the loss of N_2, when treated with:

	C_6H_5	$N(\colon N)$	$Cl\ +$		
a. Water, heat, or standing	C_6H_5	$N(\colon N)$	$Cl\ +$	H	OH
Hydrogen sulfide, heat	C_6H_5	$N(\colon N)$	$Cl\ +$	H	SH
b. Methanol, heat, light	C_6H_5	$N(\colon N)$	$Cl\ +$	H	OCH_3
Ethanol, heat, light	C_6H_5	$N(\colon N)$	$Cl\ +$	H	OCH_2CH_3
c. Benzene/ ($AlCl_3$)	C_6H_5	$N(\colon N)$	$Cl\ +$	H	C_6H_5
d. Hydrogen fluoride/ (piperidine)	C_6H_5	$N(\colon N)$	$Cl\ +$	H	F
Hydrogen chloride/ (Cu powder)	C_6H_5	$N(\colon N)$	$Cl\ +$	H	Cl
Hydrogen bromide/ (Cu powder)	C_6H_5	$N(\colon N)$	$Cl\ +$	H	Br
Hydrogen iodide/ (Cu powder)	C_6H_5	$N(\colon N)$	$Cl\ +$	H	I
e. Cuprous chloride, heat	C_6H_5	$N(\colon N)$	$Cl\ +$	Cu_2Cl	Cl
Cuprous bromide, heat	C_6H_5	$N(\colon N)$	$Cl\ +$	Cu_2Br	Br
Cuprous iodide, heat	C_6H_5	$N(\colon N)$	$Cl\ +$	Cu_2I	I
f. Metallic cyanides/ (Cu^{++} or Cu^+)	C_6H_5	$N(\colon N)$	$Cl\ +$	Na	CN
g. Mercuric nitrite / (Cu powder)	C_6H_5	$N(\colon N)$	$Cl\ +$	$HgNO_2$	NO_2

In the reaction of alcohols with benzenediazonium salts, a side reaction gives benzene, an aldehyde, nitrogen, and the inorganic acid, but the yields in this reaction are small if light is used as a catalyst.   The reactions in which powdered copper is used are known as Gattermann reactions and those in which cuprous salts are used are designated as Sandmeyer reactions.

2. By reduction to phenylhydrazine when treated with appropriate reducing agents:

The Emil Fischer classical method involves the following steps:

$C_6H_5 \cdot N\ (\colon N) \cdot Cl + Na_2SO_3 \rightarrow C_6H_5 \cdot N$ $+ NaCl$, then
$C_6H_5 \cdot N$ $\overset{..}{N} \cdot SO_2 \cdot ONa$

     $\overset{..}{N} \cdot SO_2 \cdot ONa + H_2SO_3 \rightarrow C_6H_5 \cdot N\ (SO_3H) \cdot NH \cdot SO_2 \cdot ONa$, and

$C_6H_5 \cdot N\ (SO_3H) \cdot NH \cdot SO_2 \cdot ONa + H_2O \rightarrow C_6H_5 \cdot NH \cdot NH \cdot SO_2 \cdot ONa + H_2SO_4$, then

$C_6H_5 \cdot NH \cdot NH \cdot SO_2 \cdot ONa + HCl/H_2O$, hot, conc. $\rightarrow C_6H_5 \cdot NH \cdot NH_2 \cdot HCl + H_2SO_4$, and

$C_6H_5 \cdot NH \cdot NH_2 \cdot HCl + NaOH$, aq. $\rightarrow C_6H_5 \cdot NH \cdot NH_2 + NaCl + H_2O$

3. By coupling in slightly acid, neutral, or alkaline solution with (a) aromatic amines, (b) phenols, (c) enolic compounds, and (d) aliphatic aci-nitro compounds:

a. 1. $C_6H_5 \cdot N(\colon N)$ ☐$Cl + H$☐ $NH \cdot C_6H_5$, 0-10°C. →
$C_6H_5 \cdot N\colon N \cdot NH \cdot C_6H_5 + HCl$

The acid liberated in the reaction is usually neutralized by sodium acetate. The product diazoaminobenzene, when treated with hydrochloric acid or aniline hydrochloride and heated, rearranges to *p*-aminoazobenzene hydrochloride ($C_6H_5 \cdot N\colon N \cdot C_6H_4 \cdot NH_2 \cdot HCl$).

2. $C_6H_5 \cdot N(\colon N)$ ☐$Cl + H$☐ $N(R)C_6H_5$, 0-10°C. →
$C_6H_5 \cdot N\colon N \cdot N(R)C_6H_5 + HCl$

3. $C_6H_5 \cdot N(\colon N)$ ☐$Cl + H$☐ $C_6H_4 \cdot NR_2$, 0-10°C. →
$p\text{-}C_6H_5 \cdot N\colon N \cdot C_6H_4 \cdot NR_2 + HCl$

b. $C_6H_5 \cdot N(\colon N)$ ☐$Cl + H$☐ $C_6H_4 \cdot OH$, 0-10°C. → $p\text{-}C_6H_5 \cdot N\colon N \cdot C_6H_4 \cdot OH$
$+ HCl$

c. $C_6H_5 \cdot N(\colon N)$ ☐$Cl + H$☐ $C\colon C (OH) \cdot CH_3$, 0-10°C. →
$\quad\quad\quad\quad\quad\quad\quad\quad\quad\quad\quad |$
$\quad\quad\quad\quad\quad\quad\quad\quad\quad\quad O\colon\dot{C}\cdot O\cdot R$

$(C_6H_5 \cdot N\colon N \cdot C\colon C(OH) \cdot CH_3) \rightarrow C_6H_5 \cdot NH \cdot N\colon C \cdot CO \cdot CH_3$
$\quad\quad\quad\quad |\quad\quad\quad\quad\quad\quad\quad\quad\quad\quad\quad\quad\quad |$
$\quad\quad\quad O\colon\dot{C}\cdot O\cdot R\quad\quad\quad\quad\quad\quad\quad O\colon\dot{C}\cdot O\cdot R$

(β-phenylhydrazone of the ester of α, β-diketobutyric acid)

d. $C_6H_5 \cdot N (\colon N)$ ☐$Cl + H$☐ $HC\colon N(\rightarrow O) \cdot OH$, 0-10°C. →
$(C_6H_5 \cdot N\colon N \cdot HC\colon N(\rightarrow O) \cdot OH) \rightarrow$

$C_6H_5 \cdot NH \cdot N\colon CH \cdot NO_2$ (phenylhydrazone of nitroformaldehyde)

The coupling with tertiary amines and phenols takes place in the *p*-position. If the *p*-position of a phenol is occupied, coupling occurs in the *o*-position. If a compound contains both amino and hydroxyl groups, the amino group directs to the *o*-position in acid solution and the phenol group direct to the *o*-position in alkaline solution.

4. By addition of halogen to give perhalides:

$C_6H_5 \cdot N(\colon N)^+ \cdot X + X_2 \rightarrow C_6H_5 \cdot NX \cdot NX_2.$

The perbromide is relatively stable and is readily obtained by the direct addition of bromine to a cold solution of the diazonium bromide.

5. By reactions of numerous types to give a variety of condensation and/or decomposition products (cf. An Outline of Organic Nitrogen Compounds, pp. 334-63, by Ed. F. Degering, available from University Lithoprinters, Ypsilanti, Michigan).

CHAPTER XXII

HETEROCYCLIC COMPOUNDS

Compounds such as epoxyethane and 1, 2-ethanedicarboxylic anhydride which contain oxygen in the ring have been considered (cf. Ch. V. Div. III. Ch. VIII. Div. II). Compounds of these particular types, however, are comparatively reactive and readily form addition products, whereas the types of compounds to be considered in this chapter have a rather stable heterocyclic ring structure. The most of them form substitution products much more readily than they form addition products, but pyridine, quinoline, and similar compounds form addition products with acids and many metallic salts such as those of mercury. Some of the more important heterocyclic compounds belonging to this group are considered below:

I. Compounds Containing an Heterocyclic Five-Membered Ring.

1. Furan (furfuran, 1, 4-epoxy-1, 3-butadiene) occurs in the tar obtained from pine wood. The furans may be prepared synthetically by the dehydration of a 1, 4-diketone or aldehyde.

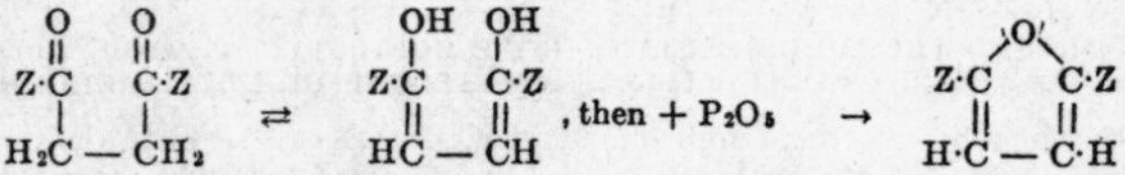

The most important derivative of furan is furfural which is produced commercially by the hydrolysis of pentosans to pentose sugars, which are subsequently dehydrated. The pentosans are widely distributed in corn cobs, oat hulls, and similar cheap materials. The sugars may exist in the form of a furanose ring, in which case they may be considered as derivatives of furan. Ordinary sucrose contains one furanose ring.

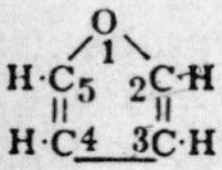

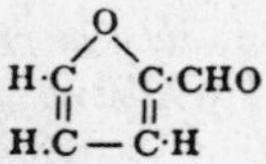

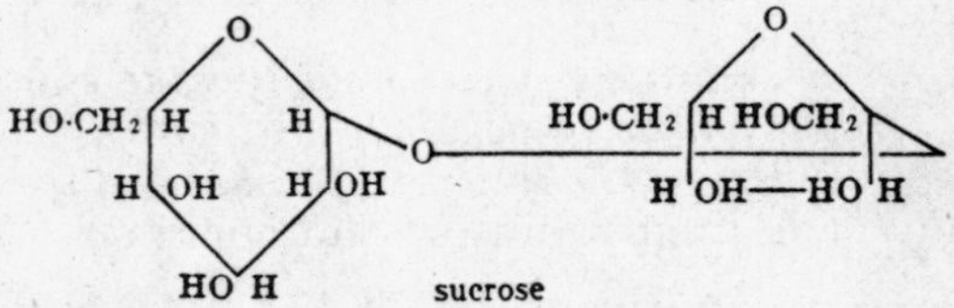

Position 2 is also indicated as α, 3 as β, 4 as β', and 5 as α'.

In those of its chemical properties which are dependent upon the carbonyl group, furfural is similar to benzaldehyde. Nuclear reactions, such as nitration or sulfonation, are not entirely analogous because of the ease with which the furan ring is oxidized. Substitution occurs in the 2- and 5-positions, the 3- and 4-positions being less active. This applies both to furan and to its derivatives. Upon mild oxidation, furfural yields 2-furancarboxylic acid (furoic acid, pyromucic acid). This, upon heating, yields furan. Catalytic hydrogenation of furan derivatives, in the presence of a nickel catalyst, yields tetrahydrofurans. Furfural is used in the preparation of synthetic resins (phenol-furfural resins) and as a solvent for the removal of aromatics, sulfur compounds, naphthenes, and other objectionable materials from lubricating oils. At present (1941), more oil is refined with furfural than with any other solvent. It is used on a smaller scale in the preparation of deodorizers, disinfectants, and paint removers.

2. Thiophene (thiofuran, 1-thia-2,4-cyclopentadiene) occurs in the benzene fraction of the coal tar distillate. Thiophenes may be prepared synthetically by the action of phosphorus pentasulfide on a 1, 4-diketone or aldehyde.

Thiophene boils at 84° C. and benzene boils at 80.08° C., hence it is difficult to effect complete separation of these compounds on a commercial scale by fractional distillation. Commercial benzene, therefore, normally contains traces of thiophene. Since thiophene sulfonates much more readily than benzene, the thiophene can be removed from benzene in the form of its sulfonate by repeated extraction with concentrated sulfuric acid. The presence of thiophene is shown by the use of its reaction with isatin in the presence of concentrated sulfuric acid to give a blue to green coloration (Indophenin Reaction).

thiophene

3. Pyrrole (azole, 1-aza-2, 4-cyclopentadiene) is obtained from coal tar and Dippel oil. Dippel oil is obtained by the destructive distillation of bones in the preparation of bone charcoal. Pyrrole may be prepared synthetically by the treatment of a 1, 4-diketone or aldehyde with ammonia.

By the addition of four hydrogen atoms, pyrrole yields pyrrolidine. Proline is a 2-carboxy derivative of pyrrolidine.

Pyrrole   Pyrrolidine   Proline

Pyrrole exhibits weakly basic properties and is easily polymerized by acids to give complicated red polymers, to which fact it owes its name. It exhibits definite acidic properties. It reacts, for example, with potassium hydroxide to yield N-potassium pyrrole. The benzoyl derivative of pyrrolidine reacts with phosphorus pentabromide to give 1, 4-dibromobutane.

The pyrroles are more stable than the furans but less stable than the thiophenes. The position taken by an entering group is the same as with furan, the 2- and 5-positions being most easily substituted.

Tetraiodopyrrole is called iodol and is used as a substitute for iodoform in medicine. Pyrrole is the essential structure in many alkaloids, in chlorophyll, in hematin, and in the bile pigment.

4. Pyrazole (1, 2-diazole) and its derivatives, some of which are shown below, are distinctly aromatic in their properties. 1, 5-Dimethyl-2-phenyl-3-pyrazolone (antipyrine) is used to combat fevers and as an analgesic. In naming the pyrazolones, the number in parenthesis indicates the position of the extra hydrogen atom.

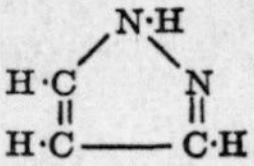

$$
\begin{array}{c}
\text{N·H} \\
\text{H·C} \diagup \quad \diagdown \text{N} \\
\text{H·C} \underline{\quad\quad} \text{C·H}
\end{array}
$$

pyrazole

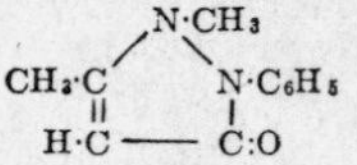

1,5-dimethyl-2-phenyl-3-pyrazolone, antipyrine

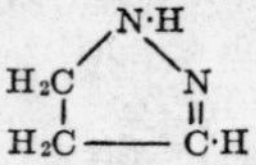

4,5-dihydropyrazole, 2-pyrazoline

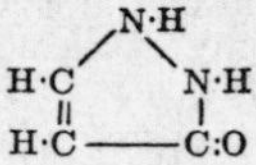

2,3-dihydro-3-oxopyrazole, 3(2)-pyrazolone

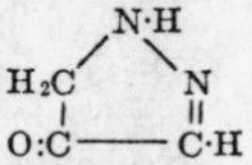

4,5-dihydro-4-oxopyrazole, 4(5)-pyrazolone

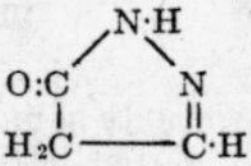

4,5-dihydro-5-oxopyrazole, 5(4)-pyrazolone

Newer compounds in this general group include Dilantin and the penicillins.

II. Compounds Containing an Heterocyclic Five-Membered Ring Fused to a Benzene Ring.

5. Coumarone (1, 2-benzofuran) occurs in coal tar along with several methyl coumarones. It is easily polymerized by acids to yield resins. It may be prepared by treating benzofuran-2-carboxylic acid with a mixture of calcium oxide and copper bronze. Benzofuran-2-carboxylic acid is prepared by treating coumarin with bromine in carbon disulfide followed by treatment with aqueous sodium hydroxide.

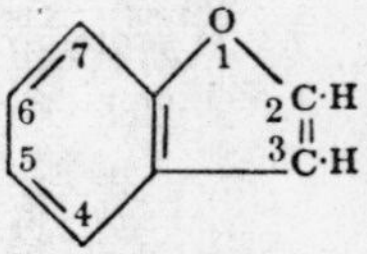

coumarone

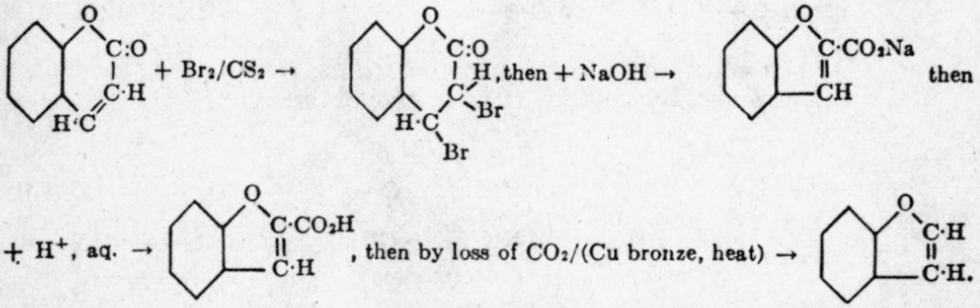

$$+ Br_2/CS_2 \rightarrow \qquad H, \text{then} + NaOH \rightarrow \qquad \text{then}$$

$$+ H^+, \text{aq.} \rightarrow \qquad , \text{then by loss of } CO_2/(Cu \text{ bronze, heat}) \rightarrow$$

Reduction with hydrogen and platinum or palladium catalysts at low temperatures and pressures yields the 2, 3-dihydro derivative. The 2-position is most easily substituted. Coumarone is easily oxidized, hence treatment with mercuric chloride results in oxidation rather than substitution. If the 2-position is blocked, substitution occurs in the 5-position.

The coumarone resins consist of a mixture of coumarones and indenes from the 160°-200° fraction from coal tar.

6. Thianaphthene (thionaphthene, benzothiophene, benzothiofuran) is of relatively slight importance except for the dyes derived from it. It reacts with sodium powder to give the 2-sodium derivative, and upon nitration the 3-nitro compound is formed. The important vat dyes derived from it are thioindigo red and various of its derivatives.

thianaphthene   thioindigo red

7. Indole [benzo (b) pyrrole, 1-azaindene] is a product of intestinal putrefaction. The crude product has a very disagreeable odor, but pure
indole is used in perfumes. It has been synthesized in a number of
ways, one of the simplest of which is by the action of sodium ethylate on
o-amino-ω-chlorostyrene.

In its chemical properties, indole is similar to the pyrroles except
for the fact that the entering group usually occupies the 3-position.

The importance of indole itself is much less than that of many of its
derivatives. Of these, 2-amino-3-indolylpropanoic acid (α-amino-β-
indolepropionic acid, tryptophan) is one of the essential amino acids,
3-indolylethanoic acid (β-indoleacetic acid, heteroauxin) appears to be
one of the growth hormones, and 3-methylindole (skatole) is a product
of protein putrefaction and has growth-promoting activity.

tryptophan   β-indoleacetic   skatole
                         acid

The most important derivative of indole is indigotin (indigo, indigo
blue) and its synthesis from indole may be accomplished by treatment
with ozone or blowing with air. It is one of the oldest known dyes and
was formerly obtained by soaking the branches and leaves of the
indigo plant in water, allowing fermentation to occur, and collecting the
blue scum which rose to the surface. It is now produced synthetically.
Of the thousand or more chemically different dyes produced commercially, indigo occupies the foremost position. The Heumann process
for the production of indigo is:

indoxyl   indigo

Another synthesis (du Pont process) is:

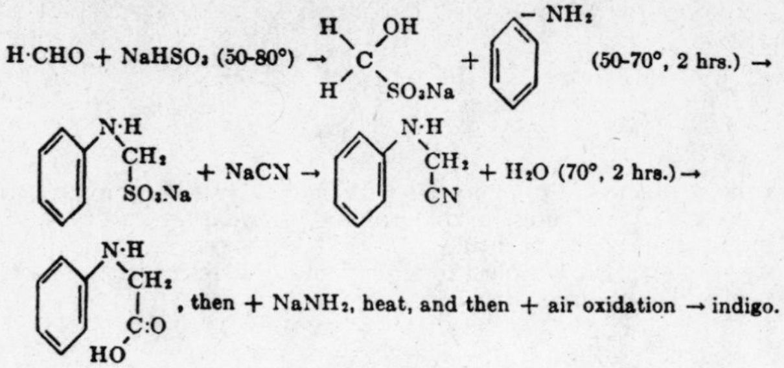

$H \cdot CHO + NaHSO_3$ (50–80°) →

+ (50–70°, 2 hrs.) →

$+ NaCN →$ $+ H_2O$ (70°, 2 hrs.) →

, then $+ NaNH_2$, heat, and then $+$ air oxidation → indigo.

The Baeyer synthesis is indicated below:

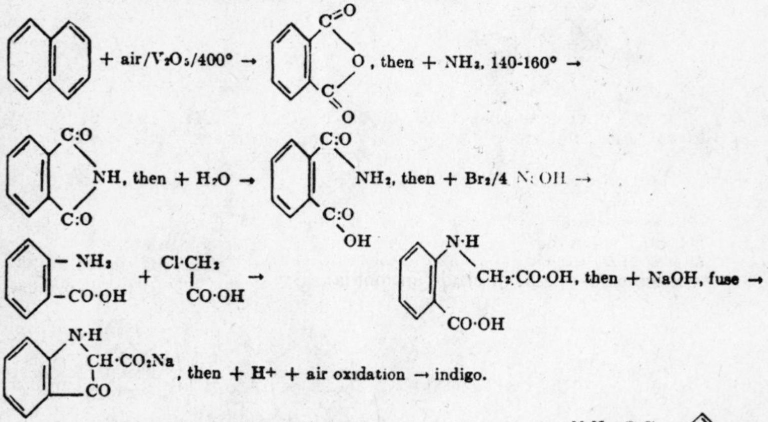

$+$ air/V_2O_5/400° → , then $+ NH_3$, 140–160° →

, then $+ H_2O →$, then $+ Br_2/4$ NaOH →

$+$ → $CH_2 \cdot CO \cdot OH$, then $+$ NaOH, fuse →

, then $+ H^+ +$ air oxidation → indigo.

Tyrian purple, the most highly prized of the dyes known to the ancients, has been shown to be 6, 6′-dibromoindigo.

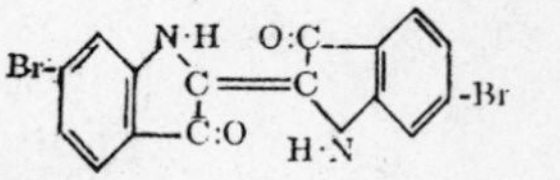

Tyrian (royal) purple

III. Compounds Containing an Heterocyclic Five-Membered Ring Fused to Two Benzene Rings.

8. Carbazole (dibenzopyrrole) is the only important member of this series and occurs in coal tar. It may be prepared by passing diphenylamine through a red-hot tube.

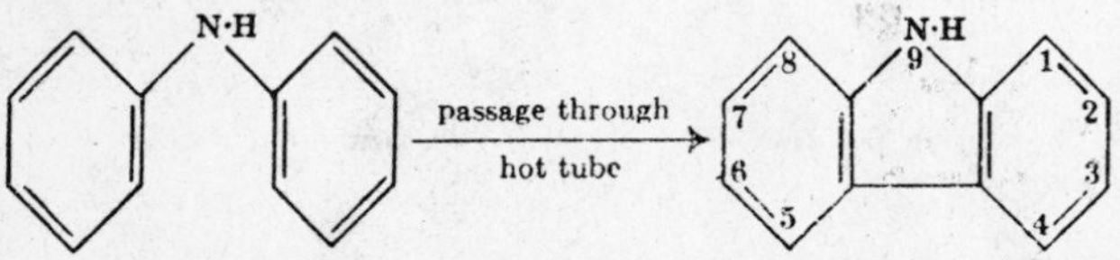

— passage through hot tube → carbazole.

Like pyrrole, it is only slightly basic. Its stability is evidenced by the fact that concentrated sulfuric acid does not attack it below 300°. Nitration yields 1- and 3-nitro compounds. The carbazole structure occurs in nature. Strychnine, for example, contains it in a partially reduced form.

IV. Compounds Containing an Heterocyclic Six-Membered Ring.

9. Pyran (γ-pyran) is of importance only by virtue of its derivatives. The γ-ketonic derivatives are found in some natural products such as meconic acid present in opium. In such products the hetero-oxygen atom is basic and forms salts with acids. The pyranoid ring structure found in sugars is so named because it is a reduced pyran ring.

10. Pyridine (azabenzene, azine), which is rather closely related to benzene in its structure, is even more stable than benzene and does not undergo substitution as readily. It is even stable toward such vigorous reagents as chromic or concentrated nitric acid. In its chemical properties it is more like nitrobenzene than benzene. Thus substitution occurs only at high temperatures and substituents introduced into the 2- and 4-positions are readily replaced. It exhibits a somewhat surprising behavior with regard to the effect of temperature on orientation. Bromine and pyridine at 300° yield 3-bromo and 3, 5-dibromopyridine in a 50 per cent yield whereas at 500° the product consists of 2-bromo and 2, 6-dibromopyridine in an 80 per cent yield. At an intermediate temperature (400°) both types of products are formed.

Pyridine is soluble in water and basic in its reactions. It has a characteristic, disagreeable odor. It occurs in the light oil fraction from coal tar, in Dippel oil, in tobacco smoke, and in crude ammonia. By the addition of six hydrogen atoms, it yields piperidine which is a constituent of piperine, which occurs in pepper. The pyridine ring is the essential ring structure of a number of alkaloids, an example of which is nicotine.

These reactions serve to establish the structure of pyridine and piperidine. Pyridine is used in organic synthesis and to some extent in the denaturing of alcohol.

This general classification includes the important group of barbitals and trioxane.

V. Compounds Containing an Heterocyclic Six-Membered Ring Fused to a Benzene Ring.

11. 1, 4-Benzopyran (benzo-γ-pyran, γ-chromene, 1, 4-coumaran) is the parent structure found in an important class of colored materials found in nature. The anthocyanins (coloring matter in the flowers, fruits, leaves, and stems of plants) are glucosides of polyhydroxy-derivatives of 2-benzopyrylium chloride. The flavones, which are yellow and are

the most widely distributed of the natural pigments which can be used as dyestuffs, are derivatives of 2-phenylbenzopyrone.

anthocyanidine structure   flavone

Another important derivative, 1, 2-benzopyrone (coumarin, *o*-coumaric acid lactone), occurs in the form of its glucoside in the tonka bean. Coumarin has an odor similar to that of new mown hay, and it is used in flavors and perfumes. It is prepared commercially by treating salicylaldehyde with acetic anhydride and sodium acetate (Perkin reaction):

coumarin

12. Quinoline (benzo[*b*]pyridine, 1-azanaphthalene, 1-benzazine) occurs in coal tar and bone oil. As would be predicted from the relative reactivities of benzene and pyridine, substitution and oxidation occur in the benzene ring whereas the pyridine ring is the easiest to reduce. It may be prepared by heating a mixture of aniline, glycerol, nitrobenzene, and sulfuric acid (Skraup synthesis). The sulfuric acid effects dehydration of the glycerol to give propenal. Aniline adds to the propenal by 1, 4-addition and the addition product loses water to give dihydroquinoline. This product is then oxidized by the nitrobenzene to give quinoline, as indicated by:

quinoline.

A number of alkaloids and synthetic medicinals possess the quinoline structure.

13. Isoquinoline (3, 4-benzopyridine) occurs in coal tar along with quinoline. The crude quinoline fraction contains about 4 per cent of isoquinoline. Its reactions are similar to those of quinoline. It may be prepared by heating the oxime of cinnamic aldehyde with phosphorus pentoxide. A Beckmann rearrangement, followed by ring closure, has been postulated.

, isoquinoline.

The isoquinoline nucleus is found in the anhalonium and the opium alkaloids.

VI. Compounds Containing an Heterocyclic Six-Membered Ring Fused to Two Benzene Rings.

14. Acridine (dibenzo[6, e]pyridine) is the most important member of this series. It is found in crude anthracene obtained from coal tar. In its reactions it shows analogies to pyridine and anthracene. It is a very stable compound, but upon vigorous oxidation either one or both the benzene rings are attacked giving quinoline-2, 3-dicarboxylic acid and pyridine-2, 3, 5, 6-tetracarboxylic acid respectively. Acridine exhibits weakly basic properties. Acriflavine and proflavine are derivatives of acridine. Both are used in dressing wounds since they are highly bacteriostatic and non-toxic. Another important derivative is atebrin, used as an antimalarial drug.

acridine   acriflavine   proflavine   atabrine

CHAPTER XXIII

MISCELLANEOUS COMPOUNDS

The compounds considered in this section have two or more functional groups. Their properties will be determined by the functional groups except as they may be modified by the presence of other substituents. These compounds may be grouped, for convenience, as derivatives of (I) acids, (II) amines, (III) phenols, and (IV) toluene.

I. Derivatives of Aromatic Acids

1. *Anthranilic acid* (*o*-aminobenzoic acid) may be prepared (a) by the reduction of *o*-nitrobenzoic acid, or (b) from phthalic anhydride:

a.

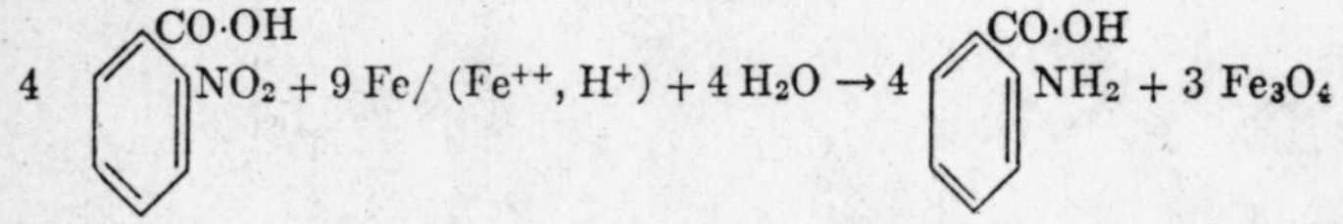

$$4 \, C_6H_4(CO \cdot OH)NO_2 + 9 \, Fe/(Fe^{++}, H^+) + 4 \, H_2O \rightarrow 4 \, C_6H_4(CO \cdot OH)NH_2 + 3 \, Fe_3O_4$$

b.

phthalimide   phthalamic acid

sodium anthranilate   anthranilic acid

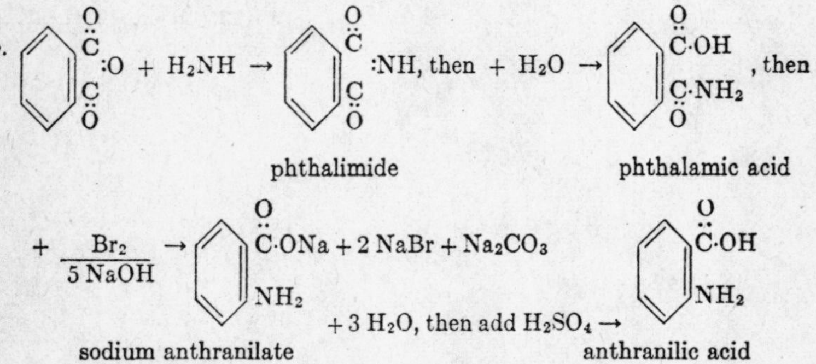

Anthranilic acid is used in the synthesis of indigo, in the manufacture of other dyes, and, in the form of its esters, in perfumes.

2. *Aspirin* (salicylic acid acetate, acetylsalicylic acid) is prepared by the acetylation of the phenolic group in salicylic acid with acetic anhydride or with acetyl chloride in the presence of sodium acetate, sulfuric acid, or zinc chloride:

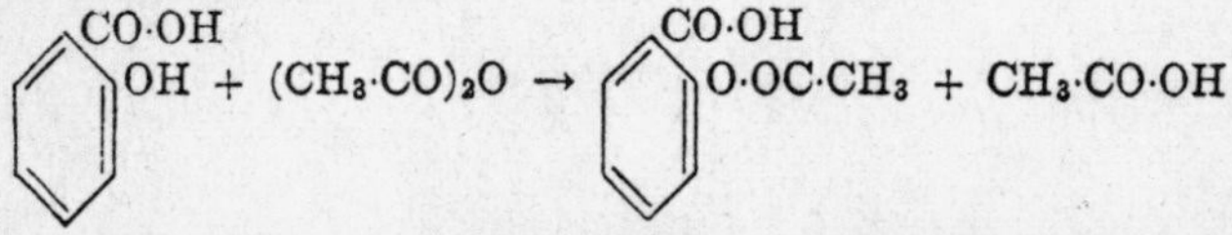

Aspirin is used extensively as an antiseptic and antipyretic.

228

3. *Gallic acid* (3,4,5-trihydroxybenzoic acid), the formula of which is shown at the right, is obtained chiefly from gall nuts. It is used as an astringent and in the preparation of inks. When subjected to dry distillation, **gallic acid** undergoes decarboxylation and yields **pyrogallic acid** (1,2,3-trihydroxybenzene).

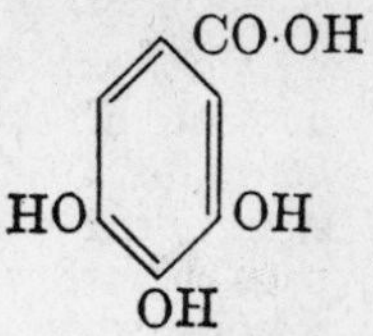

4. *o-Nitrobenzoic acid* (2-nitrobenzenecarboxylic acid), represented by the formula at the right, is obtained, along with *p*-nitrobenzoic acid, by the nitration and subsequent oxidation of toluene. The *m*-isomer is obtained by the direct nitration of **benzoic acid.** The reduction of the nitrobenzoic acids gives the corresponding aminobenzoic acids. Anesthesine, Novocaine, and Butyn are important derivatives of *p*-aminobenzoic acid.

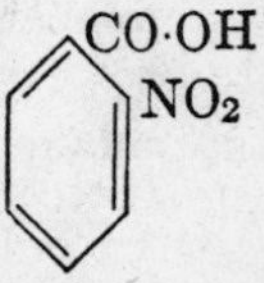

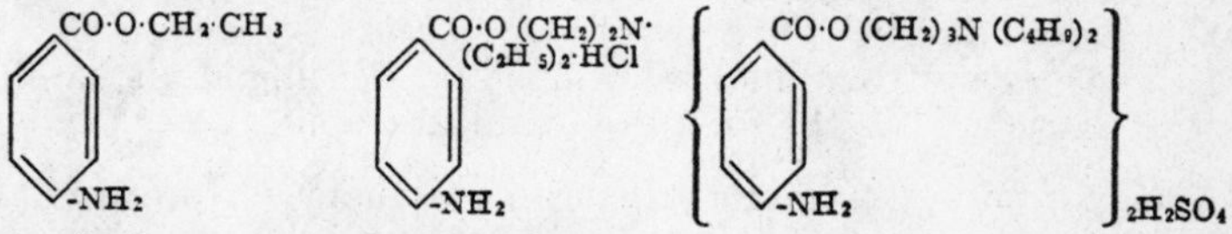

Anesthesine

Novocaine
(Procaine hydrochloride)

Butyn

Anesthesine and Novocaine are important local anesthetics, and Butyn is used in dentistry and ophthalmic surgery.

5. *Saccharin* (*o*-sulfobenzoic imide), though not related structurally to the carbohydrates, is about 500 times as sweet as cane sugar. Since it apparently undergoes no metabolic changes which yield harmful derivatives, saccharin is used somewhat in the form of its sodium salt as a sugar substitute for diabetics and, since it has no food value, by those desiring to reduce their weight. It may be prepared by the following series of reactions:

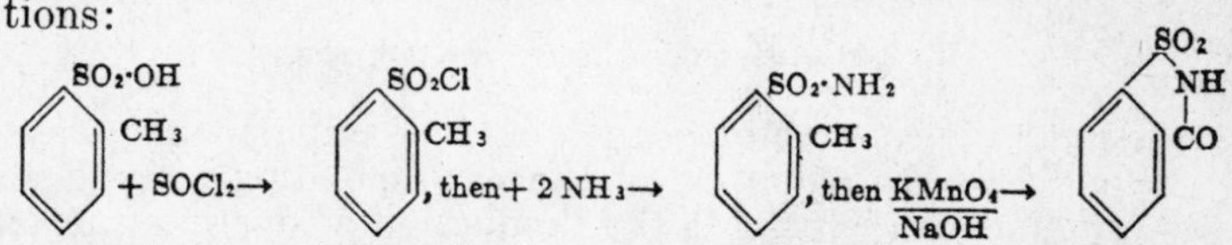

o-toluenesulfonic acid + SOCl2→ *o*-toluenesulfonyl chloride, then + 2 NH3→ *o*-toluenesulfonamide, then KMnO4 → NaOH saccharin

6. *o-Sulfobenzoic acid* (2-sulfobenzenecarboxylic acid), which has the structure shown at the right, may be prepared by the sulfonation and subsequent oxidation of toluene.

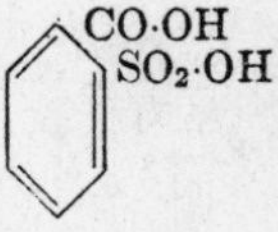

7. *Tannic acids* occur in the bark of trees. Their exact structures are unknown, but among their hydrolytic products are the hydroxy benzoic acids. The tannic acids precipitate proteins and find use in the tanning industry, as mordants in dyeing, and in the preparation of inks, as an astringent, and in the treatment of burns.

II. Derivatives of Aromatic Amines

1. *o-Chloroaniline*, and *m*-chloroaniline, are prepared by the reduction of the corresponding chloronitrobenzenes. *p*-Chloroaniline is obtained by suspending acetanilide in water and treating with chlorine to give *p*-chloroacetanilide which, upon hydrolysis, yields *p*-chloroaniline. Direct chlorination or bromination of aniline gives the symmetrical trichloro- or tribromoaniline.

o-chloroaniline

2. *o-Nitroaniline*, and *p*-nitroaniline, are prepared by nitrating acetanilide and hydrolyzing the resulting product. *m*-Nitroaniline is prepared by the partial reduction of *m*-**dinitrobenzene** with ammonium sulfide. If aniline is nitrated directly, some symmetrical trinitroaniline is obtained along with oxidized derivatives of the amino group. The explosive T.N.A. (2,3,4,6-tetranitroaniline) is obtained by treating *m*-nitroaniline with nitric acid in the presence of sulfuric acid.

3. *Sulfanilic acid* (4-aminobenzenesulfonic acid) is prepared by treating aniline with sulfuric acid to form aniline hydrogen sulfate and then heating at about 180°C. for four hours. As indicated by the formula, sulfanilic acid forms an inner salt somewhat analogous to that of the amino acids. Metanilic acid (3-aminobenzenesulfonic acid) is prepared by reducing *m*-nitrobenzenesulfonic acid.

III. Derivatives of Phenolic Compounds

1. *p-Aminophenol*, *o*-aminophenol, and *m*-aminophenol may be prepared by the reduction of the corresponding nitrophenols. *p*-Aminophenol is used in the production of dye intermediates and as a photographic developer. Other closely related compounds that are used as photographic developers are Rodinal, Amidol, and Metol:

p-aminophenol   Rodinal   Amidol   Metol

The term Rodinal is also applied to the free base and to other salts.

2. *Anethole* (1-methoxy-4-propenylben-
zene, *p*-propenylanisole) occurs in
anise seed and finds some use as an
antiseptic.

3. *Anisaldehyde* (4-methoxybenzalde-
hyde) occurs in the oil of the anise
seed and is used to a limited extent
in the perfume industry.

4. *Dulcin* (4-ethoxybenzenecarbonamide, (*p*-ethoxy-
phenyl)urea), although not related structurally
to the carbohydrates, is about two hundred times
as sweet as cane sugar.

5. *Eugenol* (4-allyl-2-methoxyphenol) is obtained
from clove oil. It is used in dentistry as an
antiseptic and as a local anesthetic. Isoeugenol
has a propenyl group in the 4-position instead of
the allyl group $(1,2,4\text{-}C_6H_4(OH)(OCH_3)\text{-}$
$HC:CH\cdot CH_3)$.

6. *Guaiacol* (2-methoxyphenol) is obtained from the
distillation of guaiac resin. Guaiacol and its
salts and esters are used as internal antiseptics.

7. *o-Nitrophenol*, and *p*-nitrophenol, are obtained by the direct
nitration of phenol under proper conditions.
m-Nitrophenol is prepared by the diazotization and subsequent hydrolysis of *m*-nitro-
aniline. Picric acid (2, 4, 6-trinitrophenol) is
prepared commercially by treating phenol
with sulfuric acid to give a mixture of the
mono- and disulfonic acid derivatives which
are then nitrated, the sulfonic acid groups being replaced
ultimately by the nitro group. Picric acid is used to determine
the glucose content of blood by the colori-
metric method, to test for creatinine, to
precipitate organic bases and proteins, in
histological work as a fixing agent, as an
antiseptic, to treat skin diseases and
burns, in the manufacture of explosives,
and in the formation of picramic acid
which is used in the synthesis of dyes.

Picric acid is also prepared by the nitration of chlorobenzene
and subsequent hydrolysis.

8. *Phenetidine* (4-ethoxyaniline) is
   used in the preparation of phenacetin   (*p*-ethoxyacetanilide)
   which in turn is used as an antipyretic and as an analgesic.

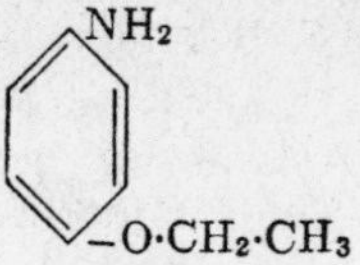

phenetidine 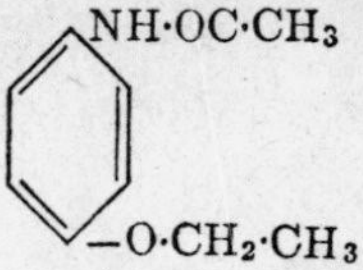   phenacetin

9. *Safrole* (1-allyl-3, 4-methylene-
   dioxybenzene) is obtained from
   oil of sassafras.   It is used as
   an anodyne.

   The formula for safrole is
   shown at the right.

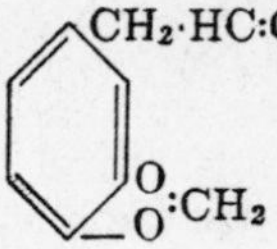

IV. Derivatives of Toluene

1. *Chloramine-T*, which was introduced by Carrel and Dakin
   during the World War,
   is used as a mouth
   wash, as an irrigating
   fluid for wounds, and
   as an active germicide.
   Dichloramine-T is used
   in the treatment of infected wounds.

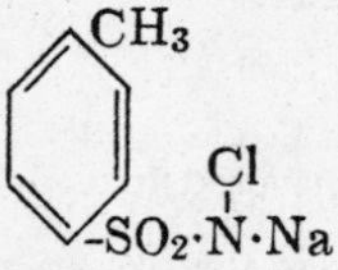

   chloramine-T 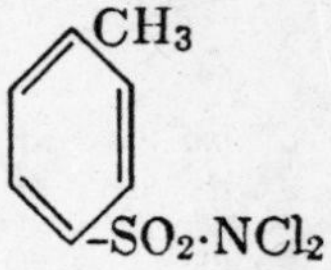   dichloramine-T

2. *o-Chlorotoluene*, and *p*-chlorotoluene, are obtained by the direct
   chlorination of toluene in the presence of a
   carrier.   *m*-Chlorotoluene is obtained by the
   diazotization of *m*-toluidine followed by treatment with a Sandmeyer reaction.

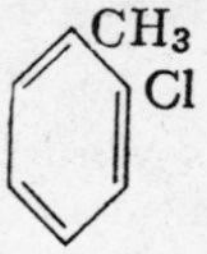

   o-chlorotoluene

3. *Trinitrotoluene* (T. N. T.) has been considered in connection
   with the derivatives of the aromatic
   hydrocarbons.   It is produced by the
   direct nitration of toluene with mixed acid
   under very carefully controlled conditions.

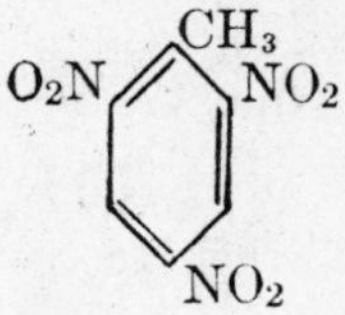

CHAPTER XXIV

I. Pronunciation

A. Accent:

1. The general trend in the American language is away from placing the accent on the final syllable. Specific illustrations are:

acetal.........ăs'ĕt-ăl
acetoacetate..ăs'ĕ-tŏ-ăs'ĕ-tāt
benzoyl.....bĕn'zŏ-ĭl
camphanic....kăm-făn'ĭk

acetaldoxime.....ăs'ĕt-ăl-dŏk'sēm
benzamide.......bĕn-zăm'ĭd
benzyl...........bĕn'zĭl
carbamide.......kär-băm'ĭd

a. Amine (ă-mēn'), arsine (är-sēn'), quinone (kwĭn-ōn'), sulfone (sŭl-fōn'), or words with these endings, and the suffix "-phenone" (fĕ-nōn') represent the principal exceptions in names of organic chemicals:

acetophenone.ăs'ĕ-tŏ-fĕ-nōn'
benzophenone bĕn'zŏ-fĕ-nōn'
ethylamine...ĕth'ĭl-ă-mēn'

allylamine.......ăl'ĭl-ă-mēn'
diethylamine.....dī-ĕth'ĭl-ă-mēn'
glucosamine......gloō'kōs-ă-mēn'

b. Biuret (bī'ů-rĕt'), buret (bů-rĕt'), cocaine (kŏ-kān'), ionone (ī'ŏ-nōn') and irone (ī-rōn'), despite other dictates, are supported by popular usage.

2. *The accent falls on the penult for adjectives ending in "ic," and the vowel preceding the suffix "ic" is usually short:*

abietic.......ăb'ĭ-ĕt'ĭk
anthranilic...ăn'thră-nĭl'ĭk
arsonic.......är-sŏn'ĭk
calorimetric...kăl'ŏ-rĭ-mĕt'rĭk
glyceric......glĭ-sĕr'ĭk
hydriodic.....hī'drĭ-ŏd'ĭk
margaric.....mär-găr'ĭk
periodic(acid).pûr'ĭ-ŏd'ĭk
pimelic.......pĭ-mĕl'ĭk

adiabatic........ăd'ĭ-ă-băt'ĭk
arsenic (acid)....är-sĕn'ĭk
asymmetric......ā'-sĭ-mĕt'rĭk
coumaric........koō-măr'ĭk
glycyrrhetic......glĭs'-ĭ-rĕt'ĭk
malic............măl'ĭk
mesoxalic.......mĕs'ŏk-săl'ĭk
phthalic........thăl'ĭk
stearic..........stĕ-ăr'ĭk

3. *In the names of salts the accent moves one syllable, sometimes more, toward the beginning of the word:*

anthranilic....ăn'thră-nĭl'ĭk
cacodylic.....kăk'ŏ-dĭl'ĭk
caproic.......kă-prŏ'ĭk
caprylic......kă-prĭl'ĭk
carbamic.....kär-băm'ĭk
cinnamic.....sĭ-năm'ĭk
glyceric......glĭ-sĕr'ĭk
salicylic.....săl'ĭ-sĭl'ĭk

vs. anthranilate......ăn-thrăn'ĭ-lāt
" cacodylate.....kăk'ŏ-dĭl-āt
" caproate.......kăp'rŏ-āt
" caprylate......kăp'rĭ-lāt
" carbamate.....kär'bă-māt
" cinnamate......sĭn'ă-māt
" glycerate.......glĭ'sĕr-āt
" salicylate.......săl'ĭ-sĭl-āt

4. *Words ending in "valent" (vā'lĕnt), take the accent on the penult:*

bivalent......bī-vā'lĕnt
quadrivalent..kwŏd'rĭ-vā'lēnt

trivalent........trī-vā'lĕnt
univalent.......ū'nĭ-vā'lĕnt

B. Vowel Sounds in Certain Suffixes:

1. *The ending "ide" should be pronounced with a long "i"* (īd):

acetamide.... ăs′ĕt-ăm′īd	acetanilide....... ăs′ĕt-ăn′ĭ-līd
amide....... ăm′ĭd	azide........... ăz′īd
benzamide.... bĕn-zăm′īd	butanolide....... bū-tăn′ŏ-līd
carbamide.... kär-băm′īd	carbanilide...... kär-băn′ĭ-līd
chloride...... klō′rīd	cyanamide....... sī′ăn-ăm′īd
formamide.... fôrm-ăm′īd	imide........... īm′īd
lipide........ lī′pĭd	phthalide........ thăl′īd
phthalimide... thăl-ĭm′īd	ureide........... ū′rĕ-īd

2. *The ending "ine" should be pronounced "ēn"* (*alkaline is an exception*):

adrenaline.... ăd-rĕn′ă-lēn	ammine.......... ăm′ēn
atropine...... ăt′rŏ-pēn	betaine.......... bē′tȧ-ēn
caffeine...... kăf′ĕ-ēn	cocaine.......... kō′kȧ-ēn, kŏ-kān′
codeine...... kō′dĕ-ēn	creatine.......... krē′ă-tēn
cysteine...... sĭs′tĕ-ēn	ephedrine........ ĕf′ĕ-drēn
heroin....... hĕr′ŏ-ēn	lutidine.......... lōō′tĭ-dēn
ptomaine..... tō′mȧ-ēn, tō′mān	thebaine........ thē′bȧ-ēn

3. *The endings "ole" and "ol" should be pronounced "ōl":*

arabitol...... ȧ-răb′ĭ-tōl	ergosterol....... ēr-gŏs′tēr-ōl
geraniol...... jĕ-rā′nĭ-ōl	linalool.......... lĭn-ăl′ŏ-ōl
pyrrole....... pĭr′ōl	skatole.......... skȧ′tōl

   a. Alcohol and sol and most words ending in "sol", are exceptions to this rule. In these words, "ol" is pronounced "ŏl":

sol........ sŏl	hydrosol........ hī′drŏ-sŏl

4. *The ending "yl" should be pronounced "ĭl":*

acetonyl...... ȧ-sĕt′ŏ-nĭl	acetyl.......... ăs′ĕ-tĭl
acyl......... ăs′ĭl	amyl........... ăm′ĭl
antimonyl.... ăn′tĭ-mŏ-nĭl	benzoyl......... bĕn′zŏ-ĭl, bĕn′zŏ-ēl
benzyl....... bĕn′zĭl	butyl........... bū′tĭl
cacodyl...... kăk′ŏ-dĭl	carbonyl........ kär′bŏ-nĭl
cetyl........ sē′tĭl	chromyl........ krō′mĭl
cresyl....... krĕs′ĭl	decyl........... dĕs′ĭl
diacetyl...... dī-ăs′ĕ-tĭl	dipropargyl...... dī′prŏ-pär′jĭl
ethyl........ ĕth′ĭl	glycyl.......... glī′sĭl
glyoxyl...... glī-ŏk′sĭl	indoxyl......... ĭn-dŏk′sĭl
linalyl....... lĭn′ă-lĭl	malonyl........ măl′ŏ-nĭl
melissyl..... mĕ-lĭs′ĭl	mesityl........ mĕs′ĭ-tĭl
methyl...... mĕth′ĭl	nitrosyl........ nī′trŏ-sĭl
nitroxyl...... nī-trŏk′sĭl	propionyl....... prō′pĭ-ŏ-nĭl
titanyl....... tī′tăn-ĭl	vinyl........... vī′nĭl

5. *The preferred pronunciation of the suffix "ile" in the names of compounds is "ĭl":*

acetonitrile... ăs′ĕ-tŏ-nī′trĭl	butyronitrile..... bū′tĭ-rŏ-nī′trĭl
nitrile........ nī′trĭl	

6. *The first vowel in the endings -acic, -alic, -anic, -aric, -elic, -enic, -eric, -etic, -idic, -ilic, -inic, -isic, -onic, -opic, and -oric, is usually short:*

abietic.......ăb′ĭ-ĕt′ĭk	arachidic.........ăr′ă-kĭd′ĭk
anisic........ă-nĭs′ĭk	behenic..........bĕ-hĕn′ĭk
benzilic......bĕn-zĭl′ĭk	camphanic.......kăm-făn′ĭk
fulminic......fŭl-mĭn′ĭk	isotropic.........ĭ′sŏ-trŏp′ĭk
itaconic......ĭt′ă-kŏn′ĭk	mandelic.........măn-dĕl′ĭk
sebacic.......sĕ-băs′ĭk	valeric...........vă-lĕr′ĭk
tartaric......tär-tăr′ĭk	terephthalic......tĕr′ĕf-thăl′ĭk

a. Some exceptions to this rule are acetic (ă-sē′tĭk), campholenic (kăm′fŏ-lē′nĭk), cetic (sē′tĭk), ceric (sē′rĭk), mesitylenic (mĕ-sĭt′ĭ-lē′nĭk), naphthenic (năf-thē′nĭk), and adjectives derived from "ene" hydrocarbons.

b. In words ending in "olic," the "o" is long (ō′lik):

alantolic...ăl′ăn-tō′lĭk	campholic........kăm-fō′lĭk
cholic.....kō′lĭk	elemolic.........ĕl′ĕ-mō′lĭk
enolic.....ĕ-nō′lĭk	glycolic..........glĭ-kō′lĭk
lauronolic..lô′rŏ-nō′lĭk	tetrolic..........tĕ-trō′lĭk

7. *"ime in oxime, at least, should be pronounced "ēm":*
acetaldoxime..ăs′ĕt-ăl-dŏk′sēm

8. *The ending "oin" is "ŏ-ĭn": Benzoin* (bĕn′zŏ-ĭn), *furoin* (fū′rŏ-ĭn), *etc.*

This division is based on "The Pronunciation of Chemical Words," by E. J. Crane, *Ind. Eng. Chem.*, News Ed., Vol. 12, No. 10, p. 202 (May 20, 1934).

II. I. U. C. Nomenclature

A. Hydrocarbons:

1. *a.* Aliphatic open-chain saturated hydrocarbons are characterized by the ending "ane." The generic name is alkanes. The names of the first four members are methane, ethane, propane, butane. Those above butane are named from Greek or Roman numerals, as pentane, hexane, heptane, octane, nonane, decane, hendecane or undecane, dodecane, etc.

Branched-chain hydrocarbons are named as substitution products of the longest continuous chain present in the formula. In the case of several side chains, their order of precedence in naming may be alphabetical or according to complexity. *Chemical Abstracts* uses the alphabetical order.

In case ambiguity results from the use of the longest chain, or if a simpler name is obtained otherwise, the chain which carries the maximum number of substituted groups may be used as the basis for naming the compound.

$CH_3 \cdot CH_2 \cdot CH (CH_3) \cdot CH_2 \cdot CH (CH_3) \cdot CH_2 \cdot CH_3$, 3, 5-dimethylheptane

$CH_3 \cdot CH_2 \cdot CH_2 \cdot CH_2 \cdot CH \cdot CH_2 \cdot CH_2 \cdot CH_2 \cdot CH_3$, 5-(1, 2-dimethylpropyl)-
$CH (CH_3) \cdot CH (CH_3) \cdot CH_3$ nonane or 4-butyl-
2, 3-dimethyloctane.

Note that the numerals should be as small as possible.

b. Aliphatic unsaturated hydrocarbons are designated by one of the following endings:

1. "Ene" where the compound contains one double bond, "diene" if two double bonds are present, and "triene" for three double bonds. The corresponding generic names are "alkenes," "alkadienes," and "alkatrienes." A numeral is used to indicate the number of the carbon atom at which each double bond starts. If the unsaturated compound contains branched chains, the *fundamental chain is selected so as to contain the greatest possible number of double bonds.*

$CH_3 \cdot HC:CH \cdot HC:CH \cdot CH_3$, 2, 4-hexadiene;

$CH_3 \cdot HC:C (CH_3) \cdot HC:C (C_2H_5) \cdot (C_4H_9) C:CH \cdot CH_3$, 3-butyl-4-ethyl-6-methyl-2, 4, 6-octatriene.

2. "Yne," "diyne," "triyne," are used to designate one, two, or three triple bonds. The corresponding generic names are "alkynes," "alkadiynes," and "alkatriynes." The requirements for indicating the position of the triple bond or bonds, and for the naming of branched-chain compounds, is the same as given under the alkenes (*b.*1.).

$CH_3 \cdot CH_2 \cdot HC(CH_3) \cdot HC(C_2H_5) \cdot C:C \cdot H$, 3-ethyl-4-methyl-1-hexyne;

$H \cdot C:C \cdot CH_2 \cdot CH_2 \cdot HC(C_2H_5) \cdot HC(CH_3) \cdot HC(C_3H_7) \cdot C:C \cdot H$, 5-ethyl-4-methyl-3-propyl-1, 8-nonadiyne.

3. "Enyne," "dienyne," "enediyne," "dienediyne," etc., are used to designate the number and types of unsaturated bonds that are present.

(The final "e" of the hydrocarbon ending is elided before vowels)

The corresponding generic names are "alkenynes," "alkadienynes," "alkenediynes," and "alkadienediynes."

$CH_3 \cdot HC:CH \cdot CH_2 \cdot C:C \cdot H$, 4-hexen-1-yne, or hexen-4-yne-1.

$CH_3 \cdot HC:CH \cdot CH_2 \cdot HC:C(CH_3) \cdot HC(C_4H_9) \cdot C:C \cdot H$, 3-butyl-4-methylnona-4,7-dien-1-yne.

$CH_3 \cdot C:C \cdot CH(CH_3) \cdot HC:CH \cdot CH(C_3H_7) \cdot HC:CH \cdot CH(C_2H_5) \cdot C:C \cdot H$, 3-ethyl-9-methyl-6-propyldodeca-4, 7-dien-1, 10-diyne.

2. The generic names for the cyclic hydrocarbons are "cycloalkanes," "cycloalkenes," "cycloalkynes," "cycloalkadienynes," etc., according to the number and types of unsaturated bonds present. Examples are:

$H_2C \cdot HCH \cdot CH_2$, cyclobutane; $H_2C \cdot CH_2 \cdot HC:CH \cdot CH_2$, cyclohexene;

$H \cdot C \cdot H$ $H \cdot C \cdot H$

$H_2C \cdot C:C \cdot CH_2$, cyclopentyne; $H_2C \cdot HC:CH \cdot C:C \cdot HC:CH \cdot CH_2$, cyclo-nona-1, 5-dien-3-yne.

$H \cdot C \cdot H$ $H \cdot C \cdot H$

In the case of partially or wholly saturated polycylic aromatic compounds, however, the prefix "hydro," preceded by "di," "tetra," etc., will be used, as in 1, 2-dihydronaphthalene and 2, 3-dihydroanthracene.

3. Many of the aromatic parent compounds have historical or commercial names.

B. Numbering:

1 Aliphatic compounds are numbered from end to end, using arabic numerals. Where more than one function is present and *ambiguity is likely to result,* the lowest numbers are given in order to (a) the *principal function,* (b) double bonds, (c) triple bonds, and (d) radicals designated by prefixes. *Whenever possible, from the standpoint of consistency, the lowest possible numbers are used.* The expression "lowest numbers" signifies those that include the lowest individual number or numbers.

$H_2C:CH \cdot CH_2 \cdot CH_3$, 1-butene; $H_2C:CH \cdot C: C \cdot H$, 1-buten-3-yne

$CH_3 \cdot HC:CH \cdot CH_2OH$, 2-buten-1-ol; $CHCl_2 \cdot HC:CH_2$, 3, 3-dichloropropene; $CH_2Cl \cdot CH (CH_3) \cdot HC:CH_2$, 4-chloro-3-methyl-1-butene.

In the last two examples, propene and 1-butene are considered as the parent compounds.

Chemical Abstracts follows the practice of placing the numbers before the functional group to which they refer. The principal chain should include the largest possible number of principal functional groups.

$H_2C:CH \cdot CH_2 \cdot C:C \cdot H$, 1-penten-4-yne or pent-1-en-4-yne.

$CH_3 \cdot CHCl \cdot CH(CH_3) \cdot CH_2 \cdot HC:CH \cdot CH_3$, 6-chloro-5-methyl-2-heptene.

In this example, 2-heptene is regarded as the parent compound. Side chains are numbered or lettered from the point of attachment. When substituted side chains are present, the numerals or letters and the name of the side chain will be placed in parentheses.

$CH_3 \cdot HC:CH \cdot CH (HC:CH \cdot CH_3) \cdot HC:C (CH_3) \cdot C:C \cdot H$, 3-methyl-5- (1-propenyl)-octa-3,6-dien-1-yne.

$CH_3 \cdot BrC:CH \cdot CH (HC:CH \cdot CH_3) \cdot HC:C (CH_2 \cdot CHI \cdot CH_2Cl) \cdot C:C \cdot H$, 7-bromo-3- (3-chloro-2-iodopropyl)-5- (1-propenyl) octa-3, 6-dien-1-yne.

As indicated elsewhere, the names of substituting groups may be arranged alphabetically or in order of increasing molecular weight. *When two equally low numberings are possible for atoms or radicals designated by prefixes, that order will be chosen which most nearly accords with the order of the prefixes in the name;* for example, $CH_2Br \cdot CH_2 \cdot CH_2 \cdot CH_2Cl$, may be 1-bromo-4-chlorobutane (alphabetical order) or 1-chloro-4-bromobutane (order of weight). For simple expressions the prefixes "di," "tri," and "tetra," are used, as in 1, 1-dichloroethane ($CH_3 \cdot CHCl_2$); 1, 2-dibromopropane ($CH_3 \cdot CHBr \cdot CH_2Br$); 3, 3-diethyl-1, 2, 4-butanetriol ($CH_2OH \cdot C [C_2H_5]_2 \cdot CHOH \cdot CH_2OH$); etc. The terms "bis," "tris," and "tetrakis" are used before complex expressions: 1, 3-Bis-(methylamino) propane ($CH_3 \cdot NH \cdot [CH_2]_3 \cdot NH \cdot CH_3$); 1, 2-bis (dimethylamino) ethane ([$CH_3]_2N \cdot CH_2 \cdot CH_2 \cdot N \cdot [CH_3]_2$); etc. The prefix "bi" is reserved to denote doubling of a radical, as illustrated by biphenyl ($C_6H_5 \cdot C_6H_5$); and 4, 4'-bibenzoic acid ($C_6H_4 \cdot CO \cdot OH)_2$.

2. Ring compounds are numbered according to the following:

a. Carbocylic systems are numbered so as to give the hydrogen atoms the lowest numbers possible (except in single rings, where double bonds are often given the lowest positions). Single rings are oriented with position "1" at the top and numbered clockwise.

b. Single heterocyclic rings are numbered so as to give position "1" to the hetero atom that is most acidic (according to its position in the periodic table), and then numbered around the ring so as to give the other hetero atoms the lowest numbers possible.

When two or more numberings conform to this rule, choose the one that gives to the hydrogen atoms the lowest numbers possible.

pyrrole

c. Complex ring formulas should be oriented so that the greatest number of rings lie in a horizontal row. If two or more possibilities are encountered, choose the one that places the remaining rings above and to the right.

(For details: cf. J. Am. Chem. Soc., **47**, 543-61 (1925)).

C. Simple Functions:

1. *Halogen derivatives* are named by prefixing the required numerals and the name of the halogen to the name of the hydrocarbon from which it is derived. $CH_3 \cdot CH_2 \cdot Cl$, chloroethane; $CH_3 \cdot CH_2 \cdot CHBr \cdot CH_2Br$, 1, 2-dibromobutane; $CH_3 \cdot CH_2 \cdot CI_2 \cdot CH_2 \cdot CH_2I$, 1,3,3-triiodopentane; $CH_2Cl \cdot HC:CH_2$, 3-chloropropene; $C_6H_5 \cdot Br$, bromobenzene; etc.

2. *Alcohols and phenols* are named by replacing the terminal "e" of the hydrocarbon name by the suffix "ol." $CH_3 \cdot CH_2 \cdot OH$, ethanol; $CH_3 \cdot CH_2 \cdot CH_2 \cdot OH$, 1-propanol; $CH_3 \cdot CHOH \cdot CH_3$, 2-propanol; and $C_6H_5 \cdot OH$, phenol (from "phene"). Cresol ($CH_3 \cdot C_6H_4 \cdot OH$) and naphthol ($C_{10}H_7 \cdot OH$) have become established by usage.

For di-, tri-, and tetra-alcohols and phenols, the ending takes the form "diol," "triol," and "tetrol," respectively. $CH_3 \cdot CHOH \cdot CH_2OH$, 1, 2-propanediol; $CH_2OH \cdot CHOH \cdot CH_2OH$, 1, 2, 3-propanetriol; o-HO$\cdot C_6H_4 \cdot OH$, 1, 2-benzenediol or 1, 2-phenediol; etc.

Mercaptans or thioalcohols are named by adding the suffix "thiol" to the name of the corresponding hydrocarbon. $CH_3 \cdot CH_2 \cdot SH$, ethanethiol; $CH_3 \cdot CH_2 \cdot CH_2 \cdot SH$, 1-propanethiol; $CH_3 \cdot CHSH \cdot CH_3$, 2-propanethiol; $C_6H_5 \cdot SH$, benzenethiol or phenethiol; etc. The suffixes "dithiol," "trithiol," and "tetrathiol," are used where two, three, or four "SH" groups occur in a given compound.

3. a. *Ethers* are named as alkoxy derivatives of the corresponding hydrocarbons. $CH_3 \cdot O \cdot CH_2 \cdot CH_3$, methoxyethane; $CH_3 \cdot O \cdot CH_2 \cdot CH_2 \cdot CH_3$, 1-methoxypropane; $CH_3 \cdot CH_2 \cdot O \cdot CH_2 \cdot CH_2 \cdot CH_3$, 1-ethoxypropane; and $CH_3 \cdot O \cdot C_6H_5$, methoxybenzene or methoxyphene. However, the common names of the symmetrical ethers, such as methyl ether ($CH_3 \cdot O \cdot CH_3$), ethyl ether ($CH_3 \cdot CH_2 \cdot O \cdot CH_2 \cdot CH_3$), propyl ether ($CH_3 \cdot CH_2 \cdot CH_2 \cdot O \cdot CH_2 \cdot CH_2 \cdot CH_3$), phenyl ether ($C_6H_5 \cdot O \cdot C_6H_5$), etc., represent good usage.

Acetals are named as 1, 1-dialkoxyalkanes. $CH_3 \cdot CH_2 \cdot CH(O \cdot CH_2 \cdot CH_3)_2$, 1, 1-diethoxypropane; $CH_3 \cdot CH_2 \cdot CH_2 \cdot CH(O \cdot CH_3)_2$, 1, 1-dimethoxybutane; etc.

b. *Inner oxides* where oxygen is linked between two carbon atoms to form a ring structure are designated by the prefix "epoxy," when it is not convenient to give the compound a cyclic name.

$H_2C \cdot \overset{O}{\cdot} \cdot CH_2$, epoxyethane; $CH_3 \cdot HC \cdot \overset{O}{\cdot} \cdot CH_2$, 1, 2-epoxypropane;

$CH_3 \cdot CH_2 \cdot HC \cdot \overset{O}{\cdot} \cdot CH_2$, 1, 2-epoxybutane; $CH_3 \cdot HC \cdot \overset{O}{\cdot} \cdot CH \cdot CH_3$, 2, 3-epoxybutane;

$H_2C \cdot CH_2 \cdot CH_2 \cdot CH_2$, 1, 4-epoxybutane or tetrahydrofuran.

4. *Sulfides, disulfides, sulfoxides,* and *sulfones* add the term "thio," "dithio," "sulfinyl," or "sulfonyl," respectively, to the name of the hydrocarbon radical. $CH_3 \cdot S \cdot CH_3$, methylthiomethane;

$CH_3 \cdot S \cdot S \cdot CH_2 \cdot CH_3$, methyldithioethane; $CH_3 \cdot CH_2 \cdot SO \cdot CH_2 \cdot CH_2 \cdot CH_3$, 1-ethylsulfinylpropane; and $CH_3 \cdot SO_2 \cdot CH_2 \cdot CH_2 \cdot CH_2 \cdot CH_3$, 1-methylsulfonylbutane.

5. *a.* 1. *Aldehydes* are named by replacing the terminal "e" of the hydrocarbon name by the suffix "al." $H \cdot CHO$, methanal; $CH_3 \cdot CHO$, ethanal; $CH_3 \cdot CH_2 \cdot CHO$, propanal; $CH_3 \cdot CH_2 \cdot CH_2 \cdot CHO$, butanal; etc. A systematic name corresponding to benzaldehyde ($C_6H_5 \cdot CHO$) is benzenecarbonal. This latter use is consistent with the use of the suffixes "-carbonamide," and "-carbonitrile." Similarly one may use pyridinedicarbonal for $C_5H_3N(CHO)_2$.

2. *Thioaldehydes* add the suffix "thial" to the name of the corresponding aliphatic hydrocarbon. $H \cdot CHS$, methanethial; $CH_3 \cdot CHS$, ethanethial; $CH_3 \cdot CH_2 \cdot CHS$, propanethial.

b. 1. *Ketones* are named by replacing the final "e" of the hydrocarbon name by the suffix "one." $CH_3 \cdot CO \cdot CH_3$, propanone; $CH_3 \cdot CH_2 \cdot CO \cdot CH_3$, butanone; $CH_3 \cdot CH_2 \cdot CH_2 \cdot CO \cdot CH_3$, 2-pentanone; $CH_5 \cdot CH_2 \cdot CO \cdot CH_2 \cdot CH_3$, 3-pentanone; etc.

In case two or more functional groups are present, the suffix takes the form of "dione," "trione," "tetrone," etc. $CH_3 \cdot CO \cdot CO \cdot CH_3$, butanedione; $CH_3 \cdot CH_2 \cdot CO \cdot CO \cdot CH_3$, 2, 3-pentanedione; $CH_3 \cdot CO \cdot CH_2 \cdot CO \cdot CH_3$, 2, 4-pentanedione; etc.

2. *Thioketones* add the suffix "thione" to the corresponding hydrocarbon name. $CH_3 \cdot CS \cdot CH_3$, propanethione; $CH_3 \cdot CH_2 \cdot CS \cdot CH_3$, butanethione; $CH_3 \cdot CH_2 \cdot CH_2 \cdot CS \cdot CH_3$, 2-pentanethione; $CH_3 \cdot CH_2 \cdot CS \cdot CH_2 \cdot CH_3$, 3-pentanethione; etc.

Use is made of the suffixes "dithione," "trithione," etc., where two or more of these functional groups occur in the compound. $CH_3 \cdot CS \cdot CS \cdot CH_3$, butanedithione; etc.

3. The term *quinone* is applied to certain ketones of "quinonoid" structure, as $p \cdot O : C_6H_4 : O$, p-benzoquinone or simply quinone;

$O : C_{10}H_6 : O$, naphthoquinone; etc.

Naphthoquinones, anthraquinone, and phenanthraquinone represent the corresponding derivatives of naphthalene, anthracene, and phenanthrene.

6. *Ketene* is $H_2C : C : O$. Its substitution products are named as derivatives. $(CH_3)_2C : C : O$, dimethylketene; etc.

7. *a.* *Carboxylic acids* are named by replacing the terminal "e" of the corresponding hydrocarbon name by the suffix "oic," and adding the word "acid." $H \cdot CO \cdot OH$, methanoic acid; $CH_3 \cdot CO \cdot OH$, ethanoic acid; $CH_3 \cdot CH_2 \cdot CO \cdot OH$, propanoic acid; $CH_3 \cdot CH_2 \cdot CH_2 \cdot CO \cdot OH$, butanoic acid; etc.

The suffix "dioic" is used for dicarboxylic acids. $HO \cdot OC \cdot CH_2 \cdot CO \cdot OH$, propanedioic acid; $HO \cdot OC \cdot CH_2 \cdot CH_2 \cdot CO \cdot OH$, butanedioic acid; $CH_3 \cdot CH (CO \cdot OH)_2$, 2-methylpropanedioic acid; etc.

The acids may also be named as alkanecarboxylic acids, especially where Geneva names are inconvenient or impossible, as, for example, $CH_3 \cdot CO \cdot OH$, methanecarboxylic acid; $HO \cdot OC \cdot CH_2 \cdot CO \cdot OH$, methanedicarboxylic acid; $HO \cdot OC \cdot CH_2 \cdot CH(CO \cdot OH) \cdot CH_2 \cdot CO \cdot OH$, 1, 2, 3-propanetricarboxylic acid; $C_6H_5 \cdot CO \cdot OH$, benzenecarboxylic acid.

b. Where *sulfur* is substituted for one or more of the oxygens in the carboxylic group, the naming is as follows: $H \cdot CO \cdot SH$ or $H \cdot CS \cdot OH$, methane*thio*ic acid, and $CH_3 \cdot CO \cdot SH$ or $CH_3 \cdot CS \cdot OH$, ethane*thio*ic acid; $H \cdot CO \cdot SH$, methane*thiol*ic acid, or $CH_3 \cdot CO \cdot SH$, ethane*thiol*ic acid; $H \cdot CS \cdot OH$, methane*thion*ic acid or $CH_3 \cdot CS \cdot OH$, ethane*thion*ic acid; and $H \cdot CS \cdot SH$, methane*thionothiol*ic acid or $CH_3 \cdot CS \cdot SH$, ethane*thionothiol*ic acid; etc.

If the sulfur acid is to be named by analogy with the "carboxylic" names, then the suffixes carbothiolic for $^-CO \cdot SH$, carbothionic for $^-CS \cdot OH$, and carbodithioic for $^-CS \cdot SH$, should be used; as, $CH_3 \cdot CO \cdot SH$, methanecarbothiolic acid.

8. *Acid derivatives* are named as such, namely:

a. *Salts*; $H \cdot CO \cdot ONa$, sodium methanoate; $CH_3 \cdot CO \cdot ONa$, sodium ethanoate; $CH_3 \cdot CH_2 \cdot CO \cdot OK$, potassium propanoate; $(CH_3 \cdot CO \cdot O)_2Ca$, calcium ethanoate, etc.

b. *Esters*; $H \cdot CO \cdot O \cdot CH_3$, methyl methanoate; $CH_3 \cdot CO \cdot O \cdot CH_3$, methyl ethanoate; $CH_3 \cdot CH_2 \cdot CO \cdot O \cdot CH_2 \cdot CH_3$, ethyl propanoate; $CH_3 \cdot CH_2 \cdot CH_2 \cdot CO \cdot O \cdot CH_3$, methyl butanoate, etc. Such names as ethanoic acid methyl ester may also be used.

c. *Anhydrides*; $H \cdot CO \cdot O \cdot OC \cdot H$, methanoic anhydride; $CH_3 \cdot CO \cdot O \cdot OC \cdot CH_3$, ethanoic anhydride; $CH_3 \cdot CH_2 \cdot CO \cdot O \cdot OC \cdot CH_2 \cdot CH_3$, propanoic anhydride;

$$\begin{matrix} H_2C \cdot C:O \\ \Big| \quad \Big\rangle O, \text{ butanedioic anhydride; etc.} \\ H_2C \cdot C:O \end{matrix}$$

d. *Acid halides*; $H \cdot CO \cdot Cl$, methanoyl chloride; $CH_3 \cdot CO \cdot Br$, ethanoyl bromide; $CH_3 \cdot CH_2 \cdot CO \cdot Cl$, propanoyl chloride; $Cl \cdot OC \cdot CH_2 \cdot CH_2 \cdot CO \cdot Cl$, butanedioyl chloride; $C_6H_5 \cdot CO \cdot Cl$, benzenecarbonyl chloride; etc.

e. *Amides*; $H \cdot CO \cdot NH_2$, methanamide; $CH_3 \cdot CO \cdot NH_2$, ethanamide; $CH_3 \cdot CH_2 \cdot CO \cdot NH_2$, propanamide; $H_2N \cdot OC \cdot CH_2 \cdot CH_2 \cdot CO \cdot NH_2$, butanediamide; $C_6H_5 \cdot CO \cdot NH_2$, benzenecarbonamide; etc.

f. *Nitriles*; $H \cdot CN$, methanenitrile (hydrogen cyanide); $CH_3 \cdot CN$, ethanenitrile; $CH_3 \cdot CH_2 \cdot CN$, propanenitrile; $CH_3 \cdot CH_2 \cdot CH_2 \cdot CN$, butanenitrile or 1-propanecarbonitrile; etc.

g. *Imides*;
$$\begin{matrix} H_2C \cdot C:O \\ \Big| \quad \Big\rangle N \cdot H, \text{ butanimide;} \\ H_2C \cdot C:O \end{matrix} \quad H_2C \Big\langle \begin{matrix} H_2C \cdot C:O \\ \Big\rangle N \cdot H, \text{ pentanimide} \\ H_2C \cdot C:O \end{matrix} \text{ or 1, 3-propanedicarbonimide.}$$

h. *Amidines*; $H \cdot C(:N \cdot H)NH_2$, methanamidine; $CH_3 \cdot C (:N \cdot H) NH_2$, ethanamidine; $CH_3 \cdot CH_2 \cdot C(:N \cdot H)NH_2$, propanamidine or ethanecarbonamidine; etc.

i. *Amidoximes*; $H \cdot C(:N \cdot OH)NH_2$, methanamidoxime; $CH_3 \cdot C(:N \cdot OH) NH_2$, ethanamidoxime; $CH_3 \cdot CH_2 \cdot C(:N \cdot OH)NH_2$, propanamidoxime or ethanecarbonamidoxime; etc.

9. *Nitrogen bases* are characterized by the ending "ine." Simple monoamines are named by prefixing the radical names to "amine"; as, $CH_3 \cdot NH_2$, methylamine; $(CH_3)_2NH$, dimethylamine; $(CH_3)_3N$, trimethylamine. Where more than one such group is present, use is made of the suffixes "diamine," "triamine," etc. $H_2N \cdot CH_2 \cdot CH_2 \cdot NH_2$, 1, 2-ethanediamine; $C_6H_4 (NH_2)_2$, benzenediamine or phenediamine; etc.

Where the nitrogen is of the quinquevalent type, the ending "ine" becomes (a) "onium" in the aliphatic series, and (b) "inium" for cyclic substances containing the nitrogen in the ring. Under similar circumstances, the ending "ole" becomes "olium." $CH_3 \cdot NH_2$, methylamine and $CH_3 \cdot NH_3I$, methylammonium iodide; pyridine and pyridinium; imidazole and imidazolium; etc.

10. *Hydroxylamine derivatives* are named as follows;

a. $CH_3 \cdot O \cdot NH_2$, methoxyamine; $CH_3 \cdot CH_2 \cdot O \cdot NH_2$, ethoxyamine; etc.

b. $CH_3 \cdot NH \cdot OH$, methylhydroxylamine; $CH_3 \cdot CH_2 \cdot NH \cdot OH$, ethylhydroxylamine; etc.

c. $CH_3 \cdot CH:N \cdot OH$, ethanal oxime; $CH_3 \cdot CH_2 \cdot CH:N \cdot OH$, propanal oxime; etc.

d. $(CH_3)_2C:N \cdot OH$, propanone oxime; $CH_3 \cdot CH_2 \cdot C(:N \cdot OH) \cdot CH_3$, butanone oxime; etc.

11. *Urea* $(H_2N \cdot CO \cdot NH_2)$, *guanidine* $(H_2N \cdot C(:N \cdot H) \cdot NH_2)$, and *carbylamine* $(H \cdot \overrightarrow{N}:C)$, as names of parent compounds used in naming their derivatives, all represent good usage.

    a. 1. $CH_3 \cdot HN \cdot CO \cdot NH_2$, methylurea; $CH_3 \cdot CH_2 \cdot HN \cdot CO \cdot NH_2$, ethylurea; $(CH_3)_2N \cdot CO \cdot NH_2$, 1, 1-dimethylurea; $CH_3 \cdot HN \cdot CO \cdot NH \cdot CH_3$, 1, 3-dimethylurea; etc.

       2. $CH_3 \cdot CO \cdot HN \cdot CO \cdot NH_2$, ethanoylurea; $CH_3 \cdot CO \cdot HN \cdot CO \cdot NH \cdot OC \cdot CH_3$, 1, 3-diethanoylurea.

    b. $CH_3 \cdot HN \cdot C(:N \cdot H) \cdot NH \cdot CH_3$, 1, 3-dimethylguanidine.

    c. $CH_3 \cdot \overrightarrow{N}:C$, methylcarbylamine or **methyl isocyanide**; $CH_3 \cdot CH_2 \cdot \overrightarrow{N}:C$, ethylcarbylamine or **ethyl isocyanide**; etc.

12. a. *Cyanate* is reserved for the true esters which, on saponification, yield cyanic acid or its hydration products $(R \cdot O \cdot CN$, not known).

    b. *Thiocyanates*; $CH_3 \cdot S \cdot CN$, methyl thiocyanate; $CH_3 \cdot CH_2 \cdot S \cdot CN$, ethyl thiocyanate; etc.

    c. *Isocyanates*; $CH_3 \cdot N:C:O$, methyl isocyanate; $CH_3 \cdot CH_2 \cdot N:C:O$, ethyl isocyanate; etc.

    d. *Isothiocyanates*; $CH_3 \cdot N:C:S$, methyl isothiocyanate; $CH_3 \cdot CH_2 \cdot N:C:S$, ethyl isothiocyanate; etc.

13. *Nitrates, nitrites*, and *nitro* and *nitroso* compounds;

    a. Nitrates; $CH_3 \cdot O \cdot NO_2$, methyl nitrate; $CH_3 \cdot CH_2 \cdot O \cdot NO_2$, ethyl nitrate.

    b. Nitrites; $CH_3 \cdot O \cdot NO$, methyl nitrite; $CH_3 \cdot CH_2 \cdot O \cdot NO$, ethyl nitrite.

    c. Nitro derivatives are always indicated by the prefix "nitro." $CH_3 \cdot NO_2$, nitromethane; $CH_3 \cdot CH_2 \cdot NO_2$, nitroethane; $C_6H_5 \cdot NO_2$, nitrobenzene; etc.

    d. Nitroso derivatives are indicated by the prefix "nitroso." $CH_3 \cdot NO$ nitrosomethane; $CH_3 \cdot CH_2 \cdot NO$, nitrosoethane; $C_6H_5 \cdot NO$, nitrosobenzene; etc.

14. *Azo, azoxy, hydrazo*, and *diazonium* compounds;

    a. Azo; $C_6H_5 \cdot N:N \cdot C_6H_5$, azobenzene.

    b. Azoxy; $C_6H_5 \cdot N(:O):N \cdot C_6H_5$, azoxybenzene.

    c. Hydrazo; $C_6H_5 \cdot HN \cdot NH \cdot C_6H_5$, 1, 2-diphenylhydrazine (hydrazobenzene).

    d. 1. Diazonium; $C_6H_5 \cdot N(:N) \cdot Cl$, benzenediazonium chloride.

       2. Diazo; $H_2C:N:N$, diazomethane; $N:N:CH \cdot CO \cdot OH$, diazoethanoic acid (**diazoacetic acid**, diazomethanecarboxylic acid); $C_6H_5 \cdot N:N \cdot OH$, benzenediazohydroxide; $C_6H_5 \cdot N:N \cdot NH \cdot C_6H_5$, diazoaminobenzene or 1, 3-diphenyltriazene.

       3. Diazoates; $CH_3 \cdot N:N \cdot OK$, potassium methanediazoate; $C_6H_5 \cdot N:N \cdot ONa$, sodium benzenediazoate; etc.

       4. Derivatives of $H_2N \cdot NH \cdot NH \cdot NH_2$ (tetrazane), of $H \cdot N:N \cdot NH \cdot NH_2$ (1-tetrazene), of $H \cdot N:N \cdot NH \cdot N:N \cdot H$, (1, 4-pentazdiene), etc.; $C_6H_5 \cdot NH \cdot NH \cdot NH \cdot NH \cdot C_6H_5$, 1, 4-diphenyltetrazane; and $C_6H_5 \cdot N:N \cdot NH \cdot NH_2$, 1-phenyl-1-tetrazene.

15. *Hydrazines, hydrazides, hydrazones*, and *semicarbazones*;

    a. Hydrazines; $CH_3 \cdot NH \cdot NH_2$, methylhydrazine; $C_6H_5 \cdot NH \cdot NH_2$, phenylhydrazine; $CH_3 \cdot HN \cdot NH \cdot CH_2 \cdot CH_3$, 1-ethyl-2-methylhydrazine; etc.

    b. Hydrazides; $CH_3 \cdot CO \cdot NH \cdot NH_2$, ethanohydrazide (ethanoic hydrazide, or methanecarbohydrazide); but $CH_3 \cdot CO \cdot HN \cdot NH \cdot OC \cdot CH_3$, 1, 2-diethanoylhydrazine.

    c. Hydrazones; $CH_3 \cdot CH:N \cdot NH \cdot C_6H_5$, ethanal (or **acetaldehyde**) phenylhydrazone; $(CH_3)_2C:N \cdot NH \cdot C_6H_5$, propanone (or **acetone**) phenylhydrazone; etc.

    d. Semicarbazones; $(CH_3)_2C:N \cdot NH \cdot CO \cdot NH_2$, propanone (or **acetone**) semicarbazone.

16. *Organometallic compounds;* $CH_3 \cdot Zn \cdot CH_3$, dimethylzinc; $(C_2H_5)_4l'b$, tetraethyllead; $CH_3 \cdot Mg \cdot Cl$, methylmagnesium chloride; $Cl \cdot Hg \cdot C_6H_4 \cdot CO \cdot OH$, chloromercuribenzoic acid; $(C_6H_5)_3Sn \cdot Na$, sodium triphenylstannide; etc.

    In order to avoid ambiguity, parentheses are used to enclose the names of complex radicals; $(CH_3)_2C_6H_3 \cdot NH_2$,(dimethylphenyl)amine, in contrast to $C_6H_5 \cdot N(CH_3)_2$, dimethylphenylamine; etc.

    For compounds possessing different functions, only the principal function will be indicated by the ending of the name. Prefixes are used to designate the other functions. (The following order is alphabetic and is *not* an order of precedence.)

Function	Prefix	Suffix
Acid..............	carboxy........	carboxylic, or "-oic"
Acid derivatives.....	....	carbonyl, "oyl," carbonamide; etc.
Alcohol...........	hydroxy.......	ol
Aldehyde, for the "O"	oxo, aldo.......	al
Aldehyde, for "CHO"	formyl.........	carbonal
Amine.............	amino.........	amine
Azo derivatives.....	azo	
Azoxy derivatives...	azoxy	
Carbonitrile (nitrile).	cyano.........	carbonitrile or nitrile
Double bond........	....	ene
Ether.............	alkoxy	
Ethylene oxide, etc...	epoxy	
Halide............	halogeno (halo)	
Hydrazine.........	hydrazino......	hydrazine
Ketone............	oxo or keto.....	one
Mercaptan........	mercapto......	thiol
Nitro derivatives....	nitro	
Nitroso derivatives..	nitroso	
Quinquevalent nitrogen	....	onium, inium (olium)
Sulfide............	alkylthio	
Sulfinic derivatives..	sulfino.........	sulfinic
Sulfone............	sulfonyl	
Sulfonic derivatives..	sulfo.........	sulfonic
Sulfoxide..........	sulfinyl	
Triple bond........	....	yne
Urea..............	ureido........	urea

D. Heterocyclic Compounds:

    The endings of customary names that do not correspond to the functional groups present in a compound, should be changed as follows:

1. "ol" to "ole"; pyrrol to pyrrole.

2. "ane" to "an"; pyrane to pyran.

    Hetero atoms occurring in the ring are indicated as follows;

1. Oxygen by "oxa"; oxadiazole.

2. Sulfur by "thia"; thiadiazole.

3. Nitrogen by "aza"; 2, 7, 9-triazaphenanthrene.

E. Radicals:

1. *a.* CH_3-, methyl; $CH_2 \cdot CH_2$-, ethyl; $CH_3 \cdot CH_2 \cdot CH_2$-, propyl; C_6H_5-, phenyl; $C_6H_5 \cdot CH_2$-, benzyl; $C_{10}H_7$-, naphthyl; $C_{14}H_9$-, anthryl; C_5H_4N, pyridyl; etc.

   b. $H_2C{:}CH$-, ethenyl (vinyl): $CH_2 \cdot HC{:}CH$-, 1-propenyl; $CH_3 \cdot CH_2 \cdot HC{:}CH$-, 1-butenyl; $C_6H_5 \cdot HC{:}CH$-, 2-phenylethenyl; etc.

   c. $H \cdot C{:}C$-, ethynyl; $CH_3 \cdot C{:}C$-, 1-propynyl; $CH_2 \cdot CH_2 \cdot C{:}C$-, 1-butynyl; $C_6H_5 \cdot C{:}C$-, phenylethynyl; etc.

2. *a.* $CH_2 \cdot CH =$, ethylidene; $CH_3 \cdot CH_2 \cdot CH =$, propylidene; $C_6H_5 \cdot CH_2 \cdot CH =$, 2-phenylethylidene; etc.

   b. $H_2C{:}C =$ ethenylidene; $CH_3 \cdot HC{:}C =$, 1-propenylidene; $C_6H_5 \cdot HC{:}C =$, phenylethenylidene; etc.

   c. $H \cdot C{:}C \cdot CH =$, propynylidene; etc.

3. *a.* $CH_3 \cdot C \equiv$, ethylidyne; $CH_3 \cdot CH_2 \cdot C \equiv$, propylidyne; etc.

   b. $H_2C{:}CH \cdot C \equiv$, propenylidyne; $H_2C{:}CH \cdot CH_2 \cdot C \equiv$, 3-butenylidyne; etc.

   c. $H \cdot C{:}C \cdot C \equiv$, propynylidyne; $H \cdot C{:}C \cdot CH_2 \cdot C \equiv$, 3-butynylidyne; etc.

4. $H_2C =$, methylene; -$CH_2 \cdot CH_2$-, ethylene; -$CH_2 \cdot CH_2 \cdot CH_2$-, trimethylene; -$CH_2 \cdot CH_2 \cdot CH_2 \cdot CH_2$-, tetramethylene; -$NH \cdot CO \cdot NH$-, ureylene; etc.

5. *a.* $CH_3 \cdot CO$-, ethanoyl; $CH_3 \cdot CH_2 \cdot CO$-, propanoyl; $C_6H_5 \cdot CO$-, benzenecarbonyl or benzoyl; etc.

   b. $C_6H_5 \cdot SO_2$-, benzenesulfonyl; $C_{10}H_7 \cdot SO_2$-, naphthalenesulfonyl; etc.

This division is based on Definitive Report of the Commission on the Reform of the Nomenclature of Organic Chemistry, *J. Am. Chem. Soc.*, *55*, 3905-3925 (1933).

For a detailed consideration of the nomenclature of the carbohydrates, the student is referred to *An Outline of the Chemistry of the Carbohydrates*, 1941, pp. 259-76, by Ed. F. Degering or to *Carbohydrate Nomenclature*, Report of the Committee on Nomenclature, Spelling, and Pronunciation, *Chemical Engineering News*, *26*, No. 22 (May 31, 1948).

CHAPTER XXV

ANTISEPTICS AND GERMICIDES

A substance capable of preventing the growth of bacteria is called an antiseptic. If it is capable of killing bacteria, it is designated as a germicide or bactericide. These terms, however, are often used indiscriminately since many compounds may fulfill the requirements of both, especially in rather concentrated solutions. Internal and external antiseptics are used, as the names imply, internally and externally, but there are some compounds that serve both purposes equally well. An antiseptic that is ingested orally and eliminated by the urine is termed a urinary antiseptic.

The antiseptic property of phenol was discovered by Joseph Lister, a young surgeon of Glasgow, in 1867. Prior to that time he had operated very skillfully, but more than half of his patients died from blood poisoning. By using phenol in disinfecting his surgical instruments and wounds, Lister found that most of his patients recovered very quickly. The caustic action of phenol on the body tissues, however, was found to be a serious disadvantage. Soon afterwards, because of their lower tissue toxicity, the cresols were introduced as substitutes for phenol, but their use was limited due to their low solubility in water. Since the introduction of phenol as an antiseptic, numerous substances have been advocated as antiseptics and germicides, and a few of them have gained considerable prominence.

A germicide is tested for its germ-killing power (bactericidal power) by determining the maximum dilution (minimum concentration) of the compound that will kill a given organism in a given length of time under specified conditions. Substances are usually compared with phenol as a standard and their killing power expressed in terms of *phenol coefficients*. To find the phenol coefficient of a compound, the reciprocal of its minimum killing concentration is divided by the reciprocal of the minimum killing concentration of phenol under the same conditions against the same organism. By definition, then, the phenol coefficient of phenol against all organisms is unity while the phenol coefficients of other substances vary with the compound and the organism employed. In this way all germicides can be compared with each other by reference to phenol as a standard.

An antiseptic, however, is tested for its growth-prevention power or bacteriostatic power by determining the minimum concentration which will prevent the growth of an organism during a given time interval (usually 48 hours) in agar medium of adjusted pH at 37 °C. The growth-prevention power is usually expressed as the reciprocal of the minimum effective concentration and not in terms of phenol coefficients. Such tests, performed in nutrient media in the bacteriological laboratory, are called *in vitro* tests. Tests performed on mammals are known as *in vivo* tests. A given substance is likely to respond differently in these two tests. Furthermore, since both the technic of bacteriologists and different strains of the same organism show marked variations, the bacteriostatic power of a compound is of little significance unless it is tested along with other substances under the same experimental conditions.

No acceptable theory has been postulated to explain the action of antiseptics or germicides. Some claim that a chemical reaction occurs between the compound and the organism while others believe that it is an adsorption phenomenon. The practical value of a germicide, especially one designated for internal use, is determined by the ratio of the toxic dose to the curative dose which is defined as the therapeutic index.

Various types of substances have been used as antiseptics and they are conveniently classified according to their chemical nature as non-metallic and metallic compounds.

I. Non-Metallic Compounds

1. Acids and Esters

The carboxylic acids such as benzoic and salicylic are weak antiseptics and find use as food preservatives and in the preparation of skin ointments. The ω-phenyl-substituted fatty acids increase in antiseptic power as the length of the aliphatic chain is increased to four or five carbon atoms. Above this point, decreasing solubility seems to lower their effectiveness. The free acids are usually better antiseptics than their sodium salts, and the unsaturated aliphatic monoacids are more effective than the corresponding saturated acids.

The sulfonic acids have no perceptible antiseptic action although sulfonic groups are often added to increase solubility. Hydrochloric acid will be considered under the halogen compounds.

Alkyl esters of *p*-hydroxybenzoic acid are effective antiseptics and their effectiveness seems to increase with the weight of the alkyl radical. They are used as preservatives in cosmetic preparations.

2. Antiseptic Dyes

Many micro-organisms are stained by adsorption of certain dyes, and various dyes have been found to be effective as antiseptics and to possess marked specificity. Many of the azo dyes such as Mallophene, Niazo, Serenium, and Picochrome are used as urinary antiseptics.

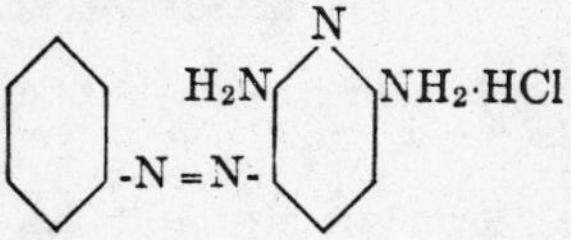

Mallophene

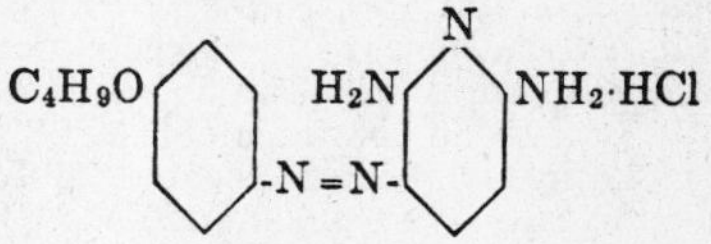

Niazo

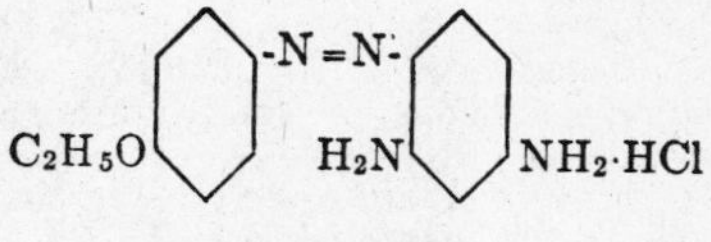

Serenium

Picochrome is a mixture of the two compounds shown at the right. Mallophene is reported to be eliminated by the urine to the extent of 60%.

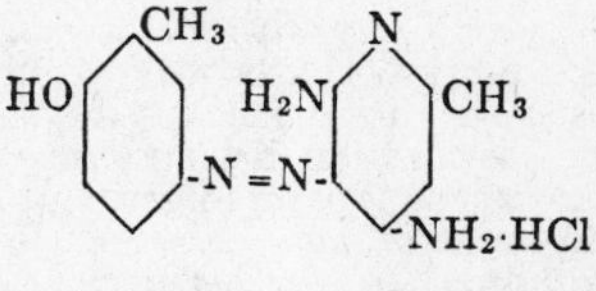

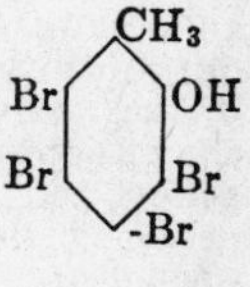

Methylene Blue (Ch. XXVI, Thiazine Dyes) stains malarial parasites *in vitro*. Malachite Green, shown at the right, was used alone and with mercuric chloride by the British during the World War as a skin disinfectant.

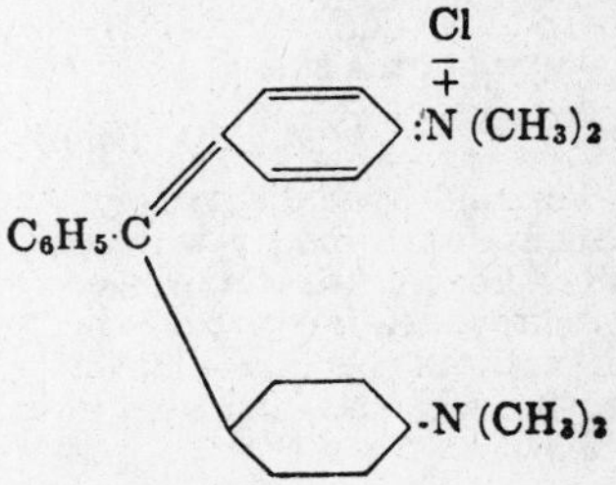

Acriflavine (Ch. XXVI, Acridine Dyes) is perhaps the most extensively used antiseptic dye. It is said to possess the exceptional property of being more active in serum than in water. Proflavine (Ch. XXII, p. 221) is also used. Mercurochrome, a phthalein dye, will be considered under the mercury compounds.

3. Halogen Compounds

Chlorine compounds such as sodium or calcium hypochlorite are used extensively for large scale disinfection. Many chlorine compounds containing so-called positive chlorine (chlorine linked to an electronegative element) have been used as antiseptics. The hypochlorites (Dakin solution) were used during the World War, but their instability led Dakin to further investigation which led to the discovery of Chloramine T and Dichloramine T. The former is soluble in water and the latter is insoluble in water. These compounds are said to be stable chlorine carriers, yet they liberate the chlorine when in contact with the wound.

Chloramine-T

Dichloramine-T

N-Chlorosuccinimide (succinochlorimide) has been shown to be quite effective in disinfecting water supplies.

$$CH_2 \cdot CO. \quad N \cdot Cl$$
$$CH_2 \cdot CO \cdot$$

N-Chlorosuccinimide

It has been found that dilute hydrochloric acid (0.005%) is a certain and definite antiseptic which appears to be superior to any other chlorine-containing solutions of the same strength. It is not painful to a wound and does not cauterize the skin.

Bromine compounds have a somewhat limited use as antiseptics. Tribromonaphthol, tetrabromo-o-cresol, and a compound of tribromophenol with bismuth have been used to some extent.

Iodine, mainly in alcoholic solution, has been used extensively as an antiseptic. Glycerol solutions of iodine are also used, and attempts have been made to prepare colloidal solutions of iodine. Iodine, however, is quite irritating to sensitive skins. Iodoform (CHI_3), iodine trichloride (ICl_3), and a number of iodine compounds have found a limited use.

4. Phenols and Substituted Phenols

Phenol and the cresols, especially m-cresol (m-$CH_3 \cdot C_6H_4 \cdot OH$), are important antiseptics. The cresols, because of their low solubility in water, are usually used in the form of emulsions. "Lysol" contains various phenols and certain aromatic hydrocarbons. Resorcinol, guaiacol, pyrogallol, the xylenols, and the naphthols have been used, although none of these is especially effective. Alkyl and halogen substituents seem to be the most effective in enhancing the germicidal action of phenols and resorcinols, the substituent being most effective when in the p-position. Among the p-alkylphenols, the germicidal power increases regularly with increase in the length of the alkyl radical until decreasing solubility becomes the controlling factor. n-Hexylphenol is more effective than n-hexylresorcinol but it is somewhat more toxic. Branched chains appear to be less effective than straight chains in intensifying germicidal action, and the alkoxy groups seem to be about as effective as the alkyl groups.

The higher alkyl cresols are more effective than the cresols, and chlorophenol and chlorothymol are much better germicides than the parent compounds. The alkylhalophenols are especially effective if the halogen atom occupies the p-position and the alkyl group is ortho to the hydroxyl group. The o-alkyl derivatives of p-chlorophenol attain their maximum germicidal power when the side chain contains 6 to 8 carbon atoms.

5. Miscellaneous Non-Metallic Compounds

Formaldehyde, or its polymers, is often used in disinfecting buildings. Its condensation product with ammonia (hexamethylenetetramine, Urotropine) finds a limited use as a urinary antiseptic.

Ichthyol (a mixture of ammonium salts of sulfonic acids formed from certain naturally occurring bitumens) is an example of a sulfur compound that has a rather wide use, and p-hydroxydiphenyl sulfide has been patented for antiseptic use.

In antiseptic amino compounds, alkylation or acetylation of the amino group seems to destroy, partially or completely, the antiseptic properties.

Alcohols, in general, are weakly antiseptic. Ethyl alcohol, glycol, and glycerol possess about the same inhibitory power. Ethyl alcohol is most efficient in a concentration of 70% as stronger solutions seem to resinify the waxy chitinous coating around the bacteria and protect them from further attack. Isopropyl alcohol in 30-50% concentration is more effective than ethyl alcohol of the same concentration.

II. Metallic Compounds

1. Antimony Compounds

Trisodium antimony-tris(3-thio-2-hydroxypropanesulfonate) ($Sb[S \cdot CH_2 \cdot CHOH \cdot CH_2 \cdot SO_2 \cdot ONa]_3$), antimony tristhioglycolamide ($Sb[S \cdot CH_2 \cdot C(:O) \cdot NH_2]_3$), antimony derivatives of aromatic thiol compounds, and salts of p-aminobenzenestibinic acid have been used in the treatment of syphilis. The antimony compounds are less effective than the arsenicals or arsenic compounds but they are also less toxic.

2. Arsenic Compounds

Since the pioneering work of Ehrlich, a number of important aromatic arsenic compounds have been synthesized. The organic arsenicals are superior to inorganic arsenicals for they are less toxic to mammals but usually as toxic and sometimes even more toxic to protozoan parasites. Atoxyl is used in the treatment of skin diseases, and Salvarsan (Arsphenamine or Ehrlich's 606), Neosalvarsan (Neoarsphenamine or "914"), and Sulfarsenol (Sulfarsphenamine) are all used in the treatment of syphilis. Neosalvarsan is more soluble than Salvarsan and Sulfarsenol is claimed to be more stable in air than is Neosalvarsan.

Atoxyl   Salvarsan   Neosalvarsan

Sulfarsenol   (Tryparsamide)

Tryparsamide has been used in the treatment of human sleeping sickness.

Ehrlich, after studying many arsenicals, formulated the general rule that only compounds containing trivalent arsenic are effective in killing trypanosomes, and that the effectiveness of compounds containing quinquevalent arsenic depends upon their reduction in the body to the trivalent form.

3. Bismuth Compounds

Bismuth compounds, although less effective than arsenic and mercury compounds, have a comparatively low toxicity and are used in the treatment of syphilis and cancer. Some of the more commonly used compounds are bismuth lactate, bismuth ammonium tartrate, bismuth potassium tartrate, bismuth salicylate, bismuth naphthenate, and sodium iodobismuthite (Na_2BiI_5, Iodobismitol).

4. Mercury Compounds

Mercury, in finely divided form, was used by the primitive people in the treatment of disease. Later, inorganic salts of mercury were used and, more recently, organic mercury compounds have attained importance in therapy.

Although the organic mercurials are effective skin disinfectants, the organic arsenicals, because of their lower toxicity, retain their superior position in the treatment of syphilis.

Mercury compounds may be divided into three classes; inorganic salts, compounds in which one valence of mercury is satisfied by an organic radical, and those in which both valences are bonded to organic radicals. Compounds of the first class are usually too toxic to be used, whereas the toxicity of the compounds containing organic radicals depends on the nature and size of the radicals. Flumerin (disodium 2-hydroxymercurifluorescein) and Aspirochyl have been used to some extent in the treatment of syphilis.

Aspirochyl

Mercurochrome-220-soluble, investigated about 1917-18, was the first organic mercurial to gain wide use. The sodium salt is soluble, but acids, blood serum, and protein material cause precipitation which has raised a question as to the effectiveness of Mercurochrome as a skin disinfectant. In concentrations of 5% or better, as used by the physician, mercurochrome is relatively effective as an antiseptic.

"Mercurochrome-220"

Mercurophen seems to be an effective germicide but its toxicity has tended to restrict its use.

Metaphen is an effective germicide of relatively low toxicity, which is used as a skin disinfectant. Tests indicate that it is stable in the presence of blood serum and protein matter, although concentrated solutions give a precipitate when treated with acids.

Mercurophen

Merthiolate is an effective germicide of relatively low toxicity which is stable, according to claims, in the presence of blood serum or protein material.

Metaphen

Basic phenylmercuric nitrate (Merphenyl Nitrate) is a most effective germicide of very low toxicity and seems to meet the requirement of being stable in the presence of blood serum or protein material. It is a double salt of the composition $C_6H_5HgOH \cdot C_6H_5Hg NO_3$. Phenylmercuric picrate, 2-hydroxyphenylmercuric chloride, and several mercury compounds of pyridine are all effective germicides of comparatively low toxicity. Phenylmercuric acetate is used in industrial microbiological control.

Merthiolate

5. Silver and Silver Compounds

Silver and its compounds have both been used as antiseptics, but evidence seems to indicate that the antiseptic action of colloidal or spongy silver is dependent upon conversion to silver ions. Some swimming pools are disinfected by the electrolytic ionization of silver.

Electrargol, prepared electrolytically, and Collargol (78% Ag) are both colloidal preparations of silver. Argyrol (20-25% Ag) and Silvol (20% Ag) are both silver-protein preparations, and Neosilvol is a colloidal preparation of silver iodide (18-22%). Silver nitrate is also used.

6. *Miscellaneous Metallic Compounds*

Compounds of aluminum, cadmium, chromium, cobalt, copper, lead, manganese, nickel, selenium, tellurium, tin, and zinc have been investigated but none of these offer promise as antiseptics.

According to published data, the phenol coefficients of some of the leading commercial preparations decrease in the following order; Merphenyl Nitrate, Metaphen, Merthiolate, Mercuric Chloride, Hexylresorcinol, Tincture of Iodine, Lysol, Mercurochrome, Dakin Solution (NaOCl), Formalin, Pepsodent Antiseptic, Listerine Antiseptic, and Hydrogen Peroxide. (Birkhaug, J. Infectious Diseases, **53**, 250 (1933)).

7. *The Antibiotics*

The antibiotics may be considered, in general, as germicidal agents and are now available in forms which are effective against both gram-positive and gram-negative microorganisms.

The antibiotics are considered in some detail in Chapter XXXIV, pages 306 to 323.

CHAPTER XXVI

DYES, PIGMENTS, STAINS AND INDICATORS

I. Dyes

A dye is usually defined as a substance, possessing a color, which is capable of dyeing fiber, with or without the aid of a mordant. Chromophore groups are responsible for the color, while auxochrome groups enable the dye to become attached to the fiber; that is, organic nucleus + chromophore → chromogen, and chromogen + auxochrome → dye.

Many colored substances, therefore, cannot be classified as dyes since they contain no auxochrome groups. Auxochrome groups, although not directly responsible for the color, often do modify the shade of a dye. The color of a dye, in general, can be deepened by adding substituents to increase the molecular weight. The relative position of the new substituent usually has a marked effect. Groups that increase color are termed "bathochromic" groups; those that diminish color are known as "hypsochromic" groups.

Important chromophore groups are $-C=C-$, $-C\equiv$, $-C=O$, $-C=N-H$, $-C=N-$, $-N=N-$, $-N=N(\rightarrow O)-$, $-N=N-N-$, $\equiv C-N=O$, $-NO_2$, $-C=S$, and $\equiv C-S-S-C\equiv$. The most important auxochromes are $-OH$ and $-NH_2$. Others are $-OCH_3$, $-NH \cdot CH_3$, $-N(CH_3)_2$, $-NH \cdot C_6H_5$, $-N(CH_3)C_6H_5$, $-NH \cdot SO_2 \cdot C_6H_5$, $-NH \cdot CO \cdot C_6H_5$, $-NH \cdot OH$, $-NH \cdot NH_2$, and sometimes $-SO_3H$. The $-O \cdot OC \cdot CH_3$, $NH \cdot OC \cdot CH_3$, and $-N(NO) CH_3$ are poor auxochrome groups.

Cotton, which is principally cellulose, has little or no acidic or basic properties. Wool and silk, however, being protein in nature, possess pronounced amphoteric properties. Hence many acidic and basic dyes will dye wool and silk directly, whereas the dyeing of cotton is usually effected by the use of mordants. Some dyes, however, such as those of the benzidine series, will dye cotton directly.

Most mordants are amphoteric substances, such as aluminum hydroxide, which act as "bridges" and link the dye to the fiber. When a dye reacts with a mordant, a lake is formed. Some colorless compounds give lakes with mordants.

Picric acid, produced in 1771, was the first artificial organic dye. Perkin, however, in 1856, pioneered the way in the synthetic dye industry by the discovery of *Mauve*, a violet dye, which he prepared by the oxidation of allyltoluene. Since that time the synthesis of dyes has developed into a large industry. Although some natural dyestuffs are still used, the commercial products are mostly synthetic. Dyes are often classified, in industry, according to their application, as follows (antiseptic dyes have been considered in Chapter XXV):

1. *Acid dyes* usually contain sulfonic, carboxylic, or other acid groups. They dye wool and silk in an acid bath, but are not suitable for cotton. Examples are picric acid, many acid azo dyes, and the eosines.

2. *Basic* or *tannin dyes* are usually the hydrochloride salts of the color bases. They dye wool and silk and dye cotton in the presence of tannin. Magenta and auramine belong to this group.

3. *Chrome dyes* are used in dyeing wool. A direct dye is applied to wool in an acid bath, followed by treatment with sodium dichromate to give an insoluble *chrome lake*.

4. *Developed dyes* are so insoluble that they must be prepared directly on the fiber. Certain azo dyes belong to this group. The fiber is treated

with a solution of a phenol or an amine and then passed through a solution of a diazonium compound to give a coupling reaction within the fiber. Para red, from β-naphthol and *p*-nitroaniline, is prepared in this way.

5. *Direct dyes* are illustrated by the sodium salts of benzidine and the primuline dyes, which have a direct affinity for cotton. Congo red is another example.

6. *Food dyes* comprise fifteen compounds certified by the U. S. Department of Agriculture for use in foods and confectioneries because of their low toxicity. Amaranth and tartrazine are examples.

7. *Mordant dyes* have acidic properties and are fixed on the fiber by the aid of a mordant. The alizarins belong to this class.

8. *Vat dyes* are insoluble compounds which, on reduction, form soluble leuco bases (sometimes colorless) which may be converted, by oxidation, into the original dye. The most important vat dye is indigo. When treated with sodium bisulfite, indigo is converted into a soluble colorless form which is applied to the fiber. This is later reconverted to indigo by blowing with air.

Dyes may be classified according to their chemical structure. Under each structural type, in the following discussion, one or more examples are given. These examples have been selected, as far as convenient, so as to include stains and indicators as well as dyes.

There is no uniformity in the naming of synthetic dyestuffs. In most cases they bear the names given them by their manufacturers. Furthermore, each dyestuff may have a number of names, or different dyes may be known by the same name. In the *Colour Index* of the Society of Dyers and Colourists, however, each dye is given an individual number. For the purpose of reference, C.I. numbers are given along with the name of the dye. Classified chemically, a dye may be one of the following:

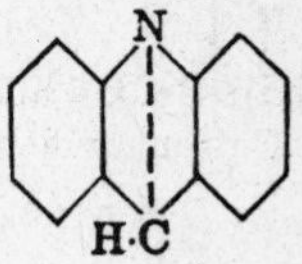

1. *Acridines,* all of which are derived from the parent substance, acridine, shown at the right. The most important member of this group, acriflavine, is prepared as follows;

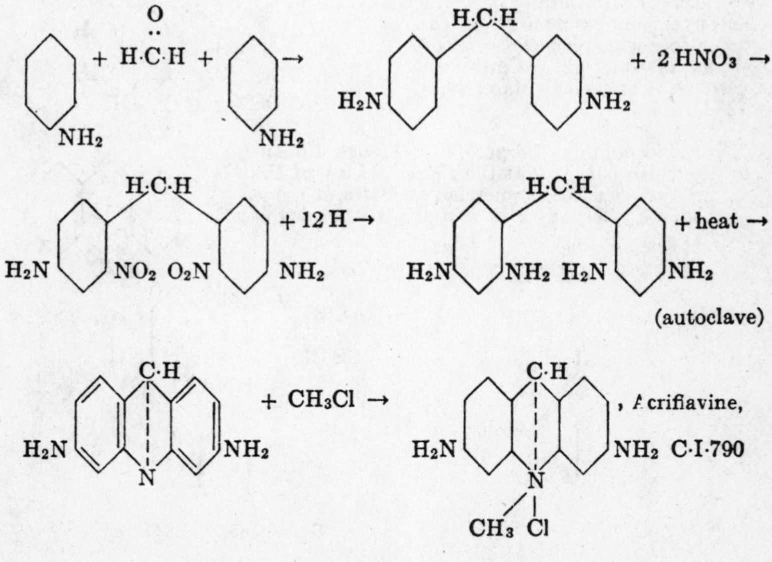

2. *Anthraquinone Dyes* are derived from the parent substance anthraquinone. Alizarin, an important member, may be prepared as follows. It is used in mordant dyeing.

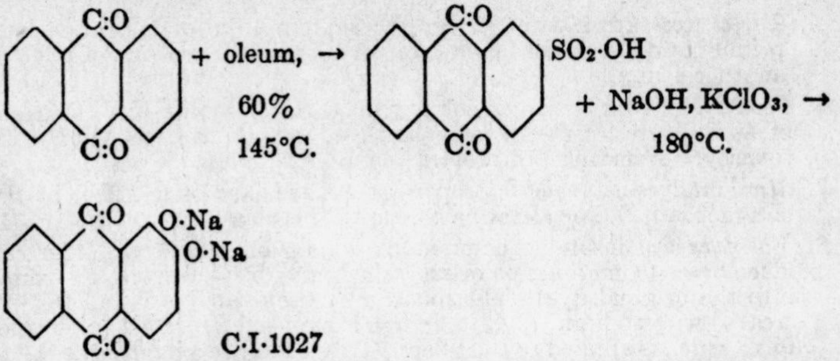

The most important anthraquinone vat dye is Indanthrene Blue R. S. Its preparation is as follows;

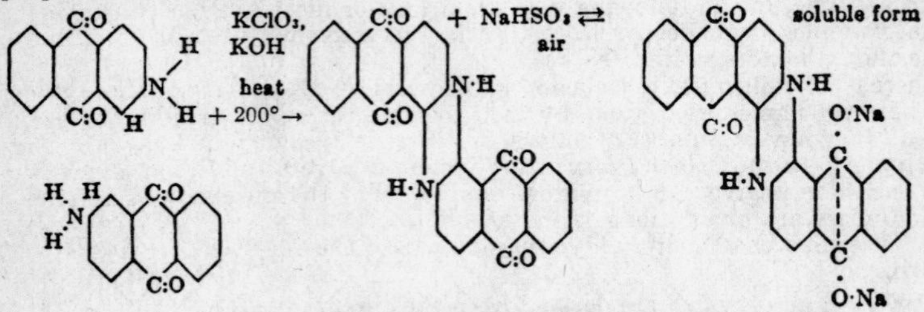

Indanthrene Blue R S, C·I·1106, Ins.

A popular blue dye, Indanthrene Blue G. C. D., has the same structure with chlorine atoms *ortho* to both of the nitrogen atoms.

Carmine, a naturally occurring red dye, consists mainly of carminic acid which has the structure shown at the right. Carmine is used as a biological stain. C.I. 1239.

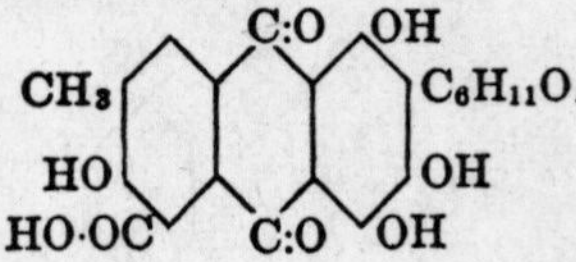

3. *Azine Dyes* contain the parent substance diphenazine, which is shown at the right. Many of the azine dyes are used commercially. Safranine, one of the fastest basic dyes known, is prepared as follows:

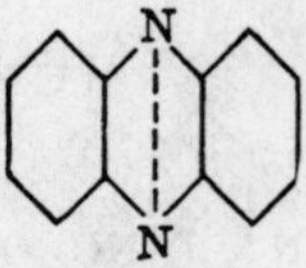

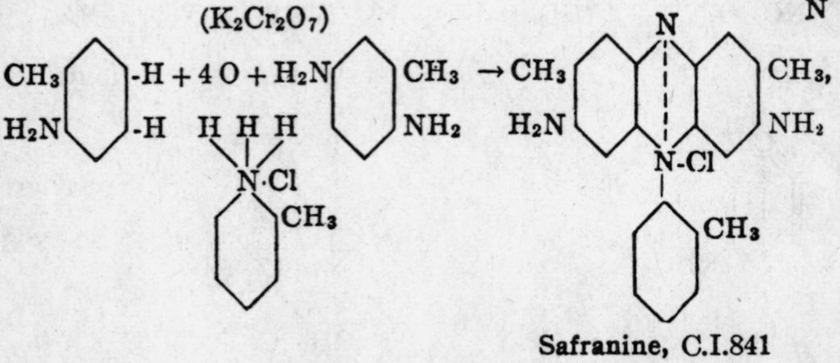

Safranine, C.I.841

Neutral Red, important as a biological stain and as an indicator, is shown at the right. C.I.825.

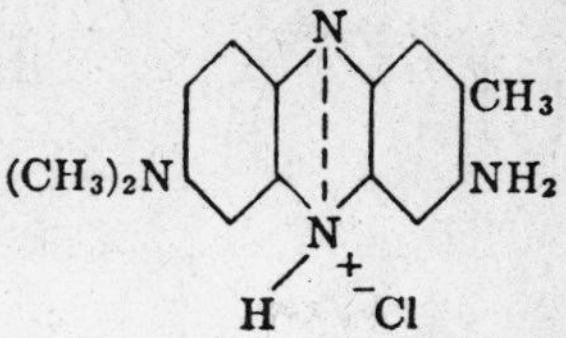

$$(CH_3)_2N \quad \text{[structure]} \quad CH_3, \ NH_2, \ N, \ \overset{+}{N}, \ H \ \bar{C}l$$

4. *Azo Dyes* are among the most important dyestuffs. Methyl orange is prepared by coupling diazotized sulfanilic acid with dimethylaniline:

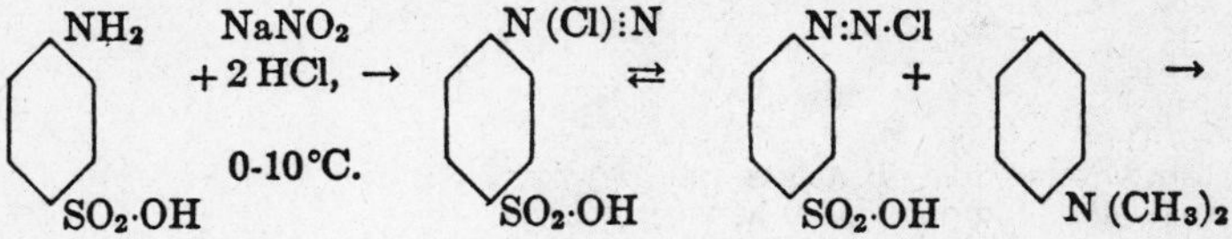

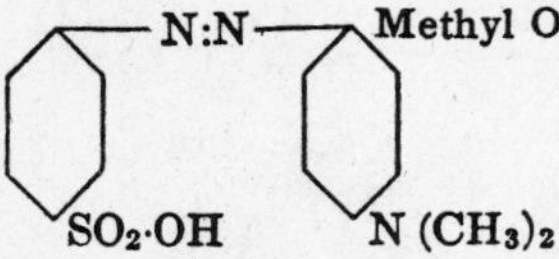

$$NH_2 \quad \xrightarrow[\text{0-10°C.}]{NaNO_2 + 2\,HCl,} \quad N(Cl){:}N \quad \rightleftarrows \quad N{:}N{\cdot}Cl \ + \ N(CH_3)_2 \quad \rightarrow$$
$$SO_2{\cdot}OH \qquad SO_2{\cdot}OH \qquad SO_2{\cdot}OH$$

$$N{:}N \quad \text{Methyl Orange, C.I. 142}$$
$$SO_2{\cdot}OH \qquad N(CH_3)_2$$

Congo Red, an important dye, stain and indicator, may be prepared by:

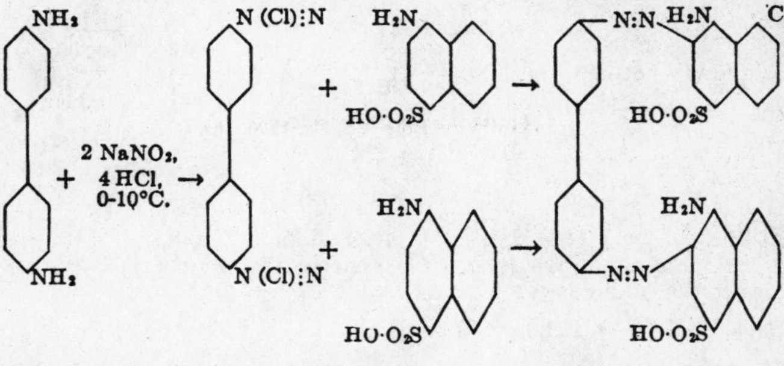

$$NH_2 \qquad N(Cl){:}N \qquad H_2N \qquad N{:}N \ H_2N \quad \text{Congo Red, C.I. 370}$$
$$+ \ HO{\cdot}O_2S \qquad HO{\cdot}O_2S$$
$$+ \quad \xrightarrow[\text{0-10°C.}]{2\,NaNO_2, \ 4\,HCl,}$$
$$NH_2 \qquad N(Cl){:}N \qquad H_2N \qquad H_2N$$
$$+ \qquad N{:}N$$
$$HO{\cdot}O_2S \qquad HO{\cdot}O_2S$$

The following azo dyes are used as stains or indicators:

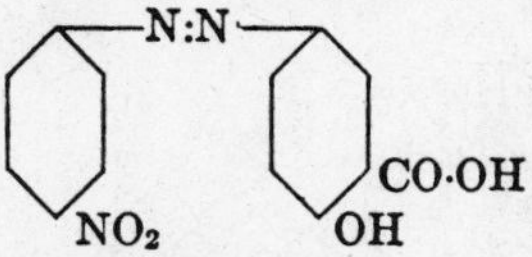

$$N{:}N \qquad\qquad N{:}N \ CO{\cdot}OH \qquad\qquad N{:}N$$
$$NO_2 \quad OH \qquad N(CH_3)_2 \qquad\qquad N(CH_3)_2$$
$$CO{\cdot}OH$$

Alizarin Yellow R **Methyl Red** **Methyl Yellow**
C.I. 40 C.I. 211 **Butter Yellow,** C.I. 19

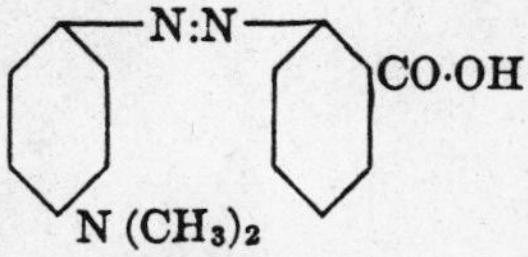

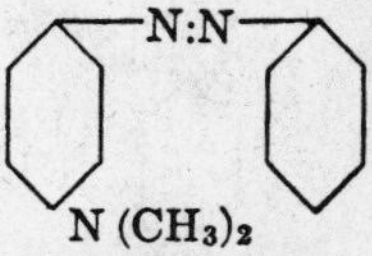

$$N{:}N \qquad SO_2{\cdot}ONa$$
$$HO \qquad SO_2{\cdot}ONa$$

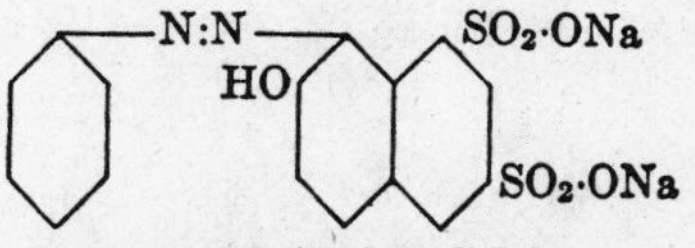

Orange G (stain), C.I.27

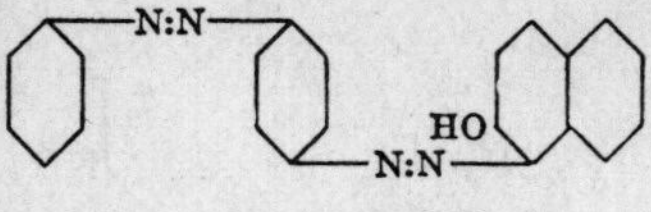

Sudan III (stain), C·I·248

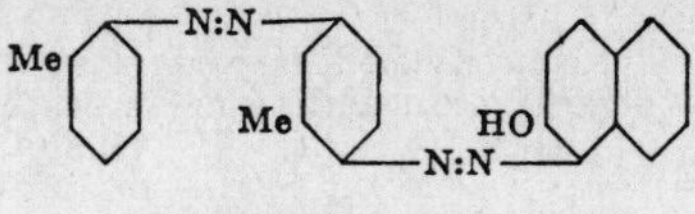

Sudan IV (stain), C·I·258

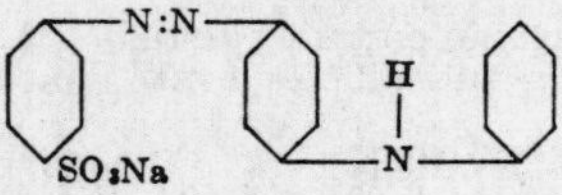

Tropeolin OO.CI.143

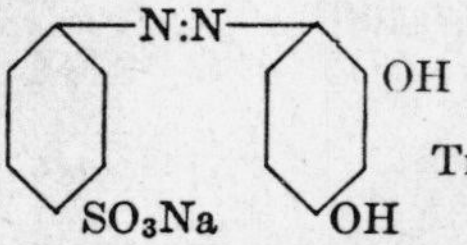

Both amines and phenols couple with diazo compounds. Phenols direct coupling to the *p*-position in alkaline solution, whereas *p*-coupling with amines is carried out in acid solution. In the benzene series, these groups direct *ortho* or preferably *para*. *beta*-Naphthol and β-naphthylamine direct coupling to the adjacent alpha position.

Tropeolin O, C.I.148

5. *Azoxy and Nitrosostilbene Dyes* may be prepared by:

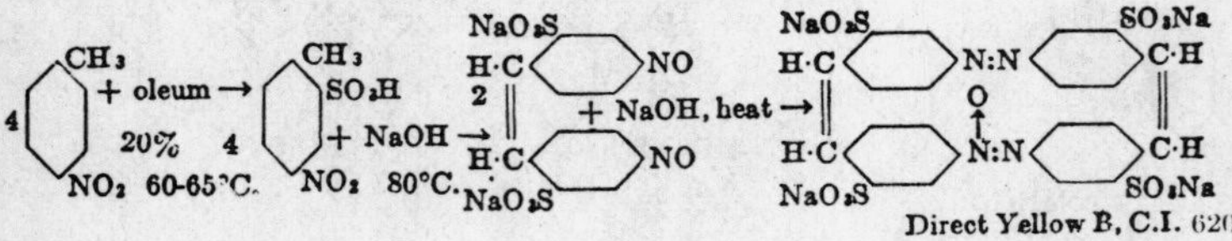

Direct Yellow B, C.I. 620

6. *Benzylidene or Azomethine Dyes* contain the linkage -N:CH-. They are not as highly colored as the azo dyes. The general method of preparation is by condensation of an aldehyde with an amine.

$$R \cdot CHO + H_2N \cdot R \rightarrow R \cdot CH:N \cdot R + H_2O$$

7. *Diphenylmethane Dyes* are of little technical importance.

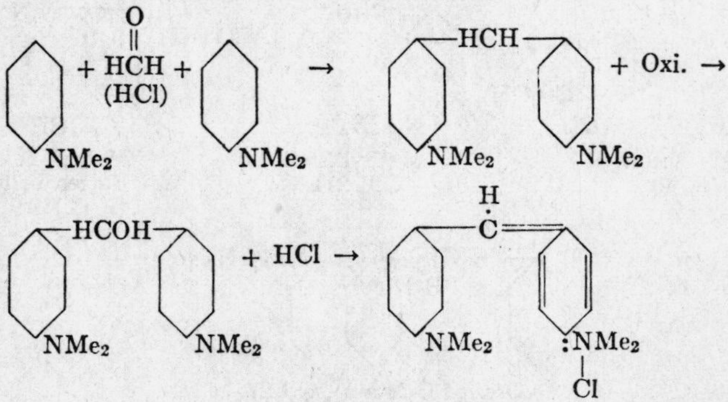

8. *Ethylenic Dyes* include many of the natural coloring pigments, such as carotene, $C_{40}H_{56}$. The effect of groups on color intensity is shown by the following structures;

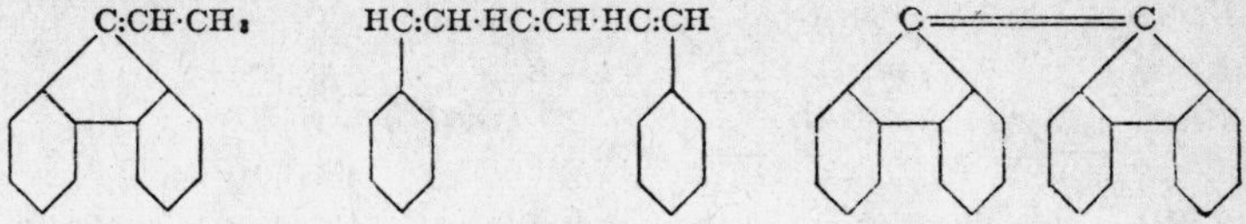
C:CH·CH₃
HC:CH·HC:CH·HC:CH
C═══C

(yellow) (bronze-yellow) (red)

9. *Flavones* comprise a series of dyestuffs, many of which are naturally occurring, containing the nucleus shown at the right. This group includes the Flavonols, Chromones, Brazilin, Hematoxylin, Benzopyranols, and Anthocyanins. Almost all of the flower pigments contain the nucleus common to this group. Hematoxylin, a naturally occurring dye, is used as a biological stain. Its formula is shown below.

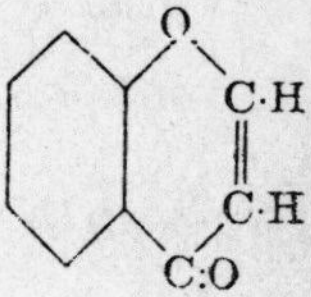
O
C·H
C·H
C:O

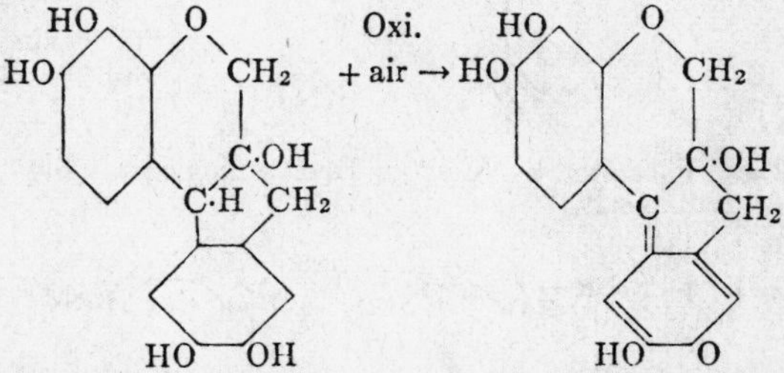
HO
O
HO
CH₂
C·OH
C·H
CH₂
HO
OH
Oxi.
+ air →
HO
O
HO
CH₂
C·OH
C
CH₂
HO
O

Hematoxylin **Hematein**, C.I. 1246

10. *Indigoid Dyes* contain the chromophoric group -OC·C:C·CO-. The most important dye of this group is Indigo. It has never been surpassed by any other blue dye. Keto form + $NaHSO_3$ → Enolic form + air → Keto form.

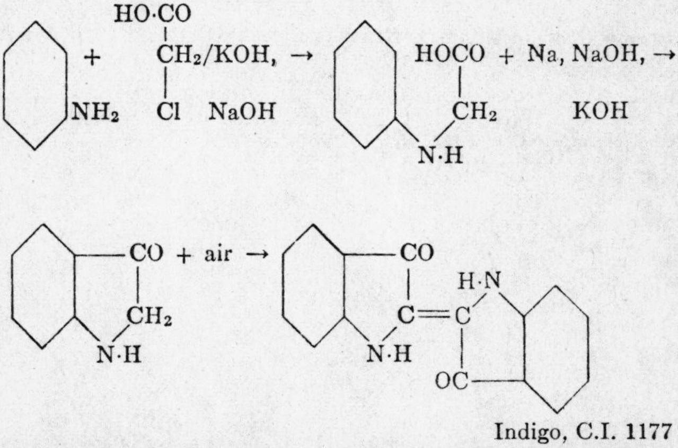
HO·CO
+
CH₂/KOH, →
NH₂
Cl
NaOH
HOCO + Na, NaOH, →
CH₂
KOH
N·H
CO + air →
CH₂
N·H
CO
H·N
C═C
N·H
OC

Indigo, C.I. 1177

The keto form, shown here, is insoluble; the enolic form is soluble.

11. *Ketoimine Dyes* are illustrated by Auramine. It is used as a basic yellow dye, although it lacks fastness.

Me₂N⟨⟩ ... Me₂N⟨⟩ ... Me₂N⟨⟩ Cl — Auramine. intensely colored. C.I.655

C:O + H₂NH, ZnCl₂ → C:N·H + HCl → C·NH₂

Me₂N⟨⟩ , Me₂N⟨⟩ ... Me₂N⟨⟩
colorless

12. *Nitro Dyes*, of which one of the simplest important members is picric acid. Nitro groups are often used in conjunction with other chromophore groups.

OH + H₂SO₄ → OH SO₃H and OH SO₃H + 3 HNO₃ ⇌ O₂N OH NO₂ → O₂N :O NO₂. — Picric acid C.I.7
NO₂ :N(O)OH

13. *Nitroso Dyes*, illustrated by Resorcin Green, which is prepared as follows:

OH OH + 2 HONO → OH NO OH NO → :O :N·OH :O :N·OH, — Resorcin Green, C.I. 1. Resorcin Green is a direct dye for animal fibers.

14. *Oxazine Dyes*, illustrated by Meldola Blue, which is usually sold as the zinc chloride double salt.

OH + ON⟨⟩NMe₂·HCl + AcOH → glacial → [structure] O ⟨⟩ NMe₂ N | Cl — Meldola Blue, C.I.909

15. *Pyrazolone Dyes*, of which Tartrazine, a certified yellow food dye, is an example.

COOH | HCOH + oleum, → (30%) | HCOH HNO₃ | COOH 40° Bé'

COOH | CO + 2 H₂N·N⟨⟩SO₃Na → | CO | COOH

COOH | H C:N·NH⟨⟩SO₃Na | C:N·NH⟨⟩SO₃Na | COOH →

O .. H·N·N:C·C·N⟨⟩SO₃Na | C = N | CO·OH   C.I. 640 SO₃Na   Tartrazine

16. *Quinoline Dyes* may be divided into (a) Cyanines, (b) Isocyanines, (c) Pinacyanols, (d) Dicyanines, (e) Carbocyanines, and (f) Dicarbocyanines. Many of the quinoline dyes have found use as photosensitizers.

a.

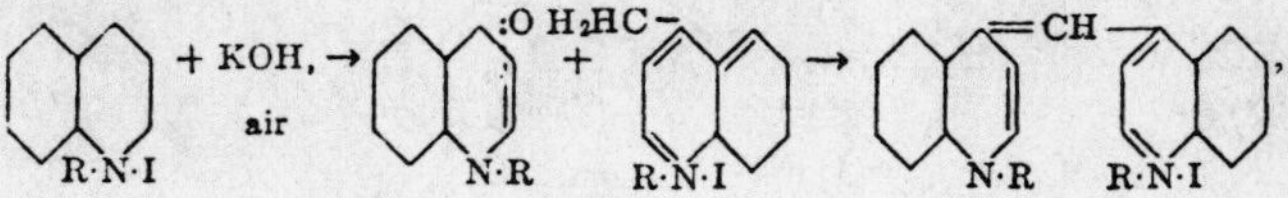

b.

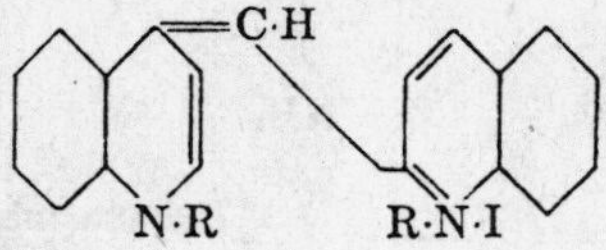

, an Isocyanine. The method of preparation is similar to that of the Cyanines.

c.

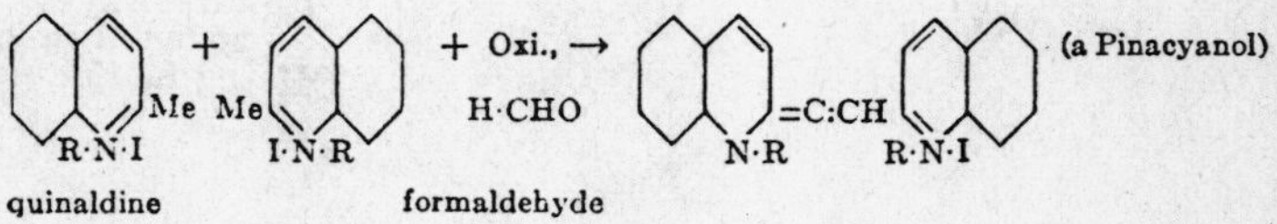

d.

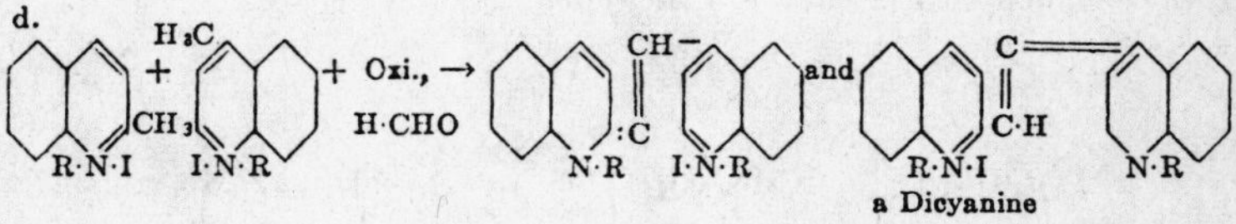

e.

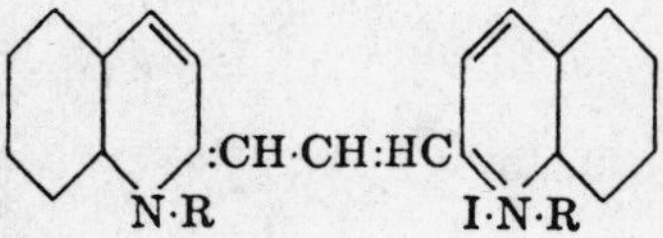

a Carbocyanine. These are also formed when two parts of quinaldine alkyl halide are condensed with formaldehyde.

f. Dicarbocyanines, having the structure indicated at the right, have also been reported.

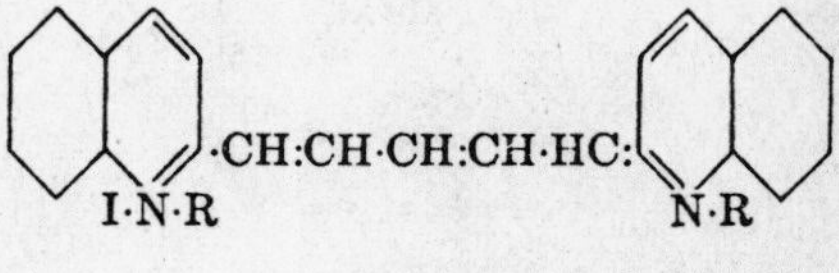

Berberine, the only natural basic dye used technically, occurs in various plants such as the *Berberis vulgaris*. Its structure is shown at the right.

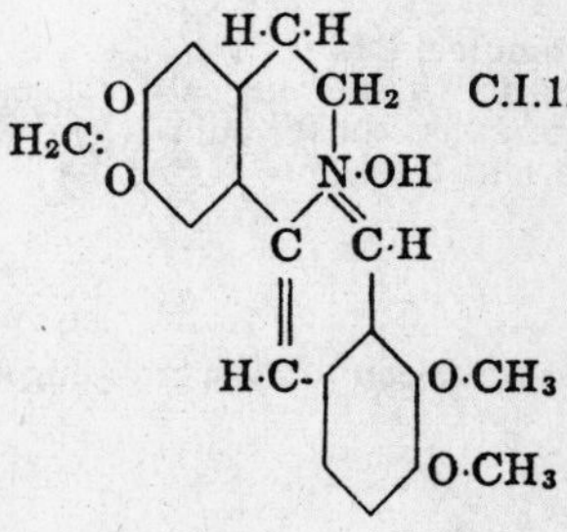

C.I. 1237

17. *Quinonimine Dyes* contain the nuclei $O:C_6H_4:NH$ and $HN:C_6H_4:NH$. The color is intensified by replacing the hydrogen atoms attached to the nitrogen.

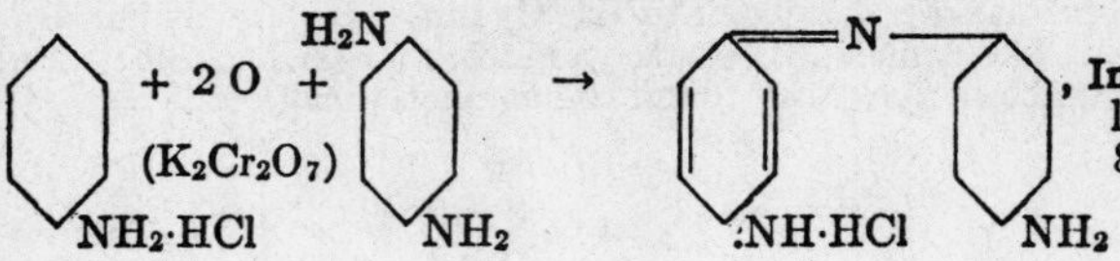

, Indamine (Phenylene Blue) C.I. 819

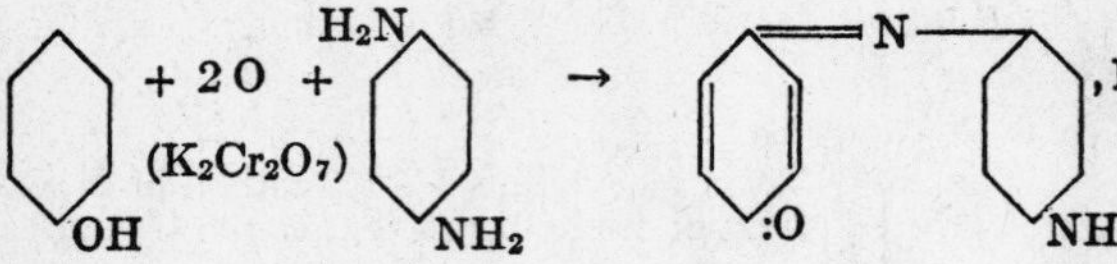

, Indoaniline

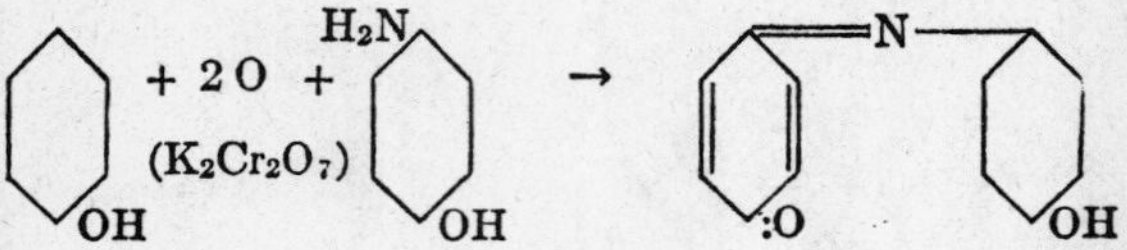

, Indophenol. Because of their lack of fastness, these dyes are not used in dyeing fabrics.

18. *Thiazine Dyes*, represented by Methylene Blue, contain the common nucleus shown at the right.

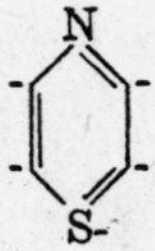

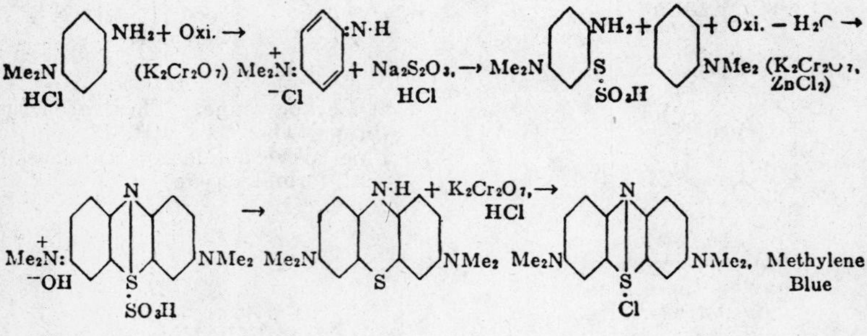

Thionine, Lauth Violet, is used as a biological stain. It has the formula shown at the right.

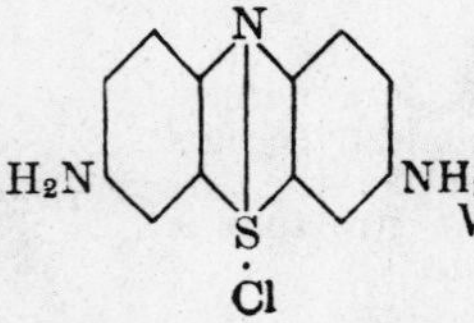

NH₂, Thionine, Lauth Violet. C.I.920.

19. *Thiazoles* contain the grouping shown at the right.

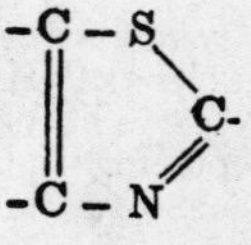

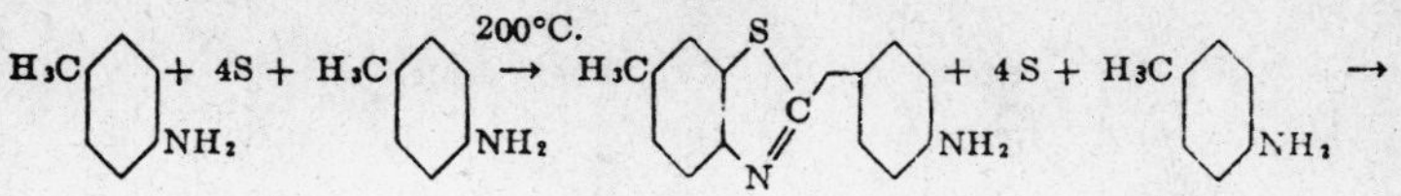

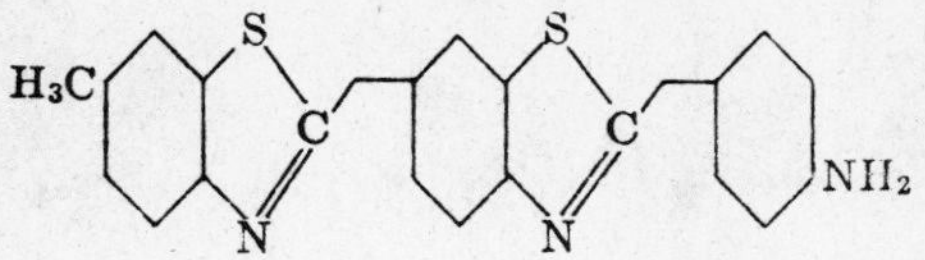

H₃C⟩ + 4 S + H₃C⟩ $\xrightarrow{200°C.}$ H₃C⟩ ... C ⟩ + 4 S + H₃C⟩ →
NH₂   NH₂   N   NH₂   NH₁

Primuline base. This can be diazotized and coupled with amines and phenols. Primuline base itself is not fast to light; so it is used as an intermediate in the preparation of other dyes.

20. *Triphenylmethane Dyes*, of which Pararosaniline is a typical example, are very important commercially. Two moles of aniline are condensed with one mole of *p*-toluidine in the presence of As_2O_5 or $C_6H_5 \cdot NO_2$, then oxidized with PbO_2, and finally converted to the hydrochloride by adding HCl.

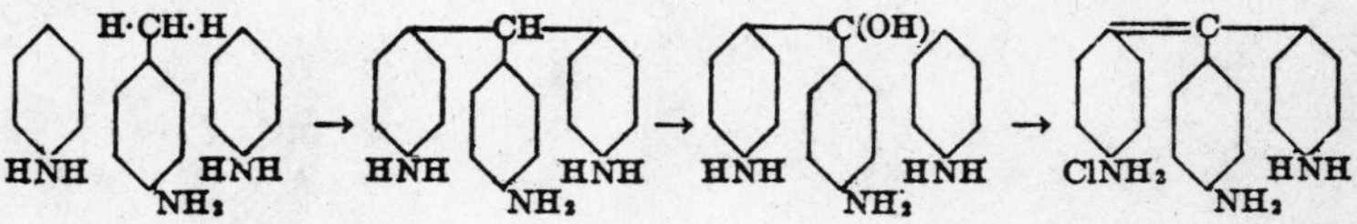

Gentian Violet is a poorly defined mixture of the violet rosanilines. Methyl Green and Methyl Violet are shown at the right. Fuchsine is closely related to Rosaniline.

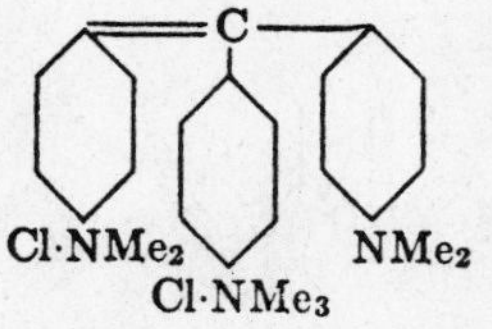

Methyl Green, C.I.684

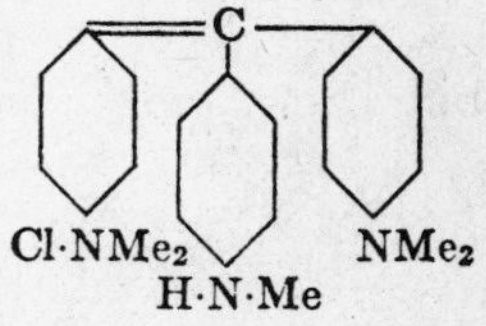

Methyl Violet, C.I.680

Mention should be made of the chromophoric properties of the triarylmethyls. They possess varying colors depending on the degree of dissociation, which in turn depends on the nature of the aryl groups.
$(C_6H_5)_3C \cdot C(C_6H_5)_3$, colorless $\rightleftarrows$ $2(C_6H_5)_3C$, yellow. $(C_6H_5 \cdot C_6H_4)_3C$ deep violet.

21. *Xanthene Dyes* include (a) Fluorones, (b) Phthaleins, (c) Pyronines, (d) Succineines, (e) Rosamines, and (f) Rhodamines. Preparations follow:

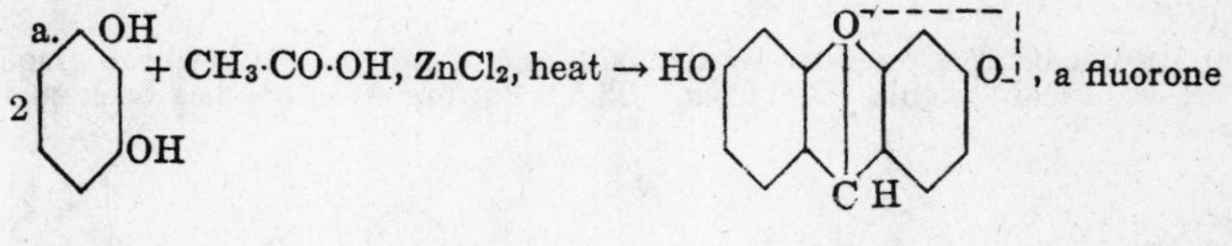

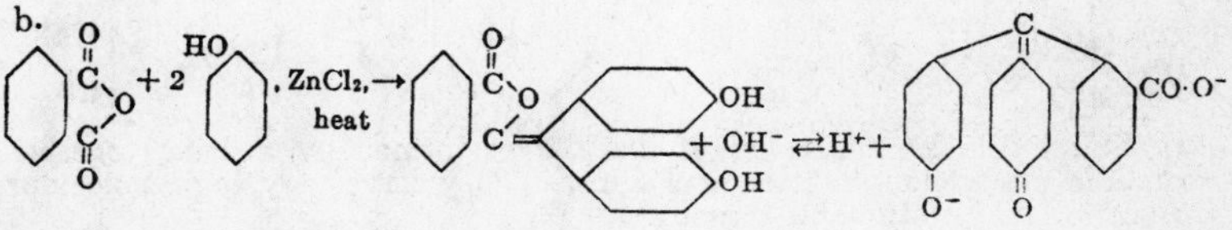

a. OH ⟩ + $CH_3 \cdot CO \cdot OH$, $ZnCl_2$, heat → HO ... O , a fluorone
2 OH

b. ... + 2 ... , $ZnCl_2$, heat → ... + OH⁻ $\rightleftarrows$ H⁺ + ...

c.

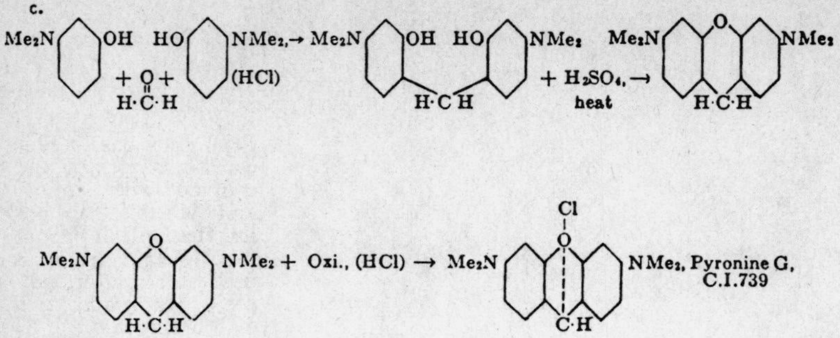

Me₂N⟨⟩OH HO⟨⟩NMe₂,→ Me₂N⟨⟩OH HO⟨⟩NMe₂   Me₂N⟨O⟩NMe₂
   + O +   (HCl)   + H₂SO₄,→
   H·C·H   H·C·H   heat   H·C·H

Me₂N⟨O⟩NMe₂ + Oxi., (HCl) → Me₂N⟨O⟩NMe₂, Pyronine G,
     H·C·H   C·H   C.I.739

d.

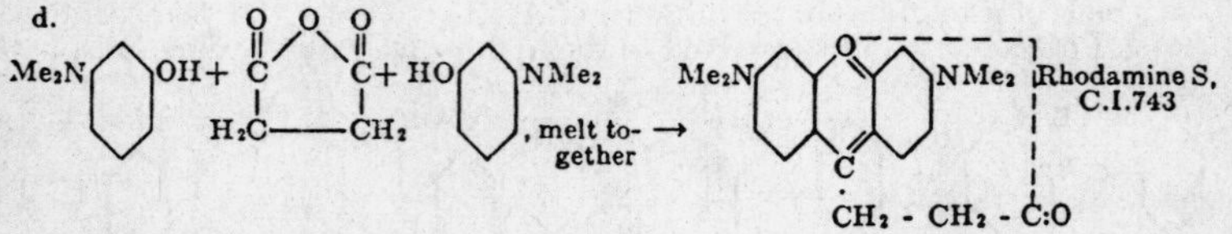

Me₂N⟨⟩OH + O O O   + HO⟨⟩NMe₂   Me₂N⟨O⟩NMe₂ Rhodamine S,
            C   C   C.I.743
           H₂C——CH₂   , melt to- →   C
                       gether
                                    CH₂ - CH₂ - C:O

e.

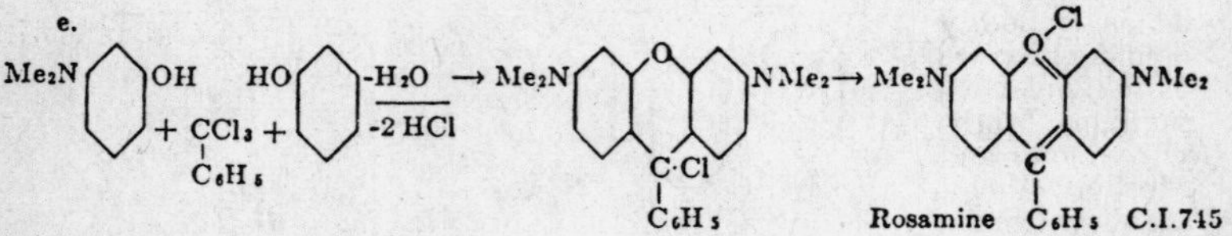

Me₂N⟨⟩OH HO⟨⟩ -H₂O → Me₂N⟨O⟩NMe₂→ Me₂N⟨O⟩NMe₂
   + CCl₃ +   -2HCl   C·Cl   C
   C₆H₅   C₆H₅
                                         Rosamine C₆H₅ C.I.745

f.

⟨⟩ O=C O + 2 HO⟨⟩NEt₂ + ZnCl₂ → Et₂N⟨O⟩NEt₂   Et₂N⟨OCl⟩Et₂N
     C=O   heat   C   C
                                       O
                                       C:O   + HCl → CO·OH
                                                         Rhoda-
                                                         mine B
                                                         C.I.749

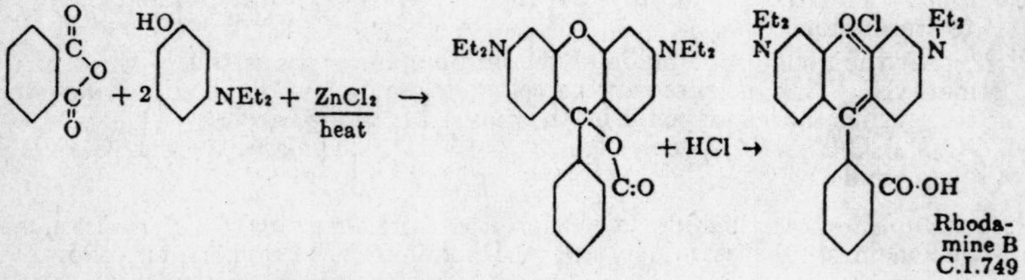

22. Dyes of Unknown Structure.

a. Aniline black is prepared by the oxidation of aniline. It has been given the empirical formula $(C_6H_5N)x$. The following structure has been suggested:

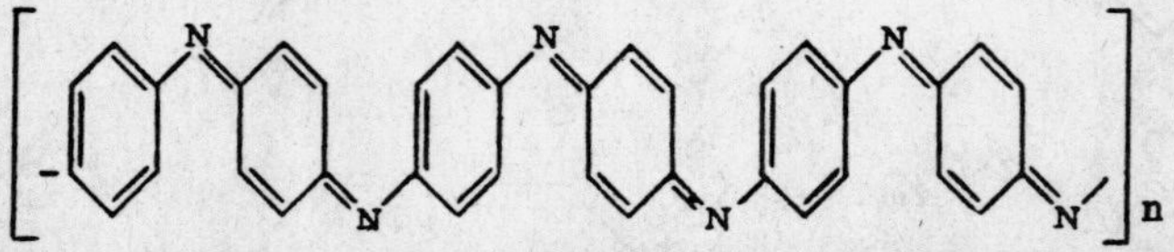

b. .Sulfur Black. A group of black dyes formed by melting together organic nitrogenous matter and sulfur. They have very high molecular weights and questionable structures.

II. Biological Stains

Biological microscopic work at the present time is greatly dependent upon the use of biological stains. These stains consist of many types of dyes which exhibit selective affinities for various cells or portions of cells. The use of dyes in microscopic work was first reported in 1838.

Both physical and chemical theories have been advanced to explain the selective dyeing of tissues. In some cases the dye is attached so firmly to the tissue that it seems to be a chemical union. On the other hand such unions can be explained on a physical basis so that some question as to the true nature still exists. It is entirely possible that in some cases the dyeing could be due to a combination of the two.

The chemical theory states that certain parts of the animal or plant cells, acidic in character, have an affinity for basic dyes. On the other hand, basic parts of the cells possess an affinity for acid dyes. It has also been found that certain tissues not stained by certain dyes are stained if used in conjunction with a mordant. On the other hand, the physical theory states that this is merely due to the precipitation of the dye by the mordant within the tissue.

Whether physical or chemical in nature, the use of dyes in staining tissues has made possible much of the biological research. Practically every type of dye has been used in biological staining. For that reason only a few of the more commonly used stains will be listed here.

1. *Nuclear stains* (acid).
   Thionine (I, 18), Methylene Blue (I, 18), Fuchsine (I, 20), Gentian Violet (I, 20), Carmine (I, 2), and Hematoxylin (I, 9).

2. *Cytoplasm stains* (acid).
   Picric acid (I, 12), Orange G (I, 4), Congo Red (I, 4), Neutral Red (I, 3).

3. *Fat stains.*
   Sudan III (I, 4), Sudan IV (I, 4).

4. *Lignified cell-wall stains.*
   Safranine (I, 3), Gentian Violet (I, 20), Methyl Green (I, 20).

5. *Cellulose wall stains.*
   Acid Fuchsine (I, 20), Eosine Y (I, 21), Hematoxylin (I, 9).

6. *Bacterial stains.*
   Methylene Blue (I, 18), Fuchsine (I, 20), Gentian Violet (I, 20).

   Numbers refer to occurrences under Part I, *Dyes.*

III. Indicators, Neutralization Type

An indicator is an acid or base whose ionizable form is different in constitution and color from that of the unionizable form. Different indicators (dyes) change from their unionized to their ionized form at different pH values. Thus, in the titration of acids and bases, in which the end point occurs at a definite and known pH value, a suitable indicator can be used to denote this end point by means of its color change. Although these color changes accompany changes from ionizable to unionizable forms, they are due to structural changes in the molecule.

Indicators belong to the general class of dyes, hence their discussion will be brief. There are certain standard indicators covering most of the entire pH range which are generally used. Below is a list of Sörensen's indicators along with their pH ranges. These dyes can be found under Part I.

		Color	
Indicator	*pH* range	*acid*	*alkaline*
Methyl Violet (I, 20)	0.1-3.22	yellow	violet
Tropeolin 00 (1, 4)	1.3-3.2	red	yellow
Methyl Yellow (I, 4)	2.9-4.0	red	yellow
Methyl Orange (I, 4)	3.1-4.4	red	orange-yellow
Methyl Red (I, 4)	4.2-6.3	red	yellow
p-Nitrophenol (I, 12)	5.0-7.0	colorless	yellow
Neutral Red (I, 3)	6.8-8.0	red	yellow
Phenolphthalein (I, 21)	8.2-10.0	colorless	red
Alizarin Yellow R (I, 2)	10.1-12.1	yellow	lilac
Tropeolin O (I. 4)	11.0-13.0	yellow	orange-brown

A number of new indicators are now available for special applications by the analytical chemist.

IV. Pigments

Among the more important pigments to be found in nature are the phthalocyanines, the detailed consideration of which is beyond the scope of this Outline. The interested student is referred to The Phthalocyanines, Miles A. Dahlen, Ind. and Eng. Chem., 31, 839 (July, 1939).

Synthetic organic pigments are of considerable interest to the paint industry for the production of special paints.

CHAPTER XXVII

TERPENES AND RUBBER

I. Terpenes

A. Occurrence:

The terpenes may be represented by the general formula C_5H_8 or some multiple thereof. They are sometimes classified as hemiterpenes (C_5H_8), true terpenes ($C_{10}H_{16}$), sesquiterpenes ($C_{15}H_{24}$), and polyterpenes (C_5H_8)n. Examples of the terpenes and some of their oxygen derivatives are:

1. Open-Chain Compounds:

  a. Citrene, $(CH_3)_2C:CH\cdot CH_2\cdot CH_2\cdot (CH_3)C:CH\cdot CH_3$

  b. Geraniol, $(CH_3)_2C:CH\cdot CH_2\cdot CH_2\cdot (CH_3)C:CH\cdot CH_2OH$

  c. Citral or geranial, $(CH_3)_2C:CH\cdot CH_2\cdot CH_2\cdot (CH_3)C:CH\cdot CHO$

2. Monocyclic Compounds:

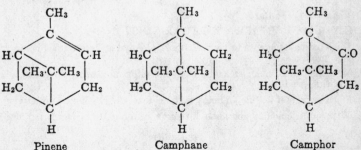

Ionone   Menthane   Menthol

3. Bicyclic Compounds:

Pinene   Camphane   Camphor

The terpenes occur in rubber, citrus fruits, lemon grass oil, and in the products of the cone bearing trees.

The terpenes are used in flavors, in perfumes, in deodorants, in inhalants, in disinfectants, and as solvents.

The terpenes may be considered as polymers of isoprene ($H_2C:C[CH_3]\cdot HC: CH_2$). The molecule may be of the open chain type as in citrene, of the monocyclic type as in menthane, or of the bicyclic type as in camphane, which represent a completely hydrogenated nucleus. Many of the terpenes, however, are unsaturated. Cymene, which may be considered as a substituted benzene, is an example of a monocyclic compound containing three olefin linkages.

Ionone, since it is formed by condensing citral with acetone in the presence of sulfuric acid, may be considered as a derivative of citrene. Still another somewhat general type of ring structure found in the terpenes is represented by dipentene, which may also be considered as a condensation product of citrene.

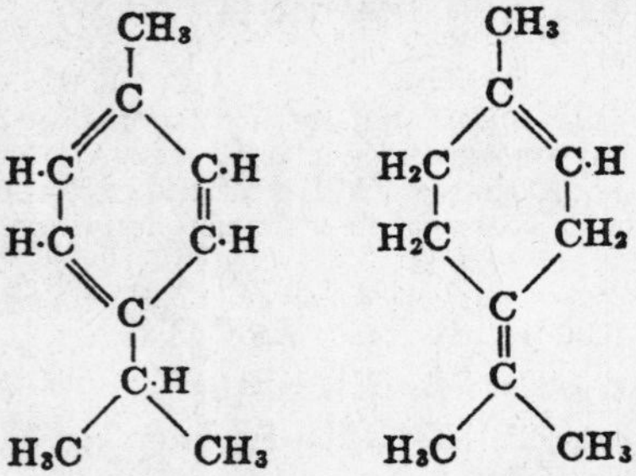

Cymene   Dipentene

B. Nomenclature:

Common names have been assigned to most of the members of the terpene series. They may be named as members of the alkenes or the alkadienes or their cyclic derivatives.

C. Preparation:

The terpenes are prepared from natural sources by steam or fractional distillation of the crude materials. Some of the terpenes have been prepared synthetically.

D. Physical Properties:

The terpenes are oily liquids or solids with low melting points. They are insoluble in water, but are soluble in alcohol, ether, and benzene. Most of them have rather pleasant odors.

E. Chemical Properties:

THEY REACT UNDER PROPER CONDITIONS:

1. With oxidizing agents to yield alcohols, aldehydes, ketones, acids, or cleavage products.

2. With halogens, halogen acids, oxygen, sulfur, and other reagents to give addition products.

3. With oxygen of the air, in many instances, to yield resinified materials.

4. With sulfuric acid, when heated, to give, in many cases, isomerization.

II. Rubber

A. Introduction:

Rubber is one of the most important naturally occurring organic chemical materials. The annual world consumption of crude rubber amounts to approximately 1,700,000 tons, and hundreds of articles essential to our modern civilization are manufactured from it.

Although rubber was probably used as early as the eleventh century, the organic chemist has not succeeded in synthesizing the natural product, and the world is still dependent on nature for its supply of this important commodity.

The term rubber was coined by Joseph Priestley who discovered that this material would *rub out* or erase pencil marks. The first technical use of rubber dates back to 1761 when rubber tubing was first produced. Following the discovery of vulcanization by Charles Goodyear in 1839, the rapid growth of the rubber industry led to the development of plantation rubber in 1876. The most important use for rubber, however, was inaugurated by the invention of the internal combustion engine and the advent of the automobile in the latter part of the nineteenth century, for about 80% of all the rubber that is fabricated makes its appearance in the form of tires and tubes.

Rubber (india rubber, or caoutchouc) is the term applied to the elastic solid obtained from the milky exudation of certain tropical trees. If it is produced from natural, uncultivated trees, shrubs, vines, or plants, the product is known as wild rubber. Nearly all of the rubber of commerce is derived from the large plantations of British Malaya, the Netherland Indies, Ceylon, British Borneo, French Indo-China, Siam, India, Burma, and Liberia. Although there are many species of latex-yielding plants, most of the plantation rubber comes from the *Hevea brasiliensis* (or hevea). Several other varieties of rubber, however, find a limited use in industry. Gutta-percha is obtained from the latex of certain species of rubber trees found chiefly in Borneo and Sumatra. Balata is obtained from the latex of a species which is found in Trinidad, the Guianas, and Venezuela. Gutta-percha and balata have a higher resin content, are tougher, more thermoplastic, and less elastic than hevea rubber. Guayule is obtained by extraction of a shrub which is found in Mexico and Texas and is cultivated in California. It has a high resin content and is softer than either gutta-percha, balata, or hevea rubber.

About 85% of all the rubber produced is for transportation purposes. Mechanical rubber goods such as engine supports, hose, belting, gaskets, and wringer rolls require considerable quantities, as also does the manufacture of footwear such as boots, shoes, heels, and soles. Other uses include sporting goods, toys, novelties, bathing apparel, sponge rubber, flooring, druggist's sundries, surgical goods, and wire insulation. Hard rubber is used in switchboards, table tops, and battery boxes. Limited quantities of rubber are used in the form of the hydrochloride for transparent wrapping paper and in the form of chlorinated rubber in corrosion-resisting paints.

Careful analytical work on very pure rubber indicates that the empirical formula for rubber is C_5H_8. The size of the rubber molecule and its configuration is not definitely known. Values for the molecular weight range from 5,000 to 200,000. The reactions of the rubber hydrocarbon show that one unsaturated linkage is present in each C_5H_8 unit. The decomposition of the ozonide of rubber to yield mostly levulinic aldehyde and levulinic acid indicates that the rubber molecule is formed by the 1, 4-polymerization of the isoprene molecule to give:

$$(-CH_2 \cdot (CH_3)C{:}CH \cdot CH_2 \cdot CH_2 \cdot (CH_3)C{:}CH \cdot CH_2 \cdot CH_2 \cdot (CH_3)C{:}CH \cdot CH_2 \cdot CH_2 \cdot (CH_3)C{:}CH \cdot CH_2 \text{-})_x$$

Even very pure rubber does not appear to be a homogenous substance. By extraction with ether or diffusion into petroleum ether, pure rubber can be separated into two modifications which are known as alpha or sol-rubber and beta or gel-rubber. X-Ray diagrams of unstretched rubber at ordinary temperatures show a broad, diffuse ring characteristic of an amorphous structure, but when rubber is stretched above 75% elongation, interference fringes appear in the X-Ray diagram which indicate a crystalline structure. These interference fringes disappear when the rubber is heated above 60°C. or when exposed to the vapors of a solvent such as benzene.

The white milky fluid (which is not the sap of the tree) as it exudes from the tree is known as latex and contains about 35% of the rubber hydrocarbon $(C_5H_8)_x$, together with varying amounts of proteins, resins, sugars, and mineral salts. The rubber particles are present in latex as a colloidal suspension, carry a negative charge, and may be coagulated by neutralization. When a current is passed through latex, the rubber particles are deposited on the anode. At the plantations, acetic or formic acid is commonly used to effect coagulation of the latex, or, if it is desirable to prevent coagulation, ammonium hydroxide may be added. Latex can be concentrated, by evaporation or by centrifuging, to a rubber content of about 60%.

B. Synthetic Rubber:

Isoprene [$H_2C:C (CH_3)\cdot HC:CH_2$] is one of the products resulting from the destructive distillation of rubber. Many attempts have been made, therefore, to synthesize rubber from isoprene. In the presence of light, acids, air, or other catalysts, isoprene does polymerize to give rubber-like products. These synthetic materials differ in many respects from natural rubber, and are usually designated as artificial rubber.

During World War I, Germany produced artificial rubber on a commercial scale from dimethylbutadiene. Depending on the method of polymerization, two products were obtained. *Methyl rubber H* was prepared by storing dimethylbutadiene in thin walled metal drums for six to ten weeks at about 30°C. in the presence of air, whereas *Methyl rubber W* was obtained by heating dimethylbutadiene under pressure for three to six months at about 70°C. Both of these products were poor substitutes for natural rubber, and their production was discontinued after the termination of the war.

In recent years there has been a marked interest in synthetic rubber in the United States. The Buna N and Buna S types of synthetic rubber, first produced in Germany, are now manufactured commercially in this country. During World War II the production of synthetic rubber in the United States was greatly increased. GR-S and cold rubber, copolymers of butadiene and styrene, are manufactured in largest tonnage as general purpose rubbers. The production of Neoprene (GR-M), Butyl rubber (GR-I), and a copolymer of butadiene and acrylonitrile, GR-A, similar to Buna N has been greatly expanded.

Butadiene, the principal raw material for the synthesis of GR-S or Buna S is obtained in Europe from alcohol or acetylene but in the United States it is produced either by cracking petroleum or by the dehydrogenation of the butane fraction of natural gas or petroleum. The accompanying chart lists the materials used in the production of the more important types of synthetic rubber together with the structural units of the polymers obtained.

TYPES OF SYNTHETIC RUBBERS

Formula and Name of Monomer	Formula and Name of Polymer
1. $H_2C:CH\cdot HC:CH_2$, 1, 3-butadiene..	($\cdot CH_2\cdot HC:CH\cdot CH_2\cdot HC:CH\cdot CH_2$)$_n$, Buna 85, Buna 115, SKA, SKB, Ker.
2. $H_2C:C$——$C:CH_2$, 2,3-dimethyl-1, 3-butadiene.............. with CH₃ CH₃ groups	($\cdot CH_2\cdot C$==$C\cdot CH_2\cdot CH_2\cdot C$==$C\cdot CH_2$)$_n$, methyl rubber H, methyl rubber W, with CH₃ CH₃ CH₃ CH₃ groups
3. $H_2C:C(Cl)\cdot HC:CH_2$, chloroprene..	($\cdot CH_2\cdot C(Cl):CH\cdot CH_2\cdot CH_2\cdot C(Cl):CH\cdot CH_2$)$_n$, Neoprene GN, GR-M, Sovprene.
4. $C:CH_2$, isobutylene............. with CH₃ / CH₃ groups	($\cdot C\cdot CH_2\cdot C\cdot CH_2\cdot C\cdot CH_2\cdot C\cdot CH_2$)$_n$, Vistanex, Oppanol. with CH₃ CH₃ CH₃ CH₃ groups
5. $H_2C:CH.Cl$, vinyl chloride.......	($\cdot CH_2\cdot CH(Cl)\cdot CH_2\cdot CH(Cl)\cdot CH_2\cdot CH(Cl)$)$_n$, Koroseal, Flamenol.

6. Cl·CH$_2$·CH$_2$·Cl, ethylene
   dichloride, with Na$_x$S$_y$ .... (·CH$_2$·CH$_2$·S·S·CH$_2$·CH$_2$·S·S·CH$_2$·CH$_2$·S·S)$_n$, Thiokol A
   Ethanite.

7. Cl·CH$_2$·CH$_2$·O·CH$_2$·CH$_2$·Cl, dichlo-
   ethylether with Na$_x$S$_y$ .... (·CH$_2$·CH$_2$·O·CH$_2$·CH$_2$·S·S·CH$_2$·CH$_2$·O·CH$_2$·CH$_2$·S·S)$_n$,
   Thiokol B, Perduren G.

8. H$_2$C:CH·C·N, acrylonitrile with
   butadiene, about 1/3 .... (·CH$_2$·HC:CH·CH$_2$·CH$_2$·CH·(CH$_2$·HC:CH·CH$_2$)x)$_n$,
   GR-A, Buna N, Perbunan, Perbunan Extra, where
   x is about 3.

9. C$_6$H$_5$·HC:CH$_2$, styrene with buta-
   diene, about 1/3 .... (·CH$_2$·HC:CH·CH$_2$·CH$_2$·CH·(CH$_2$·HC:CH·CH$_2$)x)$_n$,
   GR-S, Buna S, Buna S S.

Neoprene is prepared by the polymerization of chloroprene (H$_2$C:CCl: HC:CH$_2$, which is obtained in turn by the addition of hydrogen chloride to vinylacetylene (H·C:C·HC:CH$_2$). The vinylacetylene is produced by a controlled polymerization of acetylene (cf. p. 34). Neoprene is very resistant to abrasion and to the swelling action produced by gasoline, and lubricating and vegetable oils, on natural rubber.

Sodium polysulfide is heated with (1) ethylene dichloride to give Thiokol A or (2) dichloroethylether to give Thiokol B. The thiokols are extremely resistant to the swelling action of gasoline and lubricating oils and show good resistance to deterioration by aromatic solvents.

Vistanex is a rubbery material obtained by the polymerization of isobutylene in the presence of boron trifluoride. The polymer unlike rubber, is completely saturated and cannot, therefore, be vulcanized. Vistanex is not resistant to petroleum products, but is extremely inert to the action of acids, alkalies, and ozone.

Koroseal is, essentially, a polymer of vinyl chloride plasticized with tricresylphosphate. Koroseal cannot be vulcanized because, like Vistanex, it is completely saturated. It is very resistant to the action of acids and petroleum products.

Perbunam (or Buna N) is a copolymer of butadiene with acrylonitrile in an approximate ratio of three to one. The polymerization is effected in emulsified form in the presence of a catalyst. If the ratio of acrylonitrile is increased, the product is known as Perbunan Extra. Perbunan and Perbunan Extra are both remarkably resistant to the swelling action of gasoline, lubricating oils, and water, and are used extensively for the manufacture of mechanical rubber goods that require resistance to oil.

The co-polymerization of butadiene with styrene in a ratio of about three to one gives the product known as GR-S (Buna S in Germany). The so-called emulsion polymerization is employed in this synthesis. The GR-A (Buna N) is far more resistant to the swelling action of oils than is GR-S (Buna S rubber). For making tires, GR-S is used in this country and Buna S is used in Germany.

Butyl rubber is a copolymer of isobutylene with a small amount of diolefin such as isoprene. Butyl rubber is designated GR-I by the government.

C. Preparation and Compounding of Rubber:

The hevea tree is tapped by removing a thin shaving of bark with a sharp knife to permit the latex to ooze out into small conical shaped cups. The latex is taken to the estate factory, screened to remove bark and dirt, and the rubber coagulated in troughs by the addition of acetic or formic acid. The coagulum rises to the top as a wet, white, doughy mass, which is passed between a number of pairs of rollers to remove the serum. The rubber emerges from the last rollers in the form of ribbed sheets about a quarter of an inch thick, which are washed, dried, and smoked to render the rubber aseptic. These ribbed smoked sheets supply most of the demands of commerce for raw rubber, although crepe rubber finds a limited use. The crepe rubber is prepared by a similar process except that it is neither ribbed nor smoked.

In order to improve the mechanical properties of rubber for industrial use, it is thoroughly mixed with certain compounding ingredients before being subjected to vulcanization. Sulfur, accelerators, softeners, reinforcing agents, and antioxidants are usually added. In certain cases, coloring matter and deodorants are used to improve the attractiveness of the finished product.

The sulfur incorporated into the rubber, when subjected to heat, effects vulcanization. Accelerators greatly decrease the time of vulcanization and improve the quality of the product.

Basic inorganic materials such as calcium oxide, magnesium oxide, and litharge act as accelerators, but some of the following organic compounds are most commonly used: thiocarbanilide, diphenylguanidine, di-o-tolylguanidine, 2-mercaptobenzothiazole, tetramethylthiuram mono- and disulfide, piperidinium pentamethylene dithiocarbamate, zinc or lead dimethyldithiocarbamate, zinc butyl xanthate, and the products obtained by the action of aniline with formaldehyde, acetaldehyde, crotonaldehyde, butyric aldehyde, or heptaldehyde.

The softeners aid in the plasticization of the crude rubber and in the dispersion of inorganic fillers. Some of the commonly used softeners are stearic acid, pine tar, mineral oil, paraffin, vaseline, rosin oil, and asphaltic materials.

Reinforcing agents such as carbon black, zinc oxide, magnesium carbonate, and certain clays are used to increase stiffness.

Antioxidants increase the resistance of rubber to deterioration by heat, light, oxidation, and flexing. The most effective antioxidants are secondary aromatic amines and aromatic hydroxy compounds. Commonly used antioxidants are:

Phenyl-α-naphthylamine, phenyl-β-naphthylamine, N,N'-diphenylethylenediamine, o- and p-ditolylamine, β-dinaphthyl-p-phenylenediamine, 2,4-diaminodiphenylamine, p, p'-dimethoxydiphenylamine, diphenylnitrosoamine, hydroquinone, p-aminophenol, p, p'-diaminodiphenylmethane, p-hydroxybiphenyl, p-hydroxy-N-phenylmorpholine, and 2, 2, 4-trimethyl-1, 2-dihydroquinoline. Also, the condensation products of aniline with acetaldehyde or acetone, diphenylamine with acetone, or α-naphthylamine with aldol are sometimes used. About 1% by weight of the antioxidant is commonly used.

Large quantities of rubber are reclaimed. By the alkali process, which is most commonly used, the ground scrap is digested with a 4 to 6% solution of sodium hydroxide, in the presence of a softener such as asphalt or coal tar naphtha, at about 190°C. Fabric, saponifiable matter, soluble minerals, and sulfur are dissolved, whereas the combined sulfur and certain pigments remain with the reclaimed rubber which is replasticized.

D. Physical Properties:

Commercial crude rubber is a tough, elastic solid, sp.gr. 0.915. Cooled below 10°C., rubber becomes stiff and hard but when warmed up it again becomes soft and flexible. Above 60°C. it becomes soft and plastic, whereas at the temperature of liquid air it is very hard and extremely brittle. The distension of rubber is accompanied by evolution of heat and the contraction by the absorption of heat (Joule effect). Raw rubber exhibits a high permanent set. Mechanical action such as mastication (milling) causes rubber to become very plastic.

Rubber is insoluble in water, alcohol, and acetone, but dissolves or swells in gasoline, benzene, chloroform, carbon tetrachloride, carbon disulfide, or turpentine. It is slightly soluble in ether.

E. Chemical Properties:

THE RECURRING STRUCTURAL UNIT IN RUBBER IS $-CH_2\cdot(CH_3)$ *$C{:}CH\cdot CH_2$-CONTAINING ONE DOUBLE BOND, HENCE RUBBER REACTS UNDER PROPER CONDITIONS TO GIVE;*

1. *Addition products with (a) hydrogen in the presence of a catalyst, (b) halogens, (c) halogen acids, (d) ozone, (e) oxides of nitrogen, and (f) tetranitromethane:*

   a. $(C_5H_8)_x + x\ H_2$ (catalyst) $\rightarrow (C_5H_{10})_x$ (hydrorubber)

   b. $(C_5H_8)_x + x\ X_2 \rightarrow (C_5H_8X_2)_x$ (rubber dihalide)

      Commercial chlorinated rubber known as Tornesite is obtained by the action of chlorine on a solution of rubber in carbon tetrachloride at or above 80°C. The product contains about 65% chlorine and corresponds to the formula $(C_{10}H_{11}Cl_7)_x$. Tornesite is used in corrosion-resistant paint.

   c. $(C_5H_8)_x + x\ HX \rightarrow (C_5H_8\cdot HX)_x$ (rubber hydrohalide)

      Rubber hydrochloride containing about 30% chlorine is used commercially in the form of thin transparent sheets as a wrapping material.

   d. $(C_5H_8)_x + x\ O_3 \rightarrow (C_5H_8O_3)_x$ (rubber ozonide)

      The rubber ozonide, on decomposition with water, yields levulinic aldehyde, levulinic acid, and hydrogen peroxide.

   e. $(C_5H_8)_x + N_2O_3$ or $N_2O_4 \rightarrow$ nitrosites and nitrosates of unknown structure.

   f. $(-CH_2\cdot(CH_3)C{:}CH\cdot CH_2-)_x + x\ C(NO_2)_4 \rightarrow (-CH_2\cdot(CH_3)C\text{-}CH\cdot CH_2-)_x$
   $$\underset{ON\cdot O \quad C(NO_2)_3}{\mid \qquad\quad \mid}$$

2. *Oxidation products when treated with such reagents as oxygen, hydrogen peroxide, ozone, potassium permanganate, nitric acid, or perbenzoic acid:*

   Nitric acid yields a product with the formula $(C_5H_7NO_2)_x$, whereas perbenzoic acid gives $(C_5H_7O)_x$.

3. *Vulcanized rubber when treated with sulfur, selenium, tellurium, or sulfur monochloride:*

   a. $(C_5H_8)_x + S$, heat $\rightarrow$ soft vulcanized rubber when not more than 5% of combined sulfur is present.

   b. $(C_5H_8)_x + x\ S$, heat $\rightarrow (C_5H_8S)_x$, ebonite or hard rubber when about 32% of combined sulfur is present.

   c. Such compounds as dinitrobenzene, trinitrobenzene, organic peroxides, and persulfates may also be used to effect vulcanization.

   $2\ (C_5H_8)_x + x\ S_2Cl_2 \rightarrow 2\ (C_5H_8\cdot SCl)_x$ (cold vulcanization)

4. *Cyclo-rubber when treated with such reagents as sulfuric acid, sulfonic acids p-toluenesulfonyl chloride, stannic chloride, titanium chloride, antimony chloride, or ferric chloride:*

   The product obtained by the action of organic sulfonic acids on rubber is known as Thermoprene and is used for lining tanks. The product obtained by the action of $SnCl_4$ and HCl on rubber is known as Pliolite and is used in corrosion-resisting paints.

5. *Addition complexes when treated with chromyl chloride, selenium oxychloride, thioglycolic acid, thiocyanogen, or benzoyl chloride:*

   $C_{10}H_{16}\cdot 2\ CrO_2Cl_2$; $C_5H_8\cdot HS\cdot CH_2\cdot CO\cdot OH$; $C_5H_8(SCN)_2$

6. *Nitrones of isorubber when treated with aromatic nitroso compounds such as nitrosobenzene, nitrosotoluene, or ethyl-o-nitrosobenzoate. p-*Nitrosophenol does not react.

   This material on rubber was submitted by Dr. R. F. Dunbrook, Firestone Tire and Rubber Co., Akron, Ohio.

CHAPTER XXVIII

Synthetic Resins

In the words of Carleton Ellis,[1] "a resin may be defined as a solid or semi-solid, complex, amorphous mixture of organic substances, having no definite melting point and showing no tendency to crystallize." Synthetic resins, according to Ellis, are limited to those prepared by synthesis from non-resinous organic raw materials. Such a restriction does not permit the inclusion of ester gum and certain other derivatives of natural products, although many investigators do consider them as synthetic resins. Likewise, synthetic plastics such as nitrocellulose and cellulose acetate will not be considered as synthetic resins.

The two major fields for which synthetic resins have been developed are those of (a) molded and the associated laminated products, and (b) protective coatings such as paints, varnishes, and lacquers. The success of these developments is evidenced by the fact that the estimated 2 million pound production of synthetic resins in 1920 had increased fully fifty fold by the close of 1935. Moreover, the production during the latter year was over two times that of 1933.[2]

The two types of reaction of primary importance in connection with the synthesis of resins are *polymerization* and *condensation*. The former may be divided into (a) *additive* or *homopolymerization* (additive combination of the monomer), (b) *copolymerization* (combination of two or more different monomers, each capable of polymerizing alone), and (c) *heteropolymerization* (an additive copolymerization involving one substance which is unsaturated, but which does not readily polymerize alone). In condensation reactions, as contrasted with simple polymerizations, there is a loss of such substances as water or alcohol. In such reactions, consequently, the final product is not an exact multiple of the reacting ingredients. As shown by subsequent examples, condensation is involved in the preparation of many of the most important synthetic resins.

Phenol-Aldehyde Resins

Although Baeyer pointed out in 1872 that the reaction between phenols and aldehydes is a general one, nearly forty years elapsed before Baekeland demonstrated that the reaction could be put to practical use. This stimulated the commercial development of phenol-formaldehyde resins, and about three-fourths of all the synthetic resins now being produced are of the phenolic or modified phenolic type.

The phenol-formaldehyde reaction was considered by Baekeland to be a condensation and polymerization which takes place in three stages. According to his terminology, which is now generally accepted, the resinous products successively formed are the initial condensation product A, the intermediate product B, and the final condensation product C (Bakelite C). "A," which is soluble in many solvents, may exist in liquid, paste, or solid form. The proper heating of "A," in the presence of a suitable catalyst, will convert it into "B" and finally into "C." "B" is an insoluble solid, which merely swells in the presence of alcohol and similar solvents, but it can be melted and molded. Product "C," however, is both infusible and insoluble. In the

270

finished products, therefore, the resin is usually in the latter stage of the reaction.

Although the nature of these resins has made the experimental study of their structure very difficult, the results of certain efforts along this line are of interest. The simplest reaction product of phenol and formaldehyde is a phenol alcohol, such as saligenin. Heat will convert phenol alcohols into resins. Megson has shown the presence of dihydroxydiphenylmethane among the early reaction products when an excess of phenol is treated with formaldehyde:

$$2 \ \langle\!\!\!\bigcirc\!\!\!\rangle\, OH + H \cdot CHO \rightarrow \overset{OH}{\langle\!\!\!\bigcirc\!\!\!\rangle} - \overset{H}{\underset{H}{\overset{.}{\underset{.}{C}}}} - \langle\!\!\!\bigcirc\!\!\!\rangle\, OH + H_2O$$

The formation of the simplest branched chain has been represented by Megson as follows:[3]

$$\overset{OH}{\langle\!\!\!\bigcirc\!\!\!\rangle} + 3\,H \cdot CHO \rightarrow HOH_2C\overset{OH}{\underset{-CH_2OH}{\langle\!\!\!\bigcirc\!\!\!\rangle}}CH_2OH + 3\,C_6H_5OH \rightarrow HO\langle\!\!\!\bigcirc\!\!\!\rangle CH_2 \overset{OH}{\langle\!\!\!\bigcirc\!\!\!\rangle}\underset{\underset{\langle\!\!\!\bigcirc\!\!\!\rangle-OH}{CH_2}}{CH_2}\langle\!\!\!\bigcirc\!\!\!\rangle OH$$

Cross linkages between chains is probably necessary for the production of the final insoluble and infusible products. It is not difficult to conceive of extended molecules being built up from phenol units with connecting methylene or oxygen bridges, and of insoluble and infusible products resulting from further polymeric change.

An important development is the extension of the use of phenol-aldehyde resins from the field of molded and cast articles to that of paints, varnishes, and lacquers. The simpler phenol-aldehyde resins, which are suitable for the manufacture of molded products, are not sufficiently soluble in drying oils to be of value in the production of varnishes. One of the two general methods for obtaining the desired solubility is that developed by Behrend who fused the phenol resin with certain other products such as the esterification products of glycerol and natural rosin (ester gum). This procedure gives the so-called oil-soluble modified phenol resins which should be distinguished from the oil-soluble 100% phenol resins. In the preparation of the latter group, substituted phenols such as p-tertiary butylphenol, p-phenylphenol, and octylphenol are used. The aryl- and alkyl-substituted phenols yield resins which are markedly more soluble in drying oils than are those prepared from the simpler phenols.

Alkyd Resins

Second only to the phenol resins in the matter of production are the alkyd (or Glyptal) resins. Although they have a limited use in the production of molded products, the alkyd resins are a major center of interest to the manufacturers of paint and varnish.[4]

Alkyd resins are formed by the esterification of polyacids such as phthalic, maleic, and succinic acids with polyhydroxy alcohols. In such esterifications, simple monomeric esters being the exception, the usual products are resinous in nature and of high molecular weight. In the simple case of glycol and phthalic anhydride, the first reaction is probably of an additive nature:

$$\underset{\substack{O \\ \vdots \\ C \cdot \\ \cdot C \\ \vdots \\ O}}{\bigcirc} O + HO \cdot CH_2 \cdot CH_2 \cdot OH \rightarrow \underset{\substack{CO \cdot OH \\ CO \cdot O \cdot CH_2 \cdot CH_2 \cdot OH}}{\bigcirc}$$

Since the initial reaction product is both an acid and an alcohol, further reaction is to be expected. Upon continued heating, a long-chain molecule of the following type probably forms:

$$HO \cdot OC \cdot C_6H_4 \cdot CO(O \cdot CH_2 \cdot CH_2 \cdot O \cdot OC \cdot C_6H_4 \cdot CO)nO \cdot CH_2 \cdot CH_2 \cdot OH$$

Such linear polymers, which cannot form cross-linkages with one another, are not convertible to an infusible and insoluble form by application of heat.

The reaction which constitutes the basis for the major portion of the commercial alkyd resins is that between glycerol and phthalic anhydride. Although a monoglyceride, corresponding to the initial products shown above in the case of glycol, undoubtedly forms first, the dibasic diglyceride is also very readily produced:[5]

$$2 \underset{\substack{O \\ \vdots \\ C \cdot \\ \cdot C \\ \vdots \\ O}}{\bigcirc} O + HO \cdot CH_2 \cdot CHOH \cdot CH_2 \cdot OH \rightarrow \underset{\substack{CO \cdot OH \\ CO \cdot O \cdot CH_2 \cdot CHOH \cdot CH_2 \cdot O \cdot OC}}{\bigcirc} \underset{HO \cdot OC}{\bigcirc}$$

The continued addition of phthalic anhydride and glycerol to this product finally leads to the formation of long chains. It is observed that the secondary hydroxyl group of the glycerol molecule has not entered into the reaction, but eventually it may react with a carboxyl group in an adjacent chain and thus lead to the formation of branched chains. Resins of this type are convertible by heat into an infusible and insoluble form.

Modification of the properties of the pure alkyd resins by the addition of certain other substances, notably drying oils, greatly increase their usefulness in varnishes and lacquers. Since the drying oils are triglycerides of unsaturated monoacids such as linoleic acid, and the common alkyd resins are glycerides of diacids, it is easy to conceive of the same glycerol group being attached to the radical of an unsaturated acid and to that of phthalic acid. By suitable treatment, the drying oil fatty acids can thus be introduced directly into the resin molecule and impart to the latter the property of drying in air. Such modified alkyd resins are known as the oxygen-convertible type; that is, they are converted to a hard, insoluble form upon exposure to air at ordinary temperatures. Since they may be varied considerably in composition and physical properties without sacrificing their good resistance to weathering, these modified alkyd resins are finding wide application in paints and varnishes.

Urea-Formaldehyde Resins

Among the synthetic resins used for molded products, urea-formaldehyde resins rank next to the phenolic resins in importance. Since the issuance of the first patent on urea-formaldehyde resins to John in 1920, progress in this field has been rapid. During the early stages of their development, urea resins were thought to be suitable for the production of an "organic glass," which would be tougher and less fragile than ordinary glass. Although of satisfactory initial transparency, this "urea glass" was found to suffer from the defects of (1) softness and low resistance to abrasion, (2) inadequate water-resistance, and (3) poor aging resistance as a continuation of condensation in the molded product results in strains, cracks, and other defects. The poor aging resistance is overcome in the case of the molded products by the incorporation of substantial amounts of cellulose (wood flour). The resulting products are not transparent, but they have good strength and translucency and can be used for the manufacture of articles that require a very light color. Urea-formaldehyde resins are used considerably for impregnating papers for various purposes such as the manufacture of laminated molded products. Another important use is for crease-proofing cotton, linen, and rayon fabrics. Their utilization by the varnish and lacquer industries has been somewhat limited although the process, first disclosed by Lauter, of effecting the condensation in an organic solvent may lead to important developments in these fields.

Urea-formaldehyde resins are normally prepared by condensing urea with formaldehyde in an aqueous medium. The simplest reaction product is hydroxymethylurea (methylolurea) which results from the following reaction: $H_2N \cdot CO \cdot NH_2 + H \cdot CHO \rightarrow H_2N \cdot CO \cdot NH \cdot CH_2 \cdot OH$. Other simple reaction products, which have actually been isolated, include s-bis(hydroxymethyl)-urea ($HO \cdot CH_2 \cdot HN \cdot CO \cdot NH \cdot CH_2 \cdot OH$, dimethylolurea) and methyleneurea ($H_2N \cdot CO \cdot N:CH_2$ or its tautomeric form $O:C[NH]_2CH_2$). Further condensation of mono- or dimethylolurea, in the presence of a suitable catalyst, leads, according to Walter, Gewing, and others,[6] to the formation of a ring structure of the general type shown here. It is highly improbable, however,

$$
\begin{array}{cccc}
H \cdot N \cdot CH_2 \cdot N \cdot CH_2 \cdot N \cdot CH_2 \cdot N \cdot CH_2 \cdot OH \\
| \quad\quad | \quad\quad | \quad\quad | \\
O:C \quad\quad C:O \quad\quad C:O \quad\quad C:O \\
| \quad\quad | \quad\quad | \quad\quad | \\
H \cdot N \cdot CH_2 \cdot N \cdot CH_2 \cdot N \cdot CH_2 \cdot N \cdot CH_2 \cdot OH
\end{array}
$$

that the formula of a urea-formaldehyde resin can be represented by such a simple structure. Ramified chains or networks, such as might result from linking part of the nitrogen atoms in a given chain to carbonyl groups in another chain, undoubtedly present a more accurate picture of the final composition of these resins.

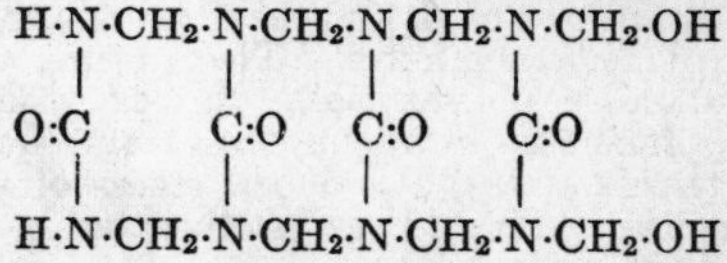

Vinyl Resins

Although the production of vinyl resins is still rather small, recent progress has justified an increase in productive capacity on the part of the manufacturers. The most important raw material is vinyl acetate, although the vinyl halides are also used. Some of the final resins may be polyvinyl alcohols or acetals, but the variety of available vinyl compounds permits the preparation of products of widely different properties. These resins are characterized by their transparency and water-white color, but they are also used in the production of translucent and opaque products. They find application primarily in the production of molded articles such as phonograph records, dentures, and large molded sheets for wall coverings. They can be used in lacquers and, despite their high cost, in special varnishes and as the intermediate layer in safety glass.

In contrast to the phenolic and urea resins which are "thermosetting" and become permanently infusible under the influence of heat, the vinyl resins are "thermoplastic." Although sufficiently rigid at room temperature, the vinyl resins can be remolded when desired by the application of heat and pressure. Since vinyl alcohol ($H_2C:CH \cdot OH$) is not known, acetylene is used as the source for the vinyl radical:

$CH_3 \cdot CO \cdot OH$ (anhydrous) $+ H \cdot C:C \cdot H$ (catalyst) $\rightarrow CH_3 \cdot CO \cdot O \cdot HC:CH_2$ (vinyl acetate)

Vinyl acetate appears to be stable if it is kept in a cool place and not exposed to light, but, upon heating in the dark to 100°C. or above or upon exposure to light, polymerization results in the formation of polyvinyl acetate. Oxidizing agents such as benzoyl peroxide greatly accelerate the formation of the resin. It has been suggested that the polymerization of vinyl acetate takes the following course:[7]

x $CH_3 \cdot CO \cdot O \cdot HC:CH_2 \rightarrow CH_3 \cdot CO \cdot O \cdot CH_2 \cdot CH_2 \cdot C(:CH_2) \cdot O \cdot OC \cdot CH_3 \rightarrow$
$CH_3 \cdot CO \cdot O \cdot CH_2 \cdot CH_2 \cdot CH(O \cdot OC \cdot CH_3) \cdot CH_2 \cdot C(:CH_2) \cdot O \cdot OC \cdot CH_3$, etc.

Koroseal is a polymer of vinyl chloride, which is obtained in turn by the addition of hydrogen chloride to acetylene or by caustic or pyrolytic dehydrohalogenation of ethylene chloride. Vinyon is a copolymer of vinyl acetate and vinyl chloride. Vinylidene chloride ($H_2C{:}CCl_2$), which may be obtained by the chlorination of vinyl chloride or the dehydrohalogenation of acetylene tetrachloride, polymerizes to give a product of the general type, ($\cdot CH_2\cdot CCl_2\cdot CH_2\cdot CCl_2)_n$, which is resistant to chemicals and has nearly the same refractive index as water. It seems particularly adapted to use for fish lines and fish nets.

Coumarone Resins

Coumarone resins are prepared by the polymerization of a mixture consisting principally of benzofuran (coumarone) and indene. The mixture for the polymerization is obtained by suitable treatment of the naphtha fractions boiling between 150–200°C. For the production of the pale-colored, odorless products demanded by the trade, very careful control of the polymerization process is necessary. This process involves the use of a catalyst, usually sulfuric acid, and the maintenance of a temperature of 20°C. or lower. These resins are used in the manufacture of varnishes, rubber products, chewing gum, linoleum, mastic flooring, and floor tiles. The low tensile strength and brittleness of the coumarone resins have militated against their use in the production of molded articles.

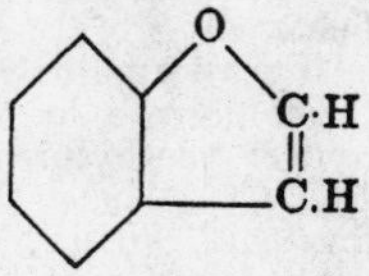

coumarone

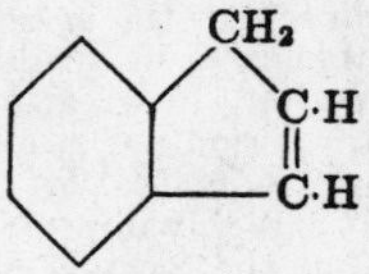

indene

Staudinger has proposed the formula given below for the polyindenes, and suggests that the unsaturated valences may be satisfied by the union of two or

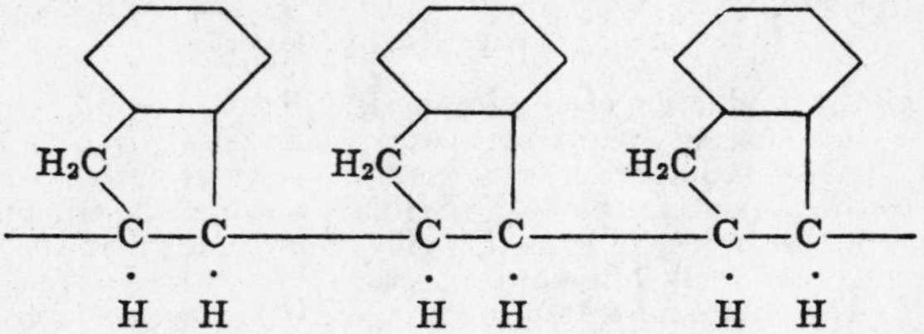

more such chains to form a ring. Whitby and Katz,[8] however, believe that hydrogen atoms wander and that the final polymer is a chain of the type shown here. The polymerization of coumarone has been less extensively

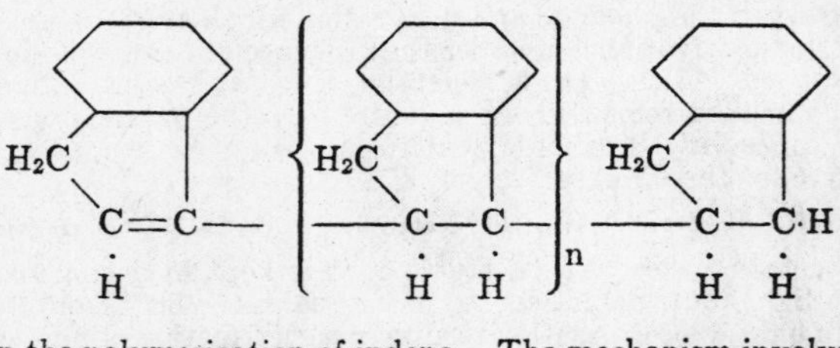

studied than the polymerization of indene. The mechanism involved is probably the same.

Styrene Resins

Although polystyrene was first produced nearly one hundred years ago and the study of its chemical constitution has contributed to the understanding of resins in general, the commercial development of this resin has not been very extensive. According to Staudinger,[9] liquid styrene ($C_6H_5 \cdot HC:CH_2$) polymerizes to form a chain of the type shown at the right. It has been estimated that the more highly polymerized products contain several thousand styrene units. The nature

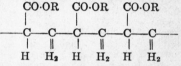

of the final polymer is dependent upon the temperature of the reaction, the absence or presence of a catalyst, and the nature of the catalyst. The properties of polystyrene are such as permit of its use in lacquers and in thermoplastic molded products.

Acrylic Resins

The acrylic resins constitute another class which has been studied for many years, but which has only recently assumed industrial production.[10] The most promising products are prepared from the esters of acrylic acid ($H_2C:CH \cdot CO \cdot OH$) or methacrylic acid ($H_2C:C[CH_3] \cdot CO \cdot OH$). In common with the monomers of some of the other resins discussed above, these esters possess a conjugated system of double bonds which is favorable for resin formation. The polymerized acrylates probably resemble polystyrene or polyvinyl esters in having a structure of the general type indicated here.

Acrylic resins are characterized by their colorless transparency, adhesive qualities, and resistance to many reagents. Their elasticity, great optical clarity, and other properties have led to their use as substitutes for glass in airplane windows and as the intermediate layer in safety glass. They are also suitable for use in clear lacquers, electrical insulators, and transparent molded articles.

Miscellaneous Resins

There are certain other resins which should be mentioned for the sake of completeness, but which are either of smaller commercial importance or are not true synthetic resins. The esterification product of glycerol and rosin is produced in very large quantities for the paint and varnish industries. Chlorinated rubber and the reaction products of rubber with sulfuric acid, chlorostannic acid, and other reagents, are being used in paints, molded products, and various other articles. Among the substitutes for rubber are the olefin polysulfide resins and α-polychloroprene which is a polymer of chloroprene [$H_2C:C(Cl) \cdot HC:CH_2$]. The successful production of synthetic rubber by the polymerization of butadiene and dimethylbutadiene has been announced in Germany. Ketone-formaldehyde resins have been commercialized on the European markets for several years. Considerable success has attended the various efforts to prepare resins of commercial importance from isobutylene and other petroleum products. Finally, resins made by the condensation of formaldehyde with aniline, p-toluenesulfonamide, or other sulfonamides have found a number of practical applications.

The material in this division was contributed by Dr. V. N. Morris, Industrial Tape Corporation.

[1] Ellis, "The Chemistry of Synthetic Resins," Vol, 1, p. 11, 1935

[2] Chem. Met. Eng. *43*, 90, 1936; Ellis, loc. cit. Ind. Eng. Chem., News ed., *28*, 265, 1936

[3] Megson, J. Soc. Chem. Ind., *49*, 251, 1930; *52*, 418, 421T, 1933

[4] Fuller and Armstrong, Chem. Met. Eng. *43*, 4, 1936

[5] Kienle and Hovey, J. Am. Chem. Soc. *51*, 509, 1929

[6] Kolloid-Beihefte, *34*, 163, 1931

[7] Allen, Meharg, and Schmidt, Ind. Eng. Chem., *26*, 663, 1934

[8] Whitby and Katz, J. Am. Chem. Soc., *50*, 1163, 1928

[9] Ann., *517*, 35, 1935

[10] Neher, Ind. Eng. Chem., *28*, 267, 1936

CHAPTER XXIX

NATURAL AND SYNTHETIC FIBERS

Fibers may be classed according to their nature and chemical composition as (I) animal, (II) vegetable, (III) regenerated, and (IV) synthetic.

I. Animal Fibers

The most important animal fibers are silk and wool. These are protein substances which, because of their amphoteric properties, combine with both acids and bases and are readily dyed.

The main protein of wool is keratin, which is found also in hair, hoof, horn, and nail. The macromolecular structure of protein has been worked out from a study of physical properties and x-ray diagrams. It is reasonably certain that keratin is not a definite chemical entity, but a composite of amino acid chains which are bound together by cross linkages of a physico-chemical type. Among the more important amino acids that have been isolated from keratin are arginine, aspartic acid, cystine, glutamic acid, lysine, tryptophan, and tyrosine.

The protein **fibroin,** which is cemented together by the protein **sericin** or silk gum, is the principal constituent of the silk fiber. During the preparation of the silk yarn, the sericin is almost entirely removed to leave comparatively pure fibroin.

In the silk fibroin molecule, every other amino acid is glycine ($H_2N \cdot CH_2 \cdot CO_2H$), every fourth amino acid is alanine ($CH_2 \cdot CHNH_2 \cdot CO_2H$), every sixteenth amino acid is tyrosine ($p\text{-}HO \cdot C_6H_4 \cdot CH_2CHNH_2 \cdot CO_2H$), and every two-hundred-sixteenth amino acid is arginine

$$[H_2N \cdot C(:N \cdot H) \cdot NH \cdot CH_2 \cdot CH_2 \cdot CH_2 \cdot CHNH_2 \cdot CO_2H].[1]$$

This gives a long chain molecule of the general type:

$$\begin{array}{cccccc} H & H\,H & H & H\,H & H & H\,H & H & H\,H \\ (\cdot \dot{N} \cdot CH_2 \cdot CO \cdot \dot{N} \cdot \dot{C} \cdot CO \cdot \dot{N} \cdot CH_2 \cdot CO \cdot \dot{N} \cdot \dot{C} \cdot CO \cdot \dot{N} \cdot CH_2 \cdot CO \cdot \dot{N} \cdot \dot{C} \cdot CO \cdot \dot{N} \cdot CH_2 \cdot CO \cdot \dot{N} \cdot \dot{C} \cdot)_n, \\ & \dot{C}H_3 & & \dot{R}_p & & \dot{C}H_3 & & \dot{R}_p \end{array}$$

where R_p is a substituent of a type to give the amino acid concerned.

II. Vegetable Fibers

The most important of the vegetable fibers are cotton, flax, hemp, and jute. Since they all consist of comparatively pure cellulose, they are practically neutral.

The principal building unit in most of the vegetable fibers is $\beta\text{-}D$-glucose. By the loss of water between two molecules of $\beta\text{-}D$-glucose, the disaccharide cellobiose is obtained (cf. p. 111). Cellobiose molecules in turn lose water to build up a long chain molecule which is known as cellulose.

The essential differences in the vegetable fibers must be attributed to substances other than $\beta\text{-}D$-glucose and to differences in the nature of the physico-chemical bonds, for they all contain the same building unit.

---

[1] Bergmann, *J. Biol. Chem.*, *109*, 317 (1935): *110*, 471 (1935): *113*, 341 (1936): *122*, 569 (1938): *Chem. Rev.*, *22*, 423 (1939).

III. Regenerated Fibers

These fibers are produced by dissolving cellulose or protein material in a suitable solvent and then extruding the viscous mass through a spinneret into a precipitating bath to produce fine threads. In such a process, the characteristics of the original material are usually retained by the regenerated threads, but slight chemical decomposition such as hydrolysis and oxidation may occur.

The process of manufacture is dependent upon the type of material that is to be regenerated. With casein, for example, an alkali solution of the protein is treated with strengthening agents such as sodium aluminate and with flexing agents such as oleic acid. The solution is then extruded through spinnerets into an acid bath containing about twenty per cent of glucose in order to increase the setting-up of the fiber. This solution also contains chemicals that will tend to increase the strength of the fiber and decrease its solubility during passage through the bath. Formaldehyde is frequently used for this purpose. The fibers, upon emerging from the bath, receive treatment with oil emulsions in order to increase their softness and flexibility. Such fibers have been regenerated from casein, peanut, soy bean, and other proteins as a substitute for wool.

The two important methods used in the regeneration of cellulose fibers are the cuprammonium and the viscose processes. In the cuprammonium process, cotton linters are dissolved in an ammonical copper salt solution, which is extruded through spinnerets into a coagulating bath to regenerate the cellulose. If an alkaline bath is used, which is usually preferred, an after treatment in an acid bath is required to remove the copper. This effects both decoloration of the filament and an increase in its tensile strength.

In the viscose process, which accounts for over eighty per cent of regenerated cellulose fibers, sheets of cellulose are treated with sodium hydroxide and then shredded and submerged in carbon bisulfide to give cellulose xanthate. This product is dissolved in sodium hydroxide and allowed to age or ripen. The product is then extruded through spinnerets into an acid bath to regenerate the cellulose in the form of a filament.

The first method used for the regeneration of cellulose fibers was the nitrocellulose process, which has now been supplanted by the two processes just considered. In this process the cellulose is nitrated, dissolved in an alcohol-ether mixture, and the viscous mass extruded counter-currently to a stream of air to remove the solvent sufficiently to cause film formation. The filaments are then denitrated by treatment with a solution of sodium bisulfide. Since complete denitration is difficult, the fibers are quite flammable. The flammability of the fibers and the cost of production have both militated against the continued operation of this process.

IV. Synthetic Fibers

The fibers of this group are essentially different from those of the three groups just considered in that they are produced by the processes of condensation and polymerization of simple molecules. The properties of the fibers may be varied, accordingly, by a choice of the reagents and control of the amount of condensation or polymerization.

Cellulose acetate, which is to be differentiated from regenerated cellulose, was one of the first synthetic fibers to be produced. The acetate silk, moreover, differs from viscose silk with respect to its properties in that the former absorbs only small amounts of moisture, retains its strength when wet, and does not mold.

In the manufacture of cellulose acetate, some form of suitable cellulose is dissolved in an acetic acid-acetic anhydride mixture that contains a trace of sulfuric acid or other catalyst. When this solution is poured into water, the cellulose acetate is precipitated, filtered off, washed, and dried. The cellulose acetate is then dissolved in acetone and extruded through spinnerets against a counter-current stream of warm air to effect removal of the solvent and formation of a filament.

The structure of the protein molecule has been emulated in the synthesis of nylon, which is a fiber-forming polymeric amide with a protein-like linkage. Nylon is made by the reaction between a primary or secondary diamine and either a dicarboxylic acid or an amide-forming derivative of such an acid. One nylon fiber that is used in the manufacture of textiles may be made from adipic acid and hexamethylene-diamine, in accordance with the equation:

$HO \cdot OC \cdot (CH_2)_4 \cdot CO \cdot OH + H \cdot NH \cdot (CH_2)_6 \cdot HN \cdot H \rightarrow HO \cdot OC \cdot (CH_2)_4 \cdot CO \cdot NH \cdot (CH_2)_6 \cdot HN \cdot H +$

H_2O, then $+ HO \cdot OC \cdot (CH_2)_4 \cdot CO \cdot OH \rightarrow HO \cdot OC \cdot (CH_2)_4 \cdot CO \cdot NH \cdot (CH_2)_6 \cdot HN \cdot OC \cdot (CH_2)_4$

$\cdot CO \cdot OH + H_2O$, then $+ H \cdot NH \cdot (CH_2)_6 \cdot HN \cdot H \rightarrow HO \cdot OC \cdot (CH_2)_4 \cdot CO \cdot NH \cdot (CH_2)_6 \cdot HN \cdot OC \cdot$

$(CH_2)_4 \cdot CO \cdot NH \cdot (CH_2)_6 \cdot HN \cdot H + H_2O$, and similarly, until a long chain molecule is built up.

The condensation is effected at a temperature of 200 to 300°, and the product formed into icy-white ribbons of a convenient size. These ribbons are then broken into small chips, the chips melted to give a water-clear liquid, and the liquid extruded through spinnerets to give filaments that are wound onto spools which rotate at a sufficient speed to keep the filament under proper tension. The tension on the filament tends to allign the molecular chains into parallel formation and thus bring them close enough together to have intermolecular attraction.

The nylon filaments may vary in diameter from that of a cobweb to as much as three or more inches. Aside from its use in the production of fibers, nylon appears in the form of bristles for tooth brushes, sutures for surgery, and in numerous other products.

Another important synthetic fiber is of the vinyl type. This may be made by the polymerization of vinyl chloride ($H_2C:CH \cdot Cl$), by the polymerization of vinyl acetate ($CH_3 \cdot CO \cdot O \cdot HC:CH_2$), or by the copolymerization of vinyl chloride with vinyl acetate, in accordance with the general equation:

$$\begin{matrix} H\,H & H\,H & H\,H & H\,H & H\,H & H\,H & & H\,H\,H\,H\,H\,H\,H\,H\,H\,H\,H\,H \\ C:C + C:C + C:C + C:C + C:C + C:C, \text{etc.} \rightarrow & H \cdot C \cdot C \cdot C \cdot C \cdot C \cdot C \cdot C \cdot C \cdot C \cdot C \cdot C \cdot C, \text{etc.,} \\ H\,W & H\,W & H\,W & H\,W & H\,W & H\,W & & H\,W\,H\,W\,H\,W\,H\,W\,H\,W\,H\,W \end{matrix}$$

where the W represents either a chlorine atom or an acetate group, depending on the particular type of product that is being produced. A product that is made by using about 80% vinyl chloride to 20% vinyl acetate is marketed under the trade name of vinyon resin for the production of vinyon fiber.

In the production of vinyon fiber, the vinyon resin, in the form of a fluffy powder, is dispersed in acetone and extruded through spinnerets to give a filament.

Vinyon is thermoplastic and softens above a temperature of 65°. It is comparatively stable, is unaffected by alcohol and the lower aliphatic hydrocarbons, and burns with difficulty. It has high elasticity, high tensile strength, and good dielectric properties.

Vinyl acetate polymer has been extruded also to give filaments, which may be spun and woven into practically any desired weave. Such a product has been marketed under the trade name of Duraklad, and used for acid-proof thread, filter cloth, and anode bags. It is thermoplastic, and resistant to both acids and alkalies although it is attacked by concentrated nitric acid.

Vinyl chloride may be polymerized, under proper conditions, to give a rubber-like product which is marketed under the trade name of Koroseal. This product is thermoplastic and may be milled on hot rolls or extruded through dies. Unlike rubber, Koroseal does not deteriorate from contact with air, light, oxidizing chemicals, or ozone. It can be made glass-clear, but is not as resilient as rubber and does not wear as well. Koroseal has been used rather extensively in the waterproofing of fabrics, although one of its principal uses is in lining pickling tanks for use in the steel industry and in the lining of tank cars for the transportation of corrosive chemicals.

Saran is the trade name of the new (1940) vinylidene chloride resins, which may be (1) polymers of vinylidene chloride as:

$$(\cdot CH_2 \cdot CCl_2 \cdot CH_2 \cdot CCl_2 \cdot CH_2 \cdot CCl_2 \cdot CH_2 \cdot CCl_2 \cdot CH_2 \cdot CCl_2)_n,$$

or (2) copolymers of vinylidene chloride with vinyl chloride or other vinyl monomer, as:

$$(\cdot CH_2 \cdot CWY \cdot CH_2 \cdot CWY \cdot CH_2 \cdot CWY \cdot CH_2 \cdot CWY \cdot CH_2 \cdot CWY \cdot CH_2 \cdot CWY)_n,$$

where W and Y may be chlorine atoms or some of the Y-groups may be either hydrogen atoms or other vinyl substituents such as the acetate group. Saran has special uses because it is tough, resistant to corrosion, water-clear, and has nearly the same refractive index as does water.

Butadiene and its homologs now serve as another raw material for the production of synthetic fibers. The butadiene is polymerized, the product dissolved in a suitable solvent, and the solution extruded through spinnerets into an appropriate bath to give filaments. Suitable solvents for this polymer are benzene, dioxan, morpholine, and quinoline. The precipitating bath may contain an alcohol, an aldehyde, a carboxylic acid, a ketone, or water. Heat treatment of the filament effects hardening.

The condensation product obtained from formaldehyde and urea may be used also in the production of synthetic fibers. An aqueous solution of formaldehyde, when treated with urea, gives a viscous solution, which is extruded into a saline bath to produce a filament. The urea, in this condensation, may be replaced by cyanamide or its derivatives.

Ethyl cellulose is produced by treating sodium cellulose with ethyl chloride in the presence of acetone. When the ethyl cellulose is dissolved in a suitable solvent and extended through a spinneret it gives a filament that is superior in some respects to that obtained from cellulose acetate. Ethyl cellulose is more resistant to the action of alkalies than is cellulose acetate, hence it withstands laundrying operations much better. The ethyl cellulose of commerce has from 2.2 to 2.6 of the three available hydroxyl groups per glucose unit replaced by ethoxy groups.

Rubber, although not a fiber, has found an important use in the textile industry. Rubber latex may be extruded through a spinneret into a coagulating bath to give a rubber thread which is marketed as **Lastron**. When such a thread is wound with some type of fiber such as cotton or silk, a highly elastic fiber is produced which is known as **Lastex**. This process makes available a flexible yarn, which is used in the knitting of foundation garments, bathing suits, and other type of clothing.

Glass also plays an important role in the production of synthetic fibers. When glass marbles are fed into an electric furnace and the melt drawn through a special type of spinneret in a metal bushing, there may be obtained filaments of almost any desired degree of fineness. These filaments are very strong, being about three time as strong as rayon, but are somewhat inferior to rayon with respect to resilence. These filaments may be produced in almost any desired shade, or the colorless fabrics produced from the filaments may be printed with colored laquers.

Orlon acrylic fiber, essentially a polyacrylonitrile, has outstanding resistance to both chemicals and weathering.

Dynel is processed from extruded vinyl chloride and acrylonitrile resins and possesses wool-like properties.

CHAPTER XXX

SURFACE-ACTIVE AGENTS

The use of soap as a detergent and wetting agent is being supplanted to some extent by new synthetic organic chemicals. These new surface-active agents are similar to soaps in that they are polar and contain a hydrocarbon chain of variable length and a solubilizing group. The hydrocarbon portion of the molecule, depending on its length, is more or less hydrophobic or water-repelling. The solubilizing group, on the other hand, is hydrophilic or water-attracting. The combination of these two groups gives a molecule which may be active at interfaces by effecting reduction of the surface tension.

At the interface between two immiscible substances, the hydrocarbon portion of these surface-active molecules becomes oriented toward the hydrophobic media whereas the other portion of the molecule is aligned toward the hydrophillic material. The behavior, of the molecule, however, will depend on the components of the interface. If the interface is made up of two hydrophobic compounds, the direction of orientation is not so readily predictable.

It may be seen, consequently, that the properties of the surface-active agent can be varied at will by changing the length of the hydrocarbon chain and/or modifying the nature of the solubilizing group. Patents have been issued for the manufacture of a number of materials with such variable combinations. When the hydrocarbon portion of the molecule is aliphatic, the chain is not branched in most cases because of the ready availability of the continuous chain compounds. When the hydrocarbon part of the molecule is aromatic, the alkyl and aryl substituents on the aromatic nucleus offer possible variations.

Examples of solubilizing or hydrophilic groups are carboxyl ($\cdot CO_2H$), bisulfate ($\cdot O \cdot SO_2 \cdot OH$), dihydrogenphosphate [$\cdot O \cdot P (\rightarrow O) \cdot (OH)_2$], hydroxyl (OH), sulfonic acid [$\cdot S(\rightarrow O)_2 OH$], and others, all of which have a solubilizing effect somewhat characteristic of the group.

The surface-active agents may be classified, for convenience, according to general types, as shown in Table I.

The A.1. type is illustrated by ordinary soap. Such compounds are readily obtained by the alkaline hydrolysis or saponification of animal or vegetable fats and oils (cf. p. 45). The use of such a

TABLE I. CLASSIFICATION OF SURFACE-ACTIVE AGENTS

Type	*General Formula*	*Name or Description*
A.1.	$R \cdot CO \cdot O^- {}^+Na$	Fatty acid salt
2.	$C_nH_{2n-2}(O \cdot SO_2 \cdot O^- {}^+Na)CO \cdot O^- {}^+Na$	Disodium salt of sulfated fatty acid
3.	$R \cdot CO \cdot O \cdot CH_2 \cdot CHOH \cdot CH_2 \cdot OH$	Glyceryl ester of a fatty acid
4.	$R \cdot CO \cdot O \cdot CH_2 \cdot CH_2 \cdot SO_2 \cdot O^- {}^+Na$	Sodium fatty acid ester sulfonate
5.	$R \cdot CO \cdot NH \cdot CH_2 \cdot CH_2 \cdot SO_2 \cdot O^- {}^+Na$	Sodium fatty acid amide sulfonate
6.	$R \cdot CO \cdot NH \cdot CH_2 \cdot CH_2 \cdot NR_2'$	Fatty acid amido ethyl dialkyl amine
B.1.	$R \cdot O \cdot SO_2 \cdot O^- {}^+Na$	Sodium alkyl sulfate where R is large
2.	$R'_2CH \cdot O \cdot SO_2 \cdot O^- {}^+Na$	Sodium secondary alkyl sulfate
3.	$R' \cdot O \cdot OC \cdot CH_2 \cdot$	Sodium salt of the bisulfate of a
	$R' \cdot O \cdot OC \cdot \overset{\mid}{C}H \cdot O \cdot SO_2 \cdot O^- {}^+Na$	dialkyl dicarboxylate
C.1.	$R \cdot SO_2 \cdot O^- {}^+Na$	Sodium alkyl sulfonate where R is large
2.	$Ar \cdot SO_2 \cdot O^- {}^+Na$	Sodium aryl sulfonate
3.	$R' \cdot O \cdot OC \cdot CH_2 \cdot$	Sodium salt of the sulfonic acid deri-
	$R' \cdot O \cdot OC \cdot \overset{\mid}{C}H \cdot SO_2 \cdot O^- {}^+Na$	vative of a dialkyl dicarboxylate
D.	$R \cdot N^+$ (pyridinium ring)	An alkyl pyridinium salt where R is a long aliphatic chain

Legend: R represents a long hydrocarbon chain of the aliphatic type,

    Ar represents an aryl or aromatic nucleus, and

    R′ represents a primary or secondary alkyl group of the short chain type.

Note: This is not intended as a complete classification, but serves to indicate the more important types of surface-active agents.

soap ($R \cdot CO \cdot O^- {}^+Na$) is limited in hard water because of the formation of insoluble salts of aluminum, calcium, magnesium, iron, and other heavy metals. The precipitation of such salts results in the formation of the so-called bath-tub ring. Most of the other surface-active agents, however, are equally effective in both soft and hard water as a consequence of the greater solubility of their salts of the metals normally found in hard water.

The A.2. type may be illustrated by the sodium salt of the sulfated derivative of ricinoleic acid $[CH_3(CH_2)_5 \cdot CHOH \cdot CH_2 \cdot HC{:}CH \cdot (CH_2)_7 \cdot CO_2H]$. The effective agent obtained from this acid has

$$CH_3(CH_2)_5 \cdot CH \cdot CH_2 \cdot HC{:}CH \cdot (CH_2)_7 \cdot CO \cdot O^- {}^+Na$$

$$O \cdot SO_2 \cdot O^- {}^+Na$$

sodium salt of the sulfated derivative
of ricinoleic acid

the formula shown above. The production and use of this compound represents one of the first attempts to synthesize surface-active agents. This product finds wide application, especially in the textile industry. The ricinoleic acid for the production of this sulfated derivative is obtained from castor oil.

The A.3. type may be represented by glyceryl oleate, which has the formula: $CH_3(CH_2)_7 \cdot HC:CH \cdot (CH_2)_7 \cdot CO \cdot O \cdot CH_2 \cdot CHOH \cdot CH_2 \cdot OH$. This group of surface-active agents is distinctive in that its members are non-ionic and may be used advantageously over a rather wide pH range.

In the commercial production of these compounds, the glycerol is usually replaced by a polyglycerol of approximately the pentaglycerol variety. When glycerol is heated in the presence of alkali at 200–300°, the polyglycerol is obtained as the condensation product. Both open chain and cyclic derivatives are obtained. Experience has shown that the most satisfactory polyglycerol for this synthesis is that of the penta-derivative. This pentaglycerol is then esterified with fatty acids, obtained from cocoanut oil, under conditions that give the mono-ester as the principal product. These pentaglycerol esters of the fatty acids have unusually good emulsifying and detergent properties and surface-activity in hard water.

The A.4. type is characterized by Igepon A ($C_{17}H_{33} \cdot CO \cdot OCH_2 \cdot CH_2 \cdot O \cdot SO_2 \cdot O^- {}^+Na$), which is produced commercially by the esterification of fatty acids, principally oleic acid, with isethionic acid (β-hydroxyethanesulfonic acid), as indicated by:

$CH_3(CH_2)_7 \cdot HC:CH \cdot (CH_2)_7 \cdot CO \cdot OH + HO \cdot CH_2 \cdot CH_2 \cdot SO_2 \cdot OH$, then $+ NaOH \rightarrow$ Igepon A, which has the formula: $CH_3(CH_2)_7 \cdot HC:CH \cdot (CH_2)_7 \cdot CO \cdot O \cdot CH_2 \cdot CH_2 \cdot SO_2 \cdot O^- {}^+Na + 2 H_2O$.

The isethionic acid for the synthesis is obtained by the treatment of ethyl alcohol with sulfur trioxide as shown by the equation:

$CH_3 \cdot CH_2 \cdot OH + SO_3 \rightarrow HO \cdot O_2S \cdot CH_2CH_2 \cdot OH$, isethionic acid.

Igepon A undergoes slow hydrolysis in acid solution, hence it is used most successfully in the presence of traces of alkali.

The A.5. type is a modification of the A.4. type and was produced to obtain a compound with detergent properties that would be reasonably stable in the presence of acid. The representative member of this group is Igepon T ($C_{17}H_{33} \cdot CO \cdot NH \cdot CH_2 \cdot CH_2 \cdot SO_2 \cdot O^- {}^+Na$), which is produced commercially by condensing oleyl chloride with 2-chloroethylamine and then treating the condensation product with sodium sulfite as shown by:

$CH_3(CH_2)_7 \cdot HC:CH \cdot (CH_2)_7 \cdot CO \cdot Cl + H \cdot NH \cdot CH_2 \cdot CH_2 \cdot Cl$, then $+ 2 Na_2SO_3 \rightarrow$ $CH_3(CH_2)_7 \cdot HC:CH \cdot (CH_2)_7 \cdot CO \cdot NH \cdot CH_2 \cdot CH_2 \cdot SO_2 \cdot O^- {}^+Na + 2NaCl + NaHSO_3$.

These compounds contain an amide and sulfonic acid linkage,

both of which are more resistant to acid hydrolysis than is the ester linkage.

The A.6. type may be represented by the Sapamines ($C_{17}H_{33}$-$CO\cdot NH\cdot CH_2\cdot CH_2\cdot NEt_2$), which are surface-active cations. They are effective in both acid and neutral solutions and may be made active in the presence of alkali by further alkylation of the nitrogen atom to give the corresponding substituted alkyl trialkylammonium salt. The Sapamines are not effected by hard water.

One member of this group may be prepared by heating oleic acid with 1-amino-2-N,N-diethylaminoethane, as:

$$CH_3(CH_2)_7\cdot HC:CH\cdot(CH_2)_7\cdot CO\cdot OH + H\cdot NH\cdot CH_2\cdot CH_2\cdot N(CH_2\cdot CH_3)_2, \text{ heat} \rightarrow$$

$$CH_3(CH_2)_7\cdot HC:CH\cdot(CH_2)_7\cdot CO\cdot NH\cdot CH_2\cdot CH_2\cdot N(CH_2\cdot CH_3)_2 + H_2O.$$

The B.1. type is represented by sodium cetyl sulfate ($C_{16}H_{33}\cdot O$-$SO_2\cdot O^- {}^+Na$), which is obtained by treating cetyl alcohol with sulfuric acid to give cetyl bisulfate, which is then converted to the salt by treatment with alkali. The cetyl alcohol for the preparation is obtained by the catalytic hydrogenation of esters of the corresponding acid. High pressures are usually required as well as the use of copper chromite or other catalyst. It has been found that the usefulness of these compounds as wetting agents and detergents shows some correlation with the length of the hydrocarbon chain. Sodium cetyl sulfate ($C_{16}H_{33}\cdot O\cdot SO_2\cdot O^- {}^+Na$), for example, is a better wetting agent and detergent than is sodium dodecyl sulfate ($C_{12}H_{25}\cdot O\cdot SO_2\cdot O^- {}^+Na$). These compounds are marketed under such trade names as *Avirol, Tide, Gardinol*, and **hymolal salts,** and should not be confused with the **sulfonates** ($R\cdot SO_2\cdot O^- {}^+Na$) under type C.

Type B.2. compounds are not as effective as the B.1. type on account of (1) a decrease in the length of the hydrocarbon chain as indicated in the preceding paragraph is attended with a decrease in surface-active properties, and (2) the alkyl sulfates prepared from the secondary alcohols are not as effective surface-active agents as are those obtained from the corresponding primary alcohols.

The type B.3. compounds may be made, presumably, by the sulfation of an unsaturated diester of a dicarboxylic acid such as dioctyl maleate. Such compounds have not found an important place on the market.

The C.1. type of compounds differ from the B. type in that the former are sulfates whereas the latter are sulfonates. Among the more general methods available for the preparation of the alkyl sulfonic acids or their salts, are: (1) the oxidation of mercaptans, (2) the interaction between a halide and sodium sulfite, or (3) the treatment of branched chain compounds with chlorosulfonic acid, the equations being:

1. $C_{12}H_{25} \cdot X + Na \cdot SH \rightarrow C_{12}H_{25} \cdot SH$, then + oxidation $\rightarrow C_{12}H_{25} \cdot SO_2 \cdot OH$, or

2. $C_{12}H_{25} \cdot X + Na_2SO_3 \rightarrow C_{12}H_{25} \cdot SO_2 \cdot O^{-+}Na$, and

3. $R(R') (R'') C \cdot H + Cl \cdot SO_2 \cdot OH \rightarrow R (R') (R'') C \cdot SO_2 \cdot OH + H \cdot Cl$.

The sulfonates of this group are comparatively stable but less desirable as detergents and wetting agents than are the alkyl sulfates of type B.

The type C.2. compounds may be obtained by the direct sulfonation of aryl hydrocarbons with either fuming sulfuric acid or sulfur trioxide, as:

$$Ar \cdot H + HO \cdot SO_2 \cdot OH/SO_3 \text{ (or } SO_3) \rightarrow Ar \cdot SO_2 \cdot OH + H_2SO_4.$$

Still more important products are obtained by effecting substitution of alkyl groups on the aryl nucleus so as to obtain alkylarylsulfonic acids as their salts. Such alkylation is usually effected by the use of a Fittig, a Grignard, or a Friedel-Crafts reaction. Partial hydrogenation of the aromatic nucleus, subsequent to sulfonation, may also be used as a means of modifying the properties of surface active agents of this general type.

The active ingredient in most of the newer surfactants such as *Breeze*, *Dreft*, *Fab*, *Surf*, *Swerl*, *Trend*, and *Vel* is the sodium salt of an alkylphenylsulfonic acid.

The C.3. group is best exemplified by Aerosol O. T. Dry, which is made by the esterification of maleic acid with octyl alcohol and subsequent sulfonation by use of sodium bisulfite, as indicated by:

$$\begin{matrix} H \cdot C \cdot CO_2H \\ \| \\ H \cdot C \cdot CO_2H \end{matrix} + \begin{matrix} HO \cdot C_8H_{17} \\ HO \cdot C_8H_{17} \end{matrix} /(H^+) \rightarrow \begin{matrix} H \cdot C \cdot CO \cdot O \cdot C_8H_{17} \\ \| \\ H \cdot C \cdot CO \cdot O \cdot C_8H_{17} \end{matrix} + 2 H_2O, \text{ then}$$

$$\begin{matrix} H \cdot C \cdot CO \cdot O \cdot C_8H_{17} \\ \| \\ H \cdot C \cdot CO \cdot O \cdot C_8H_{17} \end{matrix} + NaHSO_3 \rightarrow \begin{matrix} Na^+ {}^-O \cdot O_2S \cdot CH \cdot CO \cdot O \cdot C_8H_{17} \\ \\ CH_2 \cdot CO \cdot O \cdot C_8H_{17}, \text{ Aerosol O. T. Dry.} \end{matrix}$$

This is one of the most effective wetting agents that has been placed on the market, and has been advertised through the medium of the sinking duck photographs in popular picture journals.

The D. type compounds give surface-active cations in solution, and such agents are used as **reversed soap**. They possess the disadvantage, however, of forming precipitates with ordinary soap and with other long chain anion-containing compounds.

CHAPTER XXXI

ISOMERISM

Two compounds are said to be isomeric when they are composed of the same elements in the same proportion by weight, but have different properties. Isomerism is usually considered under three general types:

A. Ordinary Structural Isomerism:

Ordinary structural isomerism may be defined as that type of isomerism which involves compounds with the same empirical formulas and the same molecular formulas, but which have distinctly different structures and, consequently, different properties. Five types will be considered:

1. Chain isomerism is due to different arrangements of the carbon atoms in the hydrocarbon nucleus. Three types may be recognized:

   a. Ordinary chain isomerism:

      Examples are pentane ($CH_3 \cdot CH_2 \cdot CH_2 \cdot CH_2 \cdot CH_3$), 2-methylbutane ($[CH_3]_2 CH \cdot CH_2 \cdot CH_3$), and 2, 2-dimethylpropane ($[CH_3]_4 C$, tetramethylmethane).

   b. Side chain isomerism:

      Examples are propylbenzene ($C_6H_5 \cdot CH_2 \cdot CH_2 \cdot CH_3$), isopropylbenzene ($C_6H_5 \cdot CH[CH_3]_2$, 2-phenylpropane or [1-methylethyl]benzene), ethyl methyl benzene ($CH_3 \cdot CH_2 \cdot C_6H_4 \cdot CH_3$), and trimethylbenzene ($C_6H_3(CH_3)_3$).

   c. Ring-chain isomerism.

      Examples are cyclohexane (C_6H_{12}) and methylcyclopentane ($C_5H_9 \cdot CH_3$).

2. Position isomerism is due to different possible allocations, in any given carbon chain, of the functional group or groups. Five types may be recognized:

   a. In the nucleus:

      Examples are 1-butene ($CH_3 \cdot CH_2 \cdot HC:CH_2$) and 2-butene ($CH_3 \cdot HC:CH \cdot CH_3$), or 1-pentyne ($CH_3 \cdot CH_2 \cdot CH_2 \cdot C:C \cdot H$) and 2-pentyne ($CH_3 \cdot CH_2 \cdot C:C \cdot CH_3$). In this particular type, only the position of the unsaturated linkage is involved.

   b. On the chain:

      Examples are 1-chloropropane ($CH_3 \cdot CH_2 \cdot CH_2Cl$) and 2-chloropropane ($CH_3 \cdot CHCl \cdot CH_3$), 1, 1-dichloroethane ($CH_3 \cdot CHCl_2$) and 1, 2-dichloroethane ($CH_2Cl \cdot CH_2Cl$), or 2-pentanone ($CH_3 \cdot CH_2 \cdot CH_2 \cdot CO \cdot CH_3$) and 3-pentanone ($CH_3 \cdot CH_2 \cdot CO \cdot CH_2 \cdot CH_3$).

   c. On the side chain:

      Examples are (3-chloropropyl)benzene ($C_6H_5 \cdot CH_2 \cdot CH_2 \cdot CH_2Cl$), (2-chloropropyl)benzene ($C_6H_5 \cdot CH_2 \cdot CHCl \cdot CH_3$), and (1-chloropropyl)-benzene ($C_6H_5 \cdot CHCl \cdot CH_2 \cdot CH_3$).

   d. On the ring:

      Examples are *o-, m-,* and *p*-dichlorobenzene; *o-, m-,* and *p*-xylenes; and α- and β-naphthol. In this type the relative positions of the substituted groups tend to determine the properties of the compound.

e. On the ring or side chain:

Examples are o-, m-, or p-chlorotoluene ($CH_3 \cdot C_6H_4 \cdot Cl$) and benzyl chloride ($C_6H_5 \cdot CH_2 \cdot Cl$, α-chlorotoluene); o-, m-, or p-toluidine ($CH_3 \cdot C_6H_4 \cdot NH_2$) and N-methylaniline ($C_6H_5 \cdot NH \cdot CH_3$); or α-chloro- α-toluic acid ($C_6H_5 \cdot CHCl \cdot CO \cdot OH$, **phenylchloroacetic acid**) and p-chloro- α-toluic acid ($Cl \cdot C_6H_4 \cdot CH_2 \cdot CO \cdot OH$, **p-chlorophenylacetic acid**).

3. Functional isomerism is due to different isomeric functional groups. Some of the more important types are:

a. Acids and esters:

Examples are ethanoic acid ($CH_3 \cdot CO \cdot OH$) and methyl methanoate ($H \cdot CO \cdot O \cdot CH_3$), or propanoic acid ($CH_3 \cdot CH_2 \cdot CO \cdot OH$) and methyl ethanoate ($CH_3 \cdot CO \cdot O \cdot CH_3$) and ethyl methanoate ($H \cdot CO \cdot O \cdot CH_2 \cdot CH_3$).

b. Alcohols and ethers:

Examples are ethanol ($CH_3 \cdot CH_2 \cdot OH$) and methoxymethane ($CH_3 \cdot O \cdot CH_3$), 1-propanol ($CH_3 \cdot CH_2 \cdot CH_2 \cdot OH$) and methoxyethane ($CH_3 \cdot O \cdot CH_2 \cdot CH_3$), or 1-butanol ($CH_3 \cdot CH_2 \cdot CH_2 \cdot CH_2 \cdot OH$) and ethoxyethane ($CH_3 \cdot CH_2 \cdot O \cdot CH_2 \cdot CH_3$).

c. Aldehydes, ketones, and epoxy compounds:

Examples are propanal ($CH_3 \cdot CH_2 \cdot CHO$), propanone ($CH_3 \cdot CO \cdot CH_3$), and 1, 2-epoxypropane ($CH_3 \cdot HC \overset{O}{\cdot} CH_2$), or phenylethanal ($C_6H_5 \cdot CH_2 \cdot CHO$), acetophenone ($C_6H_5 \cdot CO \cdot CH_3$), and phenylepoxyethane ($C_6H_5 \cdot HC \overset{O}{\cdot} CH_2$). Some prefer to regard these as position isomers.

d. Amines; primary, secondary, and tertiary:

Examples are propylamine ($CH_3 \cdot CH_2 \cdot CH_2 \cdot NH_2$), ethylmethylamine ($CH_3 \cdot CH_2 \cdot NH \cdot CH_3$), and trimethylamine ($(CH_3)_3N$).

e. Monoacids and hydroxyaldehydes:

Examples are **acetic acid** ($CH_3 \cdot CO \cdot OH$) and hydroxyethanal ($HO \cdot CH_2 \cdot CHO$). Also, **lactic acid** ($CH_3 \cdot CHOH \cdot CO \cdot OH$) and **glyceraldehyde** ($CH_2OH \cdot CHOH \cdot CHO$).

f. Multiple-bond isomerism:

Examples are butadiene ($H_2C:CH \cdot HC:CH_2$) and 1- or 2-butyne ($CH_3 \cdot CH_2 \cdot C \vdots C \cdot H$, or $CH_3 \cdot C \vdots C \cdot CH_3$), propanal ($CH_3 \cdot CH_2 \cdot CHO$) and 2-propen-1-ol ($H_2C:CH \cdot CH_2OH$) which is sometimes described as saturated-unsaturated isomerism, or propanone ($CH_3 \cdot CO \cdot CH_3$) and 2-propen-1-ol ($H_2C:CH \cdot CH_2OH$). The lactam and lactim forms of **uracil** and similar compounds afford additional illustrations.

4. Metamerism is due to the attachment of different groups to a multivalent atom. Some of the more important types are:

a. Amines:

Examples are diethylamine ($CH_3 \cdot CH_2 \cdot NH \cdot CH_2 \cdot CH_3$) and methylpropylamine ($CH_3 \cdot NH \cdot CH_2 \cdot CH_2 \cdot CH_3$), or triethylamine ($CH_3 \cdot CH_2 (CH_3 \cdot CH_2) N \cdot CH_2 \cdot CH_3$) and ethylmethylpropylamine ($CH_3 \cdot CH_2 (CH_3) N \cdot CH_2 \cdot CH_3$). Although Schmidt and Richter both place **primary, secondary, and tertiary amines** ($CH_3 \cdot CH_2 \cdot CH_2 \cdot NH_2, CH_3 \cdot CH_2 \cdot NH \cdot CH_3$, and $[CH_3]_3N$) under this classification, they seem to fit more logically under functional isomerism.

b. Ethers:

Examples are ethoxyethane ($CH_3 \cdot CH_2 \cdot O \cdot CH_2 \cdot CH_3$) and methoxypropane ($CH_3 \cdot O \cdot CH_2 \cdot CH_2 \cdot CH_3$), or ethoxypropane ($CH_3 \cdot CH_2 \cdot O \cdot CH_2 \cdot CH_2 \cdot CH_3$) and methoxybutane ($CH_3 \cdot O \cdot CH_2 \cdot CH_2 \cdot CH_2 \cdot CH_3$). The alcohols, however, are more logically considered as functional isomers of the ethers.

c. **Esters:**

Examples are ethyl ethanoate ($CH_3 \cdot CO \cdot O \cdot CH_2 \cdot CH_3$) and methyl propanoate ($CH_3 \cdot CH_2 \cdot CO \cdot O \cdot CH_3$). The acids are functional isomers of the esters.

5. *Tautomerism* (dynamic isomerism) is due to isomeric forms which are capable, theoretically at least, of changing into each other. Dynamic isomerization is enhanced when a hydrogen atom is linked to a carbon atom which, in turn, is linked to two carbonyl ($C{:}O$), carboxyl ($CO \cdot OH$), or nitrile (CN) groups, or any two of these groups. Acetoacetic ester, for example, exists in a dynamic equilibrium involving the following forms:

(keto form) $CH_3 \cdot CO \cdot CH_2 \cdot CO \cdot O \cdot Et \rightleftarrows CH_3 \cdot C(OH){:}CH \cdot CO \cdot O \cdot Et$(enol form)

Three types of tautomerism will be considered:

a. **Desmotropism** represents an equilibrium system in which the isomers can be isolated and differ only in the position of a hydrogen atom within the molecule. Examples are:

1. **Acetoacetic ester,** $CH_3 \cdot CO \cdot CH_2 \cdot CO \cdot O \cdot Et \rightleftarrows CH_3 \cdot C(OH){:}CH \cdot CO \cdot O \cdot Et$. Tribenzoylmethane, $(C_6H_5 \cdot CO)_3C \cdot H \rightleftarrows (C_6H_5 \cdot CO)_2C{:}C(OH) \cdot C_6H_5$

   Malonic ester, $Et \cdot O \cdot OC \cdot CH_2 \cdot CO \cdot O \cdot Et \rightleftarrows Et \cdot O \cdot OC \cdot HC{:}C(OH) \cdot O \cdot Et$
   Aldoses (or ketoses, cf. p. 108) $\rightleftarrows$ cyclic structures.

2. **Phenylnitromethane,** $C_6H_5 \cdot CH_2 \cdot NO_2 \rightleftarrows C_6H_5 \cdot HC{:}N(O) \cdot OH$

   Due to the rather pronounced acidity of the one form of phenylnitromethane in contrast to that of the other form, systems with these characteristics are sometimes described by "*ionic isomerism.*"

In each of these specific cases, the pair of isomeric forms are known as desmotropes, and the pure isolated substances are termed desmotropic compounds.

b. **Pseudomerism** represents a specific type of tautomerism in which only one of the theoretical isomers is actually known. **Vinyl alcohol** ($H_2C{:}C(OH)H$), for example, has never been isolated although some of its derivatives are known. Attempts to prepare vinyl alcohol always yield **acetaldehyde,** hence it seems that the system is displaced almost quantitatively in the one direction ($H_2C{:}C[OH]H \rightarrow CH_3 \cdot CHO$). Other examples are the epoxyalkanes, the phenols, the oximes, the pyrroles, and the pyrazoles.

c. **Kryptomerism** represents a specific type of tautomerism in which the specific form of the common isomer is not known. **Hydrocyanic acid** ($H \cdot CN$ or $H \cdot NC$) and quinone are examples.

B. Stereoisomerism:

Stereoisomerism has been defined as "that type of isomerism which involves substances of the same constitution, but different configurations." Two distinct types are recognized:

1. *Geometrical isomerism* (*cis-trans*, olefin) is that particular type of stereoisomerism in which the substances have the same constitutional formulas, but differ in all of their physical properties and most of their chemical properties although they do not affect polarized light unless they contain an asymmetric grouping. Four important series are:

$$1. \quad \begin{array}{cc} a & a \\ C{:}C & \\ b & b \end{array} \quad \text{and} \quad \begin{array}{cc} a & b \\ C{:}C & \\ b & a \end{array}$$

$$2. \quad \begin{array}{cc} a & b \\ C{:}C & \\ b & c \end{array} \quad \text{and} \quad \begin{array}{cc} a & c \\ C{:}C & \\ b & b \end{array}$$

a. The "ene type:"

1. Carbon-to-carbon double bond linkage with different substituent groups may be represented by one of the three general types shown at the right. Examples of the first type are **maleic** and **fumaric** acid (allo isomers); of the second type, **crotonic** and **isocrotonic** acids, and of the third type, the 2-chlorobutenoic acids.

$$3. \quad \begin{matrix} a & & c \\ & C{:}C & \\ b & & d \end{matrix} \quad \text{and} \quad \begin{matrix} a & & d \\ & C{:}C & \\ b & & c \end{matrix}$$

2. Carbon-to-nitrogen double bond linkage with different substituent groups. The *cis*-form is also known as *n-* or *syn-*; the *trans-* or *iso-* as *anti-*. The general type is abC:Nc.

$$\begin{matrix} C_6H_5{\cdot}C{\cdot}H \\ \| \\ N{\cdot}OH \end{matrix} \quad \text{and} \quad \begin{matrix} C_6H_5{\cdot}C{\cdot}H \\ \| \\ HO{\cdot}N \end{matrix}$$

cis-benzaldoxime   *trans*-benzaldoxime

3. Nitrogen-to-nitrogen double bond linkage with different substituent groups. Specific examples are afforded by the diazo derivatives.

$$\begin{matrix} & N{:}N \\ a & \quad b \end{matrix} \quad \text{and} \quad \begin{matrix} & & b \\ N{:}N & \\ a & \end{matrix}$$

cis-form   *trans*-form

b. The "epoxy type:"

This may be regarded as a modification of the "ene type." The three general forms shown under B.*1.a.*1. may be modified to represent "epoxy types."

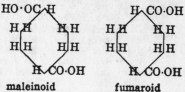

c. The "diene type:"

General examples are abC:C:Cab and abC:C:Cba, abC:C:Cbc and abC:C:Ccb, or abC:C:Ccd and abC:C:Cdc. Recently, the isolation of "diene isomers" has been reported.

d. The "cycloalkane type:"

Examples are maleinoid and fumaroid. The general type is ab·C_x·H_y·ba and ab·C_x·H_y·ab.

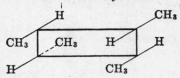

maleinoid   **fumaroid**

2. *Optical isomerism* is that particular type of stereoisomerism in which the substances have the same structural formulas and the same general properties, but differ in their action on polarized light. Optical activity is dependent upon the presence of an asymmetric structure in the configuration of the molecule, that is, the molecule must be asymmetric.

A molecule is *not* asymmetric if it has (a) a point of symmetry, (b) a line of symmetry, (c) a plane of symmetry, or (d) a rotating axis of symmetry. These are illustrated, respectively, by types like methane (CH_4), chloromethane ($CH_3{\cdot}Cl$), 2-bromo-2-butenoic acid ($CH_3{\cdot}HC{:}CBr{\cdot}CO{\cdot}OH$), and the form shown at the right which is one form of 1, 2, 3, 4-tetramethylcyclobutane. Type (b) is a special case of (d).

The common crucial test of symmetry and asymmetry is whether or not a model of the molecule and its mirror image are superposable. If they are not superposable, the molecule is asymmetric.

In order to clarify this concept, the following cases in which the molecule is asymmetric (and is, therefore, optically active) will be considered:

a. Where the molecule contains one or more asymmetric carbon atoms unless the arrangement is such that internal compensation results. Examples are afforded by the **alanines,** the **lactic acids,** the **tartaric acids** (with the exception of the meso form), and the **sugars.**

b. Where the molecule contains an *even* number of consecutive double bonds of the general type abC:C:Cab, abC:C:Cbc, abC:C:Ccd, abC:C:C: C:Cab, abC:C:C:C:Cbc, or abC:C:C:C:Ccd. 4- α(?)-Naphthyl-2,4-diphenylbutadienoic (**diphenyl-naphthyl-allene-carboxylic**) acid, which is shown at the right, has been prepared.

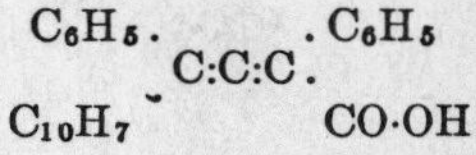

$$C_6H_5 \cdot \qquad \cdot C_6H_5$$
$$C:C:C \cdot$$
$$C_{10}H_7 \qquad CO\cdot OH$$

c. Where the molecule contains an even number of consecutive double bonds and rings. **4-Methylcyclohexylidene-acetic acid** is shown at the right, but it has not been prepared and resolved.

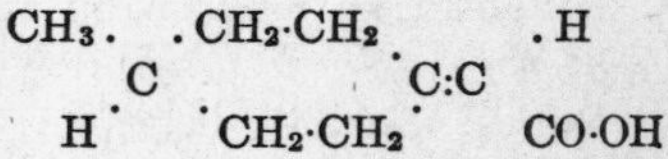

$$CH_3 \cdot \quad \cdot CH_2 \cdot CH_2 \qquad \cdot H$$
$$C \qquad\qquad C:C$$
$$H \cdot \quad \cdot CH_2 \cdot CH_2 \quad\cdot CO\cdot OH$$

d. Where the molecule contains an even number of consecutive rings. **Carone,** which is shown at the right, exists in nature in four active forms.

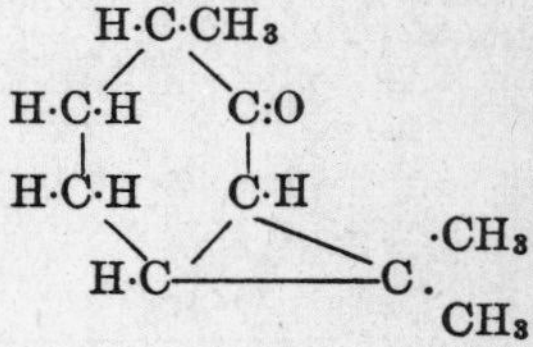

$$H\cdot C\cdot CH_3$$
$$H\cdot C\cdot H \qquad C{:}O$$
$$H\cdot C\cdot H \qquad C\cdot H$$
$$H\cdot C\underline{\qquad\qquad} C \cdot \begin{matrix} \cdot CH_3 \\ CH_3 \end{matrix}$$

e. Where the molecule is of the aliphatic diazo type shown at the right, lower.

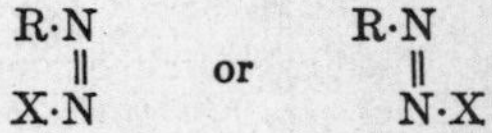

f. Where the molecule is of the *o*-substituted biphenyl type in which the substituents prevent free rotation. In these systems it is believed that the rings are coaxial but not coplanar as *o*-substituted groups would tend to force the rings into different planes. Hence if both rings are asymmetric, the molecule itself should be asymmetric and optically active, provided the substituent groups are sufficiently large to prevent free rotation.

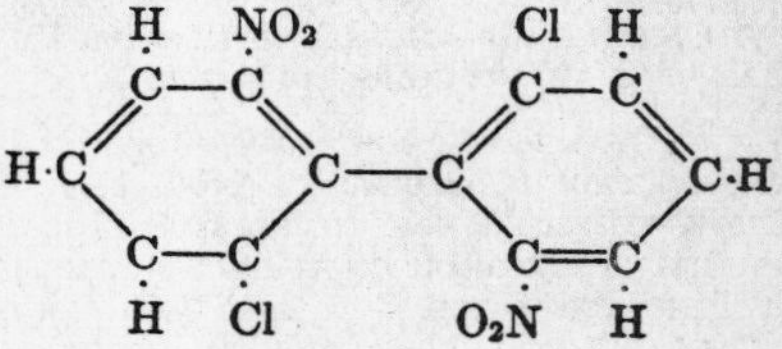

$$\begin{matrix} R\cdot N \\ \parallel \\ X\cdot N \end{matrix} \quad or \quad \begin{matrix} R\cdot N \\ \parallel \\ N\cdot X \end{matrix}$$

g. Where the molecule is of the α, α'-disubstituted naphthalene type, provided an asymmetric grouping is involved and the substituents are of sufficient size to prohibit free rotation. The type shown below has been reported

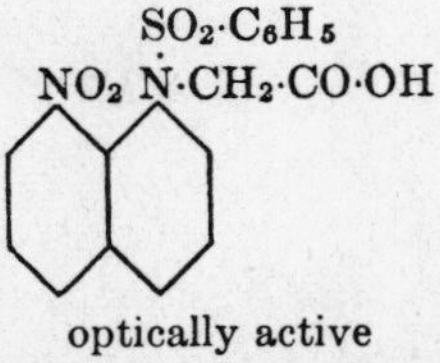

optically active

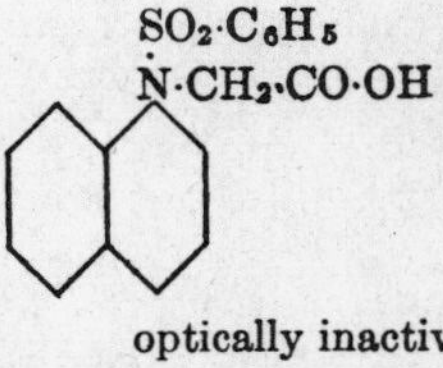

optically inactive

by Mills and Elliott, and serves as another example of restricted rotation.

h. Where the molecule is of the dimethyldiketopiperazine type. The *cis* form, shown at the left, is active and the *trans* form, shown at the right, is inactive.

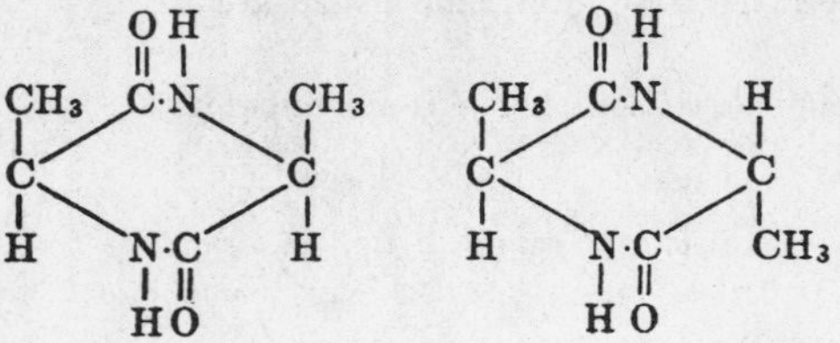

i. Where the molecule is an asymmetric derivative of chromium, cobalt, iridium, iron, nitrogen, phosphorus, rhodium, selenium, silicon, sulfur, or tin.

Optical inactivity may exist, even though optically active components are present, provided there is *external compensation.* If the *d* and *l* forms are present in equal molar concentration (racemic mixture), *external compensation* results and the mixture is optically inactive. An illustration is afforded by the tartaric acids:

CO·OH	CO·OH	CO·OH
H·C·OH	HO·C·H	H·C·OH
HO·C·H	H·C·OH	H·C·OH
CO·OH	CO·OH	CO·OH
L^{+}-tartaric acid	*D*$^{-}$-tartaric acid	*meso*-tartaric acid

racemic mixture

The term *internal compensation* represents the case where the molecule is symmetrical with respect to a point, a line, a plane, or a rotating axis of symmetry. An example is afforded by *meso*-tartaric acid in which the upper half of the molecule is the mirror image of the lower half of the molecule.

The van't Hoff-LeBel formula, 2^n, where "*n*" is the number of asymmetric carbon atoms, is used to predict the number of possible isomers in those cases where the two terminal groups are different. Since the two terminal groups in the tartaric acids are identical, the number of isomers is reduced from four to three.

C. Polymerism:

This type of isomerism exists where a given compound condenses with itself, without loss, to give a compound whose molecular weight is an even multiple of the molecular weight of the original compound. Examples are afforded by cyanic acid (HO·CN) and cyanuric acid $(HO{\cdot}CN)_3$, acetylene (HC:CH) and benzene (C_6H_6), or by glyceraldehyde $(C_3H_6O_3)$ and glucose $(C_6H_{12}O_6)$. Numerous other examples might be cited.

CHAPTER XXXII

ALKALOIDS*

A. Introduction:

1. Definition, Occurrence, Isolation.

Alkaloid means "alkali-like," and the alkaloids (morphine, 1805) were the first organic bases to be recognized. The term is applied to a class of basic, nitrogen-containing, compounds mostly found in plants, and usually in the seed-bearing varieties. With few exceptions, the alkaloids have the nitrogen in cyclic structures, and most alkaloids show high physiological activity.

The alkaloids are generally present in the plant in the form of salts of the common organic or inorganic acids, as acetic, malic, lactic, citric, phosphoric, or sulfuric. They may be distributed through all parts of the plant, but often are found concentrated in the seed, hulls, bark, leaves, or roots. It is rare that an alkaloid is found unaccompanied by others, and often as many as ten to twenty related alkaloids occur together in a single genus or species.

The procedure employed for the isolation and purification of the alkaloids depends in large degree upon the physical and chemical properties of the desired alkaloid. Liquid, volatile bases, like those of the nicotine and coniine groups, are liberated from their salts with alkali and distilled off with steam. Many others are removed from the mass of plant tissue in the form of water-soluble salts, by simple extraction with water or dilute acids, and then are precipitated with ammonia or alkali and extracted into organic solvents, or are converted to sparingly soluble salts that permit easy purification.

2. Nomenclature.

Alkaloid names all end in "ine" (ēn) to indicate the presence of the amine group. The complexity of the structures involved has made further systematization impractical, and the wide use of alkaloids as medicines make simple nomenclature a necessity. Some names refer to physiological action; examples, morphine (from Morpheus, the god of dreams), narcotine, emetine. Some have historical associations, as quinine (from the Inca word for bark, kina), codeine (from the Greek word for poppy seed-capsule), pelletierine (Pelletier, French alkaloid chemist), but the greatest number are derived from plant names; examples, papaverine (*Papaver somniferum*), hyoscyamine (*Hyoscyamus niger*), cinchonine (*Cinchona* species). To designate the many related alkaloids found in the same plant, prefixes and suffixes are often combined with the stem-name; example, cinchonine, cinchonidine, hydrocinchonine, hydrocinchonidine.

*This material on alkaloids was submitted by Dr. Lyndon Small. National Institute of Arthritis and Metabolic Disease, National Institute of Health, Bethesda 14, Maryland.

3. *Uses.* The alkaloids and their derivatives constitute some of the most valuable therapeutic agents. The importance of morphine, codeine, quinine, ephedrine, cocaine, and the alkaloids of ergot needs scarcely be mentioned. In addition, alkaloids form the active ingredient of most of the pleasure-producing drugs used by man, with the exception of alcohol and hashish. Predominant among these are tobacco, coffee and tea, betel (claiming hundreds of millions of devotées), opium, and coca, all of which owe their stimulating, sedative, or narcotic action largely to their alkaloidal content.

Well Known Useful Alkaloids

Alkaloid	Formula	Source	Action, or Field of Use
Morphine	$C_{17}H_{19}NO_3$	poppy	relief of pain
Codeine	$C_{18}H_{21}NO_3$	poppy	cough control, relief of pain
Cocaine	$C_{17}H_{21}NO_4$	coca leaf	local anesthesia
Apomorphine*	$C_{17}H_{17}NO_2$	morphine	causes vomiting
Arecoline	$C_8H_{13}NO_2$	betel nuts	worm expeller
Atropine	$C_{17}H_{23}NO_3$	belladonna	ophthalmology (mydriatic)
Scopolamine	$C_{17}H_{21}NO_4$	Datura species	narcotic
Caffeine	$C_8H_{10}N_4O_2$	coffee	stimulant, diuretic
Ephedrine	$C_{10}H_{15}NO$	ephedra	vasoconstrictor
Ergonovine	$C_{19}H_{23}N_3O_2$	ergot	obstetrics (oxytocic)
Nicotine	$C_{10}H_{14}N_2$	tobacco	insecticide
Quinine	$C_{20}H_{24}N_2O_2$	cinchona bark	malaria
Strychnine	$C_{21}H_{22}N_2O_2$	Strychnos nuts	pest exterminator
Physostigmine	$C_{15}H_{21}N_3O_2$	Calabar bean	ophthalmology, intestinal paresis

The tendency of alkaloids to form well-crystallized salts with acids is often utilized in the resolution of racemic modifications of optically active (p. 288) acids. For example, a racemic acid, d, l A, is neutralized with a levorotatory base, l B. The crystalline salts formed, $(d$ A $\cdot$ l B) and $(l$ A $\cdot$ l B) are not optical opposites, have different physical properties, and can be separated by crystallization. After the separation, the salts are treated with a mineral acid, by which pure d A is liberated from d A $\cdot$ l B, and pure l A from l A $\cdot$ l B. Morphine, brucine, strychnine, and cinchonine have been especially useful for such resolutions.

4. *Physical Properties.* As a class, the alkaloids are well crystallized, optically active, colorless compounds, soluble in most organic solvents, sparingly soluble in petroleum ether and ligroin. Berberine (page 257) is yellow. Some of the alkaloids of the nicotine and coniine groups are liquid; the liquid alkaloids generally contain no oxygen, but not all oxygen-free alkaloids are liquid Almost all alkaloids taste exceedingly bitter and are poisonous.

5. *General Chemical Features.* The alkaloids, even the liquid ones, are characterized by their tendency to form crystalline salts with both organic and inorganic acids. Only a few alkaloids are so weakly basic that they form no salts, or unstable ones.

* Alkaloid derivative, does not occur in nature.

Like other amines, the alkaloids combine with alkyl halides to form quaternary ammonium salts (page 142), for example:

$$R_3N + CH_3I \rightarrow [R_3N \cdot CH_3] \, I \rightleftharpoons R_3NCH_3^+ + I^-.$$

The alkaloid methiodides are useful for identification, and are important starting materials for the Hofmann exhaustive methylation. Most alkaloids are tertiary amines, and react with 30% hydrogen peroxide to yield amine oxides:

$$R_3N + H_2O_2 \rightarrow R_3N{\rightarrow}O + H_2O.$$

The alkaloid is readily regenerated from the oxide by mild reducing agents, as sulfur dioxide.

Alkaloids have the general property of forming sparingly-soluble precipitates with the so-called alkaloid reagents. The most common of these are potassium mercuric iodide, phosphomolybdic or phosphotungstic acids, tannic acid, and picric acid, and such reagents are useful for the detection of minute amounts of alkaloids.

Classification of the alkaloids is based on the type of nitrogen ring system present. The most important of these are pyridine, piperidine, pyrrolidine and combinations of them, quinoline, isoquinoline, indole, imidazole, and purine (see Chapter XXII). To what may be regarded as the parent ring-system are attached other rings and characteristic groups, variations that differentiate the many individual alkaloids. The most common peripheral groups encountered are methoxyl (OCH_3), methylenedioxyl ($CH_2{<}^O_O$), phenolic and alcoholic hydroxyl, N-methyl, and, less frequently, carboxyl, ester, lactone, lactam, amido, and carbonyl. The chemical reactions of the different alkaloids, beyond the general similarity due to the presence of the basic nitrogen atom, are in large degree the reactions of the peripheral groups and unsaturated systems present.

6. Structure Determination. When the empirical formula of a new alkaloid has been established by analysis and molecular weight determination, the structural question involves identification of the nucleus present, and the nature and location of the peripheral groups. The function of oxygen is generally easy to ascertain, through the reactions of the typical oxygen-containing groups mentioned above. Unsaturated linkages are detected by reduction, usually catalytic, and the unsaturation may sometimes be located through its influence on other groups, or by ozonolysis, permanganate oxidation, or bromine addition.

The manner of linkage of nitrogen is of great importance, and is established in several ways. A few alkaloids contain primary amine groups (mescaline, for example) but most contain nitrogen in a cyclic structure, and hence must be secondary or tertiary amines. Those alkaloids that fail to give secondary amine reactions (p. 142) are considered to be tertiary.

Hofmann exhaustive methylation reactions (p. 346) are widely used to demonstrate whether or not nitrogen is in a ring, and to eliminate nitrogen to give simple products that may be recognized. By another procedure, the Emde degradation, the methyl chloride of the alkaloid is boiled in alcohol or water with sodium amalgam. The nitrogen ring often opens at a different point by this method, and the Emde degradation may succeed where the Hofmann method fails.

The fundamental nucleus of the alkaloid sometimes is revealed by more drastic degradations. Vigorous oxidation, or fusion with alkali often yields recognizable fragments of the nucleus still carrying the peripheral groups, whereby the location of these in the alkaloid can be deduced. Zinc dust distillation breaks the molecule down to the most resistant aromatic nuclei, and gives valuable clues to the nuclear system present. Many of the alkaloids whose structures have been determined by these methods have been synthesized, but the syntheses have only rarely been commercially practicable.

B. Representative Alkaloids:

1. Ephedrine (ĕf'ĕ-drēn). The crude plant containing this base has been used by the Chinese for thousands of years under the name "Ma Huang." Ephedrine belongs to the phenylalkylamine class of alkaloids. Several other closely related alkaloids are present in the plant. Ephedrine finds medical application because of its power to raise blood - pressure through contraction of the blood vessels, and is especially useful in the relief of hay fever and asthma.

Ephedrine

2. Coniine (kō'nĭ-ēn). Coniine, an exceedingly poisonous alkaloid from the hemlock herb or spotted cowbane, is one of the piperidine (p. 225) type alkaloids. Together with four other poisonous alkaloids of related structure it formed the toxic principle of the hemlock drink that was once used to inflict the death penalty.

Coniine   α-Picolinic acid

Coniine is dextro-α-*n*-propylpiperidine, and on oxidative degradation yields α-picolinic acid. It was the first alkaloid to be synthesized (Ladenburg, 1886).

3. Nicotine. Nicotine is a levorotatory, strongly basic liquid alkaloid which is present, together with others, in tobacco. It is very poisonous, and finds its chief use as an insecticide. Nicotine undoubtedly contributes to the physiological effects of tobacco smoke, which contains many other basic and odorous compounds resulting from the thermal decomposition of the tobacco alkaloids.

Nicotine   Nicotinic acid

Structurally, nicotine consists of pyridine linked to an N-methylpyrrollidine ring, and has been synthesized in several ways (first synthesis, Pictet, 1904). On degradation, it yields, among other products, nicotinic acid, which has recently become important in the treatment of pellagra.

4. Atropine (ăt′rô-pēn). This alkaloid is an important medicinal drug, found in belladonna or henbane. It is the racemic (*d, l*) form of the levorotatory alkaloid hyoscyamine (hī-ŏ-sī′ăm-ēn) and is usually prepared by racemization of the latter, more abundant, compound. Hyoscyamine is an ester of tropine with tropic acid. The heterocyclic system present is a fusion of a piperidine ring with a pyrrolidine ring, the two nuclei having two carbon atoms and nitrogen in common.

Tropine   Tropic acid   Tropyltropeine
                               (Hyoscyamine, Atropine)

The tropine portion of the molecule is symmetrical and hence optically inactive; the optical activity of hyoscyamine is due to the tropic acid asymmetry. Tropic acid (Ladenburg, 1889) and tropine (Willstätter, 1903), and hence atropine, have been synthesized.

5. Cocaine (kô′kà-ēn), another member of the piperidine-pyrrollidine group, is obtained from the leaves of the coca shrub (Peru and Java). It is a valuable local anesthetic, especially in eye, nose, and throat surgery. Because of its rather high toxicity and habit-forming properties, it has been replaced in many of its uses by such synthetic local anesthetics as procaine (novocaine), anesthesine, and butyn (page 229).

Cocaine is an ester of *levo*-ecgonine (ĕk'gŏn-ēn), a carboxylic acid related to tropine, with benzoic acid and methanol.

$$\begin{array}{ccc}
CH_2-CH-CHCOOCH_3 & CH_2-CH-CHCO_2H & \\
| \quad\quad | \quad\quad\; H & | \quad\quad | & CH_3OH \\
N-CH_3\; C{<}\quad\quad \rightarrow & N-CH_3\; CHOH & + \\
| \quad\quad | \quad OCOC_6H_5 & | \quad\quad | & C_6H_5CO_2H \\
CH_2-CH-CH_2 & CH_2-CH-CH_2 &
\end{array}$$

Cocaine   l-Ecgonine

Cocaine was synthesized by Willstätter (1923), but all the cocaine of commerce is obtained from the natural source.

6. *Quinine* (kwĭn'ēn), one of the many alkaloids present in the bark of the cinchona tree (South America, East Indies) was for centuries the only effective drug for the treatment of malaria, but has in recent years been largely replaced by synthetic compounds. Quinine contains a quinoline nucleus linked to an 8-membered condensed ring-system called the quinuclidine group. *Quinidine*, identical with quinine excepting the configuration of the CHOH group, is useful in heart irregularities. Hydroquinine (ethyl group in place of the vinyl group of quinine) was synthesized by Rabe in 1931, and the synthesis of quinine itself was completed by Woodward and Doering in 1945.

$$\begin{array}{l}
CHOH-CH-N-CH_2 \\
\quad\quad\quad\quad | \\
\quad\quad\quad\quad CH_2 \\
\quad\quad\quad\quad | \\
\quad\quad\quad\quad CH_2 \\
CH_2-CH-CH-CH=CH_2
\end{array}$$

Quinine

7. *Cinchonine* (sĭn'kŏ-nēn) and cinchonidine, which are also present in considerable amounts in cinchona bark, are related structurally like quinine and quinidine, but lack the methoxyl group in the quinoline ring.

8. *Morphine* (môr'fēn) is the most important of the numerous alkaloids found in opium, the coagulated juice from the unripe seed capsules of the opium poppy. The chief opium-producing lands are India, Iran, Turkey, Yugoslavia, Japan, and China, legitimate opium production being two million kg.; the illegitimate production is estimated to be at least six times as great.

Morphine contains a partly hydrogenated phenanthrene nucleus fused with a 6-membered nitrogen-containing ring, and a 5-membered oxygen-containing ring. It is generally classed with the isoquinoline type alkaloids.

Morphine   Apomorphine

The methyl ether of morphine, codeine (kō'dĕ-ēn), is also found in opium, but most of the codeine used is prepared from the more abundant morphine by methylation of the phenolic hydroxyl group with reagents that do not react with the tertiary nitrogen atom (phenyltrimethylammonium hydroxide, $C_6H_5N(CH_3)_3OH$). Morphine has not been synthesized.

Morphine is of great importance for the alleviation of severe pain, codeine for the control of cough and for the relief of mild pain. The continued use of morphine or opium results in addiction, a psychic and physical slavery to the drugs.

Treatment of morphine with hot concentrated hydrochloric acid results in a complex structural change, yielding apomorphine. Apomorphine is one of the most effective drugs known to cause vomiting.

9. *Papaverine* (păp-ăv'ēr-ēn). In addition to the morphine group, opium has yielded a series of isoquinoline alkaloids, most of which are derivatives of 1-benzylisoquinoline. Papaverine, the first opium alkaloid whose structure was completely known (Goldschmiedt, 1888), may be regarded as the parent substance of the group. It is one of the few known optically inactive alkaloids.

Papaverine

Papaverine is used to some extent medically for its relaxing effect on smooth muscle. It was first synthesized in 1909 (Pictet and Gams), and commercially practicable syntheses have been developed more recently.

10. *Indole Alkaloids.* The alkaloids containing the indole nucleus (p. 223) are very numerous and the structures of many of them (for example, the strychnine, yohimbine, and ergot groups) have not been completely elucidated. Two of the simpler members will be discussed.

Harmine (här′mēn), an alkaloid from the seeds of the African rue, contains an indole nucleus fused with a nitrogen-containing ring in a structure known as the "4-carboline" system.

4-Carboline   Harmine

Harmine (also known as banisterine) is sometimes used in the treatment of Parkinson's disease (shaking palsy). It has been synthesized in several ways.

11. Physostigmine (fī-sō-stĭg′mēn) is found in Calabar or Esère beans, the fruit of an African vine. Although it is one of the most poisonous alkaloids known, it nevertheless finds medical use in minute doses. The nuclear system present in physostigmine is a condensation of an indole ring with a pyrrolidine ring. The aromatic ring carries a very unusual substituent, a urethane type of group (urethane is $H_2NCOOC_2H_5$).

Physostigmine

Physostigmine was synthesized in 1935 by Julian and Pikl.

CHAPTER XXXIII

GENERAL ANESTHESIA*

HISTORICAL INTRODUCTION

"Nothing in the whole realm of human effort has ever contributed so much to human comfort as the discovery of modern anesthesia."

Pain and discomfort are the archenemies of man. To escape them and effectually combat them, man has ransacked the entire earth to find drugs to bring him a surcease of pain. During the middle of the sixteenth century Ambroise Paré operated without anesthesia, except for the administration of French wines, which would produce an alcoholic stupor. It has been only about 120 years since Ephraim McDowell removed an ovarian cyst from Mrs. Jane Crawford in Danville, Kentucky, without any anesthetic agent. She was then forty-seven years old, and she lived to see her seventy-eighth birthday. It is difficult for man today to appreciate the excruciating pain suffered by surgical patients in the preanesthetic days; and, furthermore, no one can with certainty estimate the impediment to surgical progress that the absence of anesthesia would confer upon man.

Joseph Priestly, the discoverer of oxygen, prepared the first generally accepted anesthetic. Priestly was a Unitarian minister in Birmingham, England. In the congregation of this brilliant scientist-clergyman were three illustrious men; James Watt, who discovered the power of steam and holds the admiration of men in all walks of life; Erasmus Darwin, brilliant scientist and skilled clinician whose grandson, Charles, established a new order in biology; and William Withering, "Flower of English Physicians" and discoverer of the use of the purple foxglove in edema of cardiac origin.

In the year 1773, Joseph Priestly made nitrous oxide. To him it was a new chemical compound, a gas whose chemical and physical properties should be investigated. Priestly was unconcerned with its biological effects and died in Northumberland County, Pennsylvania, not knowing that nitrous oxide would confer a blessing of inestimable magnitude upon man.

A quarter of a century passed. Sir Humphry Davy, brilliant English chemist and physicist, made "Priestley's gas." It was then designated in chemical reports as "Dephlogisticated Nitrous Gas." Davy inhaled nitrous oxide and observed a period of great exhiliaration with an increase of pulse rate. In a letter to one of his friends he wrote: "I danced around my laboratory like a madman." But further than this Davy observed that continued inhalation of the gas would produce insensibility to pain. In fact, Davy anesthetized certain of his friends to unconsciousness with nitrous oxide. On July 3, 1798, Mr. Wedgewood called on Davy and permitted Davy to use nitrous oxide on him. He recorded in great detail his experiences, which read, in most respects, like a patient's account of losing consciousness under nitrous oxide. Davy suggested the use of nitrous oxide in medicine, but nothing was done about it.

Time marched on and the scene shifted to America. In Hartford, Connecticut, on December 10, 1844, G. Q. Colton was delivering a lecture on popular science. Among the experiments performed by Colton was the apparent hypnotism of certain members of the audience, presumably by means of his gesticulations. Meanwhile one of Colton's associates engulfed the individual in nitrous oxide. This made effective the hypnotic art of Colton. That afternoon, a dentist whose name was Horace Wells, was in the audience. He saw

This chapter was contributed by Dr. John C. Krantz, Jr., University of Maryland, School of Medicine, Baltimore, Maryland.

one of the people swoon, fall, and hit his shin violently against a bench, without apparent sensation of pain. Through his scintillating intellect flashed the era of painless dentistry. The next day Wells persuaded one of his dental colleagues, Dr. Riggs, to extract one of his teeth, while under the influence of "laughing gas." Wells did not whimper. The first step in man's redemption from pain had been taken. Wells did not succeed in establishing the wide-spread use of his new anesthetic agent. In Boston, where he endeavored to employ it, the gas bag failed most inopportunely; and Wells was hissed out of the room as a mountebank and charlatan. When death came, as it did prematurely, to Wells, he did not realize what a tremendous and far-reaching influence his observations would have upon the comfort and even the destiny of the race.

Crawford W. Long of Georgia used ether as a general anesthetic in 1842. He was familiar with some of the pharmacologic effects of ether, and in Jefferson there were many "ether frolics" which resembled modern parties of inebriates. Long knew ether beyond the period of excitation which its inhalation produces first. He used it to deaden pain in the reduction of fractures and on Mr. James W. Venable to permit the surgical removal of a tumor on the back of his neck. Unfortunate it is indeed that Long did not publicize his observations, for apparently the first paper published by Long on ether appeared in 1849, five years after Well's work with nitrous oxide, and three years after Morton's demonstration of the use of ether in Boston.

In Boston, a chemist named Jackson suggested the use of ether to a dentist named W. T. G. Morton, who was a pupil of Horace Wells. Morton persuaded Dr. J. C. Warren, relative of Dr. J. M. Warren, associated with General Putnam at the Battle of Bunker Hill, to permit him to use ether on one of his patients. On October 16, 1846, in Massachusetts General Hospital, Morton began the administering of ether to Dr. Warren's patient. From that operating room reverberated the memorable statement which has echoed down through the decades, "Dr. Warren, your patient is now ready." Dr. Warren commented that this was no humbug, for Mr. Abbott, the patient, was fast asleep. Ether had found its place.

Shortly after this Oliver Wendell Holmes in a letter to Morton conveyed the fact that he had assigned a generic name to ether and all such agents. He commented on the importance of a proper selection of a name, for Holmes held that it would be on the lips of every person of all races who in time to come would dwell on this planet. He coined the word "anesthesia" from the Greek $\alpha\iota\sigma\theta\eta\sigma\iota\varsigma$ perception and the $\alpha\nu$ negative, namely, without perception

The news of this important discovery soon bridged the Atlantic and James Simpson of Edinburgh began an assiduous search for substances as good as, or perhaps better than, the American Ethyl Ether. It seems strange indeed that scientists in England did not precede those in America in the use of ether, because Michael Faraday, distinguished pupil of Sir Humphry Davy, was one of the first chemists to produce ether. Simpson's experiments were fraught with many failures. Ether was better as a general anesthetic than most every substance that he and his associates tried. One day, as he was fumbling through the papers on his desk, he found a vial of a colorless liquid which had been prepared first by the German apothecary, Justus von Liebig. Simpson tried it — the liquid was chloroform. Several times did he anesthetize himself and his associates, Keith and Duncan, to unconsciousness with at least apparent impunity. It is recorded that immediately Simpson recommended the use of chloroform to alleviate the pain of childbirth. To this the clergy of England objected. They contended that this pain was a penalty pronounced upon Eve for transgressions in the Garden of Eden and that in consequence all subsequent generations of women should endure it with patience and complacence. Simpson was a careful investigator, but also was astute at repartee. To this criticism he very aptly replied, "The Lord caused Adam to fall into a deep sleep before appropriating his rib; out of which he created Eve." God administered the first anesthetic. Queen Victoria, that pioneer of English customs, broke the spell of superstition once and for all by permitting Simpson's chloroform to be used upon her in confinement.

There followed several barren decades in the field of general anesthesia and at the turn of the century, the armamentarium of the anesthetist contained only nitrous oxide, ether, and chloroform augmented in a small measure by ethyl chloride. Through the first two decades of the present century no substantial gains were made. In 1922, however, ethylene was introduced by Luckhardt of the University of Chicago. It had been observed that traces of ethylene caused the bleaching of pigmented flowers and this observation had come to Luckhardt's attention. He was curious about it. He wondered what effect ethylene would have upon animal protoplasm. Systematically he tested the gas on lower animals and observed its anesthetic effects. Ascending in the scale of development, he observed that the anesthetic properties held for monkeys; and finally he permitted himself to be anesthetized many times to unconsciousness. Ethylene subsequently, took its place among the general anesthetics. Perhaps this discovery illustrates the characteristics of a scientist, "one who has the simplicity to wonder, the ability to question, the power to generalize, and the capacity to apply." High concentrations of ethylene are required to produce anesthesia (85 to 90 per cent); and, to avoid hypoxia, the gas must be administered with oxygen. The gas mixture is extraordinarily explosive, and many tragic accidents have occurred owing to the explosion of the gas through ignition by static electric sparks. Undoubtedly this has militated against the widespread use of the gas in many places.

In 1927, Willstätter prepared a general anesthetic, tribromoethanol, marketed and employed as Avertin, dissolved in amylene hydrate. The principle involved in this discovery is based upon the theory of narcosis announced by Meyer and Overton in 1900. Essentially this theory holds that the greater the oil/water solubility is, the more potent is its activity on the central nervous system. Alcohol has anesthetic properties. The alcohols of the aliphatic series of hydrocarbons of higher molecular weight such as amyl and octyl alcohols are less water-soluble and more oil-soluble, and their potencies as anesthetic agents are greater than is that of ethyl alcohol. In Avertin, three of the hydrogen atoms of the ethyl alcohol molecule, having a combined atomic weight of 3, have been replaced by three bromine atoms with a total weight of approximately 240. This increase in molecular weight increases the oil/water coefficient and simultaneously enhances the anesthetic potency of the compound. Avertin is administered rectally. Its anesthetic index or safety margin is narrow; i.e., the anesthetic and fatal doses do not vary by a great degree of magnitude. Therefore, most anesthetists prefer to use the drug in amounts equal to three-quarters of its anesthetic dose as a basal anesthetic and to complete the relaxation with nitrous oxide or ether. The drug is contraindicated in patients suffering with hepatic or kidney diseases. It is unfortunate that all of our data on the efficacy and safety of tribromoethanol are befogged by the fact that it is employed dissolved i another anesthetic agent; namely, amylene hydrate. Furthermore, Avertin is a fixed anesthetic, and threatened collapse under agents of this kind is much more difficult to combat than it is under volatile anesthetics. Under the latter, removal of the mask initiates the immediate course of removal of the agent from the circulating blood. Obviously, when a fixed anesthetic agent is used, this safety factor is unavailable.

It occurred to Chauncey Leake of the University of California in 1930 that it would be of interest to prepare a hybrid molecule between ethyl ether and ethylene—i.e., a molecule which contained the essential features of the molecules of each of these anesthetics. Following this suggestion, Major and Ruigh prepared vinyl ether, *Vinethene.* The relation of these compounds to ethyl alcohol can be seen from the following formulas:

$CH_3.CH_2.O.H$ $CH_3.CH_2.O.CH_2.CH_3$ $H_2C:CH_2$ $H_2C:CH.O.HC:CH_2$
ethyl alcohol   ethyl ether   ethylene   vinyl ether

Vinyl ether is more powerful than ether. It is a liquid of very low boiling point. With it, anesthesia is rapidly induced, but owing to hepatic injury which may occur upon the prolonged inhalation of this anesthetic agent, its use is confined to operations of short duration. One must not pass over the production of this new agent without paying due tribute to the fertility of the mind that conceived it. In its conception a molecule was designed, synthetized, and the anticipated properties were later discovered to be inherent in it.

In 1930 Lucas and Henderson of the University of Toronto announced the anesthetic properties of the hydrocarbon cyclopropane, which is represented graphically as:

$$H_2C \underset{\diagup \diagdown}{\overset{CH_2}{\text{------}}} CH_2.$$

The gas is more potent than ethylene and hence permits the admixture during anesthesia of a larger percentage of oxygen. Relaxation of abdominal musculature is complete during cyclopropane anesthesia. The induction period is rapid and it has a safe anesthetic index. During the decade of its use the gas is now established as an important and dependable anesthetic agent. Recognition of it in the Second Supplement to the *United States Pharmacopoeia XI* bespeaks its growing field of usefulness. A brochure by Robbins, of Vanderbilt University, on Cyclopropane reviews the entire field.

At the Medical School of the University of Maryland in 1939, Krantz, Drake, Carr, and Forman succeeded in developing a chemical reaction for the convenient preparation of aliphatic cyclopropyl ethers. Four of these ethers have been prepared already and one of them has had preliminary trial. This new anesthetic agent is cyclopropyl methyl ether (*cyprome ether*), which is represented graphically above.

$$H_2C \underset{\diagup \diagdown}{\overset{CH_2}{\text{------}}} CH.O.CH_3, \quad cyprome\ ether$$

The pharmacologic studies conducted in the University of Maryland show Cyprome Ether to be more potent than ethyl ether and possibly safer. Its boiling point is 10 degrees C. higher than ethyl ether which should be a distinct advantage for anesthesia in the tropics. Black, Shannon, and Krantz in 1940 reported their first 25 human cases of anesthesia with cyprome ether in *Anesthesiology*. The compound appears to be promising.

Other new anesthetics which have been produced and are under study at the present time include:

$$H_2C \underset{\diagup \diagdown}{\overset{CH_2}{\text{------}}} CH.O.CH_2.CH_3, \quad \text{and} \quad H_2C \underset{\diagup \diagdown}{\overset{CH_2}{\text{------}}} CH.O.HC:CH_2.$$
Cypreth Ether   Cyprethylene Ether

An isomer of cyprethylene ether with the following formula $H_2C:C(CH_3)$.O.HC:CH_2, has been prepared and designated as *Propethylene*. Experimentally and clinically the new anesthetic appears promising. It is 3 to 4 times more potent than ether and the patients awaken rapidly from an anesthesia under propethylene with a minimum of nausea and abdominal distress.

Individual General Anesthetics

Ethyl Ether:

Ethyl ether or ether is prepared by the action of sulfuric acid on ethyl alcohol at 140°C. The reaction occurs in two stages as expressed by the following equations:

$$C_2H_5OH + H_2SO_4 \longrightarrow C_2H_5.HSO_4 + H_2O, \text{ then}$$
$$C_2H_5.HSO_4 + C_2H_5OH \longrightarrow \genfrac{}{}{0pt}{}{C_2H_5}{C_2H_5}\Big\rangle O + H_2SO_4.$$

The process for the production of anesthetic ether on a commercial basis was developed to a great extent by the late Dr. E. R. Squibb. The sulfuric acid employed theoretically will convert an unlimited amount of alcohol into ether, but in practice the acid becomes diluted with the water formed in the reaction and loses its capacity to unite with more alcohol to form ethyl sulfuric acid.

Ether for anesthetic purposes, "Aether," U.S.P. XIII, contains from 96 to 98 per cent of diethyl oxide: the remainder is alcohol and water. Ether is a transparent, colorless, mobile liquid which has a characteristic odor and a burning sweetish taste. Ether boils at 35° and has a specific gravity of 0.713 to 0.716 at 25°C. One volume of ether will dissolve in about 12 volumes of water at 25°C.

Ether when exposed to light and air has a tendency to develop peroxides. These are explosive and also serve as pulmonary irritants, when ether containing them is employed as an anesthetic. The structure of ether peroxide, according to Wieland, is represented as $CH_3.CHOH.O.O.CHOH.CH_3$, which is dihydroxydiethyl peroxide.

The Pharmacopoeia requires that ether used for anesthetic purposes must be peroxide free. The test is carried out in the following manner. "Shake 10 ml. of ether occasionally during 1 hour with 1 ml. of a freshly prepared aqueous solution of potassium iodide (1 in 10) in a 25 ml. glass-stoppered cylinder of colorless glass, protected from light: when viewed tranversely against a white background, no color is seen in the either liquid."

Chloroform:

Chloroform is prepared by heating alcohol or acetone with calcium hypochlorite. The following equations express the reactions which occur in each case.

$$4\ C_2H_5 \cdot OH + 8\ Ca(ClO)_2 \longrightarrow 2\ CHCl_3 + 3\ Ca(O \cdot OC \cdot H)_2 +$$
$$5\ CaCl_2 + 8\ H_2O, \text{ and } 2\ (CH_3)_2C{:}O + 3\ Ca(ClO)_3 \longrightarrow 2\ CHCl_3$$
$$+ 2\ Ca(OH)_2 + Ca(O \cdot OC \cdot CH_3)_2.$$

The calcium hypochlorite is mixed with water and eithe. alcohol or acetone and subjected to distillation; crude chloroform distils over. It is then purified by mixing with concentrated sulfuric acid, washing with sodium carbonate, and subsequent distillation.

Pure chloroform is a heavy, colorless, mobile liquid that possesses a characteristic engulfing odor and a burning sweet taste. The chloroform for anesthetic purposes recognized by the United States Pharmacopoeia contains between 99 and 99.5 per cent of $CHCl_3$; the remainder consists of a little alcohol. The presence of alcohol serves as antioxidant to chloroform.

Chloroform has a specific gravity between 1.474 and 1.478 at 25°C., it boils at 61°C., but has a high vapor pressure at room and body temperatures. Chloroform vapor does not form an explosive mixture with air, but the vapor will burn presenting a flame with a green mantle. Chloroform is miscible with alcohol and the ethereal solvents but very sparingly soluble in water: it requires about 210 ml. of water to dissolve 1 ml. of chloroform.

Upon exposure to light and air, chloroform gradually undergoes disintegration as expressed by the following equation:

$$2 \ CHCl_3 + O_2 \longrightarrow 2 \ Cl \cdot CO \cdot Cl + 2 \ HCl.$$

Phosgene is an extraordinary pulmonary irritant and must be absent in anesthetic chloroform.

It is of interest to note that the fluorine analogue of chloroform namely, fluoroform—is devoid of anesthetic properties. Chloroform is about 5 times more potent than ether as a general anesthetic. Its toxicity is strikingly characteristic on the heart and liver. The approximate mortality under chloroform anesthesia is of the order of magnitude of 1 in 2,500, with ether 1 in 10,000.

Ethylene:

Ethylene is the first member in the olefin series of hydrocarbons. It is prepared by dehydrating agents on ethyl alcohol or by the cracking of petroleum hydrocarbons.

Ethylene used for anesthetic purposes contains not less than 99 per cent by volume of $H_2C:CH_2$. The gas is colorless and possesses a slightly sweet odor and taste. At 25°C. 1 volume of ethylene dissolves in about 9 volumes of water. Ethylene is slowly soluble in concentrated sulfuric acid and is rapidly absorbed by fuming sulfuric acid.

Divinyl Oxide:

Divinyl oxide or Vinethene is the hybrid molecule between diethyl ether and ethylene. Divinyl oxide is prepared from β, β¹-dichlorethyl ether by the removal of 2 molecules of HCl, as:

$$Cl - CH_2 \cdot CH_2 - O - CH_2 \cdot CH_2 - Cl -$$
$$2 \ HCl \longrightarrow H_2C:CH - O - HC:CH_2.$$

This unsaturated ether is a colorless mobile liquid with a peculiar engulfing odor. It boils at 28.3°C. The compound, as it was marketed first, was very prone to polymerize. This is prevented by the addition of 0.01 per cent of phenyl d-naphthylamine. The rapid volatility of divinyl oxide and its great potency render it very useful as an anesthetic for short operations where it is desirable to have the induction period of short duration.

Tribromoethanol:

This compound was prepared first by Willstätter and Duisberg by the reduction of tribromoacetaldehyde. They used hops to bring about the reduction.

$$Br_3C \cdot CO \cdot H + hops \longrightarrow Br_3C \cdot CH_2 \cdot OH.$$

The reduction of bromal now is reported to be accomplished by means of aluminum ethoxide.

The solution of tribromoethanol in amylene hydrate (100 gm. in sufficient amount to make 100 ml.) is employed as an anesthetic under the trade name of *Avertin*.

Avertin:

Tribromoethanol is a white crystalline powder with a slight aromatic odor and taste. One gram of the alcohol is soluble in 35 ml. of water at 25°C. The crystals melt between 79 and 82°C.

Cyclopropane:

Methods of manufacture of cyclopropane and its characteristic reactions are given on pages 172 and 302. The gas which is used for anesthetic purposes contains not less than 99 per cent of cyclo trimethylene. One volume of the gas dissolves in about 2.7 volumes of water at 15°C. The United States Pharmacopoeia limits the amount of propylene, allene, and other unsaturates permitted to be present in cyclopropane. The test depends upon the reduction of standard potassium permanganate solution in the cold (2-3°C.) by the unsaturated compounds. Under these conditions cyclopropane does not reduce the permanganate solution.

Cyclopropane bids fair to retain its present place of deserved preeminence among the general anesthetics.

Pentothal Sodium:

This compound is a thiobarbituric acid derivative and is employed extensively as an intravenous anesthetic. The numerous barbiturates which are used in medicine are prepared by one general procedure; i.e., the condensation of the appropriately substituted malonic ester with urea or thiourea as the case might be. Pentothal sodium is the sodium salt of l-methylbutyl-ethylthiobarbituric acid as shown by the formula at the right.

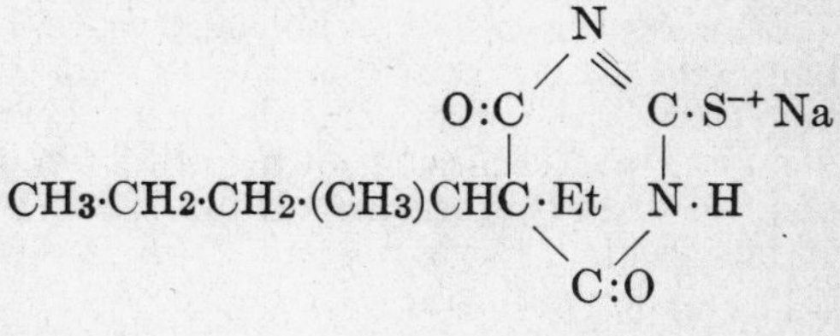

$$\text{CH}_3 \cdot \text{CH}_2 \cdot \text{CH}_2 \cdot (\text{CH}_3)\text{CHC} \cdot \text{Et} \begin{array}{c} \text{N} \\ \diagup \diagdown \diagdown \\ \text{O:C} \qquad \text{C} \cdot \text{S}^{-+}\text{Na} \\ | \qquad\qquad | \\ \qquad\qquad \text{N} \cdot \text{H} \\ \diagdown \qquad \diagup \\ \text{C:O} \end{array}$$

The compound is a white powder and very soluble in water to which, through hydrolysis, it imparts a rather strong alkaline reaction. It is used intravenously in 2.5 to 5 per cent solutions. Small volumes are slowly injected and surgical anesthesia ensues immediately. It is recommended for operations lasting no more than 30 minutes.

For a survey of anesthetics, see *Industrial Chemist*, p. 187 (May, 1938).

CHAPTER XXXIV

ANTIBIOTIC AGENTS*

Introduction

At the present time students of chemistry and biology have renewed their interest in substances capable of inhibiting the growth of pathogenic bacteria. "Antibiotic" is the term commonly applied to these substances, which are known to be widespread throughout nature. Antibiotic activity, however, is by no means a new discovery. Pasteur and Joubert, as long ago as 1877, were aware that certain organisms inhibited the growth of the Bacillus anthracis. These investigators even suggested that this phenomenon of antibiosis might be of use in the treatment of certain infections. This suggestion was based on their observation that the growth of certain bacteria could be prevented by the concomitant growth of other bacteria. It is now well established that inhibition of growth produced by certain micro-organisms is due to elaboration by the antagonistic microbe of certain products which have chemical and biologic properties. Waksman, in 1941, suggested that the term "antibiotic" be used in connection with antibacterial agents of microbial origin. The term "antibiotic," however, also has been used recently in connection with certain substances which occur in nature but which are not of microbial origin. Examples of antibiotics which are not derived from microbes include such substances as lysozyme, chlorellin, canavalin, and allicin, which will be mentioned again.

It seems fairly well established that the most important antibiotic agents yet available are those of microbial origin. These are derived from three general sources: (1) those of bacterial origin, including the earliest known antibiotic, pyocyanase (Emmerich and Loew) and tyrothricin (Dubos); (2) those derived from molds and fungi, the most important of which is penicillin (Fleming), and (3) those derived from actinomycetes, of which the most promising at the moment appears to be streptomycin (Schatz, Bugie, and Waksman).

History, Occurrence and Uses of Antibiotics of Microbial Origin

Pyocyanase and Pyocyanine:

That the organism Pseudomonas pyocyanea was antagonistic

---

* Contributed by Wallace E. Herrell, M.D., M.S. in Medicine, Division of Medicine, Mayo Clinic, Rochester, Minnesota.

to other bacterial species was discovered in 1889; however, the antagonistic substance was not named "pyocyanase" until 1899. Emmerich and Loew, who named the substance "pyocyanase," considered it to be an enzyme. The substance was known to be present in the broth in which Pseudomonas aeruginosa (Pseudomonas pyocyanea) grew. It was subsequently found that this organism produced not only the enzyme, pyocyanase, but also pyocyanine. These antibiotic agents are active against a variety of gram-positive and gram-negative pathogens but, because of their toxicity for experimental animals, neither substance has proved of any significant clinical value although each has been used locally at times in the treatment of infections of human beings.

Penicillic Acid:

Several years after the description of pyocyanase, Alsberg and Black in 1913 described another antibiotic substance to which was given the name "penicillic acid." This should not be confused with penicillin which in the free state exists as a weak acid. Penicillic acid was found to be produced by a strain of mold known as Penicillium puberulum. Sometime later it was reported that this substance was also elaborated by another species of Penicillium; namely, Penicillium cyclopium. Penicillic acid possesses considerable activity not only against gram-positive bacteria but also against members of the colon-typhoid-Salmonella group of gram-negative pathogens. Little information has been accumulated concerning its possible therapeutic value.

Actinomycetin:

This antibiotic agent was described in 1924 by Gratia and Dath. The substance was found to be elaborated by certain strains of Actinomyces and to be effective against certain gram-positive, as well as gram-negative, organisms. It is of little or no practical value.

Penicillin:

All are aware that penicillin, first described by Fleming in 1929, is the most important antibiotic agent yet described. Fleming observed that a strain of mold, Penicillium, later identified as Penicillium notatum, elaborates a substance which possesses remarkable antibacterial activity. To this substance he gave the name "penicillin." It was found in the broth in which the mold had been allowed to grow. It was evident from the earliest reports that penicillin is a substance of extremely low toxicity and yet possesses a high degree of activity against important gram-positive organisms as well as against a few important gram-negative bacteria. This substance has proved of great value in the treatment of important bacterial infections in man. It may be applied locally to infected surfaces or it may be administered intravenously or intramuscularly to man without serious toxic effects. It also may be introduced directly into infected body cavities, such as the pleural space, infected joints, or directly into the cerebrospinal canal.

Citrinin:

Hetherington, Raistrick, and colleagues, two years after the report by Fleming, obtained citrinin, another antibiotic agent, from a species of Penicillium identified as Penicillium citrinum (Thom). The material is predominantly active against gram-positive organisms. Unfortunately, although it possesses anti-bacterial activity, it is extremely toxic when administered to experimental animals either by the oral or the intraperitoneal route. Its administration to animals may be followed by ataxia, respiratory difficulty, and even convulsions. The toxicity of citrinin precludes its consideration as a therapeutic agent.

Gliotoxin:

This antibiotic agent was obtained by Weindling and Emerson in 1936 from Trichoderma lignorum. It also appears that gliotoxin can be obtained from filtrates of media in which Aspergillus fumigatus has been allowed to grow. Gliotoxin appears to possess antibacterial activity for gram-positive and certain gram-negative pathogens. As is true of many other antibiotic agents, this one is exceedingly toxic. Because of its high degree of toxicity, it holds little promise as a therapeutic agent.

Fumigatin:

This antibiotic agent was obtained in 1938 by Anslow and Raistrick in crystalline form from broth cultures in which the organism, Aspergillus fumigatus, had been allowed to grow. It was found to be especially active against certain gram-positive organisms; however, in its present form fumigatin is too toxic to be suitable for the treatment of infection.

Tyrothricin:

This substance, like the first discovered antibiotic agent, pyocyanase, is of bacterial origin. It was first described by Dubos in 1939. One source of this antibiotic agent is Bacillus brevis, an organism which was isolated from soil. This agent was first known as gramicidin but later it was evident that tyrothricin contained two substances, pure gramicidin and tyrocidine. Tyrothricin is an effective agent in the inhibition of growth of important gram-positive and a few gram-negative organisms. Because of its hemolytic property, it may not be administered parenterally in the treatment of infections. It possesses low toxicity for tissue other than blood and, therefore, has been of practical use in the local treatment of infected surfaces in man. It also has been used by means of local instillation in the treatment of bovine mastitis.

Widespread new interest and revival of interest in antibiotic agents followed the report of Dubos.

Actinomycin A and B:

Actinomycin A and B should not be confused with actinomycetin which also is derived from actinomycetes. Actinomycin A and B were obtained in 1940 by Waksman and Woodruff from Actinomyces antibioticus. These two substances were considered bacteriostatic for many gram-positive organisms and only weakly active against gram-negative pathogens. They are not identical but are closely related. Both substances have proved exceedingly toxic when given to experimental animals. Efforts to remove the toxic properties and at the same time preserve their antibacterial activity have been unsuccessful. Neither has any place in the treatment of disease of man.

Streptothricin:

Streptothricin was derived from a strain of Actinomyces, known as Actinomyces lavendulae, by Waksman and his colleagues in 1942. It appeared from certain experimental studies that this substance in addition to inhibiting the growth of certain important gram-negative, as well as some gram-positive, organisms, also inhibits the growth of certain pathogenic and saprophytic fungi. When first described, it was thought to be sufficiently safe for general or systemic administration. Later it became evident, however, that it possessed toxic features which limit its use to local application to infected surfaces.

Claviformin, Clavacin and Patulin:

These three antibiotic substances appear to be identical but were described by three different groups of investigators. Claviformin (Chain and co-workers) and clavacin (Waksman and associates) were described in 1942. Patulin (Raistrick and others) was described in 1943. Claviformin was obtained from cultures of Penicillium claviforme, clavacin from Aspergillus clavatus, and patulin from Penicillium patulum. All three substances have the same empirical formula and all three are apparently effective against the same gram-positive and gram-negative organisms. The description of the identical substance under three different names and from three microbial sources is evidence of the rapid increase in the interest which occurred following the reports on penicillin, tyrothricin, and other antibiotic substances. From a practical standpoint, this antibiotic substance (with three different names) is not suitable for subcutaneous or intravenous administration. Experimental animals die shortly after intraperitoneal administration of the substance. Whether or not the substance will prove of value as a local agent remains to be seen.

Fumigacin and Helvolic Acid:

Fumigacin was obtained from Aspergillus fumigatus by Waksman, Horning, and Spencer in 1942, and helvolic acid was reported by Chain and others in 1943 and was also obtained from a strain of Aspergillus fumigatus. Fumigacin and helvolic acid have the same empirical formula and both are active against the same gram-positive pathogens. Here, again, is an example of an antibiotic reported almost simultaneously from two different sources under two different names. Because of certain toxic effects, it seems unlikely that the identical substances, fumigacin and helvolic acid, deserve at this time serious consideration as therapeutic agents.

Aspergillic Acid:

This antibiotic agent was derived from a strain of Aspergillus flavus. The substance was described by White and Hill in 1943 and was obtained in crystalline form. Although it was found to be effective in the prevention of the growth of both gram-positive and gram-negative bacteria as well as organisms associated with gas gangrene, it is too toxic for systemic administration. Whether or not it may prove of value for local treatment has not yet been established.

Flavicin:

This antibiotic agent was isolated by Bush and Goth in 1943. It was obtained from broth filtrates of Aspergillus flavus. According to Bush and Goth, flavicin behaves in many ways like penicillin and, in fact, appears to be somewhat more active than penicillin against certain gram-positive organisms. Although the chemical nature of the substance is not well known, it appears to resemble penicillin rather closely. If this resemblance is proved on further investigation, another therapeutic agent appears to be available which may have an effect comparable to that of penicillin.

Gigantic Acid:

From Aspergillus giganteus, Philpot, in 1943, obtained gigantic acid, whose properties resemble rather closely those of penicillin. It cannot yet be stated whether or not gigantic acid is of value as a therapeutic agent. Its isolation, however, indicates clearly that substances which resemble penicillin can be produced by certain species of Aspergillus.

Subtilin:

From a strain of Bacillus subtilis, Jansen and Hirschmann, in 1944, obtained subtilin, which possesses high antibacterial activity against certain important gram-positive bacteria. The activity may be due to more than one substance. Little information is available concerning the possible clinical value of this agent.

Streptomycin:

Streptomycin was obtained by Schatz, Bugie, and Waksman in 1944 from Actinomyces griseus, which had been described previously by Waksman. This substance appears to be active against a variety of important gram-negative organisms as well as some gram-positive pathogens and, in addition, it possesses definite bacteriostatic action for Mycobacterium tuberculosis. Unlike streptothricin, this substance appears to be tolerated well and is suitable not only for local but also for systemic administration. It has been obtained in fairly pure form and what is known concerning its physical and chemical properties will be mentioned again later.

Chaetomin:

Chaetomin was described by Waksman and Bugie in 1944. Its antibacterial activity appears to be similar, but not identical, to that of penicillin. It appears to be produced by Chaetomin cochliodes. Chaetomin possesses marked bactericidal properties. It is not exceedingly toxic but unfortunately little or no protection against various bacteria occur in vivo. If the substance is active only in vitro, it is not likely to be of importance so far as clinical use is concerned.

Flavacidin:

Another substance which appears to resemble penicillin closely was obtained in 1944 by McKee, Rake, and Houck from Aspergillus flavus, which had been grown in submerged culture and was named "flavacidin." Although the chemical nature of the substance is not well known, there is considerable evidence of biologic similarity between this antibiotic agent and penicillin.

Bacitracin:

This antibiotic agent was obtained by Johnson and his colleagues in 1945 and appears to be produced by Bacillus subtilis. It is active chiefly against gram-positive organisms; however, it appears that Neisseria gonorrhoeae and Neisseria intracellularis are susceptible to its action. Bacitracin has been used clinically in the treatment of streptococcal and staphylococcal infections in man and, according to the report on its isolation, encouraging results have been obtained.

As indicated previously, antibiotics of microbial origin have in general been obtained from (1) bacteria, (2) molds and fungi, and (3) actinomycetes. The nomenclature, the origin, the author who described the agent, and the organisms against which they are effective are listed in alphabetical order in tables 1, 2 and 3.

Methods of Preparing Antibiotics

In general, the procedures for obtaining antibiotic agents of microbial origin are essentially the same. If the investigator suspects that a microbe elaborates a substance which inhibits bacteria, he grows the microbe on a suitable medium to which has been added a variety of organisms which may or may not be inhibited by the concomitant growth of the antibiotic-producing microbe. For example, sensitive bacteria will not grow in agar around a colony of Penicillium notatum. Likewise, the growth of a sensitive microbe will be inhibited frequently in broth filtrates in which the antibiotic-producing microbe has grown and from which this microbe has been removed by the simple process of filtration. Once a microbe which is thought to produce an antibiotic agent has been screened out in this fashion, certain steps are essential in obtaining the active substance. Since, as has already been mentioned, the process is similar in connection with the isolation of most antibiotic agents, the description of such a procedure will be given in some detail in regard to the isolation of only one antibiotic; namely, penicillin.

While various strains of Penicillium will produce penicillin, the one most commonly used is a strain of Penicillium notatum known as NRRL 1249.B21. This organism stems directly from the original strain of Fleming and may be obtained from the American Type Culture Collection, Washington, D. C. After the penicillin-producing mold has been obtained, the worker selects a suitable medium on which to grow the mold. The medium most frequently used is a modification of synthetic Czapek-Dox medium which consists of: Ferrous sulfate ($FeSO_4.7H_2O$), 0.01 g.; glucose, 40.00 g.; magnesium sulfate ($MgSO_4.7H_2O$), 0.50 g.; potassium acid phosphate (KH_2PO_4), 1.00 g.; potassium chloride (KCl), 0.50 g.; sodium nitrate ($NaNO_3$), 3.00 g.; and distilled water, g.s. ad., 1,000 ml.

This medium has been modified frequently by different investigators. For example, the substitution of brown sugar for glucose has at times improved the yield of penicillin. Various inorganic salts, such as zinc sulfate, likewise have been added in varying amounts. The addition of corn-steep liquor to the medium also increases the yield considerably. At all events, the person who sets out to prepare penicillin will select the medium most readily available. After the medium has been selected, spores from the master culture of the strain of Penicillium are sown on the medium. The stock culture can be grown on a slope of the Sabouraud medium and a suspension of spores made from this culture by shaking with sterile water. The spores are then implanted.

The mold may be grown in flat bottles or in flasks. The spores are seeded on the surface of a layer of culture medium which is 1.5 to 2 cm. in depth. A convenient procedure is to use 50 ml. of medium in a 200 ml. Erlenmeyer flask. If large pans are to be used, metal containers should be avoided unless they are enameled or lined with glass because heavy metals will destroy penicillin. The flasks or bottles are then incubated at temperatures between 22°C. and 25°C. The mold fails to grow at 37°C. A delicate growth will appear on the bottom of the vessel in twenty-four hours. By the third day a white growth is present on the surface of the medium. This growth becomes coalesced, green, and somewhat rigid about the fifth day. After four or five days, a few centimeters of broth under the surface growth are drawn off under sterile conditions by means of a pipette for the purpose of testing for changes in the pH and antibacterial potency. The pH of the medium which starts around 3.5 to 4.5 remains essentially the same for three or four days. After this, there is a rise in pH during which time a rapid development of the antibacterial activity of the broth containing penicillin occurs. The peak of production of penicillin occurs when the pH approaches the neutral point. Beyond the neutral point the medium rapidly loses its penicillin activity. As a rule, the maximal antibacterial titer of the medium will occur around the seventh day. The mold then is filtered off and the pH of the medium adjusted to 6.5 to 7. The filtrate may be kept at icebox temperature and will remain active for as long as seven weeks, provided the pH is kept adjusted to 6.5 to 7. If the crude penicillin broth is frozen, it may be preserved for several months. It is exceedingly important to avoid bacterial contamination by certain organisms, particularly gram-negative bacteria which rapidly inactivate the penicillin present.

If the mold is grown successfully, the broth filtrates may vary in potency from 30 to 100 Oxford units per ml. The free penicillin exists in the form of a weak organic acid which reacts chemically to form various salts. Penicillin can be extracted from the broth filtrates by use of a solvent, such as ether or amyl acetate. This extraction is best accomplished by adjusting the pH of the broth to between 2 and 3. Since penicillin is unstable at this low pH, the extraction must be done rapidly and at as low a temperature as is feasible. In the preparation of the sodium salt of penicillin, the penicillin present in the solvent is obtained by shaking the solution with aqueous sodium bicarbonate. Following this extraction the pH should be quickly readjusted to 6.5 to 7. Since aqueous solutions of sodium penicillin are rather unstable and since heat destroys the material, it is essential in the drying operation to keep the penicillin frozen and to dry it in this state. This is best accomplished by a lyophilyzing process similar to that used in the preparation of dried plasma. The final product is a powder which ranges in color from pale yellow to brown. It should be stored in sealed ampules in as near the dry state as possible.

Obviously, many refinements and modifications of extracting penicillin have been made. The foregoing surface culture method is outlined for the student interested in its preparation on a small scale.

For the large-scale production of penicillin, the submerged culture or tank method has been employed. For this work, a strain of Penicillium notatum (NRRL 832) which grows well in submerged culture is used. After the mold has been seeded in the culture in large vessels, it is necessary to aerate and agitate the culture to obtain submerged growth. This can be effected by shaking the flasks and by aerating the mixture by means of any suitable aerator. Large rotary drums equipped with aerators and agitators have been used in large-scale production.

Physical and Chemical Properties

Something is known concerning the physical and chemical properties of most of the antibiotic agents so far developed. Final information on the physical and chemical properties of some of these agents obviously is not available at this time. Much remains to be done on the chemistry (of these agents), which offers a real challenge to the student of organic chemistry. What has been published in connection with these agents will be mentioned briefly in the order shown in Tables 1, 2, and 3.

Bacitracin (Table 1):

This antibiotic agent can be extracted from the medium with normal butanol and concentrated by steam distillation in a vacuum. On further purification, the resulting substance is a grayish white powder. It is a neutral substance which cannot be precipitated from the medium by manipulating the pH. It has not yet been obtained in pure form; therefore, little is known concerning its chemical nature. It is soluble in water and withstands heating for fifteen minutes at 100°C. It is not hemolytic. It is stable in acid solution but unstable in alkaline solution of higher pH than 9.

Pyocyanase (Table 1):

Pyocyanase is a thermostable, lipoid substance. There is general agreement that the substance is an enzyme. Studies on the physical and chemical properties of the substance are not sufficiently in agreement to warrant further discussion.

Pyocyanine (Table 1):

Pyocyanine, another antibiotic derived from Pseudomonas aeruginosa, is a blue pigment. It is soluble in chloroform and is thermostable. Its chemical nature is in doubt, but it may have one of the following two structures:

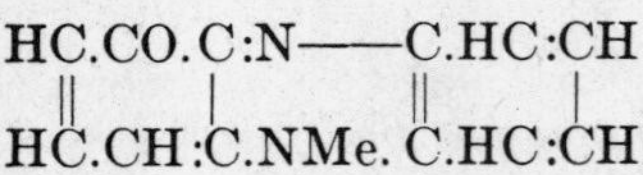

$$
\begin{array}{ll}
\text{HC.CO.C:N}\text{---}\text{C.HC:CH} & \qquad \text{HC:C}(\bar{\text{O}})\text{---}\text{C:N}\text{---}\text{C.HC:CH} \\
\text{HC.CH:C.NMe. C.HC:CH} & \qquad \text{HC:CH}\text{---}\text{C:NMe. C.HC:CH} \\
\qquad\qquad (\text{II}) & \qquad\qquad\qquad\qquad (\text{III})
\end{array}
$$

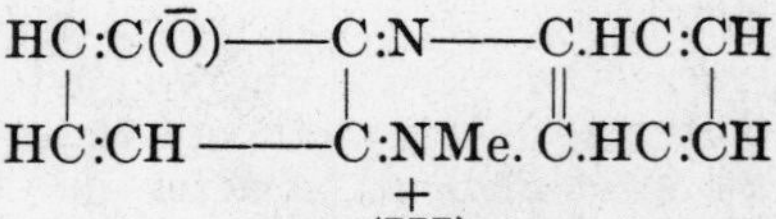

Subtilin (Table 1):

Subtilin appears to have a rather complex chemical nature. There is some evidence to suggest that it is a polypeptide. In this regard, it is like tyrothricin which is obtained from Bacillus brevis. Details of its chemical nature are not available.

Tyrothricin, Gramicidin and Tyrocidine (Table 1):

Gramicidin, one component of tyrothricin, has been obtained in crystalline form as colorless platelets. The precise empirical formula is not known but is approximately: $C_{74}H_{105}N_{15}O_{13}$ (molecular weight, 1,413). It appears to have no free carboxyl or amino groups although it is composed of amino acids. It is not soluble in many of the ordinary solvents. It is soluble and entirely stable in alcohol even when maintained at high temperatures. It is hemolytic.

In contrast to gramicidin, the other component of tyrothricin, tyrocidine ($C_{126}H_{166}N_{26}O_{26}.2HCl$? , M.W. 2,534), contains many amino acids suitable for quantitative estimation. This substance also has been obtained in crystalline form and the crystals, like those of gramicidin, appear as colorless platelets. Like gramicidin, it is hemolytic and behaves somewhat like a detergent.

Aspergillic Acid (Table 2):

Aspergillic acid has been obtained in crystalline form. The empirical formula appears to be $C_{12}H_{20}N_2O_2$. The suggested structural formula for aspergillic acid is:

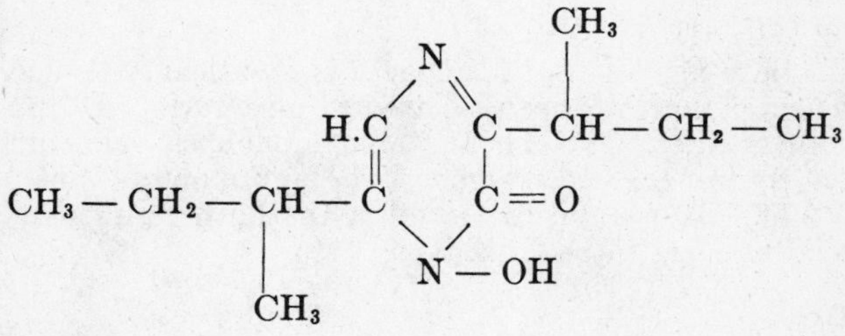

Chaetomin (Table 2):

Chaetomin is readily soluble in acetone, ethyl acetate, chloroform, benzene, dioxane, and pyridine. It is insoluble in water and petroleum ether. Chaetomin contains nitrogen and sulfur; however, it has not yet been obtained in crystalline form and neither its empirical nor structural formula can be suggested.

Citrinin (Table 2):

Citrinin is a yellow, acidic, rather stable substance which has been obtained in crystalline form. The empirical formula suggested is $C_{13}H_{14}O_5$. It is insoluble in cold water but is soluble in many of the commonly used organic solvents. It is extremely toxic. The structural formula is shown at the right.

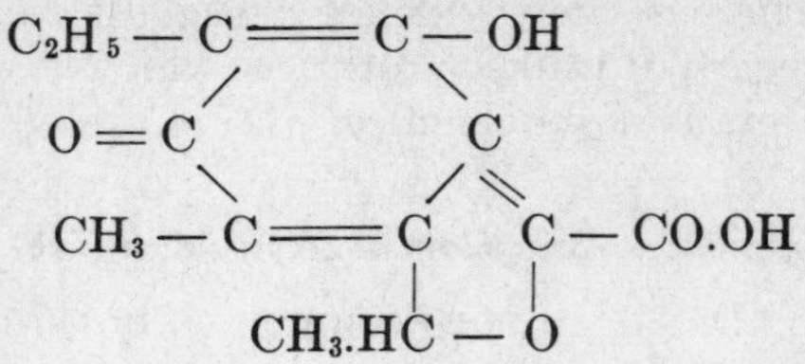

$$C_2H_5 - C = C - OH$$

Clavacin, Claviformin, and Patulin (Table 2):

Clavacin has been obtained in a fairly pure form. As was pointed out earlier, it is generally agreed that clavacin, claviformin, and patulin are identical. The empirical formula ascribed to all three is $C_7H_6O_4$ and the structural formula is shown at the right.

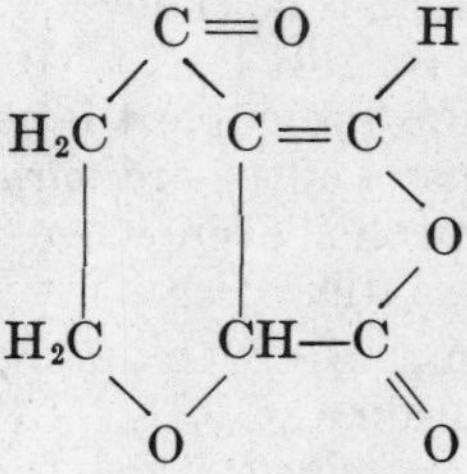

Flavacidin and Flavicin (Table 2):

Flavacidin appears to be biologically and chemically closely related to penicillin. For comments concerning the physical and chemical properties, see penicillin.

Like flavacidin, this antibiotic agent also appears to resemble penicillin rather closely. It is soluble in water and ether. It is rather unstable in acid media. It is of low toxicity and high in antibacterial activity. Details concerning its chemical nature are not available at this time.

Fumigacin and Helvolic Acid (Table 2):

It is generally agreed that fumigacin is identical with helvolic acid. The substance has been isolated in pure form. It appears as fine, white needles. It is a heat stable, monobasic acid containing carbon, hydrogen, and oxygen. The approximate molecular weight is 500. The empirical formula for both fumigacin and helvolic acid appears to be $C_{32}H_{44}O_8$.

Fumigatin (Table 2):

This antibiotic also has been obtained in crystalline form. It appears to be quinoidal in character. It is rather soluble in most solvents but is extremely toxic. The empirical formula suggested for fumigatin is $C_8H_8O_4$. The suggested structural formula is shown at the right.

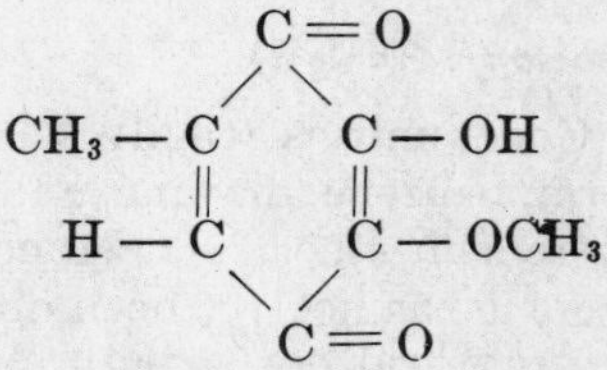

Gigantic Acid (Table 2):

All that can be definitely stated concerning the properties and chemistry of gigantic acid is that it appears to resemble penicillin closely (see penicillin).

Gliotoxin (Table 2):

This antibiotic has been obtained in crystalline form from filtrates of Aspergillus fumigatus. It appears to be a sulfur-bearing ring compound. In addition to the disulfide grouping, gliotoxin contains two other groups: the indole nucleus and the ortho ester grouping. The indole grouping occurs in several other biologically important compounds whereas the ortho ester grouping is rarely found among compounds which occur naturally. In crystalline form, gliotoxin appears as elongated plates. The empirical formula suggested for gliotoxin is $C_{13}H_{14}N_2O_4S_2$. The structural formula is shown at the upper right.

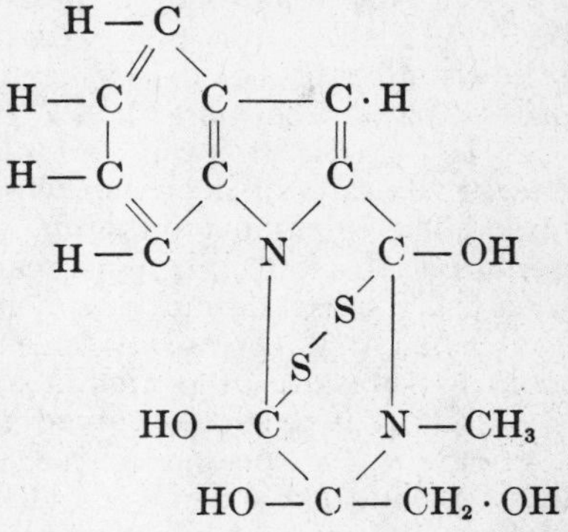

Penicillic Acid (Table 2):

The antibiotic, penicillic acid, should not be confused with penicillin although the latter substance in the free state exists as a weak acid. Penicillic acid was described nearly two decades before penicillin. The name was given to the substance long before any chemical work had been carried out on penicillin. The name only indicates that the material is acidic in nature. It is an *alpha-beta* unsaturated ketone. Its action can be inhibited in the presence of amino acids or compounds containing sulfhydryl. It is a colorless substance and is readily soluble in water. The open-chain formula is: $CH_3.C(:CH_2).CO.C(OCH_3):CH.CO.OH$. The structural formula for penicillic acid follows:

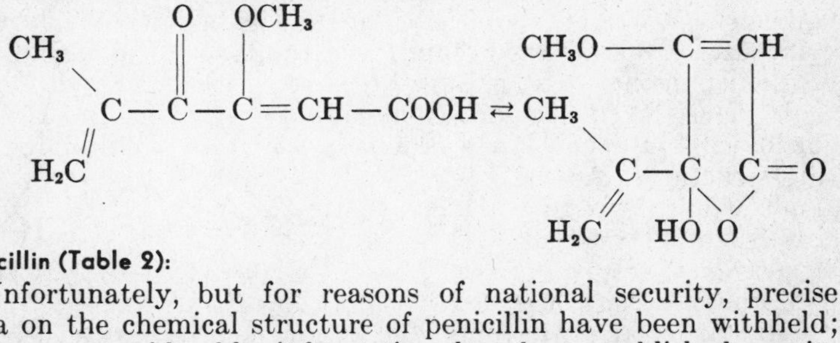

Penicillin (Table 2):

Unfortunately, but for reasons of national security, precise data on the chemical structure of penicillin have been withheld; however, considerable information has been published on its physical and chemical properties. Penicillin in the free state exists as an organic acid and reacts chemically to form various salts and esters. It is extremely unstable probably because of the presence of labile, free carboxyl groups. Free penicillin is soluble in ether, alcohol, acetone, ethyl acetate, amyl acetate, cyclo-hexanone, and dioxane. It is less soluble in carbon tetrachloride, chloroform, and benzene. Free penicillin is extremely soluble in water; as much as 5 mg. of penicillin can be dissolved in 1 ml. of water. It is unstable under treatment with dilute acid, alkalies, primary alcohols, oxidizing agents and heavy metals. At a pH of less than 5 and more than 7, it is unstable.

It is possible to prepare many salts of penicillin. Included among the salts of penicillin which have been prepared are those of sodium, calcium, potassium, ammonium, barium, silver, magnesium, and strontium. By a reaction of the free acid of penicillin with the corresponding diazo compounds, methyl, ethyl, *n*-butyl, and benzhydryl esters have been prepared. The two most commonly used salts of penicillin are the sodium and the calcium salts. Both have been obtained in highly purified crystalline form. The crystals of the sodium salt of penicillin look somewhat like fence pickets. One end is pointed and the other is straight. The sodium salt of penicillin is hygroscopic. The calcium salt of penicillin is not hygroscopic and is, therefore, more stable and easier to work with. The effect of heavy metals on the inactiva-tion of penicillin has been mentioned. It appears that the greatest inactivation occurs following its contact with copper, lead, zinc, and cadmium. Some degree of inactivation occurs after contact with nickel, mercury, and uranium. Penicillin does not appear to be inactivated to any great degree following its contact with certain bases, such as ammonia, aniline, and quinine in the ionized state. Penicillin does not completely reduce Fehling solution; however, the blue color changes to green. In the presence of potas-sium permanganate and hydrogen peroxide, there is a fairly rapid and complete oxidation of penicillin when it exists in either the form of the acid or its salts. Penicillin is less sensitive to reducing agents.

Several empirical formulas have been suggested for penicillin. They are not all in agreement. There does appear to be general agreement, however, that the compound is of fairly small mole-cular size. The following empirical formulas for the barium salt of penicillin have been suggested by Abraham and Chain: $C_{24}H_{32}O_{10}N_2Ba$ (molecular weight 645) or $C_{23}H_{30}O_9N_2Ba$. Holiday concluded that penicillin was a polysubstituted hydro-aromatic ring structure. As the result of both extensive and intensive research by various groups of persistent workers, the

$$R \cdot CO \cdot NH \cdot HC \overset{\displaystyle \diagup S \diagdown}{\underset{\displaystyle O:C \relbar N}{\relbar\relbar}} \overset{\displaystyle C \cdot H \quad C(CH_3)_2}{\underset{\displaystyle \relbar\relbar CH \cdot CO_2Na}{}}$$

The Penicillins

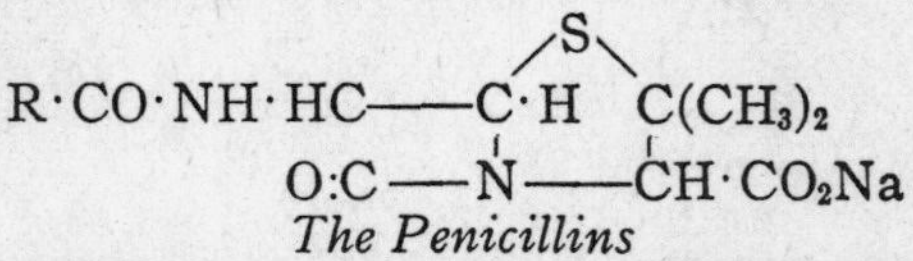

general structure of the penicillins is now believed to be correctly represented by the formulation shown here. It seems apparent that the penicillins differ from one another with respect to the R substituent, that is, the hydrocarbon residue. This accounts for the various fractions obtained from commercial penicillin, which are indicated in the next paragraph.

The work of many chemists has demonstrated that commercial penicillin is probably composed of several or many penicillins some of which have been separated from the complex and studied separately. For example, at the present time, different fractions have been described as penicillin F, G, and X.

Actinomycetin (Table 3):

Little is known concerning the chemical nature of this antibiotic. It appears to be a rather highly toxic, bacteriostatic protein. On the other hand, it has been thought by some investigators to be a fatty acid.

Actinomycin A and B (Table 3):

Actinomycin A is a bright red pigment. It appears to be a polycyclic nitrogen compound. A suggested possible empirical formula is $C_{41}H_{56}N_8O_{11}$. It is soluble in ether and alcohol but is insoluble in petroleum ether. It is thermostable but it is exceedingly toxic.

Actinomycin B is soluble in ether but is relatively insoluble in alcohol. Little information is available on the chemical nature of actinomycin B; however, it does appear that the two substances are somewhat related.

Streptomycin (Table 3):

Streptomycin has been obtained in crystalline form. Chemically it behaves as an organic base. It is thermostable and is soluble in water and acid solutions. It is relatively insoluble in ether and chloroform. According to Fried and Wintersteiner, the reineckate salt of streptomycin in crystalline form appears as thin plates. The empirical formulas suggested by them are: $(C_{14}H_{26\ 7}N_9S_4Cr)_n$ or $(C_{14}H_{26}O_8N_9S_4Cr)_n$ corresponding to $(C_{10}H_{19}O_{7-8}N_3)_n$ for the basic component.

Further studies on streptomycin hydrochloride and streptomycin helianphate, according to Kuehl and his colleagues, led them to conclude that the molecular weight of streptomycin hydrochloride was approximately 700. Streptomycin probably has the general constitution of a hydroxylated base attached through a glycosidic linkage to a nitrogen containing a disaccharide-like molecule. The present accepted empirical formula for streptomycin, according to Brink and his colleagues, is $C_{21}H_{37-39}N_7O_{12}$.

Streptothricin (Table 3):

Streptothricin, like streptomycin, behaves as an organic base. It is soluble in water and acid solutions but is not soluble in ether or chloroform. It is considerably more toxic than streptomycin.

It can be precipitated from the broth in which Actinomyces lavendulae grows by substances which precipitate protein; however, on isolation streptothricin does not appear to have proteinic characteristics. Proteolytic enzymes do not reduce its activity. It is readily soluble in water and is stable. While its exact formula is not known, nor is its molecular weight known at the present time, one suggested empirical formula based on studies of a reineckate salt is: $C_{21}H_{39}O_7N_{17}S_8Cr_2$ which corresponds to the di-reineckate of a base $C_{13}H_{25}O_7N_5$.

Chloromycetin or Chloramphenicol (Table 3):

Chloromycetin was first isolated from the broth in which had been grown Streptomyces venezuelae. It was described as a neutral compound and unique in that it contained both nitrogen and nonionic chlorine. It is stable at room temperature in aqueous solutions over the pH range of 2 to 9 for more than 24 hours. It is also unaffected by boiling for 5 hours. It was the first antibiotic of importance to be synthesized with relative ease. The chemical structure of chloramphenicol has been determined to be D-(-)-threo-2-dichloroacetamido-1-p-nitrophenyl-1, 3-propanediol.

Aureomycin (Table 3):

Aureomycin is an antibiotic produced by Streptomyces aureofaciens. It also has been obtained in crystalline form. It is a yellow substance which is a weakly basic compound. The substance forms a hydrochloride, which is the preparation most commonly used at the present time. The substance decomposes at temperatures greater than 210° C.; [a] $D^{23°}$,-240.0 (water). Its solubility in water is 14 mg. per milliliter at 25° C. The pH of the aqueous solution is 2.8 to 2.9. Analysis shows the composition, on a percentage basis, to be as follows: carbon 51.84, hydrogen 5.24, nitrogen 5.46, total chloride 13.27, ionic chloride 6.69, and oxygen 24.19 (by difference).

Terramycin (Table 3):

Terramycin is an active antibiotic which is elaborated by a strain of Streptomyces rimosus. It is similar in its antibacterial activity to aureomycin, although it is chemically slightly different. It is an amphoteric substance and forms a crystalline hydrochloride and a sodium salt. It is stable over long periods of time in aqueous solution at a pH of about 2 to 5. Chemical analysis reveals the following composition on a percentage basis: carbon 53.05, hydrogen 5.91, nitrogen 5.64, and oxygen (by difference) 35.4. The antibiotic is optically active and gives positive ferric chloride, Pauly, Friedel-Crafts, Fehling, and Molisch tests. Elemental analysis indicates the empirical formula to be $C_{22}H_{22-24}N_2O_92.H_2O$.

Antibiotics from Sources Other than Microbes

Few antibacterial substances from sources other than microbes. have yet been of great practical importance, but a number of them have been investigated.

Lysozyme:

An enzyme belonging to the class of carbohydrases, lysozyme occurs in various body fluids and in egg albumen. It is soluble in water but insoluble in alcohol, chloroform, ether, toluol, xyol, and acetone, and is of little practical importance.

Chlorellin:

Little is known about the exact physical or chemical characteristics of this substance obtained from unicellular algae.

Canavalin:

This substance is obtained from soy beans or jack beans. It is reported to have been obtained in crystalline form. It is probably a rather complex substance and appears to be a mixture of an enzyme and a co-enzyme. Its chemical structure is not known.

Allicin:

This substance has been isolated from garlic. It is a rather unstable compound and is toxic. The empirical formula ascribed to allicin is: $C_6H_{10}OS_2$ with a molecular weight of 162. The empirical formula suggested is $C_3H_5.SO.S.C_3H_5$.

TABLE 1. ANTIBIOTIC AGENTS OF BACTERIAL ORIGIN

Antibiotic agent	Organism from which derived	Author who described agent and date of report	Organisms sensitive to agent
Bacitracin	Bacillus subtilis	Johnson, Anker, and Meleney, 1945	Gram-positive; some gram-negative
Pyocyanase	Pseudomonas aeruginosa	Emmerich and Loew, 1899	Gram-positive and gram-negative
Pyocyanine	Pseudomonas aeruginosa	Wrede and Strack, 1924	Mainly gram-positive
Subtilin	Bacillus subtilis	Jansen and Hirschmann, 1944	Gram-positive
Tyrothricin Gramicidin Tyrocidine	Bacillus brevis Bacillus brevis Bacillus brevis	Dubos, 1939 Hotchkiss and Dubos, 1940 Hotchkiss and Dubos, 1940	Gram-positive Gram-positive Gram-positive; some gram-negative

TABLE 2. ANTIBIOTIC AGENTS DERIVED FROM MOLDS
AND FUNGI

Antibiotic agent	Organism from which derived	Author who described agent and date of report	Organisms sensitive to agent
Aspergillic acid	Aspergillus flavus	White and Hill, 1943	Gram-positive and gram-negative
Chaetomin	Chaetomin cochliodes	Waksman and Bugie, 1944	Gram-positive; some gram-negative
Citrinin	Penicillium citrinum	Hetherington, Raistrick, and others, 1931	Gram-positive
Clavacin*	Aspergillus clavatus	Waksman, Horning, and Spencer, 1942	Gram-positive and gram-negative
Claviformin*	Penicillium claviforme	Chain, Florey, and Jennings, 1942	Gram-positive and gram-negative
Flavacidin	Aspergillus flavus	McKee, Rake, and Houck, 1944	Gram-positive
Flavicin	Aspergillus flavus	Bush and Goth, 1943	Gram-positive
Fumigacin†	Aspergillus fumigatus	Waksman, Horning, and Spencer, 1942	Gram-positive
Fumigatin	Aspergillus fumigatus	Anslow and Raistrick, 1938	Gram-positive
Gigantic acid	Aspergillus giganteus	Philpot, 1943	Gram-positive
Gliotoxin	Trichoderma lignorum	Weindling and Emerson, 1936	Gram-positive and gram-negative
Helvolic acid†	Aspergillus fumigatus	Chain, Florey, Jennings, and Williams, 1943	Gram-positive
Patulin*	Penicillium patulum	Raistrick and others, 1943	Gram-positive and gram-negative
Penicillic acid	Penicillium puberulum	Alsberg and Black, 1913	Gram-positive and gram-negative
Penicillin	Penicillium notatum	Fleming, 1929	Gram-positive and some gram-negative

*Claviformin, clavacin and patulin are similar if not identical.
†Fumigacin and helvolic acid are similar if not identical.

TABLE 3. ANTIBIOTIC AGENTS DERIVED FROM ACTINOMYCETES

Antibiotic agent	Organism from which derived	Author who described agent and date of report	Organisms sensitive to agent
Actino-mycetin	Actinomyces	Gratia and Dath, 1924	Gram-positive and gram-negative
Actino-mycin A	Actinomyces antibioticus	Waksman and Woodruff, 1940	Gram-positive
Actino-mycin B	Actinomyces antibioticus	Waksman and Woodruff, 1940	Gram-positive
Strepto-mycin	Actinomyces griseus	Schatz, Bugie, and Waksman, 1944	Gram-negative and gram-positive
Strepto-thricin	Actinomyces lavendulae	Waksman and Woodruff, 1942	Gram-negative and gram-positive
Chloro-mycetin	Streptomyces venezuelae	Ehrlich and colleagues, 1947	Gram-negative and gram-positive Rickettsia
Aureomycin	Streptomyces aureofaciens	Duggar, 1948	Gram-negative and gram-positive Rickettsia Certain viruses
Terramycin	Streptomyces rimosus	Finlay and colleagues, 1950	Gram-negative and gram-positive Rickettsia Certain viruses

CHAPTER XXXV

PROCESSES, REACTIONS, SYNTHESES, AND TESTS

The Mechanism of Organic Reactions

A. Molecular Structure:

A discussion of atomic structure and of the forces binding atoms into molecules has been given in Chapter II, where it was shown that molecular bonds are of two extreme types. Ionic linkages are those in which one atom completely donates one or more of its electrons to another atom to give a coulombic force between the two residues, whereas the covalent type of bond has an interatomic sharing of the electrons. A variation of the covalent type is the semi-polar or dative bond. In this case, both of the electrons making up any given shared-electron single bond come from only one atom of the two so bonded.

Actually, bonds may be formed from intergradations of the above types in various proportions. It has been indicated, furthermore, that a method for the estimation of the relative contributions of the extreme types to the actual bond is available.[1] In spite of the fact that the electrons for a given bond are donated by some particular atom, these identical electrons do not necessarily maintain the same linkage. A given electron, therefore, can no longer be identified as belonging to a particular atom.

Besides the regular bond forces in the molecule, there are inductive or relative polarity effects. These result from the presence in the molecule of atoms or groups which differ in their affinity for electrons. This affinity for electrons, moreover, is dependent on two properties of an atom or group: its electronic structure, when free from the molecule, and its size.

Fluorine, which is a small molecule and lacks one electron to form its stable inert gas shell, is very electron attractive or negative in nature. Oxygen, however, which lacks two electrons of having a stable shell, is larger than fluorine so that its electronegativity is somewhat less. The size of the atom is important because of its effect on the charge distribution. The more space the charge has to spread over, the less is the tendency toward the completion of the octet by the acquisition of electrons.

The presence of every group in the molecule will, then, have an effect on the electron distribution throughout the molecule to a

[1] Cf. p. 16.

324

greater or less degree, depending upon the negativity of the group. The range of this force, except in unusual cases, extends appreciably only to immediate neighbors. There are, however, data which indicate a tendency toward alternate polarity of the carbon atoms. The benzene ring constitutes an unusual example since the effect here can be transmitted down both sides of the ring, giving a strong effect on the para carbon atom. Several groups acting in the same direction tend to extend the range of this polarity.

Such forces in the molecule affect the strengths of the bonds within their range. A negative group, by attracting electrons from immediately neighboring atoms, will weaken the bonds to these neighbors. The greatest weakening will occur on the bond nearest the negative group, since it is here that the electrons are farthest removed from their equilibrium position. When these groups act in opposition to each other, the weakest bond may be to the carbon atom midway between the two groups. An excellent example of this phenomenon is the ketonic and acid cleavage of acetoacetic acid and its substitution products. This polarity effect is also evident in the high reactivity of *beta*-hydrogen groupings in the carboxylic acid series and the relative non-reactivity of *alpha*- and *gamma*-groupings.[2] Conversely, just as an electronegative group may weaken a bond, an electropositive group may strengthen the linkage.

At this point it seems advisable to list the postulates of the theory of relative electronegativity, namely:

1. The relative distribution of the electrons in methane is assumed to represent the condition of maximum stability for a carbon-hydrogen system, that is, a hydrocarbon.

2. Replacement of a hydrogen atom in a hydrocarbon or its derivative by a **more** or **less** strongly electronegative group is attended by a shift in the relative positions of the electrons.

3. Any distortion of the electron distribution from that of methane (other than that to CO_2, CCl_4, etc.) will result in a reduction in the stability of the compound so formed; and the reactivity will be proportional to the distortion.

4. Any distortion in the relative distribution of the electrons tends to be distributed throughout the system, the effect diminishing with remoteness from the point of disturbance.

5. The influence of any atom on the relative distribution of the electrons is assumed to be a function of the relative electronegativity of the atom as indicated by the Electronegativity Map. On this basis a hydrogen atom is assigned a value of $+1$, an iodide or sulfur atom a value of 0, a bromine atom a value of -0.75, a chlorine atom a value of -1.25, a nitrogen atom a

[2] "Chlorinations with Sulfuryl Chloride. II. The Peroxide-catalyzed Chlorination of Aliphatic Acids and Acid Chlorides," by M. S. Kharasch and H. C. Brown; presented before the Division of Organic Chemistry of the A.C.S. at the Boston meeting, Sept. 11-15, 1939.

value of — 1.25, a single bonded oxygen atom a value of — 2.5, a double bonded oxygen atom a value of — 5.[3]

Perhaps the most fruitful physical picture of the molecule is obtained by considering its atoms as small balls placed in hollows. The sides of these hollows are extremely high save in the direction of the interatomic bonds. Along these directions lie passes whose height above the floor of the hollow is proportional to the bond strength. The atoms are continually moving about in their hollows but under the conditions wherein the molecule is stable never acquire enough energy to surmount one of these passes and break the bond between themselves and the next atom. Should they ever acquire the energy requisite to overcoming the "potential hill," it is obvious that the easiest way of escape is through the lowest pass, or the weakest bond, which will be the first to break.

B. Structure in Gases, Liquids, and Solids:

Gases are collections of molecules or atoms wherein the attractive energy of particle for particle is so small as to be negligible when compared with the translational energy of the particles. The particles are in rapid random motion with energies proportional to the absolute temperature. Collisions between two bodies are common, but are mostly of the elastic or billiard ball type with an infinitesimally small time of contact. At ordinary temperatures, these collisions do not possess sufficient energy to cause chemical reaction, and there are few organic reactions which occur in the gaseous phase. These gas phase reactions occur, instead, at high temperatures where the energy necessary for activation is supplied by thermal excitation.

A liquid is a collection of molecules or atoms which is kept confined to a given volume by the action of non-directional intermolecular forces leading, in general to a random arrangement of its constituents. In a liquid each molecule is surrounded by from six to twelve nearest neighbor molecules plus a great many second and third neighbors. It is the interactions between these neighbors which forces the particles to remain in the liquid state in spite of the repulsions of their electronic fields and their thermal agitations. These interactions may be thought of as ionic forces such as those which are present in inorganic salt crystals, as dipole forces (similar to magnetic forces), and as van der Waals forces which may include all other non-directional forces. There may e, moreover, directed bonding forces such as the hydrogen bond.

Ionic forces arise, of course, from coulomb attraction and repulsion between the positive and negative ions.

Molecules which contain inductive forces may be considered as

---

[3] On the Electronegativity Map, hydrogen has a value of 2.1; carbon, 2.5; iodine, 2.5; sulfur, 2.5; bromine, 2.8; chlorine, 3.0; nitrogen, 3.0; oxygen, 3.5; and fluorine, 4.0. The hydrogen atom, then, is electropositive to carbon by 0.4 units. If hydrogen is arbitrarily given a value of unity, the relative electronegative character of the other atoms is obtained by multiplying their net difference from carbon on the map by 2.5.

small magnets, the strengths of whose poles vary with the type of groups present.[4] Thus, for instance, the carbonyl group (CO) will have a strong polar effect or "dipole moment" since the electronegativities of the two atoms are quite different. The various groups and bonds present in organic compounds may be assigned values which are proportional to the strength of this effect known as their group or bond moments.[5] As in the case of magnets, such groups will attract and repel each other depending on the sign of their pole. If the molecule is symmetrical and linear, it will have no net dipole moment regardless of the relative electronegativities of the groups therein. It will, however, have a non-symmetrical distribution of charge so that it can attract dipoles or other similar molecules by its internal dipoles. Thus carbon disulfide, being linear, has no net dipole moment, but there is a strong dipole moment associated with each thiocarbonyl (C = S) bond since the carbon atom is relatively positive and the two sulfur atoms are relatively negative.

Van der Waals forces appear to result from the interaction of the electronic fields of the molecules by some such phenomena as mutual polarization upon approach. That is, the electrons in one molecule tend to make the electrons in a neighboring molecule vibrate sympathetically with them, leading to attractive forces.

The hydrogen bond, as was stated in Chapter II, is formed only by hydrogens attached to strongly electronegative atoms such as fluorine, oxygen, and nitrogen, thus uniting these atoms through the hydrogen atom to some similar electron donor atom. It is now known, furthermore, that a hydrogen atom attached to a carbon atom which is bonded to several very negative groups may also yield such a bond. The hydrogen atom in chloroform affords such an example. There is some indication that the sulfur atoms may induce hydrogen-bond formation. The hydrogen bond has a strength of from 5,000 to 8,000 calories with an average of about 6,000 calories per gram molecular weight of compound.

A solid is a collection of atoms, ions, or molecules held together in a regular arrangement by the various types of forces. In metals the exact nature of the force is not understood, but inorganic salt crystals are considered to be held together by ionic attractions. Organic crystals are thought to be held together by van der Waals forces, supplemented in some cases by dipole forces.

Since the number of organic reactions occurring in gaseous or solid phase is very limited, we shall confine our attention almost entirely to the liquid phase.

It seems that, in the liquid phase, continuous chains are actually extended in a more or less linear fashion unless some definite force causes them to be bent on themselves. The hydrocarbons, for

---

[4] Cf. Electronegativity Map, p. 17.
[5] C. P. Smyth, J. Phys. Chem. *41*, 209 (1937).

instance, tend to be extended, whereas acetoacetic acid exists in a ring form because of its internal hydrogen bond. The diesters of the dicarboxylic acids may be extended because of mutual repulsions of the carboxyl groups, but the diacids may tend to be cyclized as a consequence of internal hydrogen bonds overcoming the ordinary carboxylic repulsions.

C. Labile Hydrogen Atoms:

A high concentration of electrons at any point in space leads to strong repulsive forces among the electrons so concentrated, and any influence tending to distribute them symmetrically will meet with little resistance. Multiple linkages between atoms may be considered as such unstable structures. The relative electron affinity of the proton (hydrogen atom) is well known. Its small mass enables the proton to move from place to place in the molecule without the expenditure of a great amount of energy. If, then, a hydrogen atom in an organic molecule can, by changing its position, effect a decrease in electron concentration, such a change will tend to occur. The extent of the tautomerization will vary with the degree of super electron-saturation of the bond, and with the bonding conditions of the proton in its normal position.

The location of the proton relative to the unsaturation which promotes such a change is known as the 4-position. That is, a hydrogen is in a 4-position if it is linked to an atom adjacent to one which is multiply bonded, or the hydrogen is atom number four when the atom at the far end of the unsaturation is counted as number one.

Hydrogen atoms which can thus change their position in the molecule are known as labile hydrogen atoms. Well known examples of this type are keto-enol tautomerism and 1,4-addition. In 1,4-addition some other group or atom other than hydrogen may be labile, thus indicating that the phenomenon is not strictly limited to protons.

These labile systems will arise from various combinations of hydrogen, carbon, nitrogen and oxygen, fourteen such linkages being possible.[6]

H·C·C:C	H·C·N:O	H·O·C:O
H·C·C:O	H·C·N (→O):O	H·N·C:O
H·C·C:N	H·O·N:O	H·N·N:N
H·C·N:C	H·C·N:N	H·N·N:O
H·C·C:N	H·N·C:N	

D. The Role of the Catalyst:

Many reactions require a catalyst of some form in order to proceed at an appreciable rate. The only reactions which can be

[6] Baker, Tautomerism, Routledge and Sons (1934).

said to be independent of catalysts are those which occur in a solution of the reagents only, or in the solid or gaseous state, in the presence of no foreign matter. Many reactions which seem to proceed independent of catalysis require a trace of water. Many gas reactions, for example, take place with extreme slowness if the gases are thoroughly dried and kept free from contact with the walls of the container. Powdered glass has been found to be effective in displacing some keto-enol systems in the direction of the keto-form.[7]

A catalyst has been defined as a substance which increases the reaction rate without affecting the equilibrium constant and without itself suffering any permanent change during the reaction. It may, of course, enter into the reaction, but it must be regenerated during the process.

Catalysts may be classed into three groups: activation catalysts, addition catalysts, and solvent catalysts. The first is exemplified by the role of mercury vapor in many photochemical reactions. The mercury atom is excited by incident radiation. This excited atom then transfers its excitation energy to other atoms, thus causing them to attain the state in which it is possible for them to react.

Addition catalysts may be typified by the action of finely divided nickel or platinum in catalytic hydrogenation. In these cases the hydrogen molecules are so strongly adsorbed on the metallic surface, because of the powerful action of the unshared electrons of the metal, that the interatomic hydrogen bond may be considered as being stretched or even broken. This atomic or near atomic hydrogen is then active enough to add into any unsaturation which may be in its neighborhood. Another type of addition catalysis occurs in the benzoin condensation. Here the hydrogen cyanide adds to the aldehyde group allowing it to form hydrogen bonds with a neighboring aldehyde. These bonds bring the reacting groups together and allow the process to go to its conclusion with the ultimate regeneration of the hydrogen cyanide.

Solvent catalysts are effective because (1) they concentrate reactants, and (2) they allow the formation of a homogeneous phase.

Combination in one catalyst of all or some of these three types is also possible. Water is the most widely applicable catalyst known since it is not only a solvent, but is also capable of effecting solvation of other molecules. An intermolecular solvation or **water-bridge** between two molecules allows these molecules much more chance to react than normally. The large dipole moment of the solvated molecule leads to preferred orientation which may also affect the reaction rate.

---

[7] Bachmann, Cole and Welds, **J. Am. Chem. Soc.,** *62*, 825 (1940).

E. Molecular Fission:

It has already been stated that a molecule may be cleft by the concentration at some part of that molecule of enough energy to allow it to escape from its normal hollow over the lowest potential hill available. Such energy may come from (1) thermal excitation, (2) photonic excitation, (3) collision with a molecule having a high energy content, (4) contact with a catalyst which so affects the molecule as to effectively lower its potential hill or make it more receptive to energy transformations, or (5) external fields of force such as those already mentioned in the discussion of liquids and solids. When the molecule is thus activated, it is able to undergo chemical change, that is, to react.

Various reaction mechanisms are open to a molecule. The electrons forming the bond may undergo symmetrical fission to form two neutral particles each having an odd unpaired electron. Such particles, known as free radicals, are very highly reactive because of the odd electron. They are formed only under special conditions: as by (1) reaction between a covalent molecule and a free atom, especially an alkali metal or another free radical, or (2) photochemical or thermal decomposition. These conditions cover many of the reactions requiring presence of free metals and many oxidation and reduction mechanisms.

If the electronic bond undergoes non-symmetrical fission, two ions will be formed and each probably will carry a single charge—the one positive and the other negative. If these charges appear to be on carbon atoms, the positive ion is designated as a carbonium ion whereas the negative ion is a carbanion or an alkide ion. If, however, the charge is residual on an atom other than carbon, the change is indicated as ionic, and the ion is given its generic name. Alkide and carbonium ions seem to be about as reactive as free radicals, whereas the simple ions are capable of free existence in the liquid or solid phase.

Both the ionic and the free radical fragments can undergo secondary fission similar to the primary cleavage, with the further possibility that fission can yield a simpler neutral molecule and either an ion or free radical. The acetate radical, to illustrate, loses carbon dioxide under proper conditions to form a methyl radical (Kolbe Electrolytic Synthesis).

Neutralization of an ion by the loss or gain of an electron also yields free radicals which in turn can yield ions by reversal of the process.

Hydroxyl ion, by exerting a strong attraction for protons, tends to cause alkide or negative ion formation while positively charged particles such as protons tend to cause formation of carbonium or positive ions.

F. Molecular Addition:

The first step in many reactions is an addition. A molecule

may add another molecule, atom, or ion through (1) dipole-dipole interaction, (2) metallic forces, (3) van der Waals forces, (4) hydrogen bonds, or (5) direct addition to an unsaturated linkage. Hydrogen bonds are particularly favorable to complex formation and subsequent reaction since they are strongly directed, bring the groups together which generally undergo reaction, and may supply six kilogram calories of activation energy for each bond so formed.   In any case the complex once formed may be activated by some energy source and disintegrate or rearrange so as to yield the final products.   The fifth method of addition is that common to unsaturated compounds, and is brought about by the **electron thirst** of the multiply bonded atoms.   The approach of some substance, with surplus electrons, leads to the formation of a strong addition complex which may or may not rearrange to yield the final products.

G. Intramolecular Reactions:

A few reactions seem to occur entirely within the molecule without any apparent intervention from another molecule except, perhaps, as a source of activation energy. These reactions always take the molecule from a less stable to a more stable form. The sum of the kernel repulsions may be lowered or an inner condensation may lead to ring formation. In such cases, evidently, bond cleavage is not sufficient to remove a group from the sphere of influence of the molecule. Instead, the strong electronic field of the molecule holds the fragment in close proximity until rearrangement is complete.

H. Reduction-Oxidation or Redox Reaction:

To the fission of molecules and the transfer of groups among the molecules, there must be added a very important class of reactions in which a gain of electrons by one or more atoms is at the expense of some other part of the system. Such a gain of electrons is called reduction, while the corresponding loss is known as oxidation. The most simple form of this reaction is that which occurs in an electric cell where we have reduction at the cathode and oxidation at the anode. A neutral particle, of course, becomes an ion immediately upon the loss or gain of an electron, and can then react with the oppositely charged ions present to form some other neutral molecule. Similarly, free radicals are formed when ions lose or gain an electron. Since most redox reactions proceed by several steps, it is apparent that alternate ion and free radical formation will occur in many cases. When the electrons are being supplied or donated by compounds or atoms, instead of an electrode, the mechanism may proceed similarly through the formation of an addition compound to facilitate the electron transfer. Catalytic hydrogenation, a form of reduction, probably occurs through the formation of free radicals by addition, on the catalytic surface, of hydrogen atoms into the double bond.   Such a reaction will be classed, however, under addition complexes.

In considering the following reactions and attempting to type them, a few fundamental postulates are laid down.

1. The presence or absence in the reaction products or the reaction mixture of some constituent which may yield the final product proves neither the necessity nor uselessness of that substance in the formation of the final product.   The isolation of some of the intermediates such as free radicals in solution has not been possible, nor can the proved existence of one single mechanism outlaw the possibility of another.

2. If the mechanism is sufficiently explained in some other part of this Outline, it is not further discussed in the classification.

3. In classifying these reactions only the intermediate products leading to the final result are considered.   Thus, if both an alkide and a carbonium ion are formed and the alkide is the one which will yield the final product, the reaction will be placed under alkide ions.   It will be obvious in most cases what becomes of the rest of the molecule.   Furthermore, the most important step, and the one most difficult to comprehend, will be that under which the reaction will be classified.   Thus the Hofmann, Curtius, and Beckmann rearrangements are all classed as intramolecular although the original molecule must be changed in form before it will react.

4. When heat, free metals, or electrodes are the activators, free radical formation is assumed.   When organic or inorganic ions are the chief reagents, an ionic mechanism is postulated.   When two reactants each contain at least one oxygen or nitrogen atom, which takes part in the reaction, a hydrogen bonded intermediate is assumed if such a complex is possible.   When metal salts are the chief reagents, the complexes are held together by "residual metallic forces."   When an unsaturated linkage is the chief reagent, complex formation, through its unsatisfied "thirst" for electrons, is postulated.   When all the products are due to changes in one molecule, the reaction is classed as intramolecular.   All color reactions and tests are placed in one section as are also all the processes involving only physical changes in the molecules such as colloid discharge or formation.   Another section is reserved for the redox reactions.   All other reactions are classed together under complex formation since they proceed no doubt by this mechanism though the forces leading to reaction are not strictly definable.

Classification of Organic Reactions

I. Molecular Fissions:

    A. Free Radical Formation:

        1. By action of free atom, or

        2. By thermal excitation, or

      3. By electrolysis, or
      4. By photolysis.
  B. Ion Formation:
      1. Alkide or carbanion, or
      2. Carbonium, or
      3. Simple.
  C. Simpler Molecule Formation.

II. Molecular Addition of Molecules, Ions, or Atoms:
  A. By hydrogen bonding:
      1. Oxygen to oxygen, or
      2. Oxygen to nitrogen, or
      3. Nitrogen to nitrogen, or
      4. Otherwise.
  B. By metallic forces.
  C. By addition to unsaturated groups.
  D. By dipole-dipole or van der Waals forces, etc.

III. Intramolecular Reactions:
  A. Internal Condensation:
      1. By hydrogen bond.
      2. Through activated phenyl hydrogens.
  B. Group rearrangement or loss.

IV. Redox Reactions.

V. Color Reactions.

VI. Physical Changes or Processes.

VII. Rules.

Reactions Classified As To Types

Alphabetical List of "Named" Reactions

	Name	Type			Serial No.
1.	Acetoacetic Ester Synthesis	I	B	1	39
2.	Acyloin Condensation	I	A	1	1
3.	Adamkiewicz-Hopkins Reaction	V			217
4.	Aldol Condensation	II	A	1	69
5.	Allylic Rearrangement	I	B	2	46
6.	Angeli-Rimini Aldehyde Reactions	II	A	1	70
7.	Babcock Test	VI			248
8.	Baeyer-Drewsen Synthesis	III	A	1	188
9.	Baeyer Permanganate Test	II	C		155
10.	von Baeyer Synthesis	I	B	1	40
11.	Bamberger-Goldschmidt Synthesis	III	B		190
12.	Barbier-Wieland Degradation	II	B		136
13.	Bart Reaction	I	B	2	47
14.	Baumann-Schotten Reaction	II	A	1	71
15.	Beckmann Rearrangement of Ketoximes	III	B		191
16.	Beilstein Test for Halogen	I	A	2	13
17.	Benedict Solution	IV			208
18.	Benzidine Rearrangement	III	B		192

	Name	Type			Serial No.
19.	Benzilic Acid Rearrangement	III	B		193
20.	Benzoin Condensation	II	A	1	72
21.	Bergius Process	I	A	2	14
22.	Bergmann Degradation	III	B		194
23.	Berthelot Synthesis	II	B		137
24.	Bischler and Napieralski Reaction	III	A	1	189
25.	Biuret Synthesis and Reaction	II	A	2	98
26.	Blaise Ketone Synthesis	II	B		138
27.	Blanc Reaction	II	A	1	73
28.	Boeseken Method	VI			249
29.	Boord Olefin Synthesis	II	B		139
30.	Bouis Allene Synthesis	I	B	2	48
31.	Bouveault Aldehyde Synthesis	II	B		140
32.	Bouveault and Blanc Reduction	II	C		156
33.	von Braun Bromocyanogen Reaction	II	D		185
34.	von Braun Epimer Reagent	II	A	2	125
35.	von Braun Exhaustive Methylation	II	D		186
36.	Brooks Method	II	B		141
37.	Bucher Processes	I	A	2	15
38.	Bucherer Reaction	II	A	2	99
39.	Buchner-Curtius Reaction	II	C		157
40.	Butlerow Acetyl Chloride Reaction	I	B	2	49
41.	Cannizzaro Reaction	II	A	1	74
42.	Carbylamine Reaction	II	A	4	129
43.	Carius Determination	I	A	2	16
44.	Caro Test	V			218
45.	Carr-Price Color Reaction	V			219
46.	Castner Process	I	A	2	17
47.	Ciamician-Dennstedt Synthesis	II	A	4	130
48.	Claisen Condensation	II	A	1	75
49.	Claisen Reaction	II	A	1	76
50.	Claisen Rearrangement	III	B		195
51.	Clemmenson Reduction	II	C		158
52.	Cross-Bevan (Viscose) Process	VI			250
53.	Crum-Brown and Walker Reaction	I	A	3	34
54.	Curtius Reaction	III	B		196
55.	Dakin Isolation of Amino Acids	VI			251
56.	Darzens Reaction	II	A	1	77
57.	Decker Method	II	A	2	100
58.	Demjanow Rearrangement	I	B	2	50
59.	Dieckmann Condensation	II	A	1	78
60.	Diels-Alder Reaction	I	A	2	18
61.	Diels-Wolf Reaction	I	A	2	19
62.	Doctor Process	II	B		142
63.	Doebner-Miller Quinaldine Synthesis	II	A	2	101
64.	Dow Process	I	B	2	51
65.	Dubbs Process	I	A	2	20
66.	Dumas Determination	I	A	2	21
67.	Ehrlich Reagent	V			220
68.	Emde Degradation	II	C		159
69.	Engler-Weissberg Hypothesis	II	D		174
70.	Erlenmeyer Reaction	II	A	2	102
71.	Eschweiler Method	II	A	2	103
72.	Etard Reaction	II	D		175
73.	Fehling Test	IV			209
74.	Fenton Reagent	IV			210
75.	Ferric Chloride Test	V			221
76.	Fischer Esterification Method	VI			252
77.	Fischer Phenylhydrazine Synthesis	I	B	3	62
78.	Fischer-Dilthey Condensation	II	A	4	131

Name	Type			Serial No.
79. Fischer-Hepp Rearrangement	III	B		197
80. Fischer-Nouri Phloretin Preparation	II	A	2	104
81. Fischer-Tropsch Synthesis	I	A	2	22
82. Fittig-Erdman Synthesis	II	A	1	79
83. Fittig or Fittig-Wurtz Synthesis	I	A	1	2
84. Folin Reaction	V			222
85. Frankland Synthesis	I	A	1	3
86. Frankland-Duppa Reaction	II	B		143
87. Frasch Process	II	B		144
88. Freund Reaction	I	A	1	4
89. Friedel-Crafts Synthesis	II	B		145
90. Friedlander Synthesis	II	A	2	105
91. Fries Rearrangement	III	B		198
92. Gabriel Synthesis	I	B	2	52
93. Gattermann Aldehyde Synthesis	II	A	2	106
94. Gattermann Reaction	I	B	1	53
95. Gattermann-Koch Reaction	II	B		146
96. Gerhardt Synthesis of Quinoline	I	A	2	23
97. Gibbs Process	II	C		160
98. Gladstone Reaction	I	A	1	5
99. Goldschmidt Process	I	A	2	24
100. Griess Reactions	I	B	3	63
101. Griess Diazo Reaction	II	A	2	107
102. Grignard Reactions	II	B		147
103. Groves Process	II	D		176
104. Guerbet Reaction	II	A	1	80
105. Gustus Acid-Iodide-Ether Reaction	II	A	4	132
106. Haller-Bauer Reaction	I	B	1	41
107. Haloform Reaction	II	A	1	81
108. Hammarsten Reaction	V			223
109. Hantzsch Collidine Synthesis	II	A	1	82
110. Harries Ozonide Reaction	II	C		161
111. Haworth Methylation	II	A	1	83
112. Haworth Phenanthrene Synthesis	II	D		177
113. Heller Ring Test	VI			253
114. Hell-Volhard-Zelinsky Reaction	II	D		178
115. Heumann Synthesis	II	A	1	84
116. Hinsberg Separation	II	A	2	108
117. Hoesch Reaction	II	A	2	109
118. Hofmann Degradation	III	B		199
119. Hofmann Exhaustive Methylation	II	D		179
120. Hofmann Isonitrile Synthesis	II	A	4	133
121. Hofmann Mustard-Oil Reaction	II	A	4	134
122. Hofmann Reaction	II	D		180
123. Hofmann Rearrangement	III	B		200
124. Hopkins-Cole Reaction	V			224
125. Horbaczewski-Behrend Synthesis	II	A	2	110
126. Hydrobromic Acid Method	II	D		181
127. Indophenin Reaction	V			225
128. Ipatiev Reduction	II	C		162
129. Jacobsen Reaction	II	D		182
130. Jacobson (Semidine) Rearrangement	III	B		201
131. Janovsky Reaction	V			226
132. Japp-Klingemann Reaction	III	B		202
133. Kamlet Reaction	II	C		163
134. Karrer Aldehyde Synthesis	II	A	2	111
135. Kekulé Synthesis	I	A	1	6
136. Kiliani Synthesis	II	A	2	112
137. Kjeldahl Method	I	A	2	25
138. Knoevenagel Synthesis	II	A	1	85

	Name	Type			Serial No.
139.	Knoop *Beta*-Oxidation	IV			211
140.	Knorr Method	VI			254
141.	Knorr Reaction	II	A	3	113
142.	Kolbe Electrolytic Synthesis	I	A	2	35
143.	Kolbe Synthesis	I	B	3	64
144.	Komppa Synthesis	II	A	1	86
145.	Konigs-Knorr Synthesis	II	A	1	87
146.	Knovaloff Reaction	V			227
147.	Körner Orientation Theory	VI			255
148.	Kossel and Kutscher Method	I	B	3	65
149.	Krafft Method	I	A	2	26
150.	Ladenburg Synthesis	II	A	2	114
151.	Lederer-Manasse Synthesis	II	A	1	88
152.	Legal Test	V			228
153.	Leuckart Reaction	II	A	2	115
154.	Lieben Iodoform Reaction	II	A	1	89
155.	Liebermann Nitroso Test	V			229
156.	Liebig Combustion	I	A	2	27
157.	Lob Degradation	IV			212
158.	Lossen Rearrangement	III	B		203
159.	Lucas Test	II	D		183
160.	Lumiere and Barbier Method	II	A	2	116
161.	Malonic Ester Synthesis	I	B	1	42
162.	Mayer Reagent	I	B	3	66
163.	Mendius Reaction	II	C		164
164.	Methone Aldehyde Reagent	II	A	1	90
165.	Meyer Polymerization	II	C		165
166.	Meyer Reaction	I	B	2	54
167.	Victor Meyer Method	II	D		184
168.	Victor Meyer Synthesis	I	B	2	55
169.	Michael Synthesis	I	B	1	43
170.	Miller-Hofer Electrolysis	I	A	3	36
171.	Millon Test	V			230
172.	Molisch Test	V			231
173.	Mond Process	I	A	2	28
174.	Mustard Oil Reaction	II	C		166
175.	Nef Aldehyde and Ketone Synthesis	II	C		167
176.	Neuberg Degradation	IV			213
177.	Ninhydrin Reaction	V			232
178.	Osazone Reaction	II	A	2	117
179.	Paal-Knorr Reaction	II	A	2	118
180.	Paneth Technique	VI			256
181.	Pauly Reaction	V			233
182.	Pechmann Synthesis	II	C		168
183.	Perkin Method	I	B	1	44
184.	Perkin (Perkin-Fittig) Reaction	II	A	1	91
185.	Pettenkofer Reaction	V			234
186.	Phenylisocyanate Test	II	A	2	119
187.	Phthalic Anhydride Test	II	A	1	92
188.	Phthalein Test	V			235
189.	Pinacol Reduction	II	B		148
190.	Pinacolone Rearrangement	I	B	2	56
191.	Piria Reaction	I	B	3	67
192.	Plugge Reaction	V			236
193.	Prileschaiev Reaction	II	C		169
194.	Pschorr Synthesis	II	A	1	93
195.	Purdie (Irvine-Purdie) Methylation	II	A	1	94
196.	Reformatskii Reaction	II	B		149
197.	Reimer (Reimer-Tiemann) Reaction	II	A	4	135
198.	Riley Oxidation	II	C		170

	Name	Type			Serial No.
199.	Rimini Test	V			237
200.	Rosenmund Aldehyde Synthesis	II	B		150
201.	Ruff Degradation	IV			214
202.	Ruzicka Large Ring Synthesis	I	A	2	29
203.	Sabatier-Mailhe Aldehyde Synthesis	I	A	2	30
204.	Sabatier-Senderens Reaction	II	C		171
205.	Sakaguchi Reaction	V			238
206.	Sandmeyer Reaction	I	B	2	57
207.	Schardinger Reaction	V			239
208.	Schiff Reaction	II	A	2	120
209.	Schiff (Fuchsin-Aldehyde) Test	V			240
210.	Schmidlin Ketene Synthesis	I	A	2	31
211.	Scholl Condensation	II	B		151
212.	Schorigin Reaction	I	A	1	7
213.	Schotten-Baumann Reaction	II	A	1	95
214.	Schryver Test	V			241
215.	Schweitzer Reagent	VI			257
216.	Selivanov (Seliwanoff) Test	V			242
217.	Semidine Rearrangement	III	B		204
218.	Senderens Reaction	II	C		172
219.	Senderens Ketone Synthesis	I	A	2	32
220.	Skraup Synthesis	II	A	2	121
221.	Sörensen Formol Titration	I	B	3	68
222.	Sterol Color Reactions	V			243
223.	Strecker Reaction	II	A	2	122
224.	Sullivan Reaction	V			244
225.	Thompson Displacement Process	VI			258
226.	Thorpe Reaction	II	A	3	128
227.	Tishchenko (Tischtschenko) Reaction	II	A	1	96
228.	Tollens Test	IV			215
229.	Topfer Reagent	V			245
230.	Traube Synthesis	II	A	2	123
231.	Tschitschibabin Reaction	II	B		152
232.	Tshugaev (Tschugaeff or Chugaev) Reaction	II	C		173
233.	Twitchell Process	II	A	1	97
234.	Ullmann Reaction	II	B		153
235.	Van Slyke Method	II	A	2	124
236.	Vorlander Condensation	I	B	1	45
237.	Wagner Rearrangement	I	B	2	58
238.	Walden Inversion	I	B	2	59
239.	Walker Reaction	I	A	2	37
240.	Wallach Degradation Method	I	B	2	60
241.	Wallach Rearrangement	III	B		205
242.	Wanklyn Reaction	I	A	1	8
243.	Weermann Degradation	III	B		206
244.	Wheeler and Johnson Color Test	V			246
245.	Williamson Synthesis	I	B	2	61
246.	Wilsmore Ketene Synthesis	I	A	2	33
247.	Wislicenus Synthesis	I	A	1	9
248.	Wolff-Kishner Reduction	IV			126
249.	Wohl Degradation	IV			216
250.	Wohl-Schweitzer Electrolysis	I	A	3	38
251.	Wöhler Synthesis	III	B		207
252.	Wurtz Reaction	I	A	1	10
253.	Wurtz Synthesis	II	A	2	127
254.	Wurtz-Fittig Reaction	I	A	1	11
255.	Xanthoproteic Test	V			247
256.	Zeisel Determination	II	D		187
257.	Zerevitinov (Zerewitinoff) Method	II	B		154
258.	Zincke Reaction	I	A	1	12

In the organization of the some two hundred fifty reactions considered in the following pages, the four postulates cited on page 307 have been adhered to. Some such postulates must of necessity be adopted if any logical organization of this array of reactions is to be attained.

As a rule only one reference has been included for any given reaction. The general reference books were carefully checked, and then standard textbooks were examined in alphabetical order. Some books, consequently, have been cited as references much more than have others, but no reaction has been included unless a satisfactory reference is available. There are, accordingly, a number of reactions that might well have been included except for the fact that no attempt is made here to encourage the use of name reactions.

I. Molecular Fission

A. Free Radical Formation

1. By action of free atom.

2. By thermal excitation.

3. By electrolysis.

4. By photolysis.   (No examples given.)

B. Ion Formation

1. Alkide or carbanion.

2. Carbonium.

3. Simple.

C. Simpler Molecule Formation.

(See free radical formation by electrolysis.)

I. A. Free Radical Mechanisms

I. A. 1. Free Radicals From Action of a Free Atom

1. **Acyloin Condensation** (*J. Am. Chem. Soc. 53*, 750 (1931); Whitmore, p. 345, 1938):

The action of sodium on esters yields ketones, as:

$2 \ R \cdot CO \cdot OEt + 2 \ Na \rightarrow 2 \ R \cdot C(O^{-+}Na) \cdot OEt \rightarrow R \cdot C(O^{-+}Na) \cdot OEt$

$\qquad\qquad\qquad\qquad\qquad\qquad\qquad\qquad\qquad\quad |$

$\qquad\qquad\qquad\qquad\qquad\qquad\qquad\qquad\qquad R \cdot C(O^{-+}Na) \cdot OEt$, then

$+ \ \text{loss of 2 Et} \cdot ONa \rightarrow R \cdot C{:}O \qquad\qquad R \cdot C(O^{-+}Na)$

$\qquad\qquad\qquad\qquad\qquad\quad | \qquad\qquad\qquad\qquad \|$

$\qquad\qquad\qquad\qquad R \cdot C{:}O, \text{ then} + 2 \ Na \rightarrow R \cdot C(O^{-+}Na)$, then

$+ \ 2 \ H_2O \rightarrow R \cdot C \cdot OH \qquad\qquad\qquad R \cdot C{:}O$

$\qquad\qquad\qquad\quad \| \qquad\qquad\qquad\qquad\qquad\qquad |$

$\qquad\qquad R \cdot C \cdot OH, \text{ then} + \text{proton shift} \rightarrow R \cdot CHOH$

The sodium apparently adds to the carbonyl oxygen and donates its electron to the carbon atom to give a free radical. The union of two free radicals gives the diketone.

2. Fittig or Fittig-Wurtz Synthesis (Norris, 3rd ed., p. 372, 1931):

An aromatic halide is condensed with an aromatic or aliphatic halide by metallic sodium in the presence of an inert solvent to yield higher hydrocarbons:

$C_6H_5 \cdot X + 2 \ Na / \text{ (ether)} + X \cdot C_6H_5 \rightarrow C_6H_5 \cdot C_6H_5 + 2 \ NaX$, or

$Ar.X + 2 \ Na + X \cdot R/\text{(dry ether)} \rightarrow Ar.R + 2 \ NaX$.

Other products in the latter case are $Ar \cdot Ar$ and $R \cdot R$.

3. Frankland Synthesis (Cohen, Part I, 4th ed., p. 214-5, 1924):

Alkyl iodides react with zinc dust to give alkylzinc iodides which may be condensed, in turn, with a second mole of alkyl iodide:

$R \cdot I + Zn \ \text{dust, heat} \rightarrow R \cdot Zn \cdot I$, then

$R \cdot Zn \cdot I + R \cdot I, \text{ distil} \rightarrow R \cdot R + ZnI_2$

4. Freund Reaction (Cohen, Part I, 4th ed., p. 216, 1924):

Alpha-omega dihalogenated hydrocarbons are treated with metallic sodium in an appropriate solvent such as dry ether to effect ring closure:

$$(H_2C)_n \begin{array}{c} \diagup CH_2 \cdot Br \\ \diagdown CH_2 \cdot Br \end{array} + 2 \ Na/\text{(dry ether)} \rightarrow (H_2C)_n \begin{array}{c} \diagup CH_2 \\ | \\ \diagdown CH_2 \end{array} + 2 \ NaBr$$

5. Gladstone Reaction (Hill and Kelley, 1st ed., p. 18, 1932):

Alkyl halides by treatment with the zinc-copper couple yield the corresponding saturated hydrocarbons:

$2 \ R \cdot X + H_2O/Zn\text{-}Cu, \text{ heat} \rightarrow 2 \ R \cdot H + ZnX_2 + CuO$

6. Kekulé Synthesis (Cohen, Part I, 4th ed., p. 215, 1924):

Halogen compounds react with carbon dioxide in the presence of sodium in dry ether to yield salts of acids:

$C_6H_5 \cdot Br + CO_2 + 2 \ Na \rightarrow C_6H_5 \cdot CO \cdot ONa + NaBr$

7. Schorigin Reaction (Schmidt, English ed., p. 393, 1926):

Benzene and its homologs react with dry carbon dioxide in the presence of a mixture of sodium and diethylmercury (or diethylzinc) to give carboxylic acids:

$$C_6H_6 + CO_2 \ (Na, \ Et_2Hg) \rightarrow C_6H_5 \cdot CO \cdot OH$$

Throughout this book, reagents that appear in parenthesis are not balanced, hence benzoic acid is shown here instead of sodium benzoate, which is the product actually obtained.

8. **Wanklyn Reaction** (Schmidt, English ed., p. 154, 1926):

Alkyl sodium compounds in an inert solvent react with carbon dioxide to give salts of organic acids:

$$CH_3 \cdot Na/ \ (benzene) + CO_2 \rightarrow CH_3 \cdot CO \cdot ONa$$

9. **Wislicenus Synthesis** (Cohen, Part I, 4th ed., p. 215, 1924):

Diacids are prepared by condensing halogenated acid with silver:

$$HO \cdot OC \cdot CH_2 \cdot CH_2 \cdot I + 2 \ Ag + I \cdot CH_2 \cdot CH_2 \cdot CO \cdot OH \rightarrow$$

$$HO \cdot OC(CH_2)_4CO \cdot OH + 2 \ AgI$$

10. **Wurtz Reaction** (Chamberlain, 3rd ed., p. 17, 1934):

Alkyl halides are condensed in the presence of sodium to yield hydrocarbons:

$$R \cdot X + 2 \ Na/(dry \ ether) + X \cdot R \rightarrow R \cdot R + 2 \ NaX$$

11. **Wurtz-Fittig Reaction** (Cohen, Part I, 4th ed., p. 215, 1924):

Alkyl halides condense with aromatic halides in the presence of sodium to yield alkylated aromatic hydrocarbons:

$$C_6H_5 \cdot X + 2 \ Na/ \ (dry \ ether \ or \ benzene) + X \cdot R \rightarrow C_6H_5 \cdot R + 2 \ NaX$$

12. **Zincke Reaction** (Cohen, Part I, 4th ed., p. 224, 1924):

Reaction of aromatic hydrocarbons with halogen compounds is effected by the use of zinc dust:

$$2 \ C_6H_5 \cdot H + 2 \ Cl \cdot CH_2 \cdot C_6H_5/Zn, \ dust \rightarrow 2 \ C_6H_5 \cdot CH_2 \cdot C_6H_5 + ZnCl_2 + H_2$$

I. A. 2. Free Radicals from Thermal Excitation

13. **Beilstein Test For Halogen** (Hill and Kelley, 1st. ed., p. 360, 1932):

A copper wire is heated in a flame to volatilize any copper halide that might be present, then dipped into a halide compound or solution and heated once again to give a green coloration to the flame due to the volatilization of copper halide.

14. **Bergius Process** (Reid, 3rd. printing, p. 76, 1931):

Powdered coal is heated in the presence of hydrogen to about 450° C. under about 300 atmospheres pressure to give a mixture of hydrocarbons somewhat resembling petroleum, together with appreciable amounts of methane. The methane is treated in turn with superheated steam to give hydrogen, thus completing the cycle, as:

Coal + x H_2, 450° C./300 atms. $\rightarrow R \cdot H + CH_4$, then

$CH_4 + H_2O/(Ni, \ 600-800° \ C.) \rightarrow 3 \ H_2 + CO$, then

$CO + H_2O/(Ni, \ 600-800° \ C.) \rightarrow H_2 + CO_2$, etc.

15. Bucher Processes (Lucas, p. 308, 1935):

One part of the Bucher process for the fixation of nitrogen is to pass nitrogen into a mixture of sodium carbonate, carbon, and iron turnings at 800 to 1000°:

$N_2 + 4 C + Na_2CO_3 \rightarrow 2 NaCN + 3 CO.$

16. Carius Determination (Gattermann and Wieland, 1st. ed., p. 65–8, 1934):

An organic compound containing halogen or sulfur is heated with fuming nitric acid (and silver nitrate in the case of halogen compounds) at about 250° C. and the decomposition products analyzed by standard procedures for halogen or sulfur.

17. Castner Process (Lucas, p. 308, 1935):

Sodamide, prepared by the action of ammonia on sodium, is heated in the presence of carbon to form cyanamide, which reacts with carbon at a higher temperature to produce sodium cyanide:

$2 Na + 2 NH_3 \rightarrow 2 NaNH_2 + H_2$, then

$2 NaNH_2 + C + heat \rightarrow Na_2NCN + 2 H_2$, and

$Na_2NCN + C + heat, 800\text{-}1000° \rightarrow 2 NaCN.$

18. Diels-Alder Reaction (Waters, 1st. ed., p. 376–7, 1936):

Maleic anhydride and similar compounds undergo 1,4-addition to conjugated dienes:

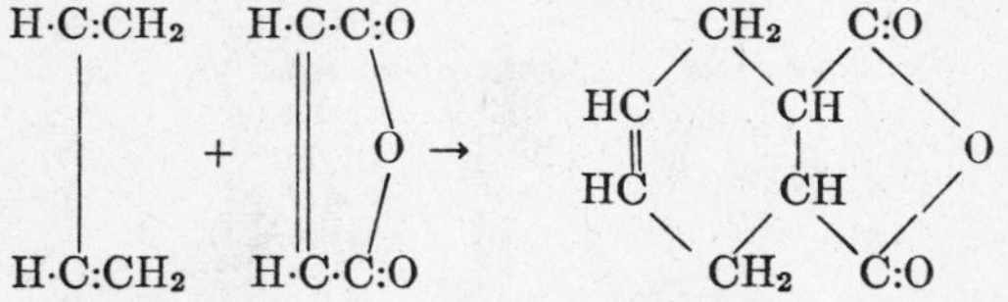

$$
\begin{array}{c}
\text{H·C:CH}_2 \quad \text{H·C·C:O} \\
\Big| \qquad\qquad \| \quad \diagdown \\
\qquad + \qquad\quad \text{O} \rightarrow \\
\Big| \qquad\qquad \| \quad \diagup \\
\text{H·C:CH}_2 \quad \text{H·C·C:O}
\end{array}
\qquad
\begin{array}{c}
\text{CH}_2 \quad \text{C:O} \\
\diagup \quad\;\; \diagdown \qquad \diagdown \\
\text{HC} \qquad \text{CH} \qquad \diagdown \\
\| \qquad\quad | \qquad\qquad \text{O} \\
\text{HC} \qquad \text{CH} \qquad \diagup \\
\diagdown \quad\;\; \diagup \qquad \diagup \\
\text{CH}_2 \quad \text{C:O}
\end{array}
$$

In certain instances this addition proceeds at room temperature, in which case the reaction is logically classed under II.C.

19. Diels-Wolf Reaction (Cohen, Part I, 4th. ed., p. 141, 1924):

Malonic acid upon heating in the presence of phosphorus pentoxide yields carbon suboxide (C_3O_2):

$CH_2(CO·OH)_2 + 2 P_2O_5, heat \rightarrow C_3O_2 + 4 HPO_3$

20. Dubbs Process (Riegel, 2nd ed., p. 370–1, 1933):

The distillation of the crude petroleum and the cracking of its heavier hydrocarbons are carried on simultaneously in the same apparatus at a temperature of about 840° F. and a pressure of about 350 pounds per square inch.

21. Dumas Determination (Chamberlain, 3rd ed., p. 809, 1934):

The nitrogen in organic compounds is converted to free nitrogen and the evolved nitrogen measured in a gas buret.

22. Fischer-Tropsch Synthesis (Ellis, 3rd ed., p. 692, 1930):

Carbon monoxide is reduced catalytically to yield alkane mixtures resembling petroleum:

$x CO + y H_2/(\text{catalyst as Co·Th}) \rightarrow \text{alkane mixtures} + x H_2O.$

The hydrocarbons, in turn, may be converted to fatty acids and subsequently to fats.

23. Gerhardt Synthesis of Quinoline (Schmidt, English ed., p. 623, 1926):

As early as 1842, Gerhardt obtained quinoline by distilling cinchonine with alkali:

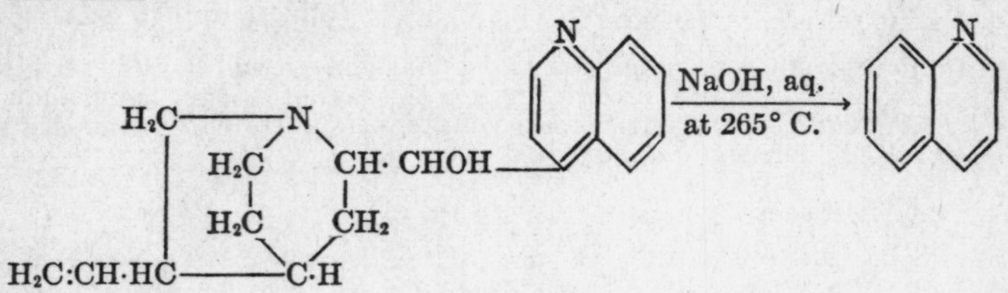

24. Goldschmidt Process (Desha, p. 386, 1936):

Oxalic acid may be produced by heating sodium formate strongly. The oxalic acid is purified by the use of its insoluble calcium salt.

$$\begin{matrix} H\cdot CO\cdot ONa \\ H\cdot CO\cdot ONa \end{matrix} + heat, 240°C. \rightarrow \begin{bmatrix} {}^*CO\cdot ONa \\ {}^*CO\cdot ONa \end{bmatrix} + H_2 \rightarrow NaO_2C\cdot CO_2Na + H_2.$$

25. Kjeldahl Method (Kamm, p. 190-1, 1932):

Organic nitrogen in its non-oxidized form is converted, in practically all cases, into ammonium sulfate upon digesting the compound with sulfuric acid. Various catalysts are used. The ammonium sulfate is treated with a non-volatile alkali to liberate ammonia, the ammonia distilled into a known volume of standard acid and the excess acid titrated.

26. Krafft Method (Whitmore, p. 306-307, 1938):

Saturated, high molecular weight, odd-numbered carboxylic acids may be prepared by distilling the barium or calcium salts of acetic acid and the naturally occurring high molecular weight acid. The mixed ketone resulting from this reaction is subsequently oxidized to produce the acid.

$$\begin{matrix} (R\cdot CO\cdot O)_2Ca, & distil \rightarrow \\ \\ (R'\cdot CO\cdot O)_2Ca, & distil \rightarrow \end{matrix} \begin{vmatrix} R\cdot CO\cdot \\ + \\ R'\cdot \end{vmatrix} + \begin{vmatrix} R\cdot \\ + \\ R'\cdot CO\cdot \end{vmatrix} \begin{matrix} \rightarrow R_2C{:}O \\ \\ \rightarrow R'_2C{:}O. \end{matrix} \quad + 2\,CaCO_3$$

Another product of the reaction is $R\cdot CO\cdot R'$. The oxidation of any one of these ketones will yield acids.

27. Liebig Combustion (Gattermann and Wieland, p. 52-61, 1934):

A weighed amount of an organic compound is oxidized in a combustion tube by oxygen in the presence of copper oxide at a dull red heat, and the products absorbed in carefully weighed absorption tubes; the water in calcium chloride or other absorbent and the carbon dioxide in a concentrated solution of potassium hydroxide. In his own words, Liebig simplified organic combustions to the extent that "even a monkey could run one."

28. Mond Process (Whitmore, p. 508, 1938):

An extraction process for nickel takes advantage of the volatility of nickel carbonyl. The carbonyl is subsequently thermally decomposed to regenerate the starting materials, as:

$Ni + 4CO$, heat $\rightarrow Ni(CO)_4$, then $+$ heat $\rightarrow Ni + 4CO$.

29. Ruzicka Large Ring Synthesis (Hill and Kelley, 1st ed., p. 324, 1932):

Cyclic ketones, containing up to 30 carbon atoms, have been prepared by vacuum distillation of the calcium, thorium or yttrium salts of diacids in the presence of copper filings. Yields are usually low, often less than 4%.

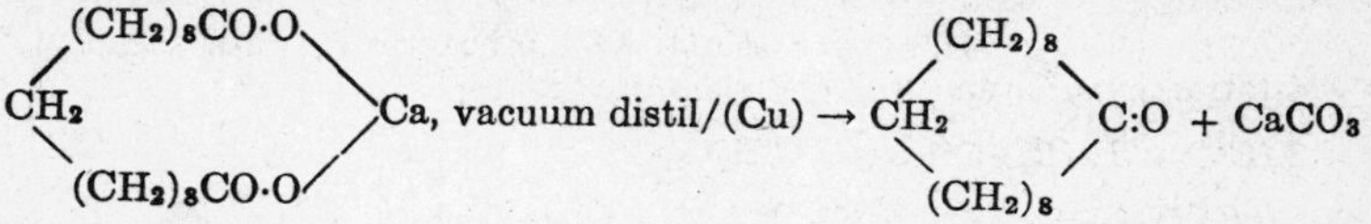

30. Sabatier-Mailhe Aldehyde Synthesis (Schmidt, English ed., p. 138, 1926):

Aliphatic acids, when mixed with the vapors of formic acid and passed over titanium oxide at 300°C., yield aldehydes:

$R \cdot CO \cdot OH + H \cdot CO \cdot OH(TiO_2, 300°C.) \rightarrow R \cdot CHO + CO_2 + H_2O$

Free radical formation may occur on the oxide surface.

31. Schmidlin Ketene Synthesis (Gattermann and Wieland, p. 117, 1932):

Acetone, by thermal decomposition, yields ketene and methane:

$CH_3 \cdot CO \cdot CH_3$, thermal decomposition $\rightarrow H_2C:C:O + CH_4$

32. Senderens Ketone Synthesis (Schmidt, English ed., p. 138, 1926):

Acids or esters, when passed over thorium or aluminum oxide at about 300-380°C., yield ketones:

$R \cdot CO \cdot OH + HO \cdot OC \cdot R/ (Al_2O_3, 300\text{-}380°C.) \rightarrow R \cdot CO \cdot R + CO_2 + H_2O$

Free radical formation may occur on the oxide surface.

33. Wilsmore Ketene Synthesis (Gattermann and Wieland, p. 117, 1932):

Acetic anhydride, on the surface of a glowing platinum wire, undergoes dehydration to yield ketene:

$CH_3 \cdot CO \cdot O \cdot OC \cdot CH_3$, by loss of $H_2O/ (Pt, hot) \rightarrow 2 H_2C:C:O$

I. A. 3. Free Radicals from Electrolysis

34. Crum-Brown and Walker Reaction (Cohen, Part I, 4th. ed., p. 233, 1924):

The electrolysis of the monoester of succinic acid yields the diester of adipic acid:

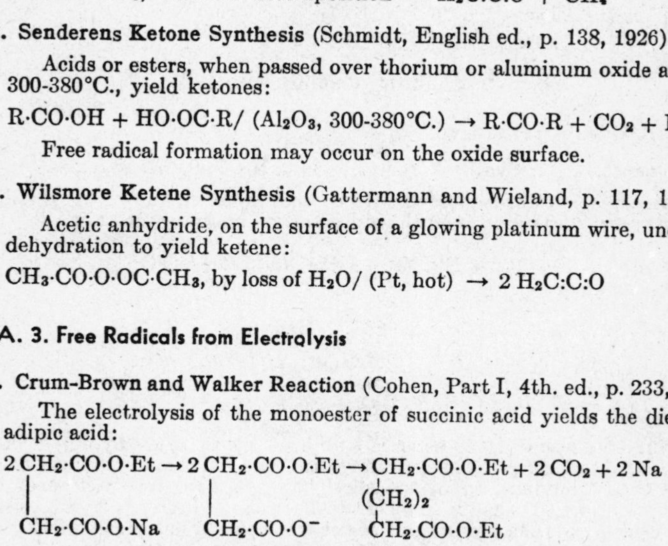

35. Kolbe Electrolytic Synthesis (Cohen, Part I, 4th ed., p. 233, 1924):

Metallic alkanoates, when subjected to electrolysis, yield saturated hydrocarbons:

$2CH_3 \cdot CO \cdot OK/2H_2O$, electrolysis $\rightarrow CH_3 \cdot CH_3 + 2 CO_2 + 2 KOH + H_2$

36. Miller-Hofer Electrolysis (Cohen, Part I, 4th ed., p. 234, 1924):

Electrolysis of mixtures of organic and inorganic salts is illustrated by the following:

$CH_3 \cdot CH_2 \cdot CO \cdot OK + K \cdot I$, electrolysis $\rightarrow CH_3 \cdot CH_2 \cdot I + CO_2 + 2 K$, and

$CH_3 \cdot CH_2 \cdot CO \cdot ONa + Na \cdot NO_2$, electrolysis $\rightarrow CH_3 \cdot CH_2 \cdot NO_2 + CO_2 + 2 Na$

37. Walker Reaction (Cohen, Part III, 4th ed., p. 268, 1924):

The electrolysis of potassium 3,3-diethoxypropanoate yields 1,1,4, 4-tetraethoxybutane:

$2 (EtO)_2 CH \cdot CH_2 \cdot CO \cdot OK + 2 H_2O$, electrolysis $\rightarrow (EtO)_2 CH \cdot CH_2 \cdot CH_2 \cdot HC$

$(OEt)_2 + 2 CO_2 + 2 KOH + H_2$

38. Wohl-Schweitzer Electrolysis (Cohen, Part I, 4th ed., p. 234, 1924):

Potassium 3,3-diethoxypropanoate (diethyl acetal of potassium malonaldehydate), when subjected to electrolysis, yields the acetal of succinic aldehyde:

$$2 CH_2 \begin{matrix} CH(OEt)_2 \\ \diagup \\ \diagdown \\ CO \cdot OK \end{matrix} + 2 H_2O, \text{ electrolysis} \rightarrow \begin{matrix} CH_2 \cdot CH(OEt)_2 \\ | \\ CH_2 \cdot CH(OEt)_2 \end{matrix} + 2 CO_2 + 2 KOH + H_2$$

Cf. Walker Reaction, No. 37.

I. B. Ionic Mechanisms

I. B. 1. Alkide Ion Formation.

39. Acetoacetic Ester Synthesis (Chamberlain, p. 253-9, 3rd ed., 1934):

Acetoacetic ester is treated with sodium ethoxide and an alkyl iodide to introduce alkyl groups on the alpha carbon grouping, as:

$CH_3 \cdot CO \cdot CH_2 \cdot CO \cdot OEt + EtONa \rightarrow CH_3 \cdot C(O^-){:}CH \cdot CO \cdot OEt + Na^+ +$ EtOH, then

$+ R \cdot X \rightarrow CH_3 \cdot CO \cdot CHR \cdot CO \cdot OEt + NaX$, then

$+ Et \cdot ONa \rightarrow CH_3 \cdot C(O^-){:}CR \cdot CO \cdot OEt + Na^+ + Et \cdot OH$, then

$+ R \cdot X \rightarrow CH_3 \cdot CO \cdot CR_2 \cdot CO \cdot OEt + NaX$.

The enol form of the ester is stabilized by an inner hydrogen bond enabling it to form easily the sodium salt. The alkide ion of this salt reacts with the carbonium ion of the alkyl halide formed by reaction of the sodium ion with the halide to yield the disubstituted ester. Hydrolysis by base then yields the sodium salt of the free acid. Such a salt will have a negative charge because of its extra electron, and the oxygens of the carboxyl group will have their electron thirst more than satisfied. This allows the carbonyl oxygen to attract the electrons of the disubstituted carbon more strongly than usual, and the simultaneous repulsion of electrons toward the disubstituted carbon by the carboxyl group weakens the bond on the carbonyl side of said carbon, causing acid cleavage. If, on the other hand, the free unionized acid is formed by acidification of the sodium salt, the carboxyl group no longer has an extra electron, and the two oxygens therein exert a much stronger pull on the electrons of the disubstituted carbon than the single carbonyl oxygen, thus weakening the bond on the carboxyl side and giving ketonic cleavage. The actual cleavage in either case is done by hydroxyl ion acting as an electron donor to the electron-deficient link, whither it has been attracted by the hydrogen bonding tendency of the oxygens.

Decomposition of this product may be by "ketonic cleavage" with dilute alkali or by "acid cleavage" with concentrated alkali:

$CH_3 \cdot CO \cdot CR_2$	CO	O·Et
H	O	H

"ketonic cleavage"

$CH_3 \cdot CO$	$CR_2 \cdot CO \cdot O$	Et
NaO	H Na	OH

"acid cleavage"

40. Von Baeyer Synthesis (Cohen, p. 220-1, Part III, 1924):

This represents the first successful synthesis of the terpenes.

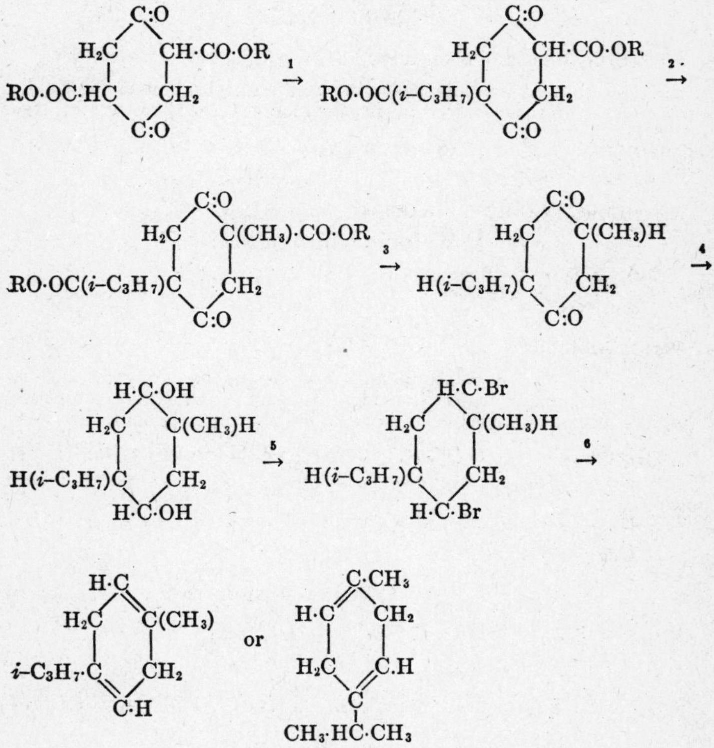

The reagents for the successive steps are: (1) NaOEt and $CH_3 \cdot CHI \cdot CH_3$, (2) NaOEt and $CH_3 \cdot I$, (3) H_2SO_4, heat, (4) H_2/catalyst, (5) HBr, and (6) NaOH, aq.

41. Haller-Bauer Reaction (Cohen, 4th ed., p. 217, 1924):

Ketones are alkylated by the use of alkyl iodides and sodamide:

$CH_3 \cdot CO \cdot CH_3 \rightleftarrows CH_3 \cdot C(OH){:}CH_2 + CH_3 \cdot I/NaNH_2 \rightarrow CH_3 \cdot C(OH)(I) \cdot$
$CH_2 \cdot CH_3 \rightarrow CH_3 \cdot CO \cdot CH_2 \cdot CH_3 + NaI + NH_3.$

The alkylation may be continued to yield hexamethyl acetone as the final product ($[CH_3]_3 C \cdot CO \cdot C[CH_3]_3$).

42. Malonic Ester Synthesis (Chamberlain, p. 253, 3rd. ed., 1934):

One or both of the *alpha*-hydrogen atoms in malonic ester may be replaced by alkyl groups by treatment with sodium ethoxide and then with an alkyl halide, as:

$EtO \cdot OC \cdot CH_2 \cdot CO \cdot OEt + Et \cdot ONa \rightarrow EtO \cdot OC \cdot HC{:}C(O^-)OEt + Na^+ +$
$Et \cdot OH,$ then

$+ R \cdot X \rightarrow EtO \cdot OC \cdot CHR \cdot CO \cdot OEt + NaX,$ then

$+ Et \cdot ONa \rightarrow EtO \cdot OC \cdot RC{:}C(O^-)OEt + Na^+ + Et \cdot OH,$ then

$+ R \cdot X \rightarrow EtO \cdot OC \cdot CR_2 \cdot CO \cdot OEt + NaX.$

43. Michael Synthesis (Cohen, Part I, 4th. ed., p. 239, 1924):

Ethyl cinnamate condenses with sodium ethyl malonate to yield, upon acidification and partial hydrolysis, the monoester of β-phenylglutaric acid:

$C_6H_5 CH{:}HC \cdot CO \cdot OEt + [(EtO \cdot OC)_2 CH^-]Na^+ \rightarrow C_6H_5 \cdot CH \cdot CH_2 (Na) \cdot CO \cdot OEt,$ then
$\qquad\qquad\qquad\qquad\qquad\qquad\qquad\qquad\qquad\qquad |$
$\qquad\qquad\qquad\qquad\qquad\qquad\qquad\qquad\qquad (Et \cdot O \cdot OC)_2 C \cdot H$

$+ HCl. aq. \rightarrow C_6H_5 CH \cdot CH_2 CO \cdot OEt,$ then $+ 3 H_2O \rightarrow$
$\qquad\qquad\qquad\qquad |$
$\qquad\qquad\qquad (EtO \cdot OC)_2 C \cdot H$

$C_6H_5 \cdot CH \cdot CH_2 \cdot CO \cdot OH + NaCl + 3 Et \cdot OH + CO_2$
$\quad |$
$H_2C \cdot CO \cdot OH$

44. Perkin Method (Gilman, p. 17, 1938):

Ethylene bromide and malonic ester condense to give the ester of 1,1-cyclopropanedicarboxylic acid. This, upon partial saponification and heating, gives a monoester of a cycloalkane carboxylic acid, as:

$Br \cdot CH_2 \cdot CH_2 \cdot Br + EtO \cdot OC \cdot HC{:}C(O^-)OEt + Na^+ \rightarrow Br \cdot CH_2 \cdot CH_2 \cdot CH$
$(CO \cdot OEt)_2,$ then

$+ Et \cdot ONa \rightarrow Br \cdot CH_2 \cdot CH_2 \cdot C[{:}C(O^-)OEt] \cdot CO \cdot OEt + Na^+ + Et \cdot OH,$ then

$+$ ring closure $\rightarrow H_2C$

$\qquad\qquad\qquad\qquad\qquad\diagdown$
$\qquad\qquad\qquad\qquad\qquad\quad C(CO \cdot OEt)_2 + NaBr,$ then
$\qquad\qquad\qquad\qquad\qquad\diagup$
$\qquad\qquad H_2C$

$\qquad\qquad\qquad\qquad\qquad\qquad\qquad H_2C$
$\qquad\qquad\qquad\qquad\qquad\qquad\qquad\quad|\diagdown$
$+$ careful hydrolysis and heat $\rightarrow \quad\quad\quad CH \cdot CO \cdot OEt + Et \cdot OH + CO_2.$
$\qquad\qquad\qquad\qquad\qquad\qquad\qquad\quad|\diagup$
$\qquad\qquad\qquad\qquad\qquad\qquad\qquad H_2C$

45. Vorländer Condensation (Cohen, Part I, 4th ed., p. 240, 1924):

Benzylideneacetone condenses with sodium malonic ester to give phenyldihydroresorcyclic ester, as:

$Bz \cdot HC:CH \cdot CO \cdot CH_3 + EtO \cdot OC \cdot HC:C(O^-)OEt + Na^+ \rightarrow$
$Bz \cdot CH \cdot CH(CO \cdot OEt)_2 + Na^+$, then

$\underset{\underset{\displaystyle H \cdot C(^-) \cdot CO \cdot CH_3}{|}}{}$

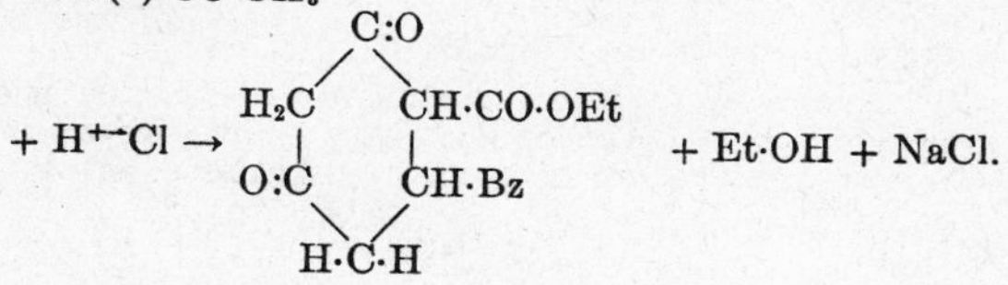

$$+ H^+{}^-Cl \rightarrow \quad \begin{array}{c} C:O \\ \diagup \quad \diagdown \\ H_2C \qquad CH \cdot CO \cdot OEt \\ | \qquad\qquad | \\ O:C \qquad CH \cdot Bz \\ \diagdown \quad \diagup \\ H \cdot C \cdot H \end{array} \quad + Et \cdot OH + NaCl.$$

I. B. 2. Carbonium Ion Formation.

46. Allylic Rearrangement (Whitmore, p. 146, 1938):

Methylvinylcarbinol undergoes reaction with halogen acid to give crotyl halide as well as the expected product, as:

$H_2C:CH \cdot CHOH \cdot CH_3 + H^+$, aq. $\rightarrow H_2C:CH \cdot CH(^+) \cdot CH_3 + H_2O$, then
$+ Br^-$, aq. $\rightarrow H_2C:CH \cdot CHBr \cdot CH_3$, or

$H_2C:CH \cdot CH(^+) \cdot CH_3 + $ rearrangement $\rightarrow H_2C(^+) \cdot HC:CH \cdot CH_3$, then
$+ Br^-$, aq. $\rightarrow Br \cdot CH_2 \cdot HC:CH \cdot CH_3$.

47. Bart Reaction (Desha, p. 333, 1936):

4-Chlorobenzenediazonium ion is treated with arsenic trioxide in the presence of alkali to yield 4-chlorophenylarsinic acid:

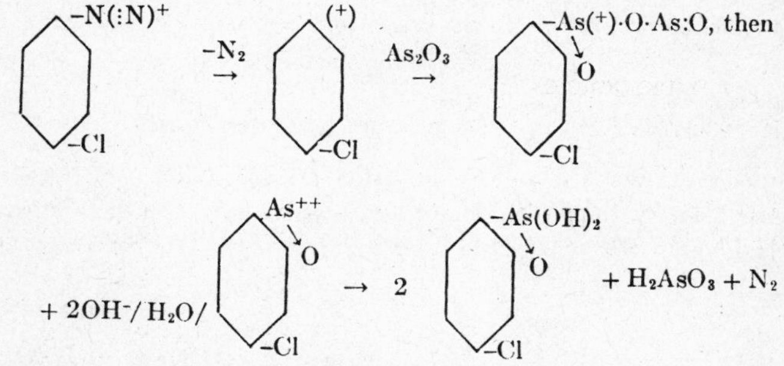

48. Bouis Allene Synthesis (Whitmore, p. 51, 1938):

An alkyl Grignard reagent undergoes reaction with acrolein to give a substituted allyl alcohol. On treatment with phosphorus tribromide the alcohol rearranges to give a 1-bromo-2-olefin. This olefin adds bromine, then hydrogen bromide is eliminated by caustic, and finally zinc removes two bromine atoms to give the allene derivative.

$H_2C:CH \cdot CHO + R \cdot Mg \cdot X/(H_2O) \rightarrow R \cdot CHOH \cdot HC:CH_2$, then
$3 R \cdot CHOH \cdot HC:CH_2 + PBr_3 \rightarrow 3 R \cdot HC:CH \cdot CH_2 \cdot Br + H_3PO_3$, then
$R \cdot HC:CH \cdot CH_2 \cdot Br + Br_2 \rightarrow R \cdot CHBr \cdot CHBr \cdot CH_2 \cdot Br$, then
$R \cdot CHBr \cdot CHBr \cdot CH_2 \cdot Br + NaOH$, alc. $\rightarrow R \cdot CHBr \cdot BrC:CH_2$, and
$R \cdot CHBr \cdot BrC:CH_2 + Zn$, heat $\rightarrow R \cdot (H)C:C:CH_2 + ZnBr_2$.

Step two is an allylic rearrangement.

49. Butlerow Acetyl Chloride Reaction (Whitmore, p. 123, 1938):

Acid halides react in certain cases as halogenation agents, one specific example being:

$$(CH_3)_3C \cdot OH + Ac \cdot Cl \rightarrow (CH_3)_3C \cdot Cl + Ac \cdot OH.$$

50. Demjanow Rearrangement (Gilman, p. 32, 1938):

Cycloalkylmethylamines rearrange on treatment with nitrous acid to give expanded ring compounds. Cyclobutylmethylamine, for example, gives cyclopentanol, cyclopentene, and methylenecyclobutane, as:

$$\begin{array}{c} H_2C - CH \cdot CH_2 \cdot NH_2 \\ | \quad\quad | \\ H_2C - CH_2 \end{array} + HONO \rightarrow \begin{array}{c} H_2C - CH \cdot CH_2 \cdot N(:N)^+ \\ | \quad\quad | \\ H_2C - CH_2 \end{array}$$

$$\downarrow -N_2$$

$$\begin{array}{c} H \cdot C \cdot OH \\ H_2C \quad CH_2 \\ | \quad\quad | \\ H_2C \text{——} CH_2 \end{array} \xleftarrow{OH^-} \begin{array}{c} H_2C - CH \cdot CH_2(^+) \\ | \quad\quad | \\ H_2C - CH_2 \end{array} \xrightarrow{OH^-} \begin{array}{c} H_2C - CH \cdot CH_2 \cdot OH \\ | \quad\quad | \\ H_2C - CH_2 \end{array}$$

$$\begin{array}{cc} \swarrow -H_2O & \nearrow -H^+ \\ \end{array} \quad\quad \begin{array}{cc} \searrow -H^+ & \nearrow -H_2O \end{array}$$

$$\begin{array}{c} C \cdot H \\ H_2C \diagup \diagdown C \cdot H \\ | \quad\quad | \\ H_2C \text{——} CH_2 \end{array} \quad\quad\quad \begin{array}{c} H_2C - C:CH_2 \\ | \quad\quad | \\ H_2C \text{——} CH_2 \end{array}$$

51. Dow Process (Whitmore, p. 718, 1938):

Chlorobenzene is treated with (a) alkali and heated under pressure in the presence of a catalyst to yield phenol, or (b) with ammonium hydroxide and heated under pressure in the presence of a catalyst to yield aniline:

a. $C_6H_5 \cdot Cl + 2 \text{ NaOH}$, aq., heat, pressure, catalyst $\rightarrow C_6H_5 \cdot ONa + NaCl + H_2O$, and

b. $C_6H_5 \cdot Cl + 2 \text{ H} \cdot NH_2$, aq., heat, pressure, catalyst $\rightarrow C_6H_5 \cdot NH_2 + NH_4Cl$.

52. Gabriel Synthesis (Hill and Kelley, 1st ed., p. 437, 1932):

Pure primary aliphatic amines are obtained by treating potassium phthalimide with an alkyl iodide and subsequent hydrolysis:

$$\begin{array}{c} \text{[benzene ring]} \begin{array}{c} O \\ || \\ C \\ \diagdown \\ N \cdot K \\ \diagup \\ C \\ || \\ O \end{array} \end{array} + I \cdot R \rightarrow \begin{array}{c} \text{[benzene ring]} \begin{array}{c} O \\ || \\ C \\ \diagdown \\ N \cdot R \\ \diagup \\ C \\ || \\ O \end{array} \end{array}, \text{ then } + 2 \text{ NaOH, aq.} \rightarrow R \cdot NH_2 +$$

$$\begin{array}{c} \text{[benzene ring]} \begin{array}{c} CO \cdot ONa \\ CO \cdot ONa \end{array} \end{array}$$

53. Gattermann Reaction (Perkin and Kipping, p. 403, 1932):

Aromatic diazonium chlorides, by treatment with halogen acids in the presence of finely divided copper, yield the aromatic halides containing the halogen atom from the acid:

$$ArN(:N)Cl + HBr/(\text{finely divided Cu, } 0\text{-}5°C.) \rightarrow Ar \cdot Br + N_2 + HCl$$

54. Meyer Reaction (Gilman, p. 471, 1938):

Sodium stannite undergoes reaction with an alkyl halide to give sodium alkyl stannate, as:

$Na \cdot O \cdot Sn \cdot O \cdot Na + R \cdot X \rightarrow R \cdot O \cdot Sn \cdot ONa + NaX.$

55. Victor Meyer Synthesis (Schmidt, English ed., p. 125, 1926):

Alkyl halides react with silver nitrite to yield alkyl nitrites, as:

$R \cdot X + Ag \cdot N(\rightarrow O){:}O \rightarrow [Ag \cdot N(\cdot O \cdot R \cdot X){:}O] \rightarrow R \cdot O \cdot NO + AgX$, but

$R \cdot X + Na \cdot N(\rightarrow O){:}O \rightarrow [R^{(+)} + (^-)N(\rightarrow O){:}O + NaX] \rightarrow R \cdot N(\rightarrow O){:}O + NaX.$

If it be assumed that sodium nitrite is more ionic than silver nitrite, it may be that most of the reaction with the sodium nitrite is with the nitrite ion whereas most of the reaction with the silver nitrite is with the exposed pair of electrons on the oxygen atom of the undissociated molecule.

The higher iodides give increasing amounts of the nitrites, whereas the corresponding bromides give more nitro compounds.

56. Pinacolone Rearrangement (Richter, 2nd. English ed., p. 216, 1919):

Pinacols rearrange, in the presence of sulfuric acid, to give unsymmetrical ketones:

$(CH_3)_2C(OH) \cdot (HO)C(CH_3)_2 + (H_2SO_4) \rightarrow (CH_3)_3C \cdot CO \cdot CH_3 + H_2O$

This mechanism may involve the formation of an alkene oxide, and subsequent rearrangement.

57. Sandmeyer Reaction (Chamberlain, 3rd ed., p. 674, 1934):

Diazonium salts react with halogen acids in the presence of cuprous salts, or with sodium cyanide in the presence of cuprous cyanide, to give halogen or nitrile derivatives of benzene:

a. $C_6H_5 \cdot N(:N)^{+-}Cl + HBr/(Cu_2Br_2, \ 0\text{-}5°C.) \rightarrow C_6H_5 \cdot Br + N_2 + HCl$

b. $C_6H_5 \cdot N(:N)^{+-}Cl + NaCN/(Cu_2(CN)_2, \ 0\text{-}5°C.) \rightarrow C_6H_5 \cdot CN + N_2 + NaCl$

58. Wagner Rearrangement (Gilman, p. 34, 1938):

Certain cycloalkanols, with an adjacent carbon atom that is completely alkylated, undergo dehydration and rearrangement to give ring contraction and expansion, as:

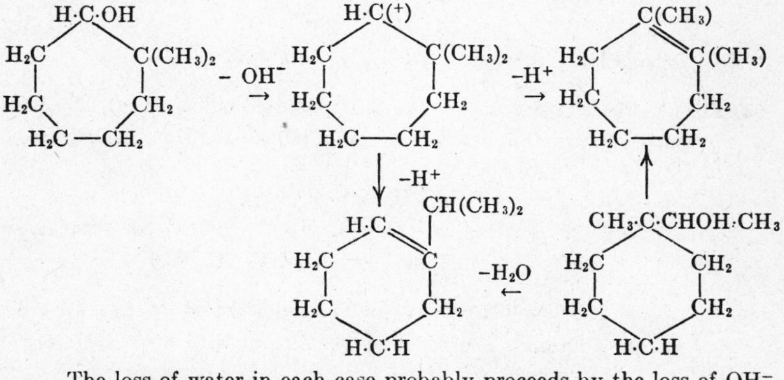

The loss of water in each case probably proceeds by the loss of OH⁻, rearrangement, and then loss of H⁺.

59. Walden Inversion (Bernthsen and Sudborough, p. 768-9, 1933):

Walden showed, by starting with *L*-chlorosuccinic acid, that either the *L*- or *D*-form of the corresponding hydroxyl derivative could be obtained by the proper choice of the reagent:

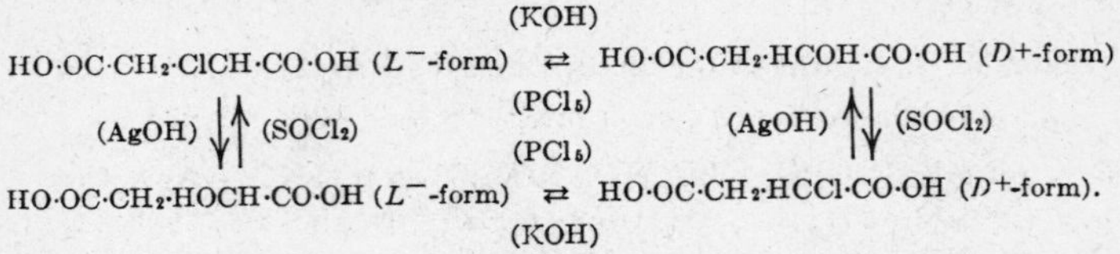

$$\text{(KOH)}$$

$$\text{HO·OC·CH}_2\text{·ClCH·CO·OH } (L^-\text{-form}) \;\rightleftarrows\; \text{HO·OC·CH}_2\text{·HCOH·CO·OH } (D^+\text{-form})$$

$$\text{(AgOH)} \;\Uparrow\;\text{(SOCl}_2) \qquad \begin{array}{c}\text{(PCl}_5)\\ \\ \text{(PCl}_5)\end{array} \qquad \text{(AgOH)} \;\Uparrow\;\text{(SOCl}_2)$$

$$\text{HO·OC·CH}_2\text{·HOCH·CO·OH } (L^-\text{-form}) \;\rightleftarrows\; \text{HO·OC·CH}_2\text{·HCCl·CO·OH } (D^+\text{-form}).$$

$$\text{(KOH)}$$

With $SOCl_2$ in pyridine, however, the L^--form is converted over to the D^+-form. With alkali, PCl_5, or $SOCl_2$/pyridine, it is postulated that an anion is formed, which can undergo an **umbrella** inversion, whereas with AgOH and $SOCl_2$ no ion is formed and no inversion results.

60. Wallach Degradation Method (Gilman, p. 35, 1938):

The 2,2-dibromocyclohexanone, upon treatment with alkali, rearranges to give a cyclopentanone, as:

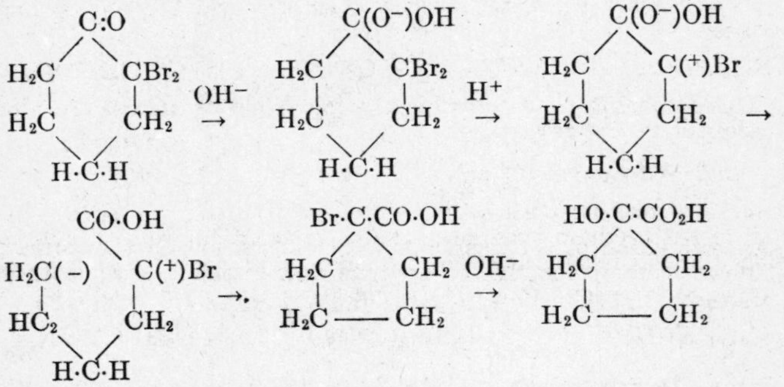

61. Williamson Synthesis (Chamberlain, p. 91, 1934):

Alkyl halides react with sodium alkoxides to yield ethers:

$$\text{R·X + Na·O·R} \rightarrow \text{R·O·R + NaX}$$

I. B. 3. Simple Ion Formation.

62. Fischer Phenylhydrazine Synthesis (Schmidt, English ed., p. 356, 1926):

Phenylhydrazine hydrochloride is obtained from benzenediazonium chloride by the following series of reactions:

$$\text{C}_6\text{H}_5\text{·N(:N)Cl + Na}_2\text{SO}_3 \rightarrow \text{C}_6\text{H}_5\text{·N:N·SO}_3\text{Na + NaCl, then}$$

$$\text{C}_6\text{H}_5\text{·N:N·SO}_3\text{Na + 2 HCl, aq./Zn} \rightarrow \text{C}_6\text{H}_5\text{·NH·HN·SO}_3\text{Na + ZnCl}_2, \text{ and}$$

$$\text{C}_6\text{H}_5\text{·NH·HN·SO}_3\text{Na + HCl/H}_2\text{O} \rightarrow \text{C}_6\text{H}_5\text{·NH·NH}_2\text{·HCl + NaHSO}_4$$

63. Griess (or Coupling) Reactions (Chamberlain, 3rd ed., p. 683, 1934):

Aromatic diazonium compounds condense with aromatic amines, phenols, and similar compounds when treated with them at 0-5°C. in the presence of sodium acetate:

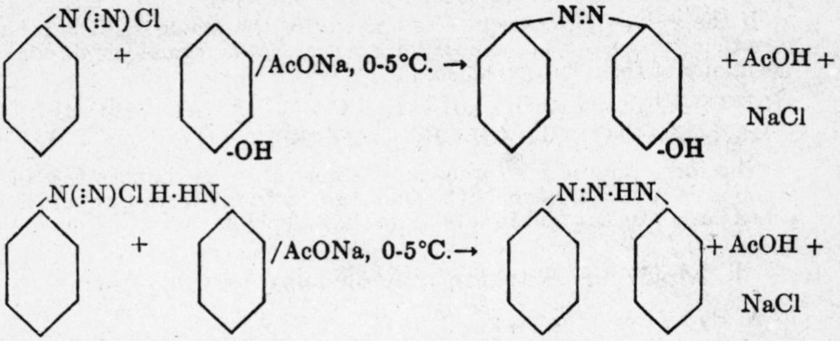

Secondary amines couple similarly, but the tertiary amines couple in the p-position. If the p-position in tertiary amimes and in phenols is occupied, they couple in the o-position.

64. Kolbe Synthesis:

a. Preparation of Nitromethane (Organic Syntheses, Col. Vol. 1., p. 393)

$$Cl \cdot CH_2 \cdot CO \cdot ONa + NaNO_2 \rightarrow O_2N \cdot CH_2 \cdot CO \cdot ONa + NaCl, \text{ then}$$
$$O_2N \cdot CH_2 \cdot CO \cdot ONa + H_2O/(H^+), \text{ heat} \rightarrow CH_3 \cdot NO_2 + NaHCO_3$$

b. Preparation of Sodium Salicylate (Chamberlain, 3rd ed., p. 558, 1934):

$$C_6H_5 \cdot ONa + CO_2, 110° C. \rightarrow C_6H_5 \cdot O \cdot CO \cdot ONa, \text{ then heat, } 130° C. \rightarrow$$
$$o\text{-}HO \cdot C_6H_4 \cdot COONa$$

65. Kossel and Kutscher Method (Schmidt, p. 210, 1938):

An estimation of the basic amino acids is made by the formation of insoluble silver compounds of arginine and histidine after sulfuric acid hydrolysis of the protein.

66. Mayer Reagent (Taylor, p. 539, 1933):

Nearly all alkaloids give precipitates with potassium mercuri-iodide. The alkaloids may be recovered by decomposing the precipitate with hydrogen sulfide.

67. Piria Reaction (Whitmore, p. 730, 732 (1938):

Aromatic nitro compounds react with sodium bisulfite to give, upon subsequent treatment with alkali, amines and salts of aminosulfonic acids, as:

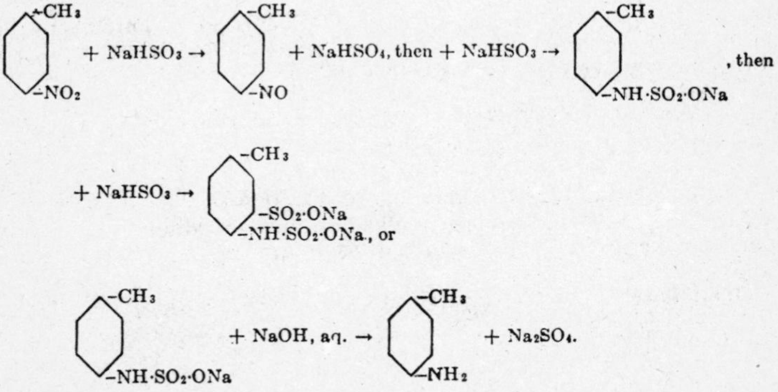

68. Sörensen Formol Titration (Schmidt, p. 191, 1938):

If the zwitter ion concept of structure for the amino acids is correct, then the titration of an amino acid in the presence of formaldehyde consists essentially of the following changes:

$$^+H_3N\cdot CZ_2\cdot CO\cdot O^- + NaOH, \text{ aq.} \rightarrow H_2N\cdot CZ_2\cdot CO\cdot O^- \, {}^+Na + H_2O, \text{ then}$$
$$H_2N\cdot CZ_2\cdot CO\cdot O^- \, {}^+Na + H\cdot CHO \rightarrow H_2C{:}N\cdot CZ_2\cdot CO\cdot O^- \, {}^+Na + H_2O.$$

The formation of the formaldehyde derivative as soon as the amino group is neutralized prevents the amino group from becoming hydrated to give a basic reaction and thus obscure the end point.

II. Molecular Addition of Molecules, Ions or Atoms.

A. By hydrogen bonding.
1. Oxygen to oxygen.
2. Oxygen to nitrogen.
3. Nitrogen to nitrogen.
4. Otherwise.

B. By metallic forces.
C. By addition into unsaturated groups.
D. By dipole-dipole or van der Waals forces.

II. A. Hydrogen Bond Mechanisms

II. A. 1. Oxygen to Oxygen Hydrogen Bonds.

69. Aldol Condensation (Chamberlain, 3rd. ed., p. 103, 1934):

A polymerization reaction which involves the carbonyl group of an aldehyde or ketone and a hydrogen atom in the alpha position to the carbonyl group of an aldehyde or ketone:

$$CH_3\cdot CHO + CH_3\cdot CHO/(\text{dilute alkali}) \rightarrow CH_3\cdot C(H)OH\cdot CH_2\cdot CHO$$

This particular product is aldol from which the reaction takes its name.

70. Angeli-Rimini Aldehyde Reactions (Whitmore, p. 227, 1938):

Aldehydes produce hydroxamic acids when heated with sodium salt of nitrohydroxylamine. These acids undergo reaction with ferric chloride to form deep purplish colors. Hydroxamic acids are also produced when an aldehyde is treated with N-benzenesulfonylhydroxylamine and a base.

$$R\cdot CHO + HO\cdot N{:}N(\rightarrow O)\cdot ONa/(OH', \text{ aq.}) \rightarrow R\cdot C({:}N\cdot OH)\cdot OH + NaNO_2.$$

These reactions fail with aldehydes having OH, NH$_2$, or CO groups in the *gamma-* or *delta-*position, which failure may be attributed to the tendency to form comparatively stable ring compounds.

71. Baumann-Schotten Reaction (Conant, 1st. ed., p. 445, 1933):

See Schotten-Baumann Reaction, No. 95.

72. Benzoin Condensation (Chamberlain, 3rd. ed., p. 576, 1934):

Benzaldehyde, in the presence of alkali cyanide as sodium or potassium cyanide, condenses with itself to give benzoin, as:

$$Bz\cdot CHO + Bz\cdot HC\overset{\nearrow}{\underset{O \rightarrow}{}} \rightarrow Bz\cdot CHOH\cdot CO\cdot Bz/(CN^-, \text{ aq.}).$$

The exact function of the cyanide is not known. It is likely that an intermediate as $Bz \cdot CH(O^-) \cdot CN$ forms a complex by hydrogen bonding with the other equivalent of aldehyde, with the subsequent loss of cyanide.

73. Blanc Reaction (Richter, p. 277, 1938):

Dicarboxylic acids of six and seven carbon atoms decompose when heated in acetic anhydride to yield cyclic ketones, as:

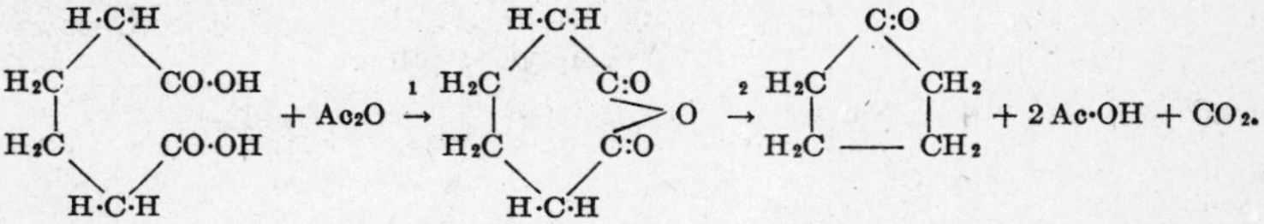

The reagents are acetic anhydride and heat. If the diacid has only four or five carbon atoms, the cyclic anhydride does not undergo decarboxylation to give the cyclic ketone. The reaction serves, consequently, to identify the number of carbon atoms in a diacid chain.

74. Cannizzaro Reaction (Conant, 1st. ed., p. 128, 1933):

Aldehydes which do not have hydrogen atoms in the α-position, undergo, in the presence of alkali, autooxidation and reduction. Benzaldehyde, for example, yields sodium benzoate and benzyl alcohol:

$C_6H_5 \cdot CHO + C_6H_5 \cdot CHO/NaOH$, aq. $\rightarrow C_6H_5 \cdot CO \cdot ONa +$

$C_6H_5 \cdot CH_2 \cdot OH + H_2O$. Similarly,

$2 (CH_3)_3C \cdot CHO + H_2O/(OH^-) \rightarrow (CH_3)_3C \cdot CH_2 \cdot OH + (CH_3)_3C \cdot CO \cdot OH$.

75. Claisen Condensation (Schmidt, Eng. ed., p. 219, 1926):

Ethyl acetate, in the presence of sodium ethoxide, condenses with itself to yield acetoacetic ester, as:

$CH_3 \cdot CO \cdot O \cdot Et + Et \cdot ONa \rightarrow H_2C:C(O^- Na^+)OEt + Et \cdot OH$, then

$+$ loss of $Et \cdot O^- {}^+Na \rightarrow H_2C:C:O + EtO^- {}^+Na$, then

$+ H \cdot CH_2 \cdot CO \cdot OEt$ addition to ketene $\rightarrow CH_3 \cdot CO \cdot CH_2 \cdot CO \cdot OEt$.

By this proposed mechanism, the sodium ethoxide is regenerated which satisfies the requirement that only a trace is needed. The classical interpretation of this mechanism is:

$CH_3 \cdot CO \cdot OEt + EtO \cdot Na \rightarrow CH_3 \cdot C(OEt)_2O^{-+}Na$, then

$+$ loss of $Et \cdot OH$ with $H \cdot CH_2 \cdot CO \cdot OEt \rightarrow CH_3 \cdot C(O^{-+}Na) (OEt) \cdot CH_2 \cdot CO \cdot OEt$, then

$+$ loss of $Et \cdot ONa \rightarrow CH_3 \cdot CO \cdot CH_2 \cdot CO \cdot OEt + Et \cdot ONa$.

76. Claisen Reaction (Cohen, Part I, 4th. ed., p. 279, 1924):

Benzaldehyde, in the presence of alkali, condenses with acetaldehyde and similar compounds. With acetaldehyde, cinnamaldehyde is obtained:

$C_6H_5 \cdot CHO + H_2CH \cdot CHO/(alkali) \rightarrow C_6H_5 \cdot HC:CH \cdot CHO + H_2O$, also

$C_6H_5 \cdot CHO + H_2CH \cdot CO \cdot CH_3/(alkali) \rightarrow C_6H_5 \cdot HC:CH \cdot CO \cdot CH_3 + H_2O$, then

$C_6H_5 \cdot HC:CH \cdot CO \cdot CH_3 + C_6H_5 \cdot CHO/(alkali) \rightarrow C_6H_5 \cdot HC:CH \cdot CO \cdot HC: CH \cdot C_6H_5 + H_2O$

77. Darzens Reaction (Richter, p. 62, 1938):

Alkyl chlorides may be obtained by the action of thionyl chloride on the corresponding alcohol dissolved in an organic base, as:

$$R \cdot OH + SOCl_2/(pyridine) \rightarrow R \cdot Cl + SO_2 + HCl.$$

78. Dieckmann Condensation (Cohen, Part I, 4th. ed., p. 266, 1924):

An inner Claisen condensation is effected by treating the esters of diacids with sodium ethoxide to give a cyclic keto ester:

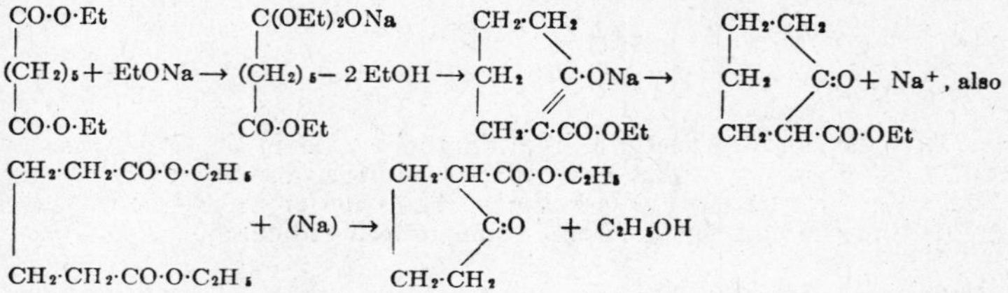

79. Fittig-Erdman Synthesis (Schmidt, English ed., p. 472, 1926):

Benzaldehyde and succinic anhydride, in the presence of sodium acetate as a catalyst, yield *alpha*-naphthol, as:

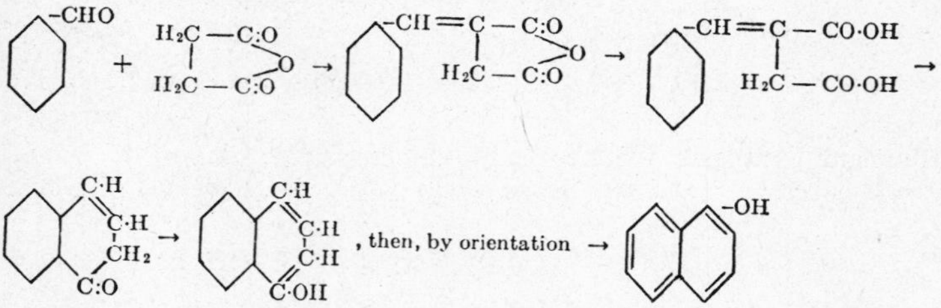

80. Guerbet Reaction (Whitmore, p. 116, 1938):

n-Propyl alcohol and sodium propylate undergo reaction at 250° C. to give 2-methyl-1-pentanol, as:

$$CH_3 \cdot CH_2 \cdot CH_2 \cdot ONa + H \cdot (CH_3)CH \cdot CH_2 \cdot OH \rightarrow CH_3 \cdot CH_2 \cdot CH_2 \cdot CH(CH_3) \cdot$$
$$CH_2 \cdot OH + NaOH.$$

Sodium hydroxide is eliminated between the *ONa* and a 2-position hydrogen atom of the alcohol unit. Guerbet considers the mechanism of the reaction to be: (1) dehydrogenation to the aldehyde, (2) aldol type of condensation, (3) loss of water, (4) hydrogenation of the olefin bond and reduction of the aldehyde group.

81. Haloform Reaction (Lucas, p. 240, 1935):

Trihalomethanes are formed when any compound containing the grouping $CH_3 \cdot CHOH \cdot Z$ or $CH_3 \cdot CO \cdot Z$, where Z may be a hydrogen atom or an alkyl group, is treated with halogen in the presence of alkali (cf. Lieben Iodoform Reaction):

$$CH_3 \cdot CH_2 \cdot OH + 4 X_2 + 6 NaOH, aq., heat \rightarrow CHX_3 + H \cdot CO \cdot ONa +$$
$$5 NaX + 5 H_2O.$$

82. Hantzsch Collidine Synthesis (Cohen, p. 289, Part III, 4th ed., 1924):

Two moles of acetoacetic ester may be condensed with one mole of an aldehyde to give a product which, upon subsequent treatment with ammonia, yields a member of the collidine series, as:

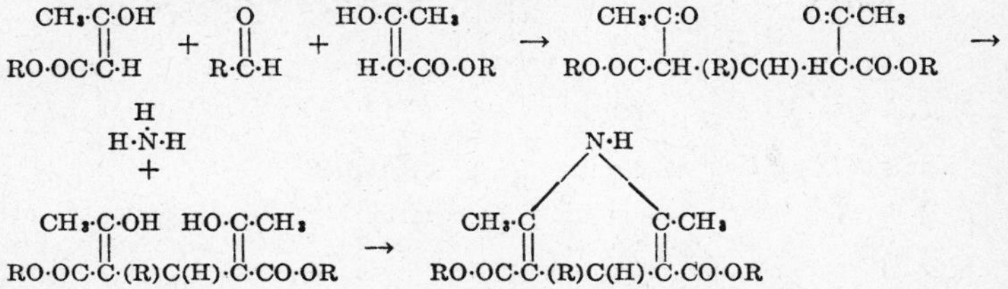

$$CH_3 \cdot C \cdot OH \atop RO \cdot OC \cdot C \cdot H \quad + \quad O \atop R \cdot C \cdot H \quad + \quad HO \cdot C \cdot CH_3 \atop H \cdot C \cdot CO \cdot OR \quad \rightarrow \quad CH_3 \cdot C : O \atop RO \cdot OC \cdot CH \cdot (R)C(H) \cdot HC \cdot CO \cdot OR \quad O : C \cdot CH_3 \quad \rightarrow$$

$$H \atop H \cdot N \cdot H \atop +$$

$$CH_3 \cdot C \cdot OH \quad HO \cdot C \cdot CH_3 \atop RO \cdot OC \cdot C \cdot (R)C(H) \cdot C \cdot CO \cdot OR \quad \rightarrow \quad N \cdot H \atop CH_3 \cdot C \qquad C \cdot CH_3 \atop RO \cdot OC \cdot C \cdot (R)C(H) \cdot C \cdot CO \cdot OR$$

83. Haworth Methylation (Cohen, Part 3, 4th ed., p. 60, 1924):

α- or β-Methyl-D-glucosides, when treated with methyl sulfate in the presence of concentrated sodium hydroxide, yield the corresponding tetramethyl-methyl-D-glucoside:

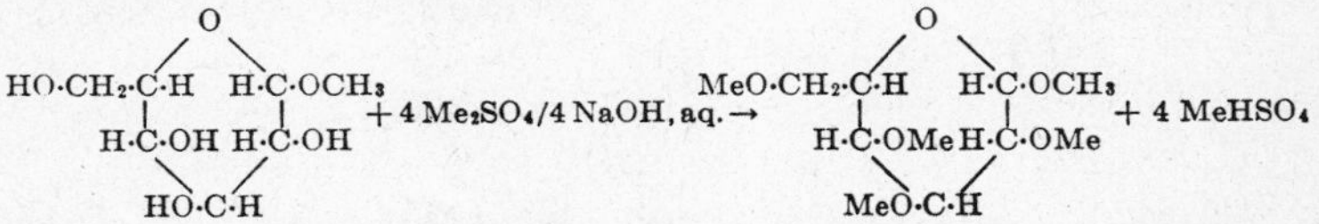

$$HO \cdot CH_2 \cdot C \cdot H \quad H \cdot C \cdot OCH_3 \atop H \cdot C \cdot OH \quad H \cdot C \cdot OH \atop HO \cdot C \cdot H \quad + 4\,Me_2SO_4/4\,NaOH, aq. \rightarrow \quad MeO \cdot CH_2 \cdot C \cdot H \quad H \cdot C \cdot OCH_3 \atop H \cdot C \cdot OMe \quad H \cdot C \cdot OMe \atop MeO \cdot C \cdot H \quad + 4\,MeHSO_4$$

Upon acid hydrolysis, the glucosidic methyl group is lost and tetramethyl-D-glucose is obtained. Irvine and Purdie methylated by the use of methyl iodide and silver oxide, and West methylated glucose directly with methyl sulfate and alkali by using chloroform as a solvent. Muskat used methyl iodide on glucose in liquid ammonia.

84. Heumann Synthesis (Chamberlain, 3rd. ed., p. 729, 1934):

2-Carboxyphenylglycine, upon fusion with sodamide, yields indoxyl which is subsequently oxidized by air to give indigo:

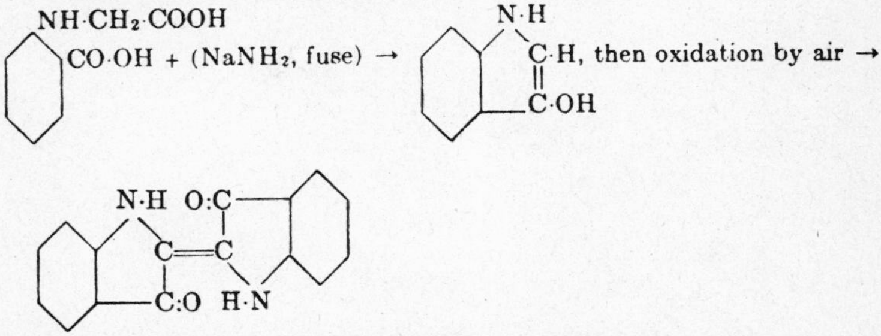

NH·CH₂·COOH
CO·OH + (NaNH₂, fuse) → N·H C·H, then oxidation by air →
C·OH

N·H O:C
C═C
C:O H·N

85. Knoevenagel Synthesis (Cohen, Part I, 4th ed., p. 282, 1924):

A series of intermolecular condensations, in the presence of ammonium bases as catalysts, that involve a carbonyl group and two hydrogen atoms that are activated by the influence of an unsaturated group, as:

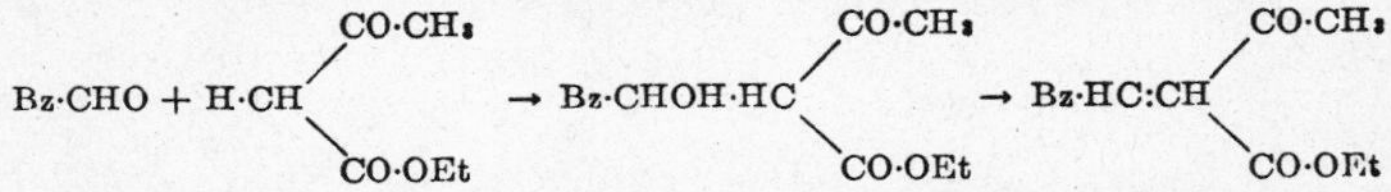

$$Bz \cdot CHO + H \cdot CH {CO \cdot CH_3 \atop CO \cdot OEt} \rightarrow Bz \cdot CHOH \cdot HC {CO \cdot CH_3 \atop CO \cdot OEt} \rightarrow Bz \cdot HC : CH {CO \cdot CH_3 \atop CO \cdot OEt}$$

In certain cases intramolecular condensations are effected.

86. Komppa Synthesis (Cohen, Part III, 4th ed., p. 261-2, 1924):

3, 3-Dialkyladipic ester is condensed with diethyl oxalate, as:

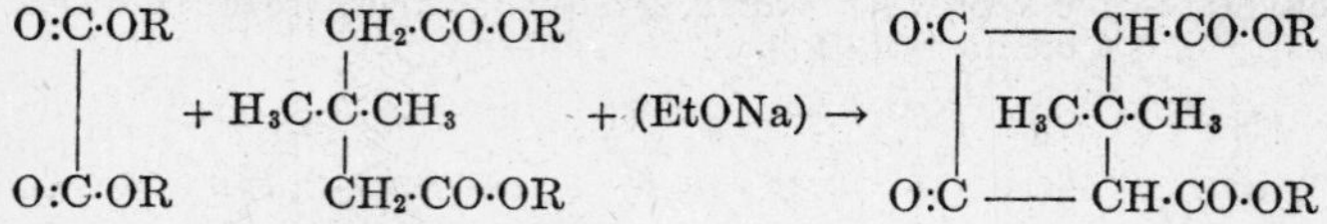

$$
\begin{array}{ll}
\text{O:C·OR} & \text{CH}_2\text{·CO·OR} \\
| & | \\
& + \text{H}_3\text{C·C·CH}_3 \quad + \text{(EtONa)} \rightarrow \\
| & | \\
\text{O:C·OR} & \text{CH}_2\text{·CO·OR}
\end{array}
\qquad
\begin{array}{ll}
\text{O:C} \longrightarrow \text{CH·CO·OR} \\
| \\
\text{H}_3\text{C·C·CH}_3 \\
| \\
\text{O:C} \longrightarrow \text{CH·CO·OR}
\end{array}
$$

See Claisen condensation, no. 75.

87. Konigs-Knorr Synthesis (Richter, p. 401, 1938):

β-Acetobromoglucose undergoes reaction with methyl alcohol in the presence of silver carbonate to give the tetraacetate of methyl β-D$^+$-glucoside. The acetate groups may be removed by an alkaline hydrolysis which does not effect the acetal linkage.

88. Lederer-Manasse Synthesis (Richter, p. 524, 1938):

Phenol and formaldehyde are condensed to give hydroxybenzyl alcohol. This is a specific example of the synthesis which is one step in the production of Bakelite resins. Formaldehyde undergoes reaction in a similar fashion with aromatic acids and nitro compounds.

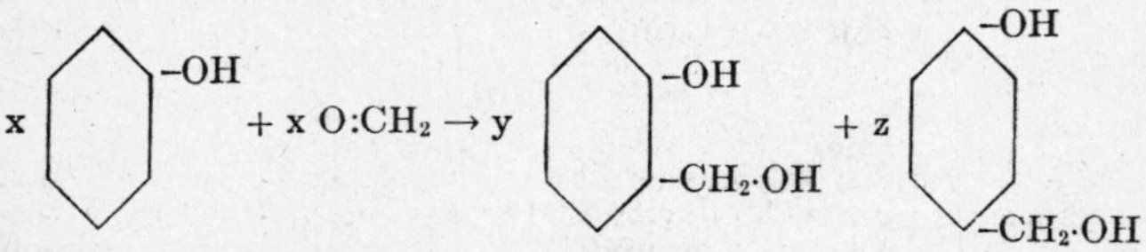

89. Lieben Iodoform Reaction (Perkin and Kipping, p. 105, 1932):

Compounds containing the $\text{CH}_3\text{·CHOH·}$ group or its first oxidized derivative, yield iodoform when treated with iodine in the presence of alkali, the postulated mechanism of the reaction being (cf. Haloform reaction no. 81):

$\text{CH}_3\text{·CH}_2\text{·OH} + \text{NaOH} \rightarrow \text{CH}_3\text{·CH}_2\text{·O}^- {}^+\text{Na} + \text{H}_2\text{O}$, then

$+ \text{I}_2/\text{NaOH} \rightarrow \text{CH}_3\text{·CHO} + 2\,\text{NaI}$, then

$+ \text{NaOH, aq.} \rightarrow \text{H}_2\text{C:CH·O}^- {}^+\text{Na} + \text{H}_2\text{O}$, then

$+ \text{I}_2/(\text{NaOH}) \rightarrow \text{I·CH}_2\text{·CHO} + \text{NaI}$, then

$+ \text{NaOH, aq.} \rightarrow \text{I·HC:CH·O}^- {}^+\text{Na} + \text{H}_2\text{O}$, then

$+ \text{I}_2/(\text{NaOH}) \rightarrow \text{I}_2\text{HC.CHO} + \text{NaI}$, then

$+ \text{NaOH, aq.} \rightarrow \text{I}_2\text{C:CH.O}^- {}^+\text{Na} + \text{H}_2\text{O}$, then

$+ \text{I}_2/(\text{NaOH, aq.}) \rightarrow \text{I}_3\text{C·CHO} + \text{NaI}$, then

$+ \text{NaOH, aq.} \rightarrow \text{I}_3\text{C·H} + \text{NaO·CO·H}$.

90. Methone Aldehyde Reagent (Whitmore, p. 657, 1938):

Aldehydes may be identified by their reaction with methone, 5,5-dimethyl-1,3-cyclohexandione. The condensation takes place between the carbonyl group of the aldehyde and the methylene group of two parts of methone, as:

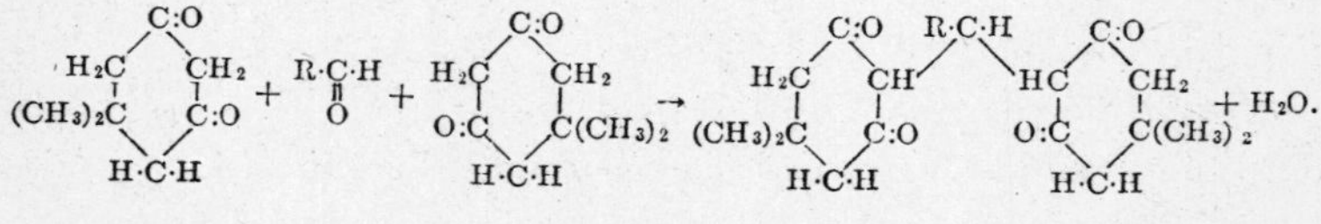

91. Perkin (Perkin-Fittig) Reaction (Chamberlain, 3rd ed., p. 546, 1934):

Benzaldehyde, when treated with sodium acetate in the presence of acetic anhydride, yields sodium cinnamate:

$$C_6H_5 \cdot CHO + H_2 \cdot HC \cdot CO \cdot ONa/Ac_2O \rightarrow C_6H_5 \cdot CH{:}HC \cdot CO \cdot ONa + 2\ AcOH$$

This reaction is not essentially different from the aldol condensation. It has been shown that any type of a mild dehydrating agent is satisfactory. See Fittig-Erdmann Synthesis, No. 79.

92. Phthalic Anhydride Test (Kamm, 2nd ed., p. 59, 1932):

Phthalic anhydride reacts with primary alcohols when the mixture is refluxed in benzene. Secondary alcohols react less readily, usually requiring a reaction temperature of 100-120° C., whereas the tertiary alcohols do not react.

93. Pschorr Synthesis (Schmidt, English ed., p. 502, 1926):

o-Nitrobenzaldehyde, when treated with phenyl acetic acid, reduced, diazotized in a sulfuric acid solution, treated with copper dust, and distilled to effect decarboxylation, yields phenanthrene:

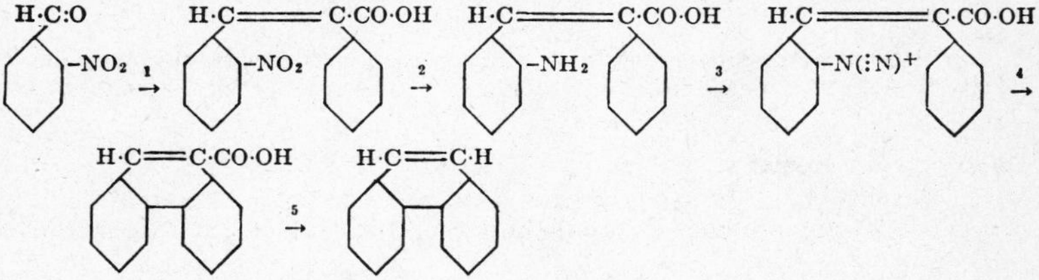

94. Purdie (Irvine-Purdie) Methylation (Cohen, Part III, 4th ed., p. 49, 1924):

Sugars are methylated by methyl alcohol in the presence of dry hydrogen chloride (Fischer method) to replace the carbonylic hydrogen atom and then refluxed with an alkyl iodide in the presence of dry silver oxide, to yield polyalkyl-α- and β-methyl-derivatives as:

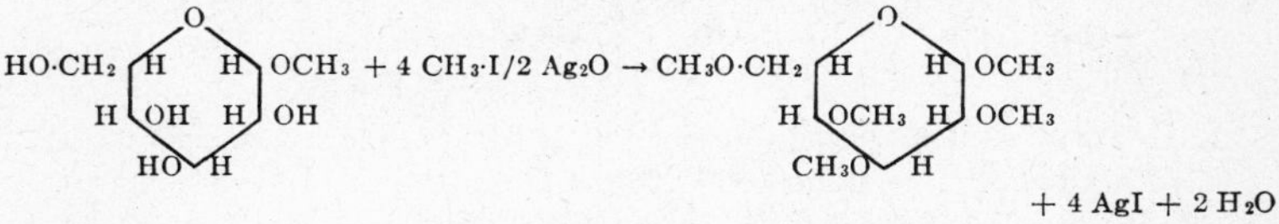

$$+ 4\ AgI + 2\ H_2O$$

95. Schotten-Baumann Reaction (Chamberlain, 3rd ed., p. 534, 1934):

Acid chlorides react with alcohols, in the presence of alkali, to yield esters as:

$$C_6H_5 \cdot CO \cdot Cl + H \cdot OR/(NaOH, aq.) \rightarrow C_6H_5 \cdot CO \cdot OR + HCl$$

96. Tischenko (Tischtschenko) Reaction (Reid, p. 193, 1931):

Aldehydes, in the presence of sodium ethoxide or aluminum ethoxide, undergo condensation accompanied by auto-oxidation and reduction, as:

$$2\ CH_3 \cdot CHO + [EtONa\ or\ Al(OEt)_3] \rightarrow CH_3 \cdot CO \cdot O \cdot CH_2 \cdot CH_3.$$

The yields are about 95% or better under proper experimental conditions.

97. Twitchell Process (Riegel, p. 516-7, 1933):

Fats or vegetable oils are heated with 30% sulfuric acid in the presence of naphthalenesulfonic acid to give the free aliphatic acids and glycerol. The glycerol and sulfuric acid are leached out with water, the sulfuric acid precipitated as calcium sulfate, and the glycerol obtained by concentration of the filtrate. Most of the acids prepared by this process are used in the manufacture of soap.

II. A. 2. Oxygen to Nitrogen Hydrogen Bonds.

98. Biuret Synthesis and Reaction (Chamberlain, 3rd. ed., p. 661-2, 1934):

Two moles of urea, upon heating, unite by the loss of one mole of ammonia to yield biuret. Some cyanuric acid is formed in the reaction due to the loss of three moles of ammonia from three moles of urea:

$2 H_2N \cdot CO \cdot NH_2$, heat $\rightarrow H_2N \cdot CO \cdot NH \cdot CO \cdot NH_2 + NH_3$, and

$$3 H_2N \cdot CO \cdot NH_2, \text{ heat} \rightarrow$$

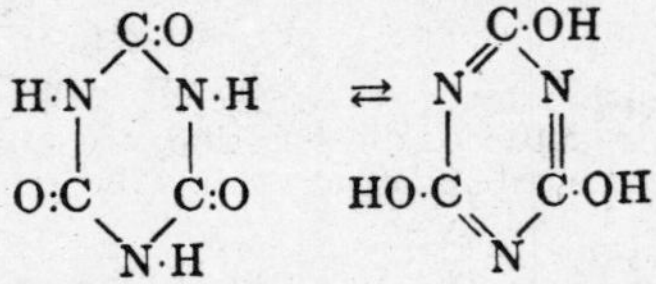

Biuret, upon treatment with a trace of cupric ion in the presence of hydroxyl ion, gives a violet coloration as a consequence, perhaps, of chelate ring formation.

99. Bucherer Reaction (Desha, p. 468, 1936):

Naphthylamines are hydrolyzed to the corresponding naphthols by water in the presence of ammonium bisulfite under pressure at about 200° C., as:

$C_{10}H_7 \cdot NH_2 + H \cdot OH/(NH_4HSO_3)$, 200° C. $\rightleftarrows C_{10}H_7 \cdot OH + NH_3$.

The reverse of this process is known also as the Bucherer reaction.

100. Decker Method (Gilman, p. 1088, 1938):

Primary amines are converted to secondary amines by condensing the primary amine with benzaldehyde, adding the desired alkyl group as an alkyl halide, and hydrolyzing.

$Bz \cdot CHO + R \cdot NH_2 \rightarrow Bz \cdot HC:N \cdot R$, then

$+ R \cdot X$, heat $\rightarrow Bz \cdot HC:NR_2^+ \,{}^-X$, then

$+ H_2O/Na^+ \rightarrow Bz \cdot CHO + R_2NH + NaX + H^+$.

101. Doebner-Miller Quinaldine Synthesis (Schmidt, Eng. ed., p. 608, 1926):

Paraldehyde, when heated with aniline in the presence of concentrated hydrochloric or sulfuric acid, yields, quinaldine, as:

$(CH_3 \cdot CHO)_3 + (H^+) \rightarrow 3 CH_3 \cdot CHO$, then

$2 CH_3 \cdot CHO + \text{(an amine)} \rightarrow CH_3 \cdot CHOH \cdot CH_2 \cdot CHO$, then

$CH_3 \cdot CHOH \cdot CH_2 \cdot CHO + \text{loss of } H_2O \rightarrow CH_3 \cdot HC:CH \cdot CHO$, then

$Bz \cdot NH_2 + CH_3 \cdot HC:CH \cdot CHO \rightarrow Bz \cdot NH \cdot HC(CH_3) \cdot HC:CH \cdot OH$, then

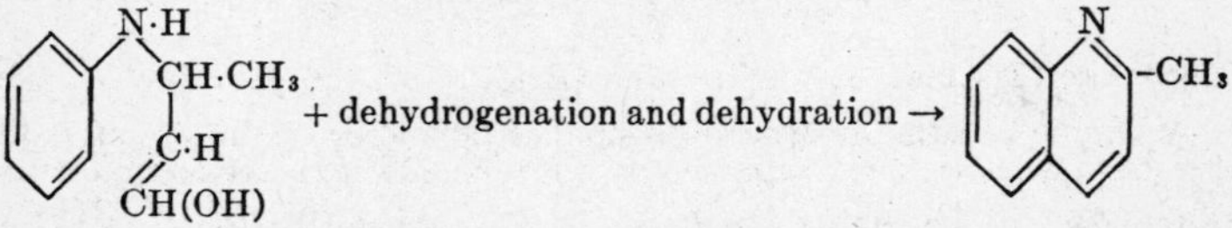

+ dehydrogenation and dehydration →

Corresponding derivatives may be obtained from other aldehydes and other amines. See the Skraup Synthesis, No. 121.

102. Erlenmeyer Reaction (Cohen, Part III, 4th ed., p. 151, 1924):

Benzaldehyde, when condensed with N-benzoylglycine and the intermediate reduced and hydrolyzed, yields an *alpha*-amino acid, as:

$Bz \cdot CHO + H_2C(CO_2H) \cdot NH \cdot OC \cdot Bz \rightarrow Bz \cdot HC:C(CO_2H) \cdot NH \cdot OC \cdot Bz + H_2O$, then

$+ 2 H^* \text{ or } H_2/(\text{catalyst}) \rightarrow Bz \cdot CH_2 \cdot CH(CO_2H) \cdot NH \cdot OC \cdot Bz$, then

$+ H_2O \rightarrow Bz \cdot CH_2 \cdot CH(NH_3^+) \cdot CO \cdot O^- + Bz \cdot CO_2H$.

103. Eschweiler Method (Gilman, p. 1041, 1938):

The methylation of primary or secondary amines is accomplished by heating the amine with formaldehyde. Excess formaldehyde furnishes the hydrogen necessary for the reduction, as:

$R \cdot NH_2 + O:CH_2$, heat $\rightarrow R \cdot NH \cdot CH_2OH$, then

$+ H \cdot CHO$, heat $\rightarrow R \cdot NH \cdot CH_3 + H \cdot CO_2H$.

104. Fischer-Nouri Phloretin Preparation (Cohen, Part I, p. 230, 1924):

The essential steps are:

$p\text{-}CH_3 \cdot CO \cdot O \cdot C_6H_4 \cdot CH_2 \cdot CH_2 \cdot CN + s\text{-}C_6H_3(OH)_3 \rightarrow p\text{-}CH_3 \cdot CO \cdot O \cdot C_6H_4 \cdot CH_2 \cdot CH_2 \cdot C(:NH) \cdot C_6H_2(OH)_3$, then $+ 2 H_2O \rightarrow p\text{-}HO \cdot C_6H_4 \cdot CH_2 \cdot CH_2 \cdot CO \cdot C_6H_2(OH)_3 + CH_3 \cdot CO \cdot ONH_4$

105. Friedlander Synthesis (Schmidt, English ed., p. 609, 1926):

o-Aminobenzaldehyde is condensed with acetaldehyde to yield quinoline. Substitution products of *o*-aminobenzaldehyde may be used along with other aldehydes or ketones containing the -$CH_2 \cdot CO$- group to give quinoline derivatives:

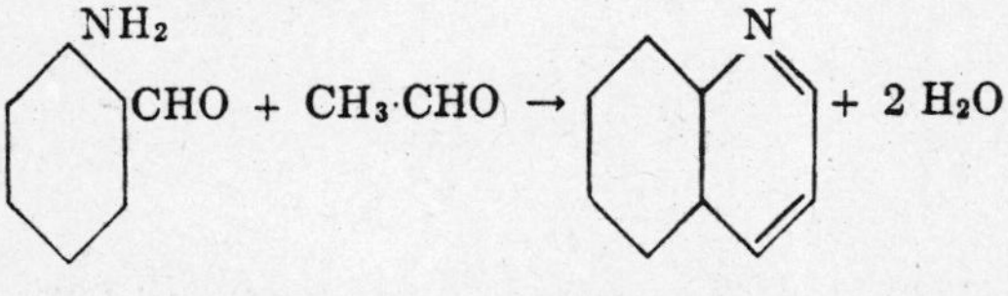

106. Gattermann Aldehyde Synthesis (Whitmore, p. 799, 1938):

When phenols are treated with hydrogen cyanide in the presence of hydrogen chloride and zinc chloride, formimides are obtained. These, upon hydrolysis, yield hydroxy aryl aldehydes, as:

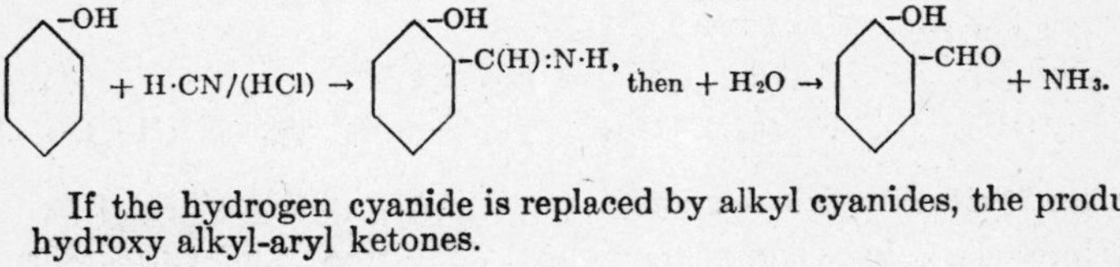

If the hydrogen cyanide is replaced by alkyl cyanides, the products are hydroxy alkyl-aryl ketones.

107. Griess Diazo Reaction (Chamberlain, 3rd ed., p. 464, 1934):

An aromatic primary amine, when treated with sodium nitrite and hydrochloric acid at 0-5°C., yields the corresponding diazonium chloride:

$Ar \cdot NH_2HCl + HONO/(NaNO_2/HCl, 0\text{-}5°C.) \rightarrow Ar \cdot N(:N)^+{}^-Cl + 2 H_2O$

108. Hinsberg Separation (Hill and Kelley, 1st ed., p. 371-2, 1932):

(a) Primary amines upon treatment with benzenesulfonyl chloride give products which are soluble in alkali, whereas (b) secondary amines by a similar treatment yield products that are insoluble in alkali:

a. $R \cdot NH \cdot H + Cl \cdot O_2S \cdot C_6H_5 \rightarrow C_6H_5 \cdot SO_2 \cdot NHR + HCl$, then

$C_6H_5 \cdot SO_2 \cdot NHR + NaOH$, aq. $\rightarrow C_6H_5 \cdot SO_2 \cdot NR(Na)$, soluble $+ H_2O$

b. $R_2N \cdot H + Cl \cdot O_2S \cdot C_6H_5 \rightarrow C_6H_5 \cdot SO_2 \cdot NR_2 + HCl$, then

$C_6H_5 \cdot SO_2 \cdot NR_2 + NaOH$, aq. $\rightarrow$ no reaction, insoluble.

109. Hoesch Reaction (Cohen, Part I, 4th. ed., p. 229-30, 1924):

Addition of resorcinol to ethanenitrile is effected in an ether solution by the use of zinc chloride and hydrogen chloride. The addition product, upon hydrolysis, yields 2, 4-dihydroxyacetophenone:

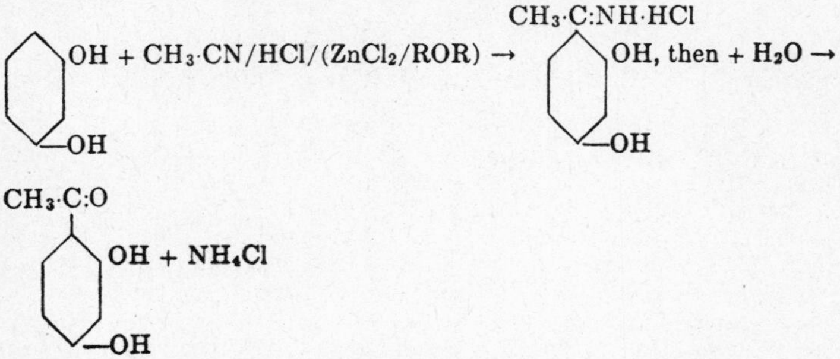

110. Horbaczewski-Behrend Synthesis (Gilman, p. 967, 1937):

The fusion of urea with trichlorolactamide yields uric acid, as:

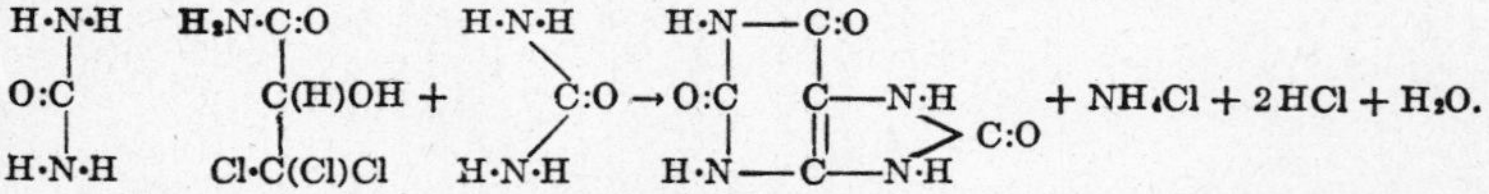

Originally, trichloroacetic acid or glycocoll was fused with urea and reported to give uric acid. Later work, however, failed to confirm this.

111. Karrer Aldehyde Synthesis (Cohen, Part I, 4th ed., p. 230, 1924):

Resorcinol, when treated with cyanogen bromide in the presence of hydrogen chloride, gives a condensation product which, upon hydrolysis, yields the aldehyde derivative:

$Br \cdot CN + HCl \rightarrow Br(H)C:N \cdot Cl$, then $+ m\text{-}C_6H_4(OH)_2 \rightarrow$

$C_6H_3(OH)_2CH:N \cdot Cl$, then $+ 2H_2O \rightarrow C_6H_3(OH)_2CHO + NH_2OH \cdot HCl$

112. Kiliani Synthesis (Chamberlain, 3rd ed., p. 301, 1934):

Aldoses and ketoses add hydrogen cyanide to give cyanohydrins which, upon subsequent hydrolysis and reduction, yield the next two isomeric sugars. See page 114.

113. Knorr Reaction (Cohen, Part I, 4th. ed., p. 307, 1924):

When phenylhydrazine is condensed with acetoacetic ester, the principal product is a heterocyclic ketone, as:

$CH_3 \cdot CO \cdot CH_2 \cdot CO \cdot OEt + Bz \cdot NH \cdot NH_2 \rightarrow CH_3 \cdot CO \cdot CH_2 \cdot CO \cdot N(NH_2) \cdot Bz + Et \cdot OH$, then

$$
\begin{array}{cc}
CH_3 \cdot C\!-\!-\!-CH_2 & \\
\ \big| \qquad \big| & \text{, by intermolecular addition} \rightarrow \\
O \qquad C{:}O & \text{and loss of water} \\
H_2N\!-\!-\!-N \cdot Bz &
\end{array}
$$

$$
\begin{array}{cc}
CH_2 \cdot C\!-\!-\!CH_2 & \qquad\qquad N \\
\quad \big| \qquad \big| & \quad\text{or } CH_3 \cdot C \diagdown\!\!\!\diagup \diagdown N \cdot Bz \\
\quad N \qquad C{:}O & \qquad\qquad \big| \\
\quad N \cdot Bz & \qquad\quad H_2C\!-\!-\!C{:}O
\end{array}
$$

The product in this case is 5-methyl-3-oxo-2-phenyl-1, 2-isodiazole.

114. Ladenburg Synthesis (Lucas, p. 622, 1935):

When α-methylpyridine is condensed with acetaldehyde, and the intermediate reduced with sodium and alcohol, the product is α-propylpiperidine, as indicated by:

$$+ CH_3 \cdot CHO \rightarrow \text{[structure]} -HC{:}CH \cdot CH_3 + H_2O, \quad \text{then}$$

$$\text{[structure]} -HC{:}CH \cdot CH_3 + 8\,H/(8\,Na/8\,Et \cdot OH) \rightarrow \text{[piperidine structure]}$$

115. Leuckart Reaction (Gilman, p. 705, 1938):

When aldehydes and ketones are treated with ammonium formate at about 160° C., the product is a formoalkylamide. These condensation products, when subjected to hydrolysis, yield primary amines, as:

$$R_2C{:}O + 2\,H \cdot CO \cdot O^{-+}NH_4, 160°\,C. \rightarrow R_2CH \cdot NH \cdot CHO + NH_3 + CO_2 + 2\,H_2O, \text{ then}$$

$$+ H_2O \text{ (hydrolysis)} \rightarrow R_2CH \cdot NH_2 + H \cdot CO_2H.$$

116. Lumiere and Barbier Method (Richter, p. 498, 1938):

Acetylation is effected by dissolving an aromatic amine in water containing an equivalent of hydrochloric acid and adding acetic anhydride. The solution, which should have been of sufficient volume to dissolve all the acetic anhydride, is treated with sodium acetate to set free the acetyl derivative, as:

$$Ar \cdot NH_2 + HCl \rightarrow Ar \cdot NH_3{}^{+-}Cl, \text{ then } + Ac_2O/Ac \cdot O \cdot Na \rightarrow Ar \cdot NH \cdot Ac + Ac \cdot OH + NaCl.$$

The hydrochloric acid is added to dissolve the amine, and the sodium acetate is added to liberate the amine in turn to react with the acetic anhydride as needed.

117. Osazone Reaction (Chamberlain, 3rd ed., p. 299, 1934):

Reducing sugars react with phenylhydrazine to give osazones which are of value in the identification of the sugars. Where two sugars produce identical osazones, they can usually be identified by the speed of formation of the osazone. The reaction is given in detail on page 122.

118. Paal-Knorr Reaction (cf. Knorr Reaction, Cohen, Part I, p. 307, 1924):

1, 4-Diketones react with ammonia or primary amines to give pyrrole derivatives, as:

$$
\begin{array}{cc}
OH \quad HO & N \cdot H \\
\ \big| \qquad \big| & \diagup \diagdown \\
R \cdot C \qquad C \cdot R & R \cdot C \qquad C \cdot R \\
\ \big|\big| \qquad \big|\big| \ \ + NH_3, \text{heat} \rightarrow & \ \big|\big| \qquad \big|\big| \ \ + H_2O. \\
H \cdot C\!-\!-\!C \cdot H & H \cdot C\!-\!-\!C \cdot H
\end{array}
$$

119. Phenylisocyanate Test (Kamm, 2nd ed., p. 58, 1932):

Phenylisocyanate reacts, under anhydrous conditions, with alcohols, phenols, ammonia, amines, and enols to give addition compounds that may be used as identification derivatives:

a. $C_6H_5 \cdot N{:}C{:}O + R \cdot O \cdot H \rightarrow C_6H_5 \cdot NH \cdot CO \cdot O \cdot R$

b. $C_6H_5 \cdot N{:}C{:}O + C_6H_5 \cdot O \cdot H \rightarrow C_6H_5 \cdot NH \cdot CO \cdot O \cdot C_6H_5$

c. $C_6H_5 \cdot N{:}C{:}O + NH_3 \rightarrow C_6H_5 \cdot NH \cdot CO \cdot NH_2$

d. $C_6H_5 \cdot N{:}C{:}O + C_6H_5 \cdot NH_2 \rightarrow C_6H_5 \cdot NH \cdot CO \cdot HN \cdot C_6H_5$

120. Schiff Reaction (Bernthsen and Sudborough, p. 399, 1933):

Aromatic aldehydes condense with primary aromatic amines, in the presence of alkali, to give an anil.

$$C_6H_5 \cdot CHO + H_2N \cdot C_6H_5/(alkali) \rightarrow C_6H_5 \cdot CH{:}N \cdot C_6H_5 + H_2O$$

Aliphatic compounds condense similarly.

121. Skraup Synthesis (Chamberlain, 3rd ed., p. 620, 1934):

Aniline, when heated with glycerol and nitrobenzene in the presence of sulfuric acid, yields quinoline, as:

1. $HO \cdot CH_2 \cdot CHOH \cdot CH_2 \cdot OH + 2 H_2SO_4$, conc. $\rightarrow H_2C{:}CH \cdot CHO + 2 H_2SO_4 \cdot H_2O$, then

2. $Bz \cdot NH_2 + H_2C{:}CH \cdot CHO \rightarrow [Bz \cdot NH \cdot CH_2 \cdot HC{:}CH \cdot OH$, then

3.

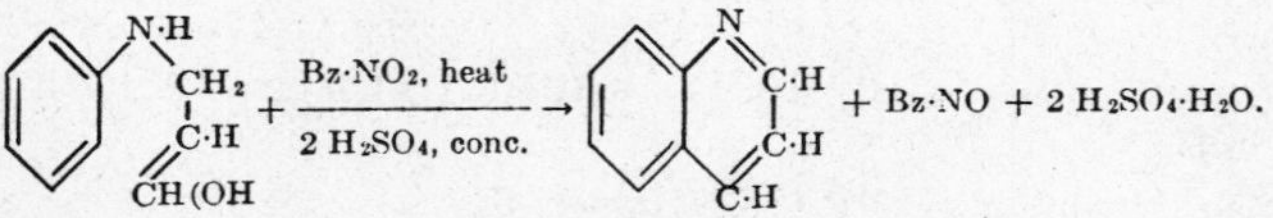

See Doebner-Miller Synthesis, No. 101.

122. Strecker Reaction (Schmidt, English ed., p. 181, 1926):

Aldehydes and ketones react with hydrogen cyanide in the presence of a trace of ammonia to yield amino nitriles which, upon subsequent hydrolysis, give α-amino acids:

$$R \cdot CHO + HCN/NH_3 \rightarrow R \cdot CH(NH_2) \cdot CN + H_2O, \text{ then}$$
$$R \cdot CH(NH_2) \cdot CN + 2 H_2O/HCl \rightarrow R \cdot CH(NH_2) \cdot CO \cdot OH + NH_4Cl$$

123. Traube Synthesis (Richter, p. 374, 1938):

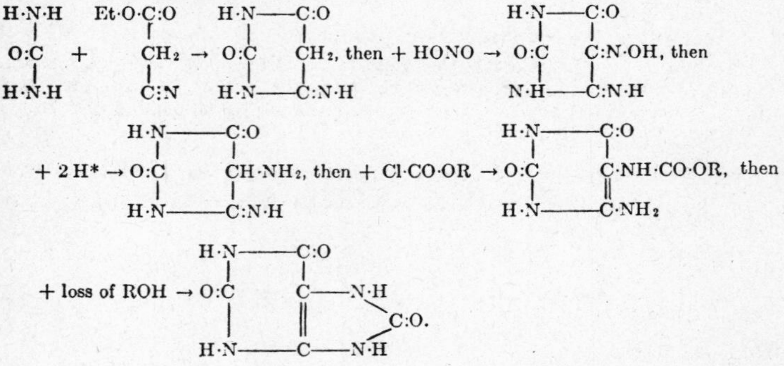

124. Van Slyke Method (Williams, 2nd ed., p. 285, 1934):

Amino nitrogen is determined by treating the compound with nitrous acid and measuring the evolved nitrogen:

$$R \cdot CH(NH_2) \cdot CO \cdot OH + HONO \rightarrow R \cdot CH(OH) \cdot CO \cdot OH + N_2 + H_2O$$

The reaction is applicable to the NH_2 group only, and in the amino acid series it proceeds most readily with the 2-amino acids, and somewhat less readily with other members of the series. This reaction is used also in the identification of some amines.

125. von Braun Epimer Reagent (Whitmore, p. 573, 1938):

Diphenylmethane-dimethyl-dihydrazine or bis- [(N-α-methylhydrazine)- 4-phenyl] -methane has been used to separate epimers as glucose and mannose. Fructose and the disaccharides do not undergo reaction with this reagent. It reacts only with those aldoses in which carbon atoms number 2, 3, and 4 have at least two adjacent hydroxyls of the same "sign," that is, on the same side of the molecule.

126. Wolff-Kishner Reduction (Richter, p. 102, 1938):

Semicarbazones, when heated with sodium ethylate, yield the corresponding hydrazones, which on continued heating with sodium ethylate give nitrogen and a hydrocarbon, as:

$$R_2C:N \cdot NH \cdot CO \cdot NH_2 + Et \cdot OH/(Et \cdot ONa, \text{ heat}) \rightarrow R_2C:N \cdot NH_2 + EtO \cdot OC \cdot NH_2, \text{ then}$$

$$R_2C:N \cdot NH_2 + (Et \cdot ONa, \text{ heat}) \rightarrow R_2CH_2 + N_2.$$

127. Wurtz Synthesis (Norris, 3rd ed., p. 211, 1931):

Alkyl isocyanates, when subjected to alkaline hydrolysis, yield primary amines:

$$R \cdot N:C:O + 2 NaOH, \text{ aq.} \rightarrow R \cdot NH_2 + Na_2CO_3$$

II. A. 3. Nitrogen to Nitrogen Hydrogen Bonds

128. Thorpe Reaction (Cohen, Part I, 4th ed., p. 296-7, 1924):

A series of condensations are effected between a cyanogen group, which is rendered acidic by the radical to which it is attached, and other cyanogen groups with a labile α-hydrogen atom:

$$C_6H_5 \cdot CN + C_6H_5 \cdot CH_2 \cdot CN/(EtONa) \rightarrow C_6H_5 \cdot C(:NH) \cdot CH(CN) \cdot C_6H_5$$

II. A. 4. Chloroform Hydrogen Bonds.

129. Carbylamine Reaction (Conant, 1st. ed., p. 245, 1933):

Primary amines react with chloroform in the presence of alkali to yield isonitriles (isocyanides):

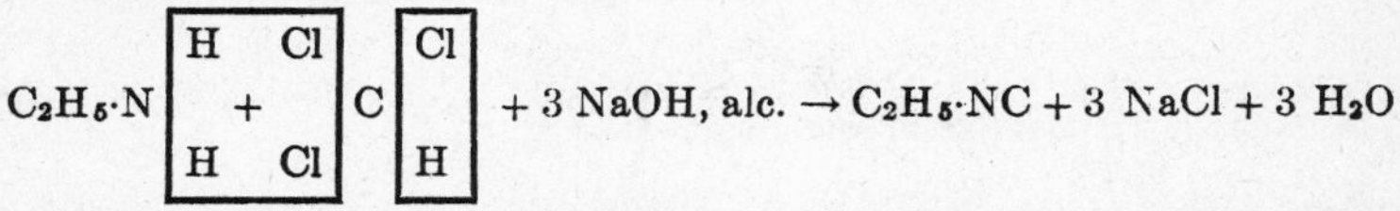

$$C_2H_5 \cdot N \begin{array}{|cc|} H & Cl \\ + \\ H & Cl \end{array} C \begin{array}{|c|} Cl \\ H \end{array} + 3 NaOH, \text{ alc.} \rightarrow C_2H_5 \cdot NC + 3 NaCl + 3 H_2O$$

130. Ciamician-Dennstedt Synthesis (Schmidt, English ed., p. 522, 1926):

Pyrrole or sodium pyrrole, upon heating with sodium ethoxide and chloroform, yields 3-chloropyridine.

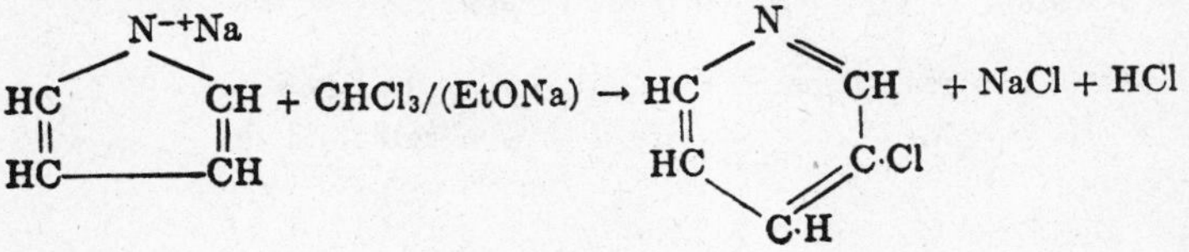

131. Fischer-Dilthey Condensation (Ann., 335, 334, 1904):

The condensation of urea and allied products with the acid halides of the diacids yields heterocyclic rings. Cf. The synthesis of barbituric acid, p. 140.

132. Gustus Acid-Iodide-Ether Reaction (Whitmore, p. 152, 1938):

Ethers are split by acid iodides, as:

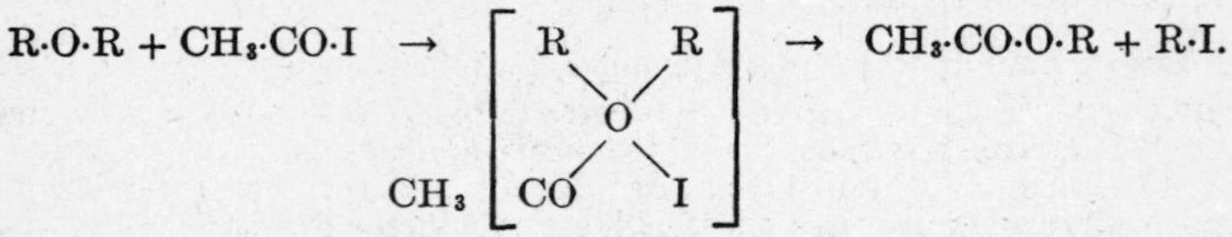

$$R \cdot O \cdot R + CH_3 \cdot CO \cdot I \rightarrow \begin{bmatrix} R \quad R \\ \diagdown O \diagup \\ \diagup \quad \diagdown \\ CH_3 \quad CO \quad I \end{bmatrix} \rightarrow CH_3 \cdot CO \cdot O \cdot R + R \cdot I.$$

Meerwein has shown that acid chlorides and acid anhydrides, in the presence of zinc chloride, effect a similar splitting of ethers.

133. Hofmann Isonitrile Synthesis (see Carbylamine reaction, No. 129).

134. Hofmann Mustard-Oil Reaction (Sidgwick, p. 25, 1937):

When primary amines are treated with carbon disulfide in the presence of mercuric chloride, and the product subjected to hydrolysis, the isothiocyanates are obtained, as:

$$R \cdot NH_2 + CS_2 + HgCl_2 \rightarrow R \cdot NH \cdot CS \cdot S \cdot HgCl + HCl, \text{ then}$$
$$+ H_2O, \text{ boil} \rightarrow R \cdot N:C:S + HgS + HCl.$$

135. Reimer (Reimer-Tiemann) Reaction (Cohen, Part I, 4th ed., p. 223 1924):

Sodium phenolate, by treatment with chloroform and sodium hydroxide and then with carbon dioxide and water, yields salicylaldehyde:

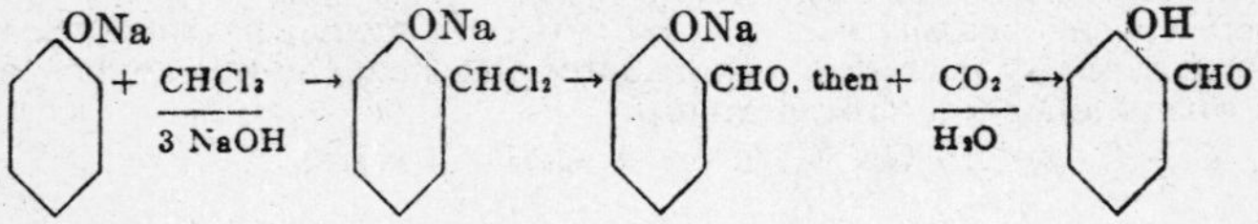

II B. Metallic Force Mechanisms

136. Barbier-Wieland Degradation (Gilman, p. 1240, 1938):

Complex acids may be degraded by treatment of their esters with two equivalents of an alkyl Grignard reagent and subsequent hydrolysis and oxidation with ozone or chromic oxide, as:

$$R \cdot CH_2 \cdot CO \cdot OEt + 2 R' \cdot Mg \cdot X/\text{dry ether} \rightarrow R \cdot CH_2 \cdot CR'_2 \cdot OMgX + EtO \cdot MgX, \text{ then}$$
$$+ HX, \text{ aq.} \rightarrow R \cdot CH_2 \cdot CHR'_2 + MgX_2, \text{ then}$$
$$+ \text{ oxidation} \rightarrow R \cdot CO \cdot OH + \text{ other products.}$$

137. Berthelot Synthesis (Chamberlain, 3rd. ed., p. 6, 1934):

Carbon disulfide is reduced with hydrogen sulfide in the presence of copper to yield methane:

$$CS_2 + 2 H_2S + 8 Cu \rightarrow CH_4 + 4 Cu_2S$$

138. Blaise Ketone Synthesis (Hickinbottom, p. 230, 1938):

Ketones may be prepared by the treatment of a Grignard reagent with zinc chloride to give an alkyl zinc chloride, which upon subsequent treatment with an acid halide, yields a ketone as:

$R \cdot Mg \cdot X + ZnCl_2 \rightarrow R \cdot Zn \cdot Cl + MgXCl$, then

$R \cdot Zn \cdot Cl + R \cdot CO \cdot Cl \rightarrow R \cdot CO \cdot R + ZnCl_2$.

139. Boord Olefin Synthesis (Whitmore, p. 47, 1938):

All of the heptenes have been obtained in pure form by the use of the following type reaction:

$R \cdot CH_2 \cdot CHO + HCl/EtOH \rightarrow R \cdot CH_2 \cdot CHCl \cdot OEt$, then

$R \cdot CH_2.CHCl \cdot OEt + Br_2 \rightarrow R \cdot CHBr \cdot CHBr \cdot OEt + HCl$, then

$R \cdot CHBr \cdot CHBr \cdot OEt + R \cdot MgX/(dry\ ether) \rightarrow R \cdot CHBr \cdot CHR \cdot OEt + MgXBr$, then

$2\ R \cdot CHBr \cdot CHR \cdot OEt + Zn\ dust \rightarrow R \cdot HC{:}CH \cdot R + ZnBr_2 + Et_2O$.

140. Bouveault Aldehyde Synthesis (Schmidt, English ed., p. 138, 1926):

Dialkylformamides, when treated with the Grignard reagent and subsequent hydrolysis, yield aldehydes:

$H \cdot CO \cdot NR_2 + R' MgX \rightarrow R' HC(OMgX)NR_2$, then $+ HX$, aq. $\rightarrow R'CHO + R_2NH + MgX_2$

141. Brooks Method (Whitmore, p. 40, 1938):

Pure 1-olefins, supposedly free from the 2-isomers, may be obtained by the reaction of a Grignard reagent on allyl bromide, as:

$R \cdot Mg \cdot X + Br \cdot CH_2 \cdot HC{:}CH_2/(dry\ ether) \rightarrow R \cdot CH_2 \cdot HC{:}CH_2 + MgXBr$.

142. Doctor Process (Read, Industrial Chemistry, 1st. ed., p. 337, 1933):

Petroleum is treated to remove thio compounds, the reactions involved in the process being essentially as follows:

$2\ R \cdot SH + Pb(ONa)_2 \rightarrow Pb(SR)_2 + 2\ NaOH$, then

$Pb(SR)_2 + S \rightarrow R \cdot S \cdot S \cdot R + PbS$

R_2S and $R \cdot S \cdot S \cdot R$ compounds are oil-soluble and are not seriously objectionable as they are comparatively stable.

143. Frankland-Duppa Reaction (Cohen, Part I, 4th ed., p. 245, 1924):

Diethyl oxalate is alkylated with diethylzinc to yield ethyl 2-ethyl-2-hydroxybutanoate as the final product:

$$
\begin{array}{ccc}
& O \cdot Zn \cdot Et & O \cdot Zn \cdot Et \\
& | & | \\
CO \cdot OEt + ZnEt_2 \rightarrow Et \cdot C \cdot O \cdot Et, & \text{then} + ZnEt_2 \rightarrow Et \cdot C \cdot Et + Et \cdot Zn \cdot OEt, & \text{and} \\
| & | & | \\
CO \cdot OEt & CO \cdot OEt & CO \cdot OEt
\end{array}
$$

$$
\begin{array}{ccc}
O\ \boxed{Zn \cdot Et + HO}\ H \rightarrow & OH & + Et \cdot Zn \cdot OH \\
| & | & \\
Et \cdot C \cdot Et & Et \cdot C \cdot Et & \\
| & | & \\
CO \cdot OEt & CO.OEt &
\end{array}
$$

144. Frasch Process (Riegel, 2nd ed., p. 368, 1933):

The mercaptans in petroleum are removed by treatment with copper oxide, the essential steps in the process being:

$2 R \cdot S \cdot H + CuO \rightarrow Cu(S \cdot R)_2 + H_2O$, then

$Cu(S \cdot R)_2$, heat $\rightarrow R \cdot S \cdot R + CuS$, or

$Cu(S \cdot R)_2 + CuO$, heat $\rightarrow R \cdot S \cdot S \cdot R + Cu_2O$

$R \cdot S \cdot R$ and $R \cdot S \cdot S \cdot R$ are oil soluble and inoffensive.

145. Friedel-Crafts Syntheses (Chamberlain, 3rd ed., p. 390, 1934):

Aromatic hydrocarbons and most of their substituted derivatives except the nitro compounds lose a hydrogen atom from the nucleus and a halogen atom from most aliphatic halogen compounds to yield condensation products. Some examples are:

a. $C_6H_5 \cdot H + X \cdot R/(AlCl_3) \rightarrow C_6H_5 \cdot R + HX$

b. $C_6H_5 \cdot H + X \cdot CH_2 \cdot C_6H_5/(AlCl_3) \rightarrow C_6H_5 \cdot CH_2 \cdot C_6H_5 + HX$

c. $C_6H_5 \cdot H + Cl \cdot CO \cdot H$ (from $CO/HCl)/(AlCl_3) \rightarrow C_6H_5 \cdot CHO + HCl$

d. $C_6H_5 \cdot H + Cl \cdot OC \cdot R/(AlCl_3) \rightarrow C_6H_5 \cdot CO \cdot R + HCl$

e. $C_6H_5 \cdot H + Cl \cdot OC \cdot C_6H_5/(AlCl_3) \rightarrow C_6H_5 \cdot CO \cdot C_6H_5 + HCl$

f. $C_6H_5 \cdot H + Cl(N:)N \cdot C_6H_5/(AlCl_3) \rightarrow C_6H_5 \cdot C_6H_5 + N_2 + HCl$

The $AlCl_3$ forms a complex with the halogen compound and likewise with the benzene, thus activating the compounds and catalyzing the reactions.

146. Gattermann-Koch Reaction (Chamberlain, 3rd ed., p. 511-2, 1934):

Aromatic hydrocarbons react with carbon monoxide and hydrochloric acid in the presence of cuprous chloride (or aluminum chloride) to yield the corresponding aldehyde (cf. Friedel-Crafts No. 145):

$C_6H_5 \cdot H + Cl \cdot CO \cdot H/(CO/HCl/Cu_2Cl_2) \rightarrow C_6H_5 \cdot CHO + HCl$

An extension of this reaction is:

$C_6H_5 \cdot H + Cl \cdot OC \cdot R/(Cu_2Cl_2) \rightarrow C_6H_5 \cdot CO \cdot R + HCl$

147. Grignard Reactions (Porter, 2nd ed., p. 375-85, 1931):

A number of important condensations utilize the Grignard reagent. A few examples are:

a. $H \cdot CHO + R \cdot MgX/(dry ether) \rightarrow R \cdot CH_2 \cdot OMgX$, then
$R \cdot CH_2 \cdot OMgX + HX$, aq. $\rightarrow R \cdot CH_2 \cdot OH$ (primary alcohol) $+ MgX_2$

b. $R \cdot CHO + R \cdot MgX/(dry ether) \rightarrow R_2CH \cdot OMgX$, then
$R_2CH \cdot OMgX + HX$, aq. $\rightarrow R_2CH \cdot OH$ (secondary alcohol) $+ MgX_2$

c. $R \cdot CO \cdot R + R \cdot MgX/(dry ether) \rightarrow R_3C \cdot OMgX$, then
$R_3C \cdot OMgX + HX$, aq. $\rightarrow R_3C \cdot OH$ (tertiary alcohol) $+ MgX_2$

d. $R \cdot CO \cdot X + R \cdot Mg \cdot X/(dry ether) \rightarrow R \cdot CO \cdot R + MgX_2$

e. $R \cdot X + R \cdot Mg \cdot X/(dry ether) \rightarrow R \cdot R + MgX_2$

f. $R \cdot CO \cdot OR' + R \cdot Mg \cdot X/(dry ether) \rightarrow R \cdot CO \cdot R + R'O \cdot MgX$

g. $R \cdot HC \overset{O}{\cdot} CH_2 + R' \cdot MgX/(dry ether) \rightarrow R \cdot CH_2 \cdot CHR' \cdot OMgX.$

h. $H \cdot CO \cdot NR_2 + R' \cdot MgX/(dry ether) \rightarrow R' \cdot CH(OMgX) \cdot NR_2.$

These reactions are applicable, in general, in both the aliphatic and the aromatic series.

148. Pinacol Reduction (Hill and Kelley, 1st ed., p. 130, 1932):

An aldehyde or ketone, when treated with magnesium and subsequent hydrolysis, yields a glycol:

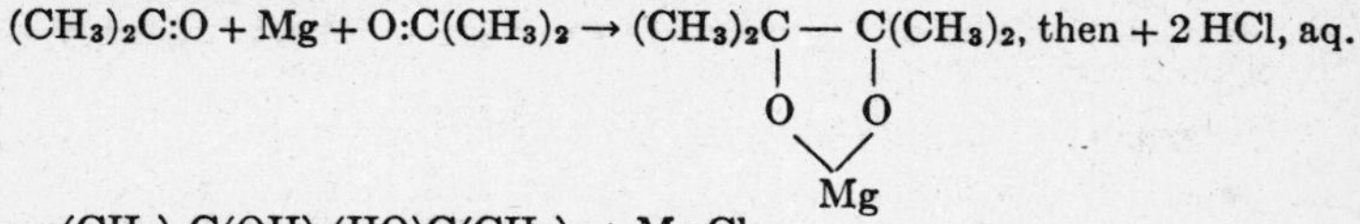

$$(CH_3)_2C{:}O + Mg + O{:}C(CH_3)_2 \rightarrow (CH_3)_2C \overset{\displaystyle |}{} \underset{\displaystyle O}{} \overset{\displaystyle |}{} C(CH_3)_2, \text{ then} + 2\,HCl, aq.$$

$$\rightarrow (CH_3)_2C(OH)\cdot(HO)C(CH_3)_2 + Mg\,Cl_2$$

For the rearrangement of this product, see Pinacolone Rearrangement (No. 56).

149. Reformatskii Reaction (Bernthsen and Sudborough, p. 272, 1933):

Hydroxy acids are prepared by treating the ester of a halogen substituted acid with zinc and then with a ketone; and, finally, with dilute acid:

$$Br\cdot CH_2\cdot CO\cdot OEt + Zn \rightarrow BrZn\cdot CH_2\cdot CO\cdot OEt, \text{ then} + CH_3\cdot CO\cdot CH_3 \rightarrow$$

$$(CH_3)_2C\cdot CH_2\cdot CO\cdot OEt, \text{ then} + HCl, aq. \rightarrow (CH_3)_2C\cdot CH_2\cdot CO\cdot OEt +$$
$$\overset{\displaystyle |}{O\cdot Zn\cdot Br} \qquad\qquad\qquad\qquad \overset{\displaystyle |}{OH}$$

$Br\cdot Zn\cdot Cl$. Subsequent hydrolysis of the ester yields the hydroxy acid.

150. Rosenmund Aldehyde Synthesis (Schmidt, English ed., p. 138, 1926):

Acid chlorides, when treated with hydrogen in the presence of colloidal or finely divided palladium or platinum, yield aldehydes:

$$R\cdot CO\cdot Cl + H_2/(Pd \text{ or } Pt) \rightarrow R\cdot CO\cdot H + HCl$$

151. Scholl Condensation (Schmidt, English, ed., p. 514, 1926):

The condensation of aromatic nuclei with the loss of hydrogen can be effected at about 100°C. by the use of $AlCl_3$ as a catalyst:

$$2\,C_{10}H_8 + (AlCl_3, \text{ at about } 100°C.) \rightarrow C_{10}H_7\cdot C_{10}H_7 + H_2$$

By continued treatment it is possible to obtain appreciable amounts of $C_{10}H_6{:}C_{10}H_6$.

152. Tschitschibabin Reaction (Whitmore, p. 138, 1938):

The symmetrical tertiary alcohols may be prepared, in general, by the treatment of ethyl carbonate with an excess of a Grignard reagent, as:

$$EtO\cdot CO\cdot OEt + R\cdot MgX/(dry\ ether) \rightarrow [EtO\cdot CR(OMgX)\,(OEt)] \rightarrow$$
$$EtO\cdot CO\cdot R + EtO\cdot MgX, \text{ then}$$
$$+ R\cdot MgX/(dry\ ether) \rightarrow [EtO\cdot CR_2(OMgX)] \rightarrow R\cdot CO\cdot R + EtOMgX,$$
$$\text{then}$$
$$+ R\cdot MgX/(dry\ ether) \rightarrow R_2CR(OMgX), \text{ then}$$
$$+ HX, aq. \rightarrow R_3C\cdot OH + MgX_2.$$

153. Ullmann Reaction (Cohen, Part I, 4th ed., p. 231-2, 1924):

Aromatic halides are condensed with themselves, with aniline, or with sodium phenolates by the use of finely divided copper as the condensing agent:

a. $C_6H_5 \cdot I$ + (Cu, powder) + $I \cdot C_6H_5$ → $C_6H_5 \cdot C_6H_5$ + I_2
b. $C_6H_5 \cdot Br$ + (Cu, powder) + $H \cdot HN \cdot C_6H_5$ → $C_6H_5 \cdot NH \cdot C_6H_5$ + HBr
c. $C_6H_5 \cdot I$ + (Cu, powder) + $Na \cdot O \cdot C_6H_5$ → $C_6H_5 \cdot O \cdot C_6H_5$ + NaI

Reactions *a* and *b* may involve free radicals, but *c* is probably ionic. These all occur, perhaps, on the catalytic-copper surface.

154. Zerevitinov (Zerewitinoff) Method (Cohen, Part I, 4th ed., p. 249, 1924):

Active hydrogen in alcohols, amines, acids, etc., is determined by reaction with methyl magnesium halide to give methane which is measured in a gas buret:

$$CH_3 \cdot Mg \cdot X + H \cdot OR → CH_4 + R \cdot O \cdot MgX$$

II. C. Unsaturated Group Mechanisms

155. Baeyer Permanganate Test (cf. Schmidt, p. 83, 1926):

A test for unsaturated linkages in which a drop of permanganate solution is added to the unknown which has been made slightly alkaline by the addition of sodium carbonate or bicarbonate. The presence of other reducing substances will obscure the test. The first steps in the reaction may be:

$$3\,R \cdot HC{:}CH_2 + 2\,MnO_4^- + 4\,H_2O/(OH^-) → (3\,R \cdot HC \overset{O}{\underset{\cdot}{\cdot}} CH_2) → 3\,R \cdot CHOH \cdot$$
$$CH_2OH + 2\,MnO_2 + 2\,OH^-$$

156. Bouveault and Blanc Reduction (Desha, p. 255, 1936):

Sodium and alcohol are used to reduce esters of the alkanoic acids to the corresponding alcohols, as:

$$CH_3(CH_2)_n CO \cdot OEt + 4\,H/(4\,Na/4\,Et \cdot OH) → CH_3(CH_2)_n CH_2 \cdot OH + Et \cdot OH.$$

157. Buchner-Curtius Reaction (Cohen, Part I, 4th. ed., p. 241, 1924):

Aldehydes react with diazomethane to yield ketones:

$$R \cdot CHO + H_2C{:}N \rightleftarrows N → R(H)C - O \quad → R(H)C \quad → R \cdot CO \cdot CH_3 + N_2$$

(with ring structures: first ring $H_2C - N$ bridging to N; second ring H_2C bridging to O)

158. Clemmensen Reduction (Conant, 1st. ed., p. 438, 1933):

Aromatic-aliphatic ketones are reduced with amalgamated zinc and hydrochloric acid to the corresponding hydrocarbons:

$$C_6H_5 \cdot CO \cdot CH_3 + 2\,Zn/(Hg) + 4\,HCl,\ aq. → C_6H_5 \cdot CH_2 \cdot CH_3 + 2\,ZnCl_2 + H_2O$$

This synthesis is used commercially for the preparation of hexylresorcinol. Caproyl chloride and resorcinol with $AlCl_3$ give $(HO)_2C_6H_3 \cdot CO \cdot (CH_2)_4 CH_3$, which is then reduced.

159. Emde Degradation (Gilman, p. 1025, 1938):

Nitrogen ring compounds, in the form of their halide salts, are opened by reduction with sodium amalgam in either alcoholic or aqueous solution, as:

$$\underset{\substack{H_2C\ \ \ \ CH_2 \\ | \ \ \ \ \ \ \ | \\ H_2C\ \ \ \ CH_2 \\ \diagdown \ \ \ \diagup \\ H \cdot C \cdot H}}{\overset{N(CH_3)_2^+}{}} + 2\,H^* \ (Na \cdot Hg/H_2O) → [CH_3 \cdot CH_2 \cdot CH_2 \cdot CH_2 \cdot CH_2N(CH_3)_2H]^+.$$

160. Gibbs Process (Conant, 1st ed., p. 453, 1933):

Naphthalene is oxidized with air at about 400-450° C. in the presence of a catalyst such as vanadium pentoxide, or a mixture of oxides, to yield phthalic anhydride. See page 204.

161. Harries Ozonide Reaction (Schmidt, English ed., p. 82, 1926):

Alkenes and other unsaturated compounds add ozone to give ozonides which, upon warming with water, decompose into carbonyl derivatives such as aldehydes and ketones:

$$-C{:}C = + O_3 \rightarrow = C{-}C{-}, \text{ then } + H_2O, \text{ warm } \rightarrow -C{:}O + O{:}C - + H_2O_2, \text{ or}$$
$$\underset{O_3}{\diagdown\diagup}$$

$$(CH_3)_2C{:}CHR + O_3 \rightarrow (CH_3)_2C{-}CHR, \text{ then } + H_2O, \text{ warm } \rightarrow$$
$$\underset{O_3}{\diagdown\diagup}$$

$$(CH_3)_2C{:}O + R{\cdot}CHO + H_2O_2$$

162. Ipatiev Reduction (Cohen, Part I, 4th ed., p. 186-7, 1924):

Hydrogen in the presence of nickel oxide as a catalyst at a temperature of about 250° C. under one to two hundred atmospheres pressure, effects reduction of benzene and other aromatic hydrocarbons, ketones, phenols, terpenes, and quinoline:

$$C_6H_6 + 3\ H_2/(NiO, 250° C., 100\ atms.) \rightarrow C_6H_{12}$$

163. Kamlet Reaction (U. S. Patent, 2,151,517, March 21, 1939):

The sodium bisulfite addition product of an aldehyde is condensed in the presence of traces of alkali or weak acids, with a primary or secondary nitroparaffin, as:

$$R{\cdot}CH(OH){\cdot}SO_3Na + R'{\cdot}CZ{:}N(\rightarrow O)ONa/(OH^- \text{ or } H^+) \rightarrow \text{(addition complex)} \rightarrow R{\cdot}CHOH{\cdot}CZ(NO_2){\cdot}R' + Na_2SO_3.$$

164. Mendius Reaction (Desha, p. 294, 1936):

Nitriles, when reduced with sodium and alcohol, yield the corresponding primary amines, as:

$$R{\cdot}CN + 4\ H^*/(4\ Na/4\ Et{\cdot}OH) \rightarrow R{\cdot}CH_2{\cdot}NH_2.$$

165. Meyer Polymerization (Schmidt, English ed., p. 697-8, 1926):

Primary nitriles polymerize, in the presence of sodium, to give cyanalkines:

$$3\ CH_3{\cdot}CN \rightarrow$$

$$=C{\cdot}H + N{:}C{\cdot}CH_3$$
$$=C{\cdot}NH_2 + CH_3{\cdot}C{:}N$$

$$\rightarrow CH_3{\cdot}C \begin{matrix} N{-}C{\cdot}CH_3 \\ \diagup\ \ \ \ \ \diagdown \\ C{\cdot}H \\ N{=}C{\cdot}NH_2 \end{matrix}$$

166. Mustard Oil Reaction (Schmidt, English ed., p. 291, 1926):

Carbon disulfide reacts with primary amines and subsequent treatment with silver nitrate to yield alkylisothiocyanates:

$$\begin{matrix} & NHR & & NHR \\ & | & & | \\ 2\ CS_2 + 4\ R{\cdot}NH_2 \rightarrow 2\ & C{:}S, & \text{then} + 2\ AgNO_3 \rightarrow 2\ & C{:}S & + 2\ R{\cdot}NH_2{\cdot}HNO_3, \text{ and} \\ & | & & | \\ & S{\cdot}NH_3R & & S{\cdot}Ag \end{matrix}$$

$$2\ RHN{\cdot}CS{\cdot}S{\cdot}Ag, \text{ heat} \rightarrow 2\ RNCS + Ag_2S + H_2S$$

This is known also as the Hofmann Mustard-Oil Reaction (cf. 134).

167. Nef Aldehyde and Ketone Synthesis (Gilman, 625-26, 1938):

A primary or secondary nitrocompound is converted to a salt by treatment with alkali and then added to sulfuric acid to give the corresponding aldehyde or ketone, as

$2\ R\cdot CH_2\cdot NO_2 + 2\ NaOH \rightarrow 2\ R\cdot CH{:}N(\rightarrow O)ONa$, then $+ 2\ H_2SO_4 \rightarrow$
$2\ R\cdot CHO + N_2O + 2\ NaHSO_4 + H_2O$, or
$2\ R_2CH\cdot NO_2 + 2\ NaOH \rightarrow 2\ R_2C{:}N(\rightarrow O)ONa$, then $+ 2\ H_2SO_4 \rightarrow$
$2\ R\cdot CO\cdot R + N_2O + 2\ NaHSO_4 + H_2O$.

168. Pechmann Synthesis (Schmidt, English ed., p. 565, 1926):

Acetylene condenses with diazomethane to yield pyrazole:

$$
\begin{array}{c}
\text{C·H} \\
\parallel \\
\text{C·H}
\end{array}
+ H_2C{:}N \rightleftharpoons N \rightarrow
\begin{array}{c}
\quad\ N\cdot H \\
\quad \diagup \diagdown \\
H\cdot C \quad\quad N \\
\parallel \quad\quad \parallel \\
H\cdot C\text{——}C\cdot H
\end{array}
$$

169. Prileschaiev Reaction (Gattermann and Wieland, p. 99, 1934):

Perbenzoic acid reacts with alkenes to give epoxyalkanes (**alkene oxides**) and benzoic acid:

$$R\cdot HC{:}CH\cdot R + C_6H_5\cdot CO\cdot O\cdot OH \rightarrow R\cdot HC\overset{\displaystyle O}{\diagup\diagdown}CH\cdot R + C_6H_5\cdot CO\cdot OH$$

170. Riley Oxidation (Whitmore, p. 234, 1938):

The oxidation of acetaldehyde with selenium dioxide yields glyoxal. This reagent seems to have a certain specificity for oxidizing the alpha-hydrogen atoms in preference to a carbonyl hydrogen atom, as indicated by:

$$CH_3\cdot CHO + SeO_2\ [\rightleftarrows H_2C{:}CH\cdot OH \rightarrow \underset{\underset{O\cdot Se\cdot O}{|\quad\ |}}{H_2C\text{——}CH\cdot OH}] \rightarrow H\cdot OC\cdot CHO + H_2O + Se.$$

171. Sabatier-Senderens Reaction (Chamberlain, 3rd ed., p. 606, 1934):

Alkenes are reduced by hydrogen in the presence of finely divided nickel at about 150-200°C., to give the corresponding alkanes:

$$R\cdot HC{:}CH_2 + H_2/(Ni,\ 150\text{-}200°C.) \rightarrow R\cdot CH_2\cdot CH_3$$

172. Senderens Reaction (Whitmore, p. 151, 1938):

When ethyl alcohol is passed over alumina at 250-300° C., the principal product is ethyl ether, as:

$$2\ CH_3\cdot CH_2\cdot OH + Al_2O_3,\ 250\text{-}300° C. \rightarrow CH_3\cdot CH_2\cdot O\cdot CH_2\cdot CH_3 + H_2O.$$

The mechanism may involve olefin formation and subsequent addition of alcohol.

173. Tshugaev (Tschugaeff, or Chugaev) Reaction (Cohen, Part III, 4th ed., p. 243-4, 1924):

Methyl alcohol reacts with carbon disulfide in the presence of alkali to give, through the loss of water, a condensation product which upon treatment with another alcohol gives an unstable intermediate that decomposes to give an alkene:

$$CH_3\cdot OH + CS_2/NaOH \rightarrow CH_3\cdot O\cdot CS\cdot SNa + H_2O,\ \text{then}$$
$$CH_3\cdot O\cdot CS\cdot S\cdot Na + HO\cdot R \rightarrow CH_3\cdot O\cdot CS\cdot SR + NaOH,\ \text{and}$$
$$CH_3\cdot O\cdot CS\cdot SR,\ \text{decomposes} \rightarrow CH_3\cdot O\cdot CS\cdot SH + C_nH_{2n}$$

This reaction gives normal olefins.

II. D. Dipole-Dipole and Van Der Waals Force Mechanisms.

174. Engler-Weissberg Hypothesis (Cohen, Part I, 4th ed., p. 134–5, 1924):

An explanation for oxidation by air which seems to agree with the known facts is exemplified by the air oxidation of benzaldehyde:

$$C_6H_5 \cdot CHO + O_2 \rightarrow C_6H_5 \cdot CO \cdot O \cdot OH, \text{ then } + C_6H_5 \cdot CHO \rightarrow 2\,C_6H_5 \cdot CO \cdot OH$$

175. Etard Reaction (Hill and Kelley, 1st ed. p. 406, 1932):

Toluene is oxidized to benzaldehyde by the use of chromyl chloride. The mechanism of the reaction is supposed to be as follows:

$$3\,C_6H_5 \cdot CH_3 + 6\,CrO_2Cl_2/(CS_2) \rightarrow 3\,(C_6H_5 \cdot CH_3 \cdot 2\,CrO_2Cl_2), \text{ then}$$

$$3\,(C_6H_5 \cdot CH_3 \cdot 2\,CrO_2Cl_2), \text{ decomposes} \rightarrow 3\,C_6H_5 \cdot CHO + 2\,H_2CrO_4 + H_2O + 4\,CrCl_3$$

176. Groves Process (Perkin and Kipping, Part I, p. 66, 1932):

Alcohols react with hydrogen chloride in the presence of anhydrous zinc chloride to yield the corresponding chloride:

$$R \cdot OH + HCl(gas)/(ZnCl_2, \text{ anhydrous}) \rightarrow R \cdot Cl + H_2O$$

177. Haworth Phenanthrene Synthesis (Richter, p. 618, 1938):

Condensation between naphthalene and succinic anhydride takes place on the *beta*-position. The oxo group is next reduced, and the resulting product treated with sulfuric acid to effect ring closure to give a cyclic ketone. Subsequent treatment with amalgamated zinc and hydrochloric acid (cf. 158) and then with selenium effects reduction and dehydrogenation to give phenanthrene.

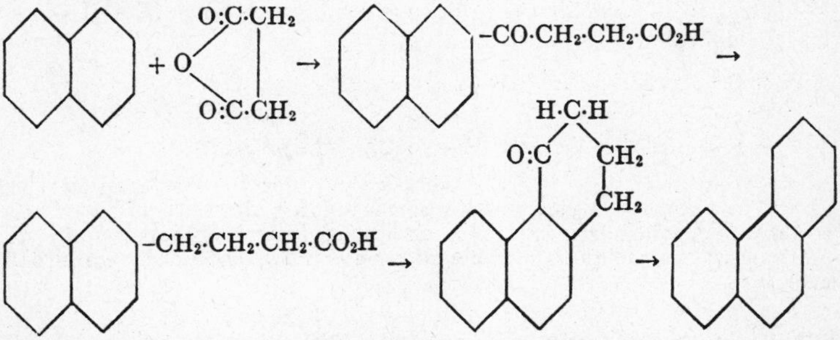

178. Hell-Volhard-Zelinsky Reaction (Hill and Kelley, 1st ed., p. 225, 1932):

Acids are treated with bromine in the presence of red phosphorus to give the acid bromides; then by additional treatment with bromine the α-bromoacyl bromides are obtained. These, upon hydrolysis, yield the α-bromoacids:

$$6\,R \cdot CH_2 \cdot CO \cdot OH + 3\,Br_2/2\,P \rightarrow 6\,R \cdot CH_2 \cdot CO \cdot Br + 2\,H_3PO_3, \text{ then}$$

$$R \cdot CH_2 \cdot CO \cdot Br + Br_2 \rightarrow R \cdot CHBr \cdot CO \cdot Br + HBr, \text{ and}$$

$$R \cdot CHBr \cdot CO \cdot Br + H_2O \rightarrow R \cdot CHBr \cdot CO \cdot OH + HBr$$

179. Hofmann Exhaustive Methylation (Cohen, Part III, 4th ed., p. 315, 1924):

Certain amines, when treated successively with methyl iodide and moist silver oxide and then distilled, yield a tertiary amine, an unsaturated hydrocarbon, and water. (Cf. von Braun, No. 186).

A specific example is:

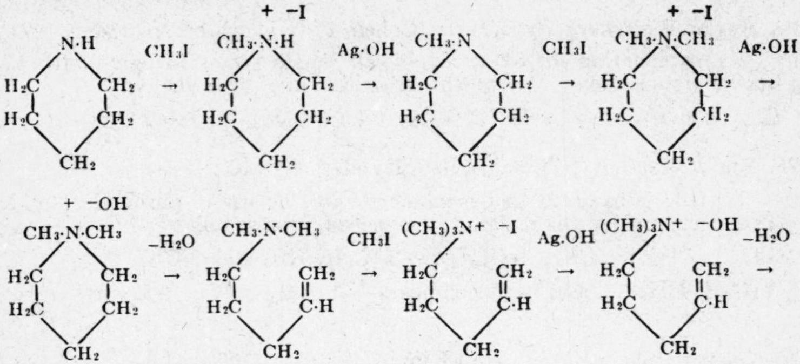

$CH_3 \cdot HC:CH \cdot HC:CH_2 + N(CH_3)_3 +$ other products.

180. Hofmann Reaction (Chamberlain, 3rd. ed., p. 44, 1934):

Primary, secondary, and tertiary amines are obtained by the action of ammonia on alkyl halides (cf. von Braun Exhaustive Methylation, No. 186):

$R \cdot X + NH_3 \rightarrow R \cdot NH_2 \cdot HX$, then $+ NH_3 \rightarrow R \cdot NH_2 + NH_4X$, and

$R \cdot NH_2 + R \cdot X \rightarrow R_2NH \cdot HX$, then $+ NH_3 \rightarrow R_2NH + NH_4X$, and

$R_2NH + R \cdot X \rightarrow R_3N \cdot HX$, then $+ NH_3 \rightarrow R_3N + NH_4X$

The tertiary amines react in turn with alkyl halides to form tetraalkyl ammonium salts:

$R_3N + R \cdot X \rightarrow R_4N \cdot X$

181. Hydrobromic Acid Method (Kamm, 2nd ed., p. 58, 1932):

Hydrobromic acid (48%) reacts very quickly with most tertiary alcohols to give good yields of the corresponding alkyl bromides, whereas secondary alcohols require refluxing and primary alcohols require the addition of one mole of sulfuric acid for every two moles of hydrobromic acid.

182. Jacobsen Reaction (Whitmore, p. 775, 1938):

1-Bromo-2,4,6-trimethylbenzene, when treated with concentrated sulfuric acid, yields, 2,4,6-trimethylbenzenesulfonic acid, 1,3-dibromo-2,4,6-trimethylbenzene and 1,3,5-tribromo-2,4,6-trimethylbenzene:

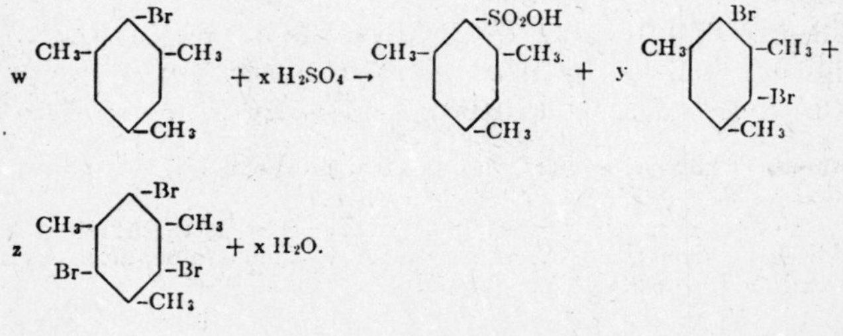

183. Lucas Test (J. Am. Chem. Soc. **52,** p. 803, 1930):

To one ml. of alcohol in a test tube are added six ml. of a hydrochloric acid-zinc chloride solution, and the tube is stoppered and shaken. Tertiary alcohols give a cloudy appearance immediately due to the formation of two phases, whereas secondary alcohols react within a few minutes, and the primary alcohols react only upon heating.

$$R \cdot OH + HCl/(ZnCl_2) \rightarrow [R \cdot OH_2^+ \ ^-Cl] \rightarrow R \cdot Cl + H_2O.$$

184. Victor Meyer Method (Kamm, 2nd ed., p. 59, 1932):

Primary, secondary, and tertiary alcohols are identified by conversion to the corresponding iodides and then to the nitro compounds.

$$R \cdot OH + HI \rightarrow [R \cdot OH_2^{+-}I] \rightarrow R \cdot I + H_2O,$$
$$R \cdot I + AgNO_2 \rightarrow R \cdot NO_2 + AgI.$$

Tertiary nitro compounds do not dissolve in alkali, whereas primary and secondary nitro compounds do. The secondary nitro compounds react with nitrous acid to form nitroso derivatives which are insoluble in alkali.

a. $R \cdot CH_2 \cdot NO_2$, alkali soluble, and

$$R \cdot CH_2 \cdot NO_2 + HONO \rightarrow R \cdot C(:N \cdot OH)NO_2, \text{ alkali soluble } + H_2O$$

b. $R_2CH \cdot NO_2$, alkali soluble, and

$$R_2CH \cdot NO_2 + HONO \rightarrow R_2C(NO)NO_2, \text{ insoluble in alkali } + H_2O$$

c. $R_3C \cdot NO_2$, insoluble in alkali.

185. von Braun Bromocyanogen Reaction (Schmidt, Eng. ed., p. 605, 1936):

The N-alkylpiperidines and certain other cyclic tertiary bases are ruptured when treated with cyanogen bromide, and yield brominated cyanamides, as:

$$\begin{array}{c} N \cdot R \\ H_2C \quad CH_2 \\ | \quad\quad | \\ H_2C \quad CH_2 \\ H \cdot C \cdot H \end{array} + Br \cdot CN \rightarrow Br \cdot CH_2 \cdot CH_2 \cdot CH_2 \cdot CH_2 \cdot CH_2 \cdot NR \cdot CN.$$

186. von Braun Exhaustive Methylation (Schmidt, Eng. ed., p. 530, 1926):

When pyrrolidine is subjected to exhaustive methylation by the phosphorus halide method, the products are butadiene, trimethylamine, and water:

$$\begin{array}{ccccccc}
N \cdot H & & N \cdot CH_3 & & N(CH_3)_2^{+-}I & & N(CH_3)_2^{+-}OH \\
H_2C \quad CH_2 & \rightarrow & H_2C \quad CH_2 & \rightarrow & H_2C \quad CH_2 & \rightarrow & H_2C \quad CH_2 & \rightarrow \\
H_2C - CH_2 & & H_2C - CH_2 & & H-C - CH_2 & & H_2C - CH_2
\end{array}$$

$$\begin{array}{ccccc}
N(CH_3)_2 & & N(CH_3)_3^{+-}OH & & \\
H_2C \quad CH_2 & \rightarrow & H_2C \quad CH_2 & \rightarrow & H_2C \quad CH_2 \\
H_2C - C \cdot H & & H_2C - C \cdot H & & H \cdot C - C \cdot H
\end{array} + (CH_3)_3N + H_2O.$$

187. Zeisel Determination (Gattermann and Wieland, p. 68-70, 1932):

Alkoxy groups, usually methoxy, are determined by treating the compounds with concentrated hydriodic acid and distilling off the alkyl iodide into an alcoholic solution of silver nitrate:

$$C_6H_5 \cdot O \cdot CH_3 + HI, \text{ heat } \rightarrow C_6H_5 \cdot OH + CH_3 \cdot I, \text{ then}$$
$$CH_3 \cdot I + AgNO_3, \text{ alc.} \rightarrow CH_3 \cdot O \cdot NO_2 + AgI$$

III. Intramolecular Reactions

A. Internal Condensation.

   1. By hydrogen bond.

   2. By activated phenyl hydrogens.

B. Group Rearrangement or Loss.

III. A. Internal Condensation Mechanisms

188. Baeyer-Drewsen Synthesis (Chamberlain, 3rd. ed., p. 619, 1934):

    2-Nitrocinnamaldehyde is reduced and dehydrated to give quinoline which is then oxidized and decarboxylated to yield pyridine:

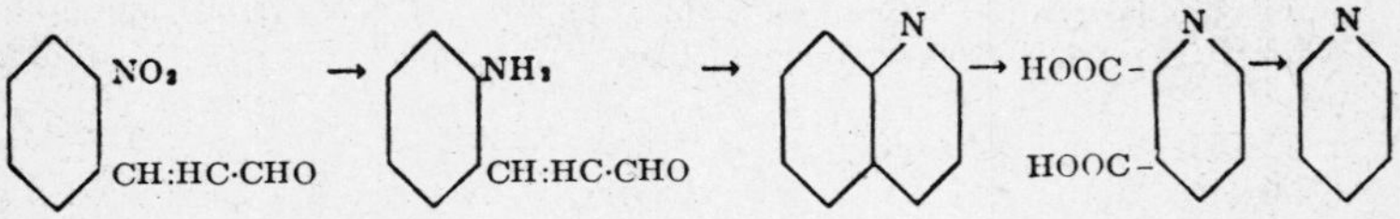

189. Bischler and Napieralski Reaction (Sidgwick, p. 566, 1937):

    When acyl derivatives of phenylethylamine are treated with powerful dehydrating agents in boiling toluene, the main product of the dehydration is 3, 4-dihydroisoquinoline, as:

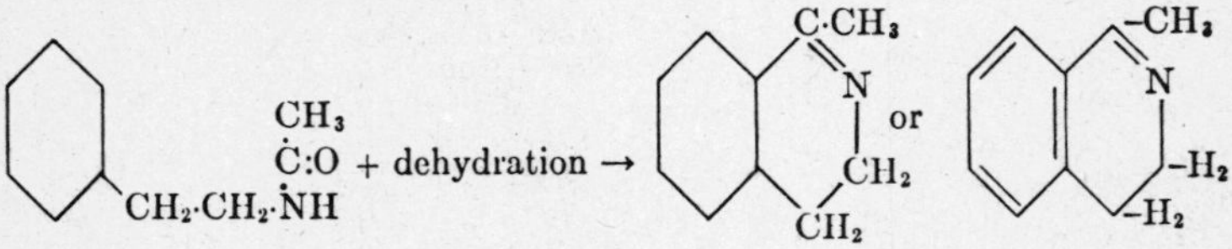

III. B. Group Migration or Loss Mechanisms

190. Bamberger-Goldschmidt Synthesis (Schmidt, English ed., p. 618, 1926):

    Cinnamaldoxime, upon heating with phosphorus pentoxide, yields isoquinoline:

$C_6H_5 \cdot HC:CH \cdot C(:N \cdot OH)H$ + $(P_2O_5,$ heat) $\rightarrow$
$C_6H_5 \cdot HC:CH \cdot N:C(OH)H$ then, $\rightarrow$ **isoquinoline**

The *trans*-isomer gives better yields.

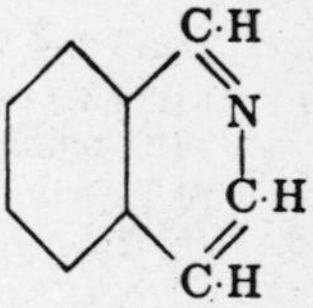

isoquinoline

191. Beckmann Rearrangement of Ketoximes (Porter, 2nd. ed., p. 416, 1931):

    A ketoxime is treated with phosphorus pentachloride to give the halogen derivative which undergoes hydrolysis and rearranges to yield a substituted amide:

$$C_6H_5 \cdot C \cdot C_6H_5$$
$$\parallel$$
$$N \cdot OH + PCl_5 \rightarrow$$

$$C_6H_5 \cdot C \cdot C_6H_5$$
$$\parallel$$
$$N \cdot Cl \rightarrow$$

$$C_6H_5 \cdot C \cdot Cl$$
$$\parallel$$
$$N \cdot C_6H_5 + H_2O \rightarrow$$

$$C_6H_5 \cdot C \cdot OH$$
$$\parallel$$
$$N \cdot C_6H_5 \rightarrow$$

$$C_6H_5 \cdot C:O$$
$$\mid$$
$$H \cdot N \cdot C_6H_5 \text{ (N-phenylbenzamide)}$$

192. Benzidine Rearrangement (Chamberlain 3rd. ed., p. 459, 1934):

Hydrazobenzene, in the presence of hydrogen ion, undergoes rearrangement to yield 4,4'-diaminobiphenyl:

$$C_6H_5 \cdot NH \cdot HN \cdot C_6H_5 + (H^+) \rightarrow H_2N \cdot C_6H_4 \cdot C_6H_4 \cdot NH_2$$

The first step of this reaction (the Semidine rearrangement) gives $C_6H_5 \cdot NH \cdot C_6H_4 \cdot NH_2$. See No. 204.

193. Benzilic Acid Rearrangement (Cohen, Part II, 4th. ed., p. 381–2, 1924):

Benzil rearranges, in the presence of strong alkali, to give the salt of diphenyl hydroxyacetic acid:

$$C_6H_5 \cdot C{:}O + NaOH \ (\text{or KOH}), \ \text{heat} \rightarrow$$

194. Bergmann Degradation (Whitmore, p. 606, 1938):

The free amino group of the polypeptide is first blocked with phenylisocyanate, and this intermediate in turn is methylated with diazomethane, converted to the hydrazide with hydrazine, then to the azide with nitrous acid, and finally to the benzyl urethane by treatment with benzyl alcohol. This product, upon catalytic reduction and hydrolysis, gives the amide of the polypeptide residue, toluene, ammonia, and an aldehyde which is identified to determine the amino acid that was split off. The steps are:

$$Q \cdots CO \cdot NH \cdot CHR \cdot CO_2H + H_2C{:}N_2 \rightarrow Q \cdots CO \cdot NH \cdot CHR \cdot CO \cdot O \cdot CH_3 + N_2,$$
then

$$+ H_2N \cdot NH_2 \rightarrow Q \cdots CO \cdot NH \cdot CHR \cdot CO \cdot NH \cdot NH_2 + CH_3 \cdot OH, \ \text{then}$$

$$+ HONO \rightarrow Q \cdots CO \cdot NH \cdot CHR \cdot CO \cdot N{:}\overrightarrow{N{:}N} + H_2O, \ \text{then}$$

$$+ Bz \cdot CH_2OH \rightarrow Q \cdots CO \cdot NH \cdot CHR \cdot NH \cdot CO \cdot O \cdot CH_2.Bz, \ \text{then}$$

$$+ 2 \ H^*/H_2O \rightarrow Q \cdots CO \cdot NH_2 + R \cdot CHO + CO_2 + Bz \cdot CH_3 + NH_3$$

Q is used here to indicate the rest of the peptide chain with the terminal free amino group blocked as $Bz \cdot NH \cdot CO \cdot HN \ldots$

195. Claisen Rearrangement (Richter, p. 521, 1938):

Unsaturated phenolic ethers rearrange on boiling to give o-substituted phenols, as:

196. Curtius Reaction (Cohen, Part III, 4th. ed., p. 147, 1924):

An azide is prepared by the reaction of hydrazine on an acid chloride with subsequent diazotization, and the azide is then converted to an amine by treatment with an alkali and hydrolysis:

$$R \cdot CO \cdot Cl + H_2N \cdot NH_2 \rightarrow R \cdot CO \cdot NH \cdot NH_2 + HCl, \ \text{then}$$

$$R \cdot CO \cdot NH \cdot NH_2 + HONO \rightarrow R \cdot CO \cdot N_3 + 2 \ H_2O, \ \text{and}$$

$$R \cdot CO \cdot N_3 + (KOH, \ aq.), \ \text{heat} \rightarrow R \cdot NCO + N_2, \ \text{and}$$

$$R \cdot NCO + 2 \ NaOH, \ aq. \rightarrow R \cdot NH_2 + Na_2CO_3$$

An ester may be used as the starting material instead of the acid chloride.

197. Fischer-Hepp Rearrangement (Cohen, Part II, 4th ed., p. 391, 1924):

Nitroso derivatives of secondary aromatic amines rearrange upon heating to give nuclear substituted nitroso compounds:

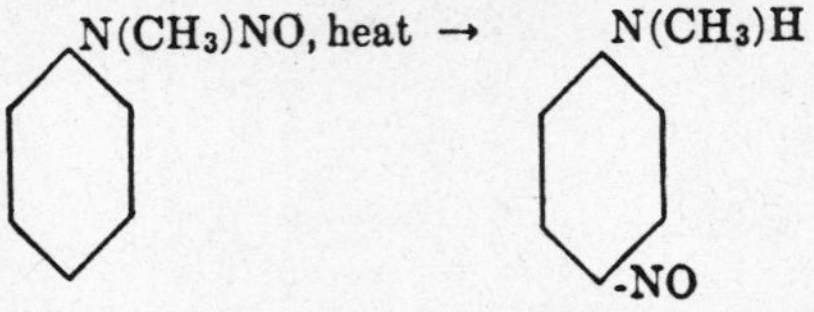

$N(CH_3)NO$, heat $\rightarrow$ $N(CH_3)H$

-NO

198. Fries Rearrangement (Gilman, p. 752, 1938):

Aromatic esters of alkanoic acids undergo rearrangement, in the presence of aluminum chloride, to yield hydroxyaromatic-aliphatic ketones, as:

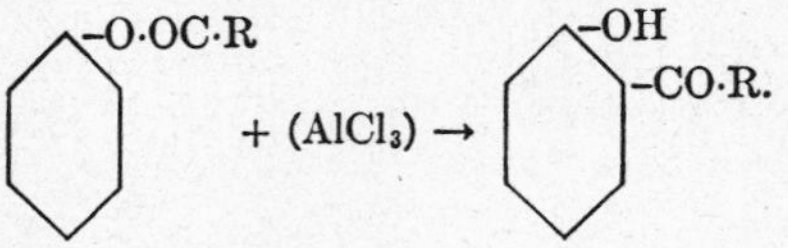

$-O \cdot OC \cdot R$ $-OH$
                         $-CO \cdot R.$
+ $(AlCl_3) \rightarrow$

199. Hofmann Degradation (Chamberlain, 3rd ed., p. 135, 1934):

Amides, upon treatment with halogen and alkali, yield primary amines:

$R \cdot CO \cdot NH_2 + Br_2/NaOH \rightarrow R \cdot CO \cdot NHBr + NaBr + H_2O$, then

$R \cdot CO \cdot NHBr + NaOH$, aq. $\rightarrow R \cdot N:C:O + NaBr + H_2O$, then

$R \cdot N:C:O + 2 NaOH$, aq. $\rightarrow R \cdot NH_2 + Na_2CO_3$

200. Hofmann Rearrangement (Porter, 2nd ed., p. 423, 1931):

N-Methylaniline, upon heating, isomerizes to o-toluidine:

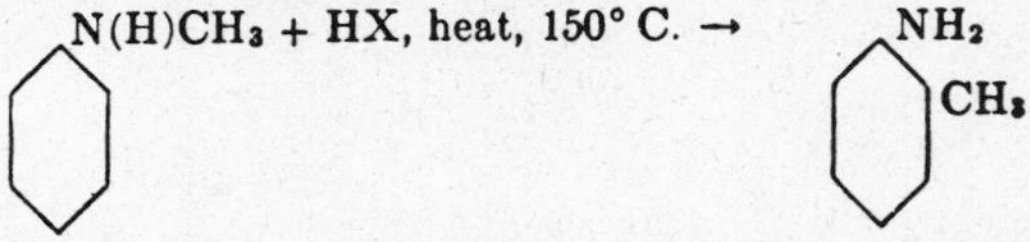

$N(H)CH_3 + HX$, heat, 150° C. $\rightarrow$ NH_2

CH_3

201. Jacobson (Semidine) Rearrangement (see Semidine Rearrangement, No. 204).

202. Japp-Klingemann Reaction (Sidgwick, p. 411, 1937):

Enolic forms of β-diketones and β-ketonic esters undergo reaction with diazo compounds to give oxygen azo compounds, which rearrange readily to carbon azo compounds and, finally, to hydrazones, as:

$R \cdot (HO)C:CH \cdot CO \cdot OEt + Bz \cdot N(:N)^+ \, ^-Cl \rightarrow R \cdot (Bz \cdot N:N \cdot O)C:CH \cdot CO \cdot OEt$
+ HCl, then

+ rearrangement $\rightarrow R \cdot CO \cdot (Bz \cdot N:N)CH \cdot CO \cdot OEt$, then

+ rearrangement $\rightarrow R \cdot CO \cdot (Bz \cdot NH \cdot N:)CH \cdot CO \cdot OEt.$

203. Lossen Rearrangement (Whitmore, p. 361, 1938):

Hydroxamic acids, in the presence of strong acid, undergo rearrangement to give an isocyanate, as:

$R \cdot C(:N \cdot OH) \cdot OH + (H^+, conc.) \rightarrow R \cdot C(:O) \cdot NHOH \rightarrow [R \cdot C(:O) \cdot N] \rightarrow$
$R \cdot N:C:O.$

The isocyanate may be hydrolyzed to give a primary amine or treated with an alcohol to give an urethane.

204. Semidine Rearrangement (Schmidt, English ed., p. 350, 1926):

If one *para* position is blocked, only half of the benzidine rearrangement occurs:

$CH_3 \cdot CO \cdot HN \cdot C_6H_5 \cdot NH \cdot HN \cdot C_6H_5$ + (H⁺, heat) →

$CH_3 \cdot CO \cdot HN \cdot C_6H_4 \cdot NH \cdot C_6H_4 \cdot NH_2$

205. Wallach Rearrangement (Schmidt, English ed., p. 348, 1926):

Azoxybenzene, when warmed with concentrated sulfuric acid, isomerizes into *p*-hydroxyazobenzene:

$C_6H_5 \cdot N{:}N (\rightarrow O) \cdot C_6H_5 / (H_2SO_4,$ heat$) \rightarrow$ *p*-$C_6H_5 \cdot N{:}N \cdot C_6H_4 \cdot OH$

206. Weermann Degradation (Cohen, Part III, 4th ed., p. 8, 1924):

The lactones of aldonic acids are treated with a saturated alcoholic solution of ammonia to give the corresponding amide, and then with hypochlorous acid (NaOCl/HCl, aq.) to yield an aldose sugar containing one less carbon atom. The next to the last step in this reaction is essentially a Hofmann Rearrangement.

207. Wöhler Synthesis (Chamberlain, 3rd ed. p. 357, 1934):

Sodium cyanate, upon heating with ammonium sulfate, yields urea:

$Na \cdot O \cdot CN + (NH_4)_2SO_4$, heat $\rightarrow$ $(NH_4 \cdot O \cdot CN) \rightarrow H_2N \cdot CO \cdot NH_2 + NaNH_4SO_4$

IV. Redox Reactions

208. Benedict Solution (Lowy and Harrow, 4th. ed., p. 177, 1936):

This is a modified Fehling solution made by mixing copper sulfate with a solution of sodium citrate and sodium carbonate to give a reagent which does not deteriorate on standing. The essential reaction is the reduction of the cupric ion to cuprous ion to give a red precipitate of hydrated cuprous oxide:

Cu^{++} + reducing agent such as glucose → Cu^+

209. Fehling Test (Lowy and Harrow, 4th ed., p. 176-82, 1936):

A reducing agent is added to Fehling solution to give a red precipitate of hydrated cuprous oxide:

Cu^{++} + reducing agent → Cu^+

Fehling solution is made up in separate containers and mixed just before using. The one solution contains copper sulfate and the other contains sodium hydroxide and Rochelle salt. Aldehydes, in general, and the monosaccharides give positive results with this test. Benzaldehyde, however, does not reduce Fehling solution readily.

210. Fenton Reagent (Whitmore, p. 489, 491, 1938):

Controlled oxidation in the aliphatic diacid series may be obtained in certain cases by the use of hydrogen peroxide in the presence of ferrous sulfate. Malic acid, for example, gives oxosuccinic acid, as:

$$
\begin{array}{l}
O{:}C \cdot OH \\
| \\
H \cdot C \cdot OH \\
| \quad + H_2O_2/(FeSO_4) \rightarrow \\
H \cdot C \cdot H \\
| \\
O{:}C \cdot OH
\end{array}
\qquad
\begin{array}{l}
O{:}C \cdot OH \\
| \\
C{:}O \\
| \\
H \cdot C \cdot H \\
| \\
O{:}C \cdot OH
\end{array}
\text{, oxo-succinic acid} + 2\ H_2O.
$$

The sugars, also, may be oxidized by this reagent, as:

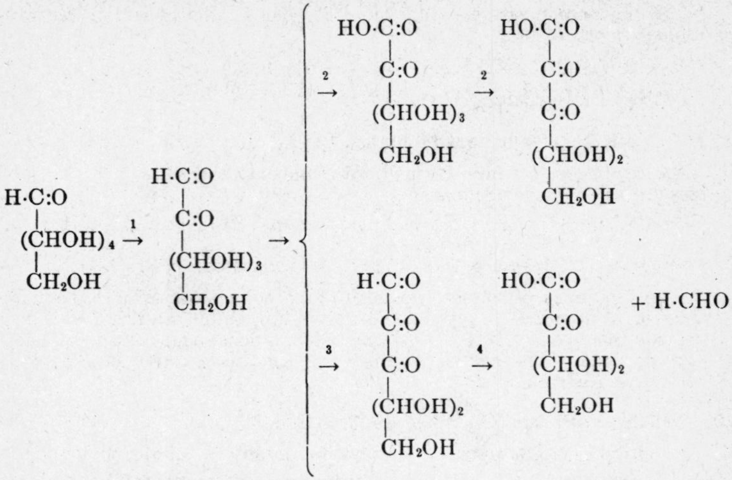

$$
\begin{array}{c}
H{\cdot}C{:}O \\
|\\
(CHOH)_4 \\
|\\
CH_2OH
\end{array}
\xrightarrow{1}
\begin{array}{c}
H{\cdot}C{:}O \\
|\\
C{:}O \\
|\\
(CHOH)_3 \\
|\\
CH_2OH
\end{array}
\rightarrow
\left\{
\begin{array}{cc}
\begin{array}{c}
HO{\cdot}C{:}O \\
|\\
C{:}O \\
|\\
(CHOH)_3 \\
|\\
CH_2OH
\end{array}
&
\begin{array}{c}
HO{\cdot}C{:}O \\
|\\
C{:}O \\
|\\
C{:}O \\
|\\
(CHOH)_2 \\
|\\
CH_2OH
\end{array}
\\
\\
\begin{array}{c}
H{\cdot}C{:}O \\
|\\
C{:}O \\
|\\
C{:}O \\
|\\
(CHOH)_2 \\
|\\
CH_2OH
\end{array}
&
\begin{array}{c}
HO{\cdot}C{:}O \\
|\\
C{:}O \\
|\\
(CHOH)_2 \\
|\\
CH_2OH
\end{array}
\quad + H{\cdot}CHO
\end{array}
\right.
$$

(with reagents 2, 2, 3, 4 over the arrows)

The reagents are: (1) H_2O_2/Fe^{++}, aq., (2) H_2O_2, dilute, (3) H_2O_2, conc., (4) H_2O, heat. Cf. Ruff Degradation, No. 214.

211. Knoop *Beta* Oxidation (Bodansky, 3rd ed., p. 358-9, 1934):

An oxidation mechanism which attempts to account for the fact that fatty acids undergo metabolic oxidation with the loss of two carbon atoms at a time. See page 160.

212. Löb Degradation (Cohen, Part III, 4th ed., p. 8, 1924):

Aldose sugars, upon electrolysis in the presence of sulfuric acid and a lead anode, yield the next lower member of the series.

213. Neuberg Degradation (Cohen, Part III, 4th ed , p: 9, 1924):

The copper salts of the aldonic acids, upon electrolysis, yield the aldose sugar containing one less carbon atom, carbon dioxide, and hydrogen.

214. Ruff Degradation (cf. Fenton Reaction, Cohen, Part III, 4th ed., p. 8, 1924):

Aldonic acids are oxidized with hydrogen peroxide in the presence of basic ferrous acetate to yield the next lower aldose sugar.

215. Tollens Test (Reid, p. 197, 1931):

Aldehydes are oxidized by an ammoniacal solution of silver nitrate, and metallic silver is deposited due to the reduction of the silver ion:

$$H{\cdot}CHO + 2\ Ag\ (NH_3)_2OH/2\ H_2O \rightarrow H{\cdot}CO{\cdot}ONH_4 + 3\ NH_4OH + 2\ Ag$$

216. Wohl Degradation (Cohen, Part III, 4th ed., p. 8, 1924):

An aldose sugar, by treatment with alcoholic ammoniacal silver nitrate, yields an aldose sugar containing one less carbon atom. See page 117.

V. Color Reactions

217. Adamkiewicz-Hopkins Reaction (cf. Hopkins-Cole, Chamberlain, 3rd, ed., p. 662, 1934):

A test for the tryptophan nucleus in protein by the use of glyoxylic acid in the presence of acetic or sulfuric acid to give a reddish violet coloration:

Protein + $H \cdot OC \cdot CO \cdot OH/H_2SO_4$ → reddish violet coloration

218. Caro Test (cf. Schiff Test, Kamm 2nd. ed., p. 53, 1932):

Schiff reagent (Fuchsin aldehyde reagent) is added in small amounts to a solution of the material to be tested. Most aldehydes produce a bluish pink coloration almost immediately, although benzaldehyde responds slowly. Most ketones give a positive test on standing or when heated with the reagent. The reagent is a dilute solution of magenta which seems to give the following equilibrium:

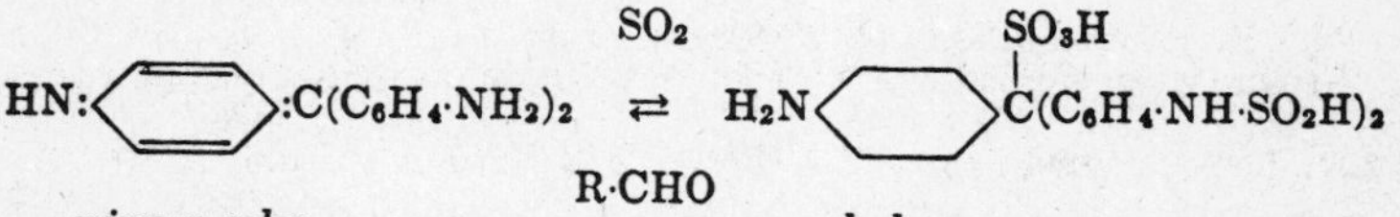

$$HN: \langle \quad \rangle : C(C_6H_4 \cdot NH_2)_2 \overset{SO_2}{\underset{R \cdot CHO}{\rightleftarrows}} H_2N \langle \quad \rangle C(C_6H_4 \cdot NH \cdot SO_2H)_2 \overset{SO_3H}{|}$$

crimson color   *colorless*

219. Carr-Price Color Reaction (SbCl$_3$) (Gilman, p. 1171, 1938):

Carotenoids, and vitamin A particularly, give blue colors when treated with an anhydrous solution of antimony trichloride.

220. Erlich Reagent (Schmidt, Carl L.A., p. 187, 1938):

Compounds containing the indole nucleus, such as tryptophan, give an intense blue color when treated with a strong hydrochloric acid solution of *p*-dimethylaminobenzaldehyde. This is a modification of the tryptophan aldehyde reaction. (Cf. Hopkins-Cole Reagent, No. 224, and Adamkiewicz-Hopkins Reaction, No. 217).

221. Ferric Chloride Test (Desha, p. 188, 416-417, 1936):

Most phenols, acetates, enols, and oximes give intense colorations when their solutions in water or dilute alcohol are treated with ferric chloride. Certain α-hydroxy acids give deep yellow colorations when treated with ferric chloride solution. The structure of the compounds formed is of a complex nature.

222. Folin Reaction (Schmidt, Carl L.A., p. 187, 1938):

1, 2-Naphthoquinone-4-sulfonic acid, in the presence of alkali, gives a deep red color with amino acids.

223. Hammarsten Reaction (Gilman, p. 1306, 1938):

Powdered cholic acid, apocholic acid, and their derivatives, give a violet color when treated with 25% hydrochloric acid. The color changes gradually green to yellow.

224. Hopkins-Cole Reaction (See Adamkiewicz-Hopkins Reaction, No. 217).

225. Indophenin Reaction (Chamberlain, 3rd ed., p. 612, 1934):

Thiophene reacts with isatin in the presence of concentrated sulfuric acid to give a green to blue coloration.

226. Janovsky Reaction (Hickinbottom, p. 322, 1938):

Acetone solutions of polynitro-derivatives of aromatic nuclei give color reactions when treated with 10% aqueous sodium hydroxide solution, as:

1, 3, 5-trinitrobenzene.....................blood red

2, 4-dinitrotoluene........................royal blue

m-dinitrobenzene..........................reddish-violet

2, 4-dinitrochlorobenzene.................red, changing to lilac

227. Konovaloff Reaction (Hickinbottom, p. 321, 1938):

Primary or secondary nitro-compounds are treated with sodium hydroxide, the salt formed is extracted with water, and ether is added to this extract. Upon the dropwise addition of ferric chloride, a red to reddish-brown color develops in the ether.

228. Legal Test (Gilman, p. 1321, 1938):

Pyridine solutions of glycosides or of genins give characteristic red colors when treated with an alkaline solution of sodium nitroprusside.

229. Liebermann Nitroso Test (Perkin and Kipping, p. 196, 1932):

Nitroso compounds, when dissolved in alcohol and treated with sulfuric acid, and then diluted and treated with alkali, give a blue to violet coloration.

230. Millon Test (Chamberlain, 3rd ed., p. 661, 1934):

When protein material or phenol is heated with a nitric acid solution of mercury (mercurous nitrate + nitrite), a brick red coloration or precipitate is produced. Tyrosine is probably the protein constituent that gives the test.

231. Molisch Test (Bodansky, 3rd ed., p. 68, 1934):

Carbohydrate material, when treated with a trace of α-naphthol and then with sulfuric acid in such a way as to form two layers, produces a violet ring at the juncture of the two layers.

232. Ninhydrin Reaction (Schmidt, Carl L.A., p. 185, 1938):

Amino acids containing a free amino group in the alpha-position give a blue color when treated with ninhydrin. (Ninhydrin is triketohydrindenehydrate).

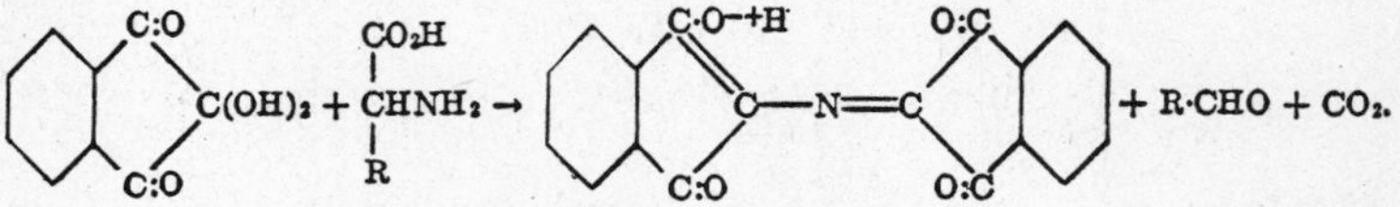

blue color

233. Pauly Reaction (Whitmore, p. 600, 1938):

Tyrosine produces a red azo dye when treated with diazotized sulfanilic acid.

234. Pettenkofer Reaction (Gilman, p. 1305, 1937):

Solutions of bile salts give red colors when treated with concentrated sulfuric acid, sugar solution, and agitated at 70°. A blue-red color finally develops.

235. Phthalein Test (Desha, p. 188, 1936):

Many phenols give colored phthalein dyes when heated with phthalic anhydride and sulfuric acid. Cf. Phenolphthalein, p. 259.

236. Plugge Reaction (Richter, p. 520, 1938):

Most phenols give red colors when treated with mercuric nitrate solutions containing some nitrous acid.

237. Rimini Test (Chem. Zentr., 1898, II, 132):

This is better known as the Angeli-Rimini Aldehyde Reaction (cf. 70).

238. Sakaguchi Reaction (Schmidt, Carl L. A., p. 188, 1938):

Guanidine derivatives, such as arginine, give an intense red color when treated with α-naphthol and sodium hypochlorite. The guanidine group is probably partially destroyed.

239. Schardinger Reaction (Gattermann and Wieland, p. 210, 1932):

Milk is treated at 50°C. with formaldehyde and methylene blue. Boiled milk gives a blue coloration, whereas unboiled milk fails to give any coloration.

Fresh cow milk contains an enzyme which greatly accelerates reduction by aldehyde of methylene blue to its leuco base. Heat destroys the enzyme, thus inhibiting the reaction. The enzyme activates two hydrogen atoms on the hydrated aldehyde to cause reduction. Finely divided platinum works similarly.

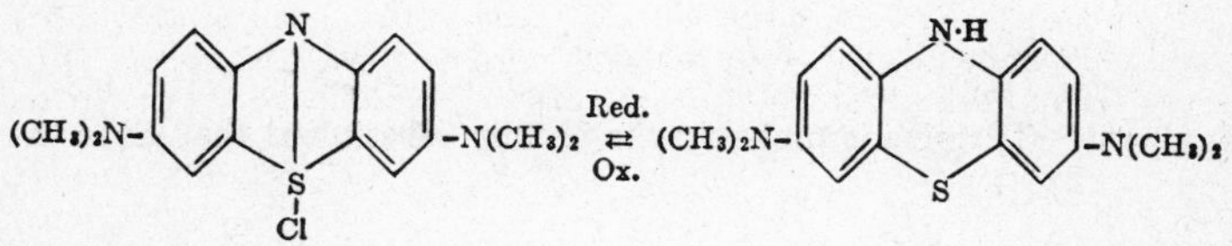

$(CH_3)_2N-$ [structure] $-N(CH_3)_2$ $\underset{Ox.}{\overset{Red.}{\rightleftarrows}}$ $(CH_3)_2N-$ [structure] $-N(CH_3)_2$

Cl
Blue

240. Schiff (Fuchsin-Aldehyde) **Test** (See Caro Test, No. 218).

241 Schryver Test (Whitmore, p. 232, 1938):

Phenylhydrazine hydrochloride, potassium ferricyanide and concentrated hydrochloric acid give a red color with formaldehyde The test has been used where formaldehyde has been suspected as a harmful preservative.

242. Selivanov (Seliwanoff) Test (Bodansky, 3rd ed., p. 68-9, 1934):

Ketose sugars give a red coloration or a red precipitate when heated with resorcinol dissolved in dilute hydrochloric acid, whereas aldose sugars give negative results.

243. Sterol Color Reactions (Gilman, p. 1271-2, 1938):

Salkowski Reaction: The sterol, in chloroform solution, is shaken with concentrated sulfuric acid. The chloroform layer is red, and the sulfuric acid layer is green.

Liebermann-Burchard Reaction: The sterol, in a chloroform solution, is treated with sulfuric acid and a few drops of acetic anhydride. A red color is produced which changes through blue to green.

Rosenheim Test: The sterol or its chloroform solution is treated with chloroacetic acid. Sterols containing a conjugated system of carbon atoms give red colors which change to blues.

244. Sullivan Reaction (Schmidt, Carl L.A., p. 189, 1938). See Folin Reaction, No. 222.

245. Topfer Reagent (Desha, p. 569, 1936):

An alcoholic solution of dimethylaminoazobenzene is used in gastric analysis for free hydrochloric acid. A red color appears if any free HCl is present when 5 ml. of 20.5% solution of the above reagent is added to 10 ml. filtered stomach fluid. The acid may then be titrated to a canary-yellow endpoint.

246. Wheeler and Johnson Color Test (Gilman, p. 983, 1938):

Uracil is converted into dibromooxyhydrouracil. This derivative is converted into the purple barium salt of dialuric acid by hydrolysis with barium hydroxide.

247. Xanthoproteic Test (Chamberlain, 3rd ed., p. 662, 1934):

When protein material is heated with concentrated nitric acid, it gives a yellow coloration which turns to orange on the addition of ammonium hydroxide or to red when neutralized with sodium hydroxide.

VI. Physical Changes or Processes

248. Babcock Test (Desha, p. 655, 1936):

Equal volumes of conc. sulfuric acid and milk or cream are placed in a flask which has a long narrow neck. The sulfuric acid destroys the milk proteins, and the heat from this reaction melts the butter fat. Hot water is added, and the flask is centrifuged. A column of clear fat is brought into the neck of the bottle where it can be measured.

249. Boeseken Method (Gilman, p. 1417, 1938):

The fact that a cis-configuration of a glycol in a boric acid solution shows maximum conductivity has been used in the absolute-configuration determination of sugars.

250. Cross-Bevan (Viscose) Process (Riegel, 2nd. ed., p. 340–5, 1933):

Cellulose is treated with alkali and then with carbon disulfide to give a viscous mass which is then dissolved in dilute alkali to obtain a solution known as viscose. The viscose is forced through small openings into a coagulating bath such as sulfuric acid.

251. Dakin Isolation of Amino Acids (Schmidt, Carl L.A., p. 142, 1938):

After hydrolyzing the protein with sulfuric acid and removing the acid by treatment with barium hydroxide, the solution is concentrated and extracted with butyl alcohol.

252. Fischer Esterification Method (Schmidt, Carl L.A., p. 138, 1938):

Monoamino acids are isolated following acid hydrolysis of proteins by the fractional distillation of their esters under reduced pressure.

253. Heller Ring Test (Chamberlain, 3rd ed., p. 663, 1934):

This is a clinical test for albumin which is commonly used in urinalysis. Five ml. of urine are placed in a test tube and 5 ml. of strong nitric acid introduced into the bottom of the tube by means of a pipette. At the juncture of the two liquids a cloud of precipitate, after standing a few minutes, indicates the presence of albumin. The test is quite sensitive.

254. Knorr Method (Lucas, 518, 1935):

By the use of low temperatures, acetoacetic ester is separated into its isomeric forms.

255. Körner Orientation Theory (Conant, p. 356, 1933):

When a third group is substituted on a benzene nucleus, the *p*-disubstitution product yields only one derivative, the *o*-disubstitution product yields two derivatives, and the *m*-disubstitution product yields three derivatives. By this means, a disubstitution product of benzene can be assigned an *o*-, *m*-, or *p*-structure.

256. Paneth Technique (Gilman, p. 520, 1938):

By comparing the relative rates at which standard metallic mirrors were removed at a definite distance from the source of the free radicals, Paneth arrived at values for the half-life period for the ethyl and methyl radicals.

257. Schweitzer Reagent (Chamberlain, 3rd ed., p. 335, 1934):

A specially prepared solution of ammoniacal cuprous chloride disintegrates cellulose to give an apparent solution. It has been shown, however, that the reagent yields a colloidal suspension of cellulose fibers.

258. Thompson Displacement Process (Chamberlain, p. 645, 1934):

In the nitration of cellulose, the cotton is placed in a nitric-sulfuric acid mixture and allowed to remain for about two hours. Cold water is then admitted slowly at the top over a period of about three hours while the acid mixture is withdrawn at the bottom. The replacement of the water for the acid takes place with little or no heat production.

CHAPTER XXXVI

ORGANIC CATALYSIS*

A. Introduction:

A catalyst may be defined as a substance which changes the rate of a chemical reaction without undergoing any net change itself. Many times only a trace of the catalytic material is sufficient to bring about a manifold change in the rate of a chemical reaction. A true catalyst neither adds energy to a reaction nor changes the equilibrium point of a reaction. Although the phenomenon of catalysis was first clearly recognized by Berzelius in 1835, our knowledge of catalysis is still largely uncoordinated and catalysts are generally sought by trial and error rather than by the application of basic concepts. Practically every chemical reaction known is influenced by catalysis in one way or another.

Over 40 per cent of all petroleum products and 50 per cent of all organic chemicals are dependent upon catalysts for their commercial production. The widespread use of catalysts in organic processes is largely due to their amenability to continuous processes, with the corresponding decrease in labor costs. Oftentimes, higher purity and greater uniformity of a product also result from the choice of selective catalysts.

Catalytic reactions may be classified under two headings— homogeneous catalysis and heterogeneous catalysis. Although there are several types of homogeneous catalytic reactions, acid-base catalysis in the liquid phase is the only one of technical importance in organic chemistry. A similar situation exists in heterogeneous catalysis where the important types are liquid and gaseous reactions with solid catalysts.

B. Catalytic Theory:

How does a solid catalyst speed up a reaction? The answer probably lies in the ability of the catalyst to adsorb and react with the reactants in such a way as to dissociate certain bonds. Naturally, to be a good catalyst, only the bonds which need to be broken to cause the desired reaction should be influenced by the catalyst. This requires special arrangements of atoms on the catalyst's surface and an ability to form bonds with one or more of the reactants. Molecules thus bonded to the surface are said to be adsorbed. If the material is adsorbed too strongly, it monopolizes the surface and slows down the reaction. Thus, a

---

*The material for this chapter was contributed by Dr. John N. Pattison, Batelle Memorial Institute, Columbus, Ohio.

successful catalyst must have a certain surface structure and be capable of bonding with one or more of the reactants but not too tightly with any of them.

The hydrogenation of ethylene by a transition-metal such as Ni, Pt, Pd, Fe, or Rh catalysts illustrates these points. The catalyst adsorbs the hydrogen and dissociates it into atoms. The closer the spacing of the metal atoms is to 3.7 angstroms, the better the catalyst, because this is the optimum distance of separation of hydrogen atoms for reaction with ethylene. Ethylene reacts with two hydrogens which are properly spaced without itself being adsorbed. Some ethylene is adsorbed, but it is so tightly bound that it merely decreases the surface available for hydrogenation.

C. Catalysts:

The method of preparation of solid catalysts has an important influence on the activity obtained. In general, it may be said that a catalyst should be prepared by a chemical reaction rather than by physical means. It is generally desired to have as much surface as possible and fine grinding does not compare with chemical methods. Usual techniques are precipitation as a gel, co-precipitation, depositing a solid from solution on a material of high surface area, dissolving one metal from an alloy, decomposition of metallic salts such as a formate, acetate, carbonate, or nitrate, and reduction of the oxide.

Many supported catalysts are made in which the active catalytic material is only present to the extent of 2 to 10 per cent. Promoters present, from traces to 1 per cent, often greatly influence the activity of the catalyst, as does the type of support which is used.

D. Applications:

In this section, some of the more important types of catalytic reactions will be discussed. Examples of important industrial applications and laboratory syntheses will be listed to illustrate typical conditions. This should serve as a guide to what can be accomplished by catalytic reactions and as an indication of the importance of catalysis to chemical industry and in organic synthesis.

1. Hydrogenation and Dehydrogenation.

Hydrogenation is one of the oldest and most important types of catalytic reaction and refers to the catalytic addition of hydrogen atoms to a molecule. Many reactions of this type cannot be carried out without a catalyst. Reductive cleavage of a molecule such as the reduction of an ester to two alcohols is called hydrogenolysis. Dehydrogenation is the removal of hydrogen. These three reaction types are grouped together because they have much in common. Metals of the eighth group of the periodic table (Fe, Co, Ni, Rh, Pd, Pt) are generally the preferred catalysts for these reactions. Other prominent catalysts are copper chromite, which is especially useful for the hydrogenation of carbonyl compounds such as aldehydes, ketones and acids and molybdenum sulfide. Copper, chromium, and molybdnum oxides are prominent dehydrogenation catalysts.

In the laboratory, platinum (Adams catalyst, PtO_2, which is reduced in situ to Pt), Raney nickel (prepared by dissolving the Al from a NiAl alloy with NaOH), and copper chromite (CuO + $Cu_2Cr_2O_2$, prepared by decomposition of the nitrates) are the most widely used.

Temperature, pressure, and poisons are important factors in hydrogenation catalysis. Most hydrogenation catalysts also act as dehydrogenation catalysts at higher temperatures. Although the temperature at which this change occurs differs from one catalyst to another, it is generally best to operate hydrogenation reactions below 300°C. Small amounts of arsenic or sulfur compounds are very harmful to the activity of platinum catalysts, and nickel is poisoned by sulfur and halogen compounds. Hydrogenations of sulfur-containing materials may be carried out in the presence of molybdenum sulfide.

Some typical hydrogenation and dehydrogenation reactions are listed in Table 1.

TABLE 1.   CATALYTIC HYDROGENATION, HYDROGENOLYSIS, AND DEHYDROGENATION REACTIONS

Reaction	Catalyst	Temp., °C.	Press., Atm.
Hydrogenation:			
$CO + H_2 \rightarrow$ Hydrocarbons	Fe or Co	300	20
$CO + 2H_2 \rightarrow CH_3OH$	$Cu\text{-}ZnO\text{-}Cr_2O_3$	340	200
Linoleic acid $+ H_2 \rightarrow$ oleic acid	Ni on kieselguhr	165	2
$C_6H_6 + H_2 \rightarrow C_6H_{12}$	Raney Ni	220	28
$CH_3 \cdot HC = CH \cdot CHO + H_2 \rightarrow$ $CH_3CH = CHCH_2OH$	Cu chromite	150	3
$CH_3CH_2CH_2CH_2 \cdot HC = CH_2 \cdot CHO +$ $H_2 \rightarrow CH_3(CH_2)_5CH_2OH$	Raney Ni	150	3
$C_6H_5CN + H_2 \rightarrow C_6H_5CH_2CH_2NH_2$	Raney Ni	125	100
Hydrogenolysis:			
$CH_3CH_2CH_2CH_2COOC_2H_5 + H_2 \rightarrow$ $CH_3(CH_2)_3CH_2OH + C_2H_5OH$	Cu chromite	250	100
$C_6H_5NO_2 + H_2 \rightarrow C_6H_5NH_2$	Cu chromite	175	3
Dehydrogenation:			
$C_6H_5C_2H_5 \rightarrow C_6H_5 \cdot HC = CH_2 + H_2$	ZnO	600	1
$C_4H_{10} \rightarrow C_4H_8 + H_2$	Cr_2O_3 on Al_2O_3	600	1
$C_7H_{16} \rightarrow C_6H_5CH_3 + H_2$	MoO_3 on Al_2O_3	500–550	1
$CH_3OH \rightarrow HCHO + H_2$	Cu	350	1

2. Oxidation.

Oxidation is one of the most valuable tools of the synthetic chemist. The cheapest reagent, and the one always employed when possible, is air. The reactions are carried out in both the liquid and vapor phase using a variety of catalysts. The reactions are highly exothermic and the problem of heat removal is an important one if the reaction is to be kept under control. The most important oxidation catalysts are vanadium, silver, copper, and manganese oxides. Good oxidation catalysts generally have the ability to change readily and reversibly from one oxidation state to another. This fact has led to a theory of alternate oxidation and reduction reactions to explain oxidation catalysis. Some commercial catalytic oxidation reactions are listed below as a guide to the choice of catalysts and conditions.

TABLE 2. SOME CATALYTIC OXIDATION REACTIONS

Reaction	Catalyst	Temp., °C.	Press., Atm.
naphthalene + air → phthalic anhydride	V_2O_5	400–460	1
benzene + air → maleic anhydride	V_2O_5	450	1
$CH_3OH + air \longrightarrow HCHO + H_2O$	Cu gauze	550–600	1
$CH_3CHO + air \longrightarrow CH_3COOH$	$Mn(CH_3COO)_2 + CH_3COOH$	55	5
$H_2C=CH_2 + air \longrightarrow H_2C\overset{O}{-}CH_2$	Ag + BaO	275	1
$C_2H_5SH + air \longrightarrow C_2H_5SO_3H$	Oxides of N	40–60	1

3. Hydration and Dehydration.

Catalytic dehydration is a very important reaction in organic synthesis, whereas hydration has only been employed to a limited extent. Alumina (Al_2O_3) is a very good catalyst for dehydration reactions in the range of 250-400°C. and has been the most widely used. Thoria (ThO_2), silica gel, and mineral acids, such as sulfuric and phosphoric, have found wide use. Catalytic dehydration is a clean and easy way to prepare ethers, esters, amides, ethylenic compounds, and many others as illustrated in Table 3.

Table 3. EXAMPLES OF CATALYTIC HYDRATION AND DEHYDRATION

Reaction	Catalyst	Temp., °C.	Pressure, Atm.	Remarks
Hydration: $(C_2H_5)_2O + H_2O \rightarrow 2C_2H_5OH$	$Al_2O_3 + H_3PO_4$	320	1	
$HC{\equiv}CH + H_2O \rightarrow CH_3CHO$	$HgSO_4 + H_2SO_4$	70–100	2	Liquid phase
Dehydration: $C_2H_5OH \rightarrow (C_2H_4)_2O$	Al_2O_3	250	1	
$CH_2OHCHOHCH_2OH \rightarrow$ $H_2C{=}CHCHO + 2H_2O$	Al_2O_3	360	1	
$C_2H_5OH + H_2S \rightarrow (C_2H_5)_2S$ $+ 2H_2O$	ThO_2	200–300	1	
$C_6H_5COOH + CH_3OH \rightarrow$ $C_6H_5COOCH_3 + H_2O$	ThO_2	350	1	
$CH_3CHO + C_2H_5OH \rightarrow$ $H_2C{=}CH{-}HC{=}CH_2 +$ $2H_2O$	$SiO_2 + Ta_2O_5$	425	1	
$CH_3CONH_2 \rightarrow CH_3CN +$ H_2O	Al_2O_3	250	1	
$2CH_3COOH \rightarrow (CH_3COO)_2O$ $+ H_2O$	ThO_2	400–500	1	

4. Halogenation.

Most halogenation reactions are not carried out catalytically. However, in the chlorination of toluene where it is desired selectively to chlorinate the ring and not the side chain, an iron catalyst (generally $FeCl_3$) is used, with chlorine as the reagent.

5. Ammonolysis.

The preparation of amino compounds by replacement reactions using ammonia is generally done catalytically. In the liquid phase, copper and its salts is most often used in the replacement of halogens. Hydroxyl groups are more readily replaced by ammonia in the presence of sulfites in the liquid phase and alumina in the gas phase. Table 4 illustrates these reactions.

TABLE 4. CATALYTIC AMMONOLYSIS

Reaction	Catalyst	Temp., °C.	Press., Atm.	Remarks
$C_6H_5Cl + NH_3 \rightarrow$ $\quad C_6H_5NH_2 + HCl$	Cu	200	60	Commercial. Catalyst essential.
$CH_3OH + NH_3 \rightarrow CH_3NH_2$ $\quad + HCl$	Al_2O_3	450	1	Commercial.
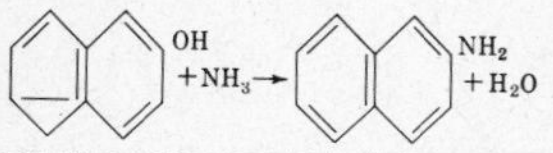 (naphthol) OH $+NH_3 \rightarrow$ (naphthylamine) NH_2 $+ H_2O$	$(NH_4)_2SO_3$	150	High	Commercial. Catalyst increases yield.

E. Catalysis in the Petroleum Industry:

Initially, the refining of petroleum consisted simply of separating the constituents already present in the crude oil by distillation. It was indeed fortunate that the first crude oil (discovered in Pennsylvania in 1859) was adaptable to such an operation. As markets were developed for more specialized products, and different types of crude oil became available, it became necessary to treat the less useful petroleum fractions chemically in order to increase the percentage of gasoline. The first step along this line was thermal cracking. In this process, the longer molecules are broken into smaller ones, some of which are suitable for gasoline. The need for more gasoline continued, and catalytic cracking was discovered to be more efficient. Other processes which followed were alkylation, isomerization, polymerization, cyclization, and hydroforming, each designed to improve the quality or yield of gasoline.

A simplified outline of how these processes are interrelated in an integrated oil refinery is shown below.

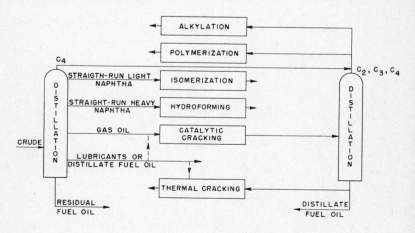

1. Catalytic Cracking.

Cracking is such a complicated process that it is difficult to define by an equation. Catalytic cracking favors the formation of aromatics and isoparaffins, thus greatly increasing the octane rating, and produces only a small amount of undesirable olefins. The most important catalysts are natural clays, such as montmorillonite, and synthetic silica-alumina gels (alumina is the minor constituent). The reaction is generally carried out at 400-600°C. and ordinary pressures. Until a few years ago, fixed catalyst beds were used, and the formation of carbon on the catalyst (undesirable side reaction) made it necessary to reactivate the catalyst every 20 or 30 minutes. This loss of time has been largely eliminated by the development of "fluidized" catalyst systems. These consist of a large clyindrical reactor containing finely divided (100-300 mesh) catalyst particles kept in suspension by the stream of oil vapor. The violent mixing which results eliminates local overheating ("hot spots"), which often developed during the regeneration, and allows the regeneration and cracking to be carried on as a continuous process.

The mechanism of catalytic cracking probably consists of a proton transfer between the catalyst (active catalysts are known to be capable of proton transfer under cracking conditions) and the reactant hydrocarbons. The fragments thus formed (possibly carbonium ions of paraffins and olefins) then are desorbed or react further on the catalyst to form aromatics which are desorbed or react further to form coke.

2. Alkylation.

Alkylation refers to the addition of an alkyl group to an organic compound. Important examples in the petroleum field are:

$$\underset{\text{isobutane}}{(CH_3)_3CH} + \underset{\text{2 butene}}{CH_3 \cdot HC = CH \cdot CH_3} \xrightarrow[\substack{25\text{-}45°C. \\ 100\text{-}125 \text{ p.s.i.}}]{HF} \underset{\text{"isooctane"}}{(CH_3)_3C\text{---}CH_2CH(CH_3)_2}$$

$$\underset{\text{benzene}}{C_6H_6} + \underset{\text{ethylene}}{H_2C = CH_2} \xrightarrow[\substack{250°C. \\ 400 \text{ p.s.i.}}]{\substack{H_3PO_4 \text{ on} \\ \text{Kieselguhr}}} \underset{\text{ethyl benzene}}{C_6H_5C_2H_5}$$

This provides a use for the petroleum gases which are a by-product of the cracking process and at the same time produces highly branched hydrocarbons which are valuable ingredients of aviation gasoline. Aklylation catalysts fall into two principal classes, both of which may be referred to as acid-acting catalysts. The first consists of anhydrous halides of the Friedel-Crafts type, such as aluminum chloride, aluminum bromide, zirconium chloride, and boron fluoride, which may be promoted by their respective hydrogen halide gases. The second type consists of acids, principally concentrated sulfuric, liquid hydrogen fluoride, and synthetic silica-alumina cracking catalyst. In either case, the alkylation reactions are carried out under sufficient pressure to maintain at least part of the reactants in the liquid phase and with an excess of the saturated hydrocarbon to minimize polymerization of the olefin.

Much controversy has resulted concerning the mechanism of the alkylation of paraffins. The simple addition shown above in the alkylation of butane is a simplification of the results obtained. Actually, a variety of products results, closely paralleling the thermodynamically predicted mixture. The addition of a proton to the double bond, and subsequent formation of a carbonium ion, apparently is the initial step. The carbonium ion then rearranges by methyl and proton shifts that tend to convert secondary carbonium ions to tertiary ions which are much more stable. Subsequent proton transfers then yield the saturated hydrocarbons as products.

3. Polymerization.

Another process which utilizes the by-product olefin gases is polymerization. The process is very similar to the alkylation reaction just described except that it takes place between two olefins. The resulting product is usually a low-molecular-weight polymer, such as a dimer, and is an olefin. Subsequent hydrogenation is necessary if the product is to be blended into aviation gasoline. The catalysts generally employed are "solid phosphoric acid," which is phosphoric acid adsorbed on kieselguhr, and silica-alumina cracking catalysts. The reaction is operated at 500 pounds per square inch pressure and at about 750°C.

4. Isomerization.

Isomerization is an important process for the final treatment of gasoline to produce more isohydrocarbons and as the source of isobutane for making alkylate (butenes plus isobutane; see Alkylation above). The latter reaction may be written:

$$CH_3 \cdot CH_2 \cdot CH_2 \cdot CH_3 \rightleftharpoons CH_3 \cdot \overset{\displaystyle CH_3}{\overset{|}{CH}} \cdot CH_3$$

These are equilibrium reactions in which the iso forms are favored at low temperatures. The same acid-type catalysts which are effective in cracking are active in isomerization. In commercial practice, aluminum chloride promoted with hydrogen chloride is most widely used. The temperature is from 80-150°C. at 9-18 atmospheres. The contact time (in the order of 10 minutes) is long enough to permit the reaction to come nearly to equilibrium. Conversions of butane to isobutane usually range from 45 to 55 per cent per pass.

5. Cyclization and Aromatization.

Cyclization, as the name implies, refers to the formation of cyclic hydrocarbons (naphthenes) from straight-chain hydrocarbons. An extension of this process which includes dehydrogenation is called aromatization or hydroforming and is an important source of toluene for explosives, alkyl aromatics for aviation fuel, and benzene.

A typical reaction is the formation of toluene from n-heptane:

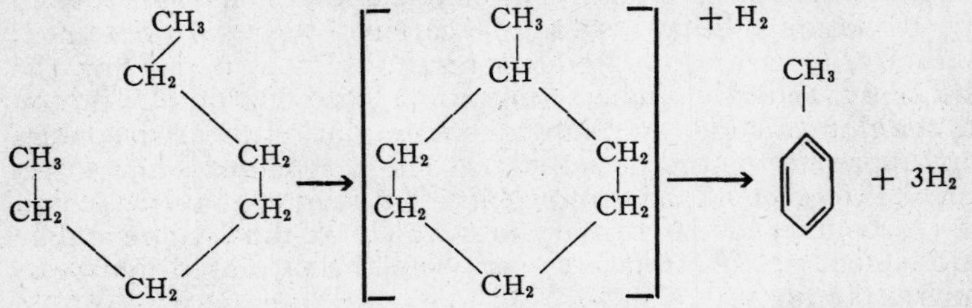

This reaction is generally carried out at 450-500°C. in the presence of molybdenum or chromium oxide catalysts supported on alumina. Hydrogen dilution is used to reduce coke formation even though this is a dehydrogenation reaction. By the proper choice of conditions, almost any desired increase in aromatic content may be achieved.

CHAPTER XXXVII

THE REFERENCE LITERATURE OF
ORGANIC CHEMISTRY

The service which an introductory organic text performs is to acquaint a prospective chemist with the theoretical foundations, the classes of compounds, and the types of reactions in this vast field of organic chemistry. Fortunately, the organic chemist has at his disposal a number of important aids in searching for additional information. Four excellent guides[1] to the use of chemical literature have been published, in which not only items of universal interest to all chemists may be found but also special pages devoted to organic chemistry alone.

It would pay real dividends for an earnest student to take a few hours' time and read at least one of these books on the literature of chemistry, particularly those pages where information concerning the general reference books, the general journals, and the specialized material for organic chemistry is located. This is an age of rapid scientific progress when an investigator cannot afford not to use the material discovered in the past. He must master the methods to locate the records of completed research, because not even a genius has time to repeat the work of others which might be found easily if he used a few moments in the library.

The author can never include in an introductory text the information concerning the thousands of individual compounds, the countless illustrations of methods to prepare them and their derivatives, and the complete survey of the experimental evidence for the theories accepted in the organic field. Nor can any book, treatise, or text include the latest material that is published day by day—material which not only adds to the present information but may overnight radically change some aspect of the science. It cannot be too strongly emphasized that both the older literature and the contemporary journal articles must be consulted by anyone who wants to do active teaching or research.

Although organic chemistry comprises a tremendous number of distinct substances, material to ascertain whether any compound has been prepared has been painstakingly organized into the re-

---

[1] These guides are: E. J. Crane and A. M. Patterson, *A Guide to the Literature of Chemistry*, John Wiley and Sons, Inc., N. Y. C., 1927; B. A. Soule, *Library Guide for the Chemist*, McGraw-Hill Book Co., Inc., N. Y. C., 1938; M. G. Mellon, *Chemical Publications, Their Nature and Use*, McGraw-Hill Book Co., Inc., N. Y. C., 1940, Ed. 2 (a new edition is in preparation); G. Malcolm Dyson, *A Short Guide to Chemical Literature*, Longmans, Green and Co., N. Y. C., 1951.

liable and indispensable empirical formula indexes. With knowledge of the system of classification used in these indexes, and a little practice, the investigator can confidently expect to find this information. A chronological search may be made by using the available formula indexes[1] for those years not covered by the general formula indexes of Beilstein.[2] The methods of listing the compounds are described in detail in the formula indexes themselves and also in the four books on chemical literature. When the arrangement of the empirical formulas in the German indexes has been learned, it is necessary only to note the different order used by the so-called Hill System for Chemical Abstracts. Richter and Stelzner's formula indexes give references to the literature and some physical constants. The *Chemisches Zentralblatt* and *Chemical Abstracts* refer to abstracts which record more information.

For a comprehensive discussion of the organic compounds, the monumental F. K. Beilstein's *Handbuch der Organischen Chemie* is the most extensive chemical treatise known. To use Beilstein is an art imperative to master. The third edition with its supplements covers the literature to July 1, 1903. Edition 3 of Richter is actually a formula index for the third edition of Beilstein, and for all organic references in the *Chemisches Zentralblatt* from 1902 to 1910. The fourth edition of Beilstein covers the literature to January 1, 1910, and consists of thirty-one volumes which include the subject and empirical formula index for the first twenty-seven volumes. The first thirty-one volume supplement embraces the period from 1910 to 1919. The second supplement, which includes material from 1919 to 1929, is complete through twenty-seven volumes and the cumulative subject index for the entire fourth edition. At present, the cumulative empirical formula index for the whole set lists compounds from C_1 to C_{17}. The remainder is to be published soon. The editors contemplate a third supplement to be finished by 1965. Meanwhile, since the editors classified all organic structures and assigned system numbers to them from 1 to 4877, compounds may be located through the Table of Contents. In addition to the explanation for the use of the treatise in Volume 1 of the main edition, there is a clear discussion by E. H. Huntress.[3] Volumes 30 and 31 are outside the scope of these general indexes and end with system number 4767. In spite of the increasing gap between the latest date covered by Beilstein and the present day, this is not serious because

[1] Listed according to years of the literature covered, the formula indexes are: M. M. Richter, *Lexikon der Kohlenstoffverbindungen*, Ed. 3, covering the period prior to 1910; R. Stelzner, *Literatur-Register der Organischen Chemie*, in five successive volumes covering the period from 1910-1921; the collective formula indexes of the *Chemisches Zentralblatt*, 1922-24, 1925-29, 1930-34, followed by annual indexes of that journal except for gaps during World War II. To check the period beginning with 1920, use the Collected Formula Index of *Chemical Abstracts* for 1920-1946 and after that the annual formula index of that journal.

[2] F. K. Beilstein, *Handbuch der Organischen Chemie*, Springer Verlag, Berlin, 1919 *et seq.*

[3] E. H. Huntress, *A Brief Introduction to the Use of Beilstein's Handbuch der Organischen Chemie*, John Wiley and Sons, Inc., N. Y. C., 1938, Ed. 2.

modern abstract journals can be relied upon to be quite thorough, which was not true of the nineteenth-century publications.

For each entry, Beilstein contains the following information if it has been published: name, formula, structure, configuration, bibliography if in a book, some historical information, occurrence, formation, preparation, properties, chemical changes, physiological properties, usual analysis, addition compounds, and salts. References in Beilstein and other German publications employ German abbreviations which may be unfamiliar. Comprehensive works like Beilstein print lists of those abbreviations which they use.[1]

An English treatise not so complete as Beilstein is the three-volume publication edited by I. M. Heilbron.[2] The 1953 edition is compiled by Heilbron and H. M. Bunbury. These volumes include information to 1950 with some insertions as late as the early months of 1953. *Elsevier's Encyclopedia of Hydrocarbon Compounds*, edited by E. Josephy (deceased) and F. Radt,[3] will comprise twenty volumes with many parts. There are to be four series: (1) aliphatic compounds, (2) carboisocyclic noncondensed compounds, (3) carbocyclic condensed compounds, (4) heterocyclic compounds, and a general index. As yet, only parts of Volumes 12, 13, and 14, which cover rapidly expanding topics not treated adequately in Beilstein, are published. The general literature is covered to within ten years and structural studies to within one year of the publication date. There is distinct emphasis upon biological and physical properties. The *Encyclopedia of Hydrocarbon Compounds*, edited by J. E. Faraday, lists in expandable, loose-leaf form the pertinent data for hydrocarbons which contain C_1 to C_{14} carbon atoms. Beilstein is the source of information up to 1919, and *Chemical Abstracts, British Chemical Abstracts*, and the *Chemisches Zentralblatt* since then. At present, the literature has been scrutinized to within four years. Insertion of pages with later information will be made as they are printed.

There are several advanced books which stress theoretical aspects of the field and are neither texts nor encyclopedias. *Organic Chemistry, An Advanced Treatise*[4] covers important topics for graduate instruction. Specialists wrote individual chapters. A French publication planned by Victor Grignard as editor and carried on by P. Baud, G. Dupont, and R. Locquin is now complete in twenty-three volumes. The long-famous Victor von Richter text is now superseded by E. H. Rood's *Chemistry of*

[1] For unusual abbreviations consult Crane and Patterson's appendix on Symbols or A. M. Patterson, *German-English Dictionary for Chemists*, John Wiley and Sons, Inc., N. Y. C., 1950, Ed. 3.

[2] *Dictionary of Organic Compounds*, Oxford University Press, N. Y. C., 1934, Vol. 1, 1936, Vol. 2, 1938, Vol. 3; New Rev. Ed. 1943, Vol. 1, 1944, Vols. 2 and 3 reprinted with supplements; New Rev. Ed. 1953 in 4 volumes.

[3] Published jointly by the Elsevier Publishing Company and the Nordeman Publishing Company of N. Y. C. and Amsterdam.

[4] Edited by Henry Gilman, John Wiley and Sons, Inc., N. Y. C., 1938, 2 Vols.; 1943-1953, Ed. 2, 2 Vols.

the Carbon Compounds,[1] to be complete in five volumes. In the first volume there is a good concise historical introduction to organic chemistry.

Methods of laboratory practice are compiled in Houben-Weyl, *Methoden der Organischen Chemie*,[2] which consists of seven volumes with others in preparation. This displaces the older comprehensive four-volume work,[3] which could still be useful. These volumes include information relating to general methods and operations in organic analysis, followed by numerous examples of reactions such as oxidations, halogenations, and the more specialized types as illustrated by diazotization. Reagents, varying conditions, and starting materials are stressed. A smaller, older compilation of laboratory methods is H. J. L. Meyer's *Lehrbuch der Organischchemischen Methodik*.[4] Lassar-Cohn, similar and older, has two volumes, part of which have been translated into English by R. E. Oesper.[5] In addition to these voluminous works, other books on methods or processes have appeared from time to time.[6]

The above formula indexes, treatises, advanced texts, and books on methods do not constitute all the so-called secondary source material, that organization of known facts made easily accessible to the reader. The abstract and review journals are the means by which the research worker can ascertain contemporary progress without reading all primary source material, most of which appears in the form of articles in thousands of journals. The principal chemical societies of the United States, Great Britain, France, and Germany have sponsored the four best chemical abstract journals in the world. In *Chemical Abstracts*, *British Chemical Abstracts*, *Bulletin de la société chimique de France*, and *Chemisches Zentralblatt*, abstracts for organic chemistry are always included.

In *Chemical Abstracts*, which is the logical journal to use first, and which has been in existence since 1907, the organic chemistry section is Number 10. Abstracts of articles are followed by cross references, lists of new books, and brief notes on patents. Each abstract now indicates the title, the author(s), the laboratory where the work was done, and a brief summary which must include any new method, new compound, or measured physical constant. The most useful features of *Chemical Abstracts* are

[1] Elsevier Publishing Company, Amsterdam, 1951-1954.

[2] E. Müller (Ed.), Georg Thieme, Stuttgart, 1952-1955.

[3] Weyl-Houben, *Die Methoden der Organischen Chemie*, Georg Thieme, Leipzig, 1925-1941, 4 Vols.

[4] This contains four volumes: 1. Analysis and Determination of Constitution (1929); 2. Detection and Estimation of Organic Compounds (1933); 3. Part 1. Synthesis of Open Chain and Isocyclic Compounds (1938); and 4. Part 2. Synthesis of Heterocyclic Compounds (1939).

[5] Lassar-Cohn, *Organic Laboratory Methods*, authorized translation from the general Part of 5th Rev. Ed., by Ralph E. Oesper, Roger Adams (Ed.), Williams & Wilkins Co., Baltimore, Md., 1928.

[6] Gatterman-Wieland, *Laboratory Methods of Organic Chemistry*, The Macmillan Co., N. Y. C., 1937, and 35 German editions. P. H. Groggins, *Unit Processes in Organic Synthesis*, McGraw-Hill Book Co., Inc., N. Y. C., 1938, Rev. Ed.

the indexes. There are four decennial indexes which cover the years from 1907 to 1946. In addition to author and subject indexes, there are ring formula lists and organic radical lists by name and by formula, which list for each item the nomenclature in best usage. The annual indexes also include Patent Number lists for 1912-14 and from 1935 to the present. There are two collective Patent Number lists. One is the *Patent Index to Chemical Abstracts 1907-36*, prepared by the Science Technology Group of the Special Libraries Association, and the other, compiled by *Chemical Abstracts* editors, is the *Ten-Year Numerical Index to Chemical Abstracts* for the years 1937-1945. The latter has over 143,000 entries classified by countries in numerical order with the column and page references to *Chemical Abstracts*. Patent information can be very important to organic chemical synthetic work.

The *Journal of the Chemical Society* had unusually fine abstracts for organic chemistry from 1870 to 1926 when it was succeeded by *British Chemical Abstracts*, which in turn ceased publication as of January 1, 1954, except for some sections. The French journal which began its abstracting service in 1858 seems to have stopped its abstracts in the past few years. The *Chemisches Zentralblatt*, founded by Leibig in 1830, is the oldest of the four. Except for some irregularities during World War II, it has maintained excellent service, especially since 1919. It is worth while for complete coverage to consult all the journals, particularly prior to World War I.

There is a type of review journal which does not attempt to cover the entire field of organic chemistry but rather selects the important articles for a definite period of time, usually a year. These serials with their bibliographies give an excellent idea of who the prominent research men are and their specializations. This may prove fruitful in selection of an institution in which to do graduate work. Representative of this class are the well-known *Jahresberichte*, from Germany, the *Annual Report of the Progress of Chemistry*, from the British Chemical Society, and for a single decade the *Annual Survey of American Chemistry*, a publication authorized by the National Research Council (1925-35 only).

There are several publications of immense value to organic chemistry. *Organic Syntheses*,[1] an annual publication instituted in 1921 by a group of American chemists with Roger Adams as editor-in-chief, is designed to collect together satisfactory methods for preparation of organic compounds. *Organic Syntheses* has three collective volumes for the first twenty-nine years, in which all information is revised and re-edited. Seven volumes with Roger Adams as editor-in-chief are entitled *Organic Reactions*.[2] These contain chapters on special topics which are written by

[1] John Wiley and Sons, Inc., N. Y. C., 1921 *et seq.*
[2] John Wiley and Sons, Inc., 1942 *et seq.*

experts. W. Theilheimer is the editor of *Synthetic Methods of Organic Chemistry*,[1] of which ten volumes and two cumulative indexes have appeared since 1947. The content is selected from abstracts of articles on methods which have appeared in the *Chemisches Zentralblatt* and *Chemical Abstracts*. The index lists reactions by name, operation, starting material, and product. A. Weissberger edits *Physical Methods of Organic Chemistry*,[2] and *Technique of Organic Chemistry*.[3]

The problem of organic nomenclature is serious and important. The greatest organic chemists of their day have given time and energy to the solution of the difficulties. The Geneva Congress in 1892 was called to establish rules to eliminate confusion, and any student who reads the literature for the forty-year period succeeding that date must know the Geneva System in order to name compounds properly or interpret the names of others.[4] The increase in number and variety of compounds since 1892 necessitated further co-operation among research chemists. In 1923, the Council of the International Chemical Union appointed a committee to modify the Geneva System where needed.[5] Since *Chemical Abstracts*[6] adheres to the best usages of that report, the editors published an admirable summation of all chemical nomenclature used in their journal for the benefit of their abstractors and readers. A book by A. M. Patterson and L. T. Capell, *Organic Ring Systems*,[7] was prepared to bring order to the rapidly multiplying intricacies of such systems of organic rings as are known to exist in identifiable compounds. The position numbers are indicated to promote uniform use. There is no need with these modern aids to create the nomenclature turmoil of the nineteenth century, when there might be a dozen names for the same compound. A student must be alert and resourceful and familiarize himself with such nomenclature rules to be able to read the literature to the best advantage.

Book lists for texts and special topics are rapidly outmoded. There are recent publications for qualitative and quantitative organic analysis, theoretical relationships, and monographs on specific subjects. These can be found in publishers' lists, in book review sections, and in the library card catalogue. Examination and personal preference is then an aid in selection. A few items of unusual interest or uniqueness are the book by Hückel;[8] on theory of organic physical constants by Timmermans;[9] on or-

[1] S. Karger, N. Y. C., 1947 *et seq.*

[2] Interscience Publishers, Inc., N. Y. C., 1945 *et seq.*

[3] Interscience Publishers, Inc., N. Y. C., 1949 *et seq.*, Ed. 2.

[4] F. Tiemann, Ber. deut. chem. Ges. *26*, 1595 (1893).

[5] A. M. Patterson, J. Am. Chem. Soc. *55*, 3905-25 (1933).

[6] C. A. *39*, 5868-951 (1945).

[7] A. C. S. Monograph, Rheinhold Publishing Co., N. Y. C., 1940.

[8] W. Hückel, *Theoretische Grundlagen der Organischen Chemie*, Akademische Verlagsgesellschaft, Leipzig, 1956.

[9] J. Timmermans, *Physico-Chemical Constants of Pure Organic Compounds*, Elsevier Publishing Co., Inc., N. Y. C., 1950.

ganic analysis by Mulliken and Huntress;[1] on advice for a research program by Reid;[2] and Gysel,[3] who has prepared a book to assist in computations of composition.

The use of the chemical literature entails three steps. The first is to know the general types of source material. The second step is to examine it. The final step is to write accurate references for the information acquired. Close observation of references in the various publications, and study of the frequent "Notices to Authors," as published in the American Chemical Society journals, will show the standard methods for bibliography. *Chemical Abstracts* and the *Journal of the American Chemical Society* are good models to follow.

This chapter is not intended to be a complete bibliography for organic chemistry, since passage of time rapidly diminishes the usefulness of any list. As an introduction to the important reference works and abstract journals, it is designed to direct the search for a background which will lead to more specialized publications as they are needed.

The material in this chapter has been prepared by Dr. Virginia Bartow, Department of Chemistry and Chemical Engineering, University of Illinois.

[1] S. P. Mulliken, *A Method for the Identification of Pure Organic Compounds*, John Wiley and Sons, N. Y. C., 1904-1922, 4 Vols. S. P. Mulliken and E. H. Huntress, New Ed., 1941.

[2] E. E. Reid, *Introduction to Organic Research*, D. Van Nostrand and Co., Inc., N. Y. C., 1924.

[3] H. Gysel, *Tables of Percentage Composition of Organic Compounds*, Birkhauser, Basel, 1951.

GLOSSARY OF TECHNICAL TERMS

ACTIVATE—To put into a state of increased chemical activity.

ACYCLIC—Open chain, not ring formation.

ACYLATING AGENT—A reagent used to introduce an acyl ($R \cdot CO \cdot$) group.

ACYLATION—The reaction of introducing an acyl group.

ADDITIVE—The tendency to add.

ALCOHOLYSIS—A reaction in which the hydroxyl hydrogen of an alcohol is substituted.

ALKANE—A member of the saturated hydrocarbon series.

ALKENE—An open chain hydrocarbon containing one double bond.

ALKINE, or ALKYNE—An open chain hydrocarbon containing one triple bond.

ALKYL—A radical derived from a saturated hydrocarbon by removing one hydrogen atom.

ALKYL HALIDE—A compound consisting of an alkyl group combined with one halogen atom.

AMMONOLYSIS—A reaction by which one hydrogen is removed from a molecule of ammonia.

ANGSTROM UNIT—One-one hundred millionth of a centimeter, or 1×10^{-8} cm.

ANHYDRIDE—Without water. The residue of an acid after the elements of water are removed from the carboxyl groups.

ASTRINGENT—An agent which causes contraction of tissues.

ASYMMETRIC—Without symmetry. A carbon atom holding four different substituents.

ATOMIC NUMBER—A number indicating the location of an element in the Periodic Table.

CARBONYL—A group having the structure, $\cdot C(:O) \cdot$.

COVALENT—A sharing of electrons, non-ionic.

DEGREDATION—A stepwise disintegration of a carbon chain.

DERIVATIVE—A resultant obtained by chemical reaction.

ENDOTHERMIC—Absorption of heat energy.

EPIMER—A spatial arrangement of an hydroxyl group in a sugar opposite from one taken as a standard.

EPOXY—A group having the structure —C—C—.

ESTERIFICATION—The process of forming an ester.

EXOTHERMIC—Evolution of heat energy.

GLYCOGEN—Liver starch; animal starch.

HALIDE—A compound consisting of a alkyl radical and one halogen atom.

HYDRIDE—A compound in which hydrogen is the negative component.

HYPOHALOUS—A compound in which a hydroxyl group is combined with an halogen atom.

INULIN—A polysaccharide derived from dahlia tubers and Jerusalem artichokes.

ISOMER—A compound having the same molecular composition as another compound.

ISOMERISM—The state in which two or more compounds have the same molecular formula but different arrangements of the atoms.

Kg. CALORIE—A heat unit equal to one thousand gram calories.

LAEVO ROTATORY—The property of rotating plane of polarized light to the left.

MACERATE—To wear away; to soften by steeping in a liquid.

METAMER—A compound resembling another in properties but differing from it in structure.

METATHESIS—Double decomposition.

MOLAL—A solution containing one gram-molecular weight of solute in one thousand grams of solvent.

MOLAR—A strength of solutions involving one gram molecular weight or fraction thereof per liter of solution.

N—Nitrogen; a notation to indicate that a substituent is attached to nitrogen.

n—Normal; Preceding the name of a compound, this notation indicates that the chain is continuous and not branched.

OLEFIN—Same as alkene.

OXIDATION—Loss of electrons.

POLYMER—A compound whose percentage of composition is the same as some initial compounds but whose molecular weight is a multiple of the initial compound.

RECTIFY—To purify by fractional distillation.

REDUCTION—A gain of electrons.

SAPONIFICATION—The process of making soap; a special case of the hydrolysis of esters in which a base is added to neutralize the acid form.

SATURATED—A type of hydrocarbons in which the valences of carbon are used in holding other carbon atoms or hydrogen atoms.

SUDORIFIC—A medicinal agent used to induce sweating.

INDEX

Principal references are given in *Italic* type.